XIX Chilean Physics Symposium 2014

Journal of Physics: Conference Series Volume 720

Concepcion, Chile
26 – 28 November 2014

Editors:

Gonzalo Gutierrez
Luis Roa

ISBN: 978-1-5108-2537-6
ISSN: 1742-6588

Printed from e-media with permission by:

Curran Associates, Inc.
57 Morehouse Lane
Red Hook, NY 12571

proceedings.com

Some format issues inherent in the e-media version may also appear in this print version.

Printed by Curran Associates, Inc. (2016)

Institute of Physics
Dirac House, Temple Back
Bristol BS1 6BE UK

Phone: 44 1 17 929 7481
Fax: 44 1 17 920 0979

techtracking@iop.org

Additional copies of this publication are available from:

Curran Associates, Inc.
57 Morehouse Lane
Red Hook, NY 12571 USA
Phone: 845-758-0400
Fax: 845-758-2633
Email: curran@proceedings.com
Web: www.proceedings.com

Table of contents

Volume 720

XIX Chilean Physics Symposium 201426–28 November 2014, Concepción, Chile

Accepted papers received: 19 May 2016 Published online:

Preface

Papers

Statistical Mechanics and Complex systems

Gravitation and Cosmology

Fields, particles and Nuclear Physics

Condensed Matter and material science

Applied Physics and experimental techniques

General physics, Education and Society

The Chilean Physics Symposium is the most important meeting of Chilean physicists. It has been held every two years since 1978, in different cities of the country, gathering professional physicists, young PhDs, and postgraduate and undergraduate students. This is a meeting where it is possible to review the status of physics in Chile, to discuss the progress of the various groups and researchers, as well as discuss the educational aspects and the relationship of the discipline with the country's development.

The XIX Symposium was held in the city of Concepción, on days 25, 26 and 27 November 2014 and was attended by 374 participants, most Chileans, but also others from other Latin American countries, North America, and Europe. There were 4 invited plenary lectures, 127 oral contributions and 186 posters. The scientific works presented covered virtually all areas of physics, both experimental and theoretical, and its relations with neighboring disciplines: condensed and material physics, particles and fields, gravitation and cosmology, optical and quantum physics, non-linear and mathematical physics, statistical mechanics and fluids, atomic, molecular and nuclear physics, electrodynamics and plasma, and education, society and history.

There was also a session where three members of the physics community were rewarded for their outstanding work as researchers and educators of new generations: Professor Jaime Roessler, of the University of Chile in Santiago, Professor Iván Schmidt of the University Santa María in Valparaíso, and Dr. Jorge Zanelli, CECs in Valdivia. It also paid tribute to Professor Sergio del Campo, who died in June 2014, and served as professor at Pontificia Universidad Católica de Valparaíso.

In these Proceedings there is a sample of the works that were presented at the Symposium. All authors were invited to submit articles for publication in the Symposium Proceedings. The articles received were reviewed by members of the Scientific Committee and by invited peers. The criteria for review, as usual in our meetings, focused on the demand for a consistent piece of research, and a clear statement of results. We thank the authors who sent their works, which comprise the 58 papers in this issue.

We wish to acknowledge the commendable support received for the edition of the Proceedings from Ariana Espinoza Muñoz, University of Concepcion. We also thank Felipe González-Cataldo and Leopoldo Soto. Finally, we list below the Scientific Committee of the Symposium, the Organizing Committee and the board of the Chilean Society of Physics 2013-2015, who with the Department of Physics at the University of Concepción- that took the organization of this event in their hands - they made possible this Symposium and the editing of these Proceedings. Last, but not least, we would like to thank Journal of Physics: Conference Series, and in particular to Sarah Toms, for her enthusiasm and patience.

Gonzalo Gutiérrez
Luis Roa
Editors

Journal of Physics: Conference Series **720** (2016) 011001 doi:10.1088/1742-6596/720/1/011001

Scientific Committee

Marcel Clerc,	Universidad de Chile
Rafael Correa,	Universidad Tecnológica Metropolitana
Claudio Dib,	Universidad Técnica Federico Santa María
Rodolfo Figueroa,	Universidad de La Frontera
Gonzalo Gutiérrez,	Universidad de Chile
Luis Huerta,	Universidad de Talca
Fernando Méndez,	Universidad de Santiago
Luis Roa,	Universidad de Concepción
Leopoldo Soto,	Comisión Chilena de Energía Nuclear
Juan Alejandro Valdivia,	Universidad de Chile
Ulrich Volkmann,	Pontificia Universidad Católica de Chile
Julio Yañez,	Universidad Católica del Norte

Organizing Committee

José Aguirre,	Universidad de Concepción
Luis Braga,	Universidad de Concepción
Francisco Brito,	Universidad del Bío Bío
Juan Crisóstomo,	Universidad de Concepción
Joaquín Díaz de Valdés,	Universidad de Concepción
Claudio Faúndez,	Universidad de Concepción
Fernando Gutiérrez,	Universidad de Concepción
Fernando Izaurieta,	Universidad de Concepción
Gustavo Lima,	Universidad de Concepción
Paulraj Manidurai,	Universidad de Concepción
Julio Oliva,	Universidad Austral de Chile
Luis Roa,	Universidad de Concepción
Ximena Torres,	Universidad del Bío Bío
Asticio Vargas,	Universidad de la Frontera

Directive Board, Chilean Physical Society 2013-2015

President: Fabiola Arévalo,	Universidad de la Frontera
Secretary: Leopoldo Soto,	Comisión Chilena de energía Nuclear
Treasurer: Pedro Vega,	Universidad de La Serena
Director: Luis Huerta,	Universidad de Talca
Director: Fabián Torres,	Universidad de la Frontera

Journal of Physics: Conference Series **720** (2016) 011002 doi:10.1088/1742-6596/720/1/011002

Peer review statement

All papers published in this volume of *Journal of Physics: Conference Series* have been peer reviewed through processes administered by the proceedings Editors. Reviews were conducted by expert referees to the professional and scientific standards expected of a proceedings journal published by IOP Publishing.

Forecasting of DST index from auroral electrojet indices using time-delay neural network + particle swarm optimization

J A Lazzús, C H López-Caraballo, P Rojas, I Salfate, M Rivera and L Palma-Chilla

Departamento de Física y Astronomía, Universidad de La Serena, Casilla 554, La Serena, Chile

E-mail: `jlazzus@dfuls.cl`

Abstract. In this study, an artificial neural network was optimized with particle swarm algorithm and trained to predict the geomagnetic DST index one hour ahead using the past values of DST and auroral electrojet indices. The results show that the proposed neural network model can be properly trained for predicting of $DST(t+1)$ with acceptable accuracy, and that the geomagnetic indices used have influential effects on the good training and predicting capabilities of the chosen network.

1. Introduction

Geomagnetic activity is usually characterized by geomagnetic indices, the most common being the storm DST index and the auroral electrojet indices [1]. Most indices have long records that allow statistical studies of the causes of activity and of related phenomena. Correlations between indices and possible drivers provide the basis for empirical prediction. The $DST(t)$ is defined as the average of the disturbance variation of the H component, divided by the average of the cosines of the dipole latitudes at the observatories for normalization to the dipole equator [2]. DST index can serve as a good measure of the overall strength of the near-Earth electric currents, especially the ring current. It is obtained from selected geomagnetic observatories operating at equatorial regions. It is important to realize that the ultimate goal of space weather research should be to forecast space weather and, at least, predict geomagnetic activity in terms of the geomagnetic DST index as a function of time.

The auroral electrojet index was originally introduced by Davis and Sugiura [3] as a measure of global electrojet activity in the auroral zone. The auroral electrojet index is derived from geomagnetic variations in the horizontal component H observed from selected observatories along the auroral zone in the northern hemisphere [4]. The auroral electrojet index is usually used to represent four indices AU, AL, AE and AO. The AU and AL indices [3], are intended to express the strongest current intensity of the eastward and westward auroral electrojets, respectively [4]. The AE index is defined as $AE = AU - AL$ [3], provides an estimate of the overall horizontal current strength, and, to some extent, a rough measure of the ionospheric energy losses [5]. And the AO index is defined as $AO = (AU + AL)/2$ [3], provides a measure of the equivalent zonal current [6].

In this work, the DST one hour ahead was estimate with a hybrid neural network+particle swarm algorithm using the time series of the past values of DST and auroral electrojet (AE and AO) indices.

2. Computational method

In this study, we consider one of the most successful and frequently used types of neural networks: a multilayer feed-forward neural network with a back-propagation learning algorithm. This consists of one input layer with N inputs, one hidden layer with q units and one output layer with n outputs. The output of this model can be expressed as [7]:

$$y_n = F_n \left(\sum_{j=1}^{q} W_{nj}^{(a)} f_j \left(\sum_{l=1}^{N} W_{jl}^{(b)} x_l + W_{j0}^{(b)} \right) + W_{n0}^{(a)} \right) \tag{1}$$

where $W_{nj}^{(a)}$ are the weights between unit j and unit n of input and hidden layers and $W_{jl}^{(b)}$ are the weights between hidden layer and an output. The activation functions $F_n(x)$ and $f_j(x)$ are linear or nonlinear. We used one hidden layer with $f_j(x)$ as a tangent hyperbolic nonlinear activation functions and $F_1(x)$ as linear function in output layer. For a given set of N inputs, we define the root mean square error (RMSE) by:

$$\text{RMSE} = \sqrt{\frac{\sum_{s=1}^{N} \left(y_s^{real} - y_s^{calc}\right)^2}{N^2}} \tag{2}$$

where y^{real} denotes the actual given output and y^{calc} the neural network output. This network was trained to minimize RMSE, replacing the gradient descendent error by a paricle swarm optimization (PSO) [8].

PSO is a population-based optimization tool, where the system is initialized with a population of random particles and the algorithm searches for optima by updating generations [9]. In each iteration, the velocity of each particle j is calculated according to the following formula [10]:

$$v_j^{k+1} = \omega v_j^k + c_1 r_1 \left(\psi_j^k - s_j^k\right) + c_2 r_2 \left(\psi_g^k - s_j^k\right) \tag{3}$$

where s and v denote a particle position and its corresponding velocity in a search space, respectively. k is the current step number, ω is the inertia weight, c_1 and c_2 are the acceleration constants, and r_1, r_2 are elements from two random sequences in the range (0,1). s_j^k is the current position of the particle, ψ_j^k is the best one of the solutions that this particle has reached, and ψ_g is the best solutions that all the particles have reached. In general, the value of each component in v can be clamped to the range $[-v_{\max}, +v_{\max}]$ control excessive roaming of particles outside the search space [9]. After calculating the velocity, the new position of each particle is:

$$s_j^{k+1} = s_j^k + v_j^{k+1} \tag{4}$$

Thus, all the neurons of the ANN have an associated activation value for a give input pattern, and the algorithm continues finding the error that is presented for each neuron, except those of the input layer. After finding the output values, the weights of all layers of the network are actualized $W_{nj} \to W'_{nj}$ by PSO, using eqs. (3 and 4) [8]. The velocity is used to control how much the position is updated. On each step, PSO compares each weight using the data set. The network with the highest fitness is considered the global best. The other weights are updated based on the global best network rather than on their personal error or fitness [9]. In PSO, the inertial weight ω, the constant c_1 and c_2, the number of particles N_{part} and the maximum speed of particle summary the parameters to syntonize for their application in a given problem. An exhaustive trial-and-error procedure was applied for tuning the PSO parameters. Firstly,

Journal of Physics: Conference Series **720** (2016) 012001 doi:10.1088/1742-6596/720/1/012001

the effect of ω was analyzed for values of 0.1 to 0.9. Next, the effect of N_{part} was analyzed for values of 10 to 100 particles in the swarm. Table 1 shows the selected parameters for this hybrid algorithm.

Table 1. Parameters used in the hybrid ANN+PSO algorithm.

Section	Parameter	Value
ANN	NN-type	*feed-forward*
	Number of hidden layers	1
	Transfer function	*tansig*
	Number of iterations	1500
	Normalization range	[–1, 1]
	Weight range	[–100, 100]
	Bias range	[–10, 10]
	Minimun error	1e–3
PSO	Number of particles in swarm (N_{part})	50
	Number of iterations ($k_{\max}$)	1500
	Cognitive component (c_1)	1.494
	Social component (c_2)	1.494
	Maximum velocity ($v_{\max}$)	12
	Minimum inertia weight ($\omega_{\min}$)	0.5
	Maximum inertia weight ($\omega_{\max}$)	0.7
	Objective function	RMSE

3. Database and training

A data set of geomagnetic DST index and auroral electrojet indices were taken from World Data Center for Geomagnetism of Kyoto [11], and used to train the network. This work used a leave-20%-out cross-validation method to estimate the predictive capabilities of the model. Training and prediction sets were selected with the consideration that the DST variations are present with adequate frequency in the training database. Then, 43848 hourly data points (years 2000 to 2004) were used in the training set, and other 8760 hourly data points (year 2005) were used in the prediction set. From this database, the best input vectors obtained to solve $DST(t+1)$ were:

$$\text{N}[DST(t-3), AE(t-3), DST(t-2), AE(t-2), DST(t-1), AE(t-1), DST(t), AE(t)] \quad (5)$$

$$\text{N}[DST(t-3), AO(t-3), DST(t-2), AO(t-2), DST(t-1), AO(t-1), DST(t), AO(t)] \quad (6)$$

The selected geomagnetic indices cover wide ranges, going from -442 to 67 (nT) for DST index, from 4 to 2241 for AE index, and from -942 to 1071 for AO index. Figure 1 shows the time series used and ranges for these indices.

4. Results and discussion

Several network architectures were tested to select the most accurate scheme. The most basic architecture normally used for this type of application involves a neural network consisting of three layers [8]. The number of hidden neurons needs to be sufficient to ensure that the

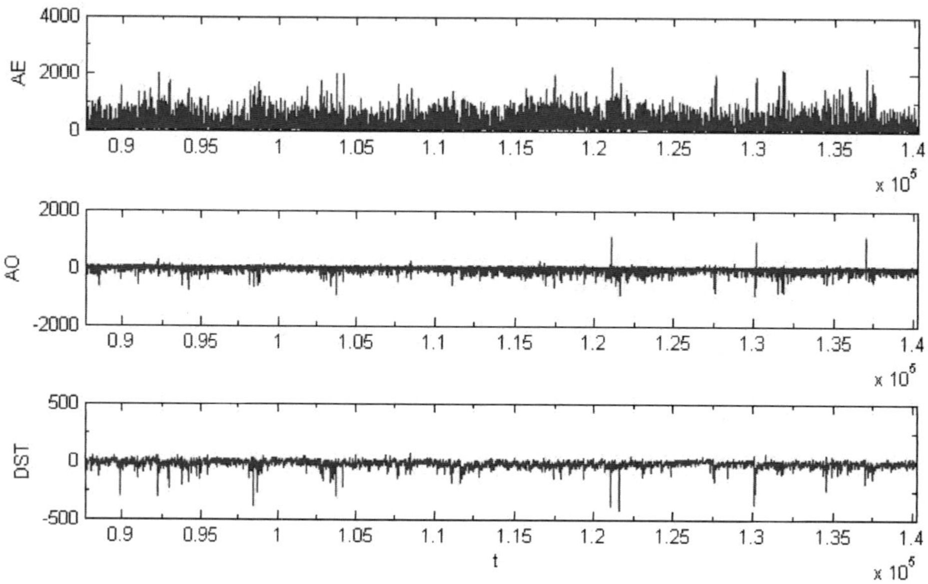

Figure 1. Time series of geomagnetic indices used in this study.

information contained in the data utilized for training the network is adequately represented. There is no specific approach to determine the number of neurons of the hidden layer, but many alternative combinations are possible. The optimum number of neurons was determined by adding neurons in systematic form and evaluating the average absolute deviations of the sets during the learning process [9]. Figure 2 shows the deviation found in the prediction of $DST(t+1)$ as a function of the number of neurons in the hidden layer. As observed in this figure, the optimum number of neurons in the hidden layer is between 6 and 9. Table 2 summarize the statistical results obtained in the estimation of $DST(t+1)$ using the selected input. In this table MAE is the mean absolute error, RMSE is the root mean square error, and R^2 is the correlation coefficient.

Table 2. Statistical results obtained in the estimation of $DST(t+1)$.

Input	Architecture	Step	MAE	RMSE	R^2
		Training	2.7070	4.2852	0.9829
Eq. (5)	8-6-1	Prediction	1.9991	2.8671	0.9661
		Total	2.5654	4.0416	0.9824
		Training	2.7564	4.3279	0.9826
Eq. (6)	8-8-1	Prediction	2.0112	2.8971	0.9653
		Total	2.6074	4.0821	0.9820

Figure 3 shows a comparison between the real (solid line) and calculated values (points) of $DST(t+1)$, obtained by the neural network models. Fig. 3a show the results obtained using

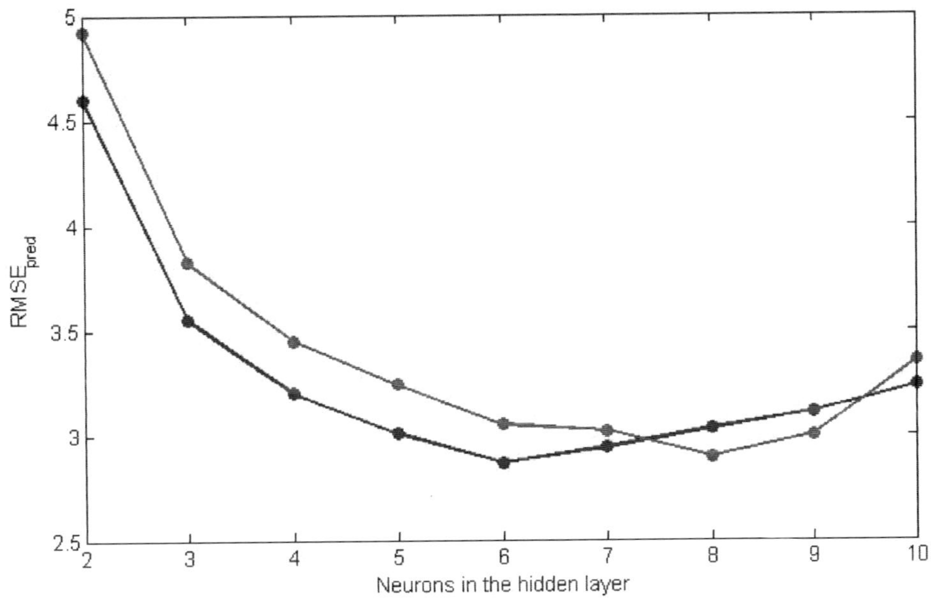

Figure 2. Time series of geomagnetic indices used in this study. $DST + AE$ (blue line), and $DST + AO$ (green line).

the past values of DST and AE (Eq. 5). For the training (years 2000–2004), the correlation coefficient R^2 is 0.9829 and the slope of the curve m is 0.9691 (expected to be 1.0) with RMSE of 4.2852. For the prediction (year 2005), R^2 is 0.9661, m is 1.0102 (expected to be 1.0) and RMSE of 2.8671. Fig. 3b show the results obtained with Eq. 6. In this case, for the training step (years 2000–2004), R^2 is 0.9826, m is 0.9769 and RMSE is 4.3279. And for the prediction step (year 2005), R^2 is 0.9653, m is 1.0095 and RMSE is 2.8971.

Currently, various models have been developed to predict DST [12]. However, comparative studies of artificial neural networks (ANNs) and the traditional regression approaches in modelling DST have also been conducted, and it has been shown that ANN methodology offers a promising alternative to the traditional approach [13]. In this case, a comparison was made with a multiple linear regression (MLR) method, and similar database. The MLR method show errors higher than 500% and accuracy of <50% with R^2 of 0.5 for the prediction of $DST(t+1)$ using $DST + AE$ versus the ANN+PSO method with accuracy of >97% and R^2 >0.96. Similar results were obtained for the prediction of $DST(t+1)$ using $DST + AO$. These results represent a tremendous increase in accuracy for estimating this important geomagnetic index and show that not only the optimum architecture obtained was crucial, also the appropriate selection of the independent parameters (Eq. 5 and 6). This is important because the auroral electrojet indices AE and AO are commonly available parameters [11]. Note that the coefficients of linear correlation of these parameters show a non-linear relationship with the DST [1], and consequently an ANN is the best alternative to model the DST for several applications. The innovative aspect of this study is the use of a neural model with only two input variables and a very limited number of neurons in the hidden layer. This ANN was optimized with a particle swarm algorithm to update the weight of the network. To the best of the authors's knowledge, there is no application for DST forecast such as the one presented here, using a hybrid ANN+PSO algorithm.

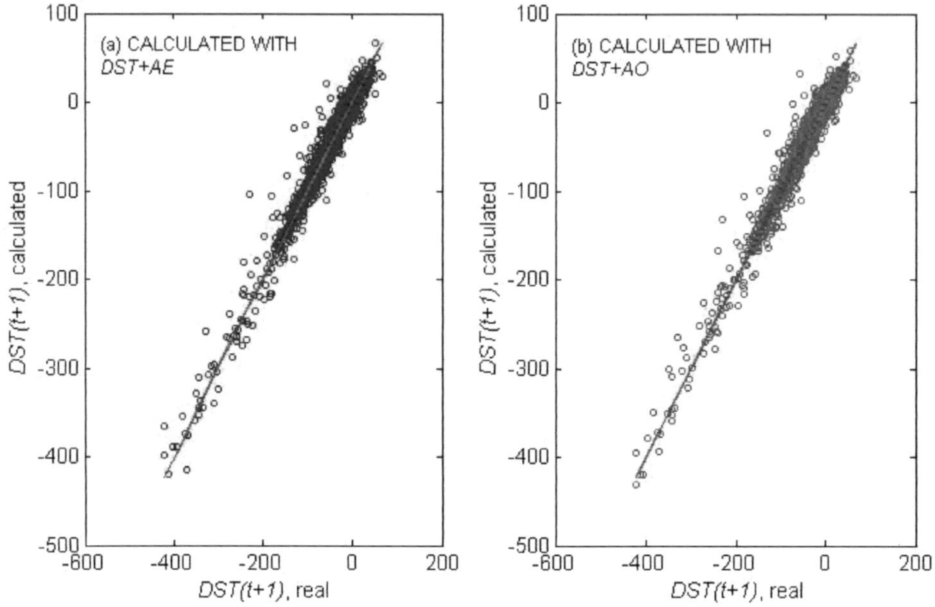

Figure 3. Comparison between calculated and real values for the estimation of $DST(t+1)$. (a) Calculated using Eq. 5, $DST + AE$. (b) Calculated using Eq. 6, $DST + AO$.

5. Conclusions

This study presents a method that includes an artificial neural network (ANN) replacing standard backpropagation with particle swarm optimization (PSO). Then, DST index one hour ahead was estimate by this hybrid algorithm using the time series of the past values of DST and auroral electrojet (AE and AO) indices.

Based on the results presented in this study, the following main conclusions are obtained: i) The results show that the proposed neural network models with eqs. 5 and 6 can be properly trained for predicting the $DST(t+1)$, with acceptable accuracy; ii) the geomagnetic indices used have influential effects, on the good training and predicting capabilities, of the chosen network; iii) The low deviations found with the proposed method indicate that it can predict the future values of DST with better accuracy than other methods available in the literature.

Acknowledgments

The authors thank the Direction of Research of the University of La Serena (DIULS) for the special support that made possible the preparation of this paper.

References

[1] Plotnikov I Y and Barkova E S 2007 *Adv. Space Res.* **40** 1858
[2] Sugiura M and Kamei T 1991 *IAGA Bulletin No. 40*
[3] Davis T N and Sugiura M 1966 *J. Geophys. Res.* **71** 785
[4] Pallocchia G, Amata E, Consolini G, Marcucci M F and Bertello I 2008 *J. Atmos. Sol.-Terr. Phys.* **70** 663
[5] Ahn B H, Akasofu S I and Kamide Y 1983 *J. Geophys. Res.* **88** 6275
[6] Menvielle M, Iyemori T, Marchaudon A and Nosé M 2011 *Geomagnetic Observations and Models*, ed M Mandea and M Korte (New York: Springer) chapter 8 pp 183–228
[7] Lazzús J A 2009 *Chin. J. Chem. Phys.* **22** 19
[8] Lazzús J A, Salfate I and Montecinos S 2014 *Neural Netw. World* **24** 601

[9] Lazzús J A 2013 *Math. Comput. Model.* **57** 2408
[10] Lazzús J A 2010 *Comput. Math. Appl.* **60** 2260
[11] WDC 2014 *World Data Center for Geomagnetism of Kyoto* (Database)
[12] Watanabe S, Sagawa E, Ohtaka K and Shimazu H 2003 *Adv. Space Res.* **31** 829
[13] Stepanova M, Antonova E and Troshichev O 2005 *J. Atmos. Sol.-Terr. Phys.* **67** 1658

Mackey–Glass noisy chaotic time series prediction by a swarm-optimized neural network

C H López-Caraballo, I Salfate, J A Lazzús, P Rojas, M Rivera and L Palma-Chilla

Departamento de Física y Astronomía, Universidad de La Serena, Avda. J. Cisternas 1200, Casilla 554, La Serena, Chile.

E-mail: clopez@dfuls.cl,jlazzus@dfuls.cl

Abstract. In this study, an artificial neural network (ANN) based on particle swarm optimization (PSO) was developed for the time series prediction. The hybrid ANN+PSO algorithm was applied on Mackey-Glass noiseless chaotic time series in the short-term and long-term prediction. The performance prediction is evaluated and compared with similar work in the literature, particularly for the long–term forecast. Also, we present properties of the dynamical system via the study of chaotic behaviour obtained from the time series prediction.

Then, this *standard* hybrid ANN+PSO algorithm was complemented with a Gaussian stochastic procedure (called *stochastic* hybrid ANN+PSO) in order to obtain a new estimator of the predictions that also allowed us compute uncertainties of predictions for noisy Mackey-Glass chaotic time series. We study the impact of noise for three cases with a white noise level (σ_N) contribution of 0.01, 0.05 and 0.1.

1. Introduction

The prediction of times series (TS) has played an important role in many science fields of practical application as engineering, biology, physics, meteorology, etc. In particular, and due to their dynamical properties, the analysis and prediction of chaotic time series have been of interest for the science community. In the literature, we found many methods focused on the prediction of chaotic time series, for example, those based on artificial neural network (ANN) models as the back-propagation algorithm [1], radial basic function [2], recurrent network [3], genetic algorithms [4], fuzzy system application [5], and wavelet approach [6], etc. These systems are usually modeled by delay-differential equations; standard examples are the Mackey–Glass equation [7], the Ikeda equation [8], and equation for an electronic oscillator with delayed feedback [9] (for more examples see [10]). The main goal is to forecast future values of a series known (or observed) up to a specific time ($x_{(t+T)}$, with T the ahead prediction) using past values. Note that the non-linear dependence level of a series on previous data points is of interest, particularly, because of the possibility of a chaotic behaviour of the input series.

Up to now, several short-term prediction methods have proven to be efficient in the forecast of chaotic time series and therefore in the characterization of dynamical system (for example, see [5, 11, 12, 6, 13] and references therein). In contrast, the long-term prediction still requires methods to improve the forecast effectiveness. Furthermore, in the literature the long-term forecast has not been widely studied.

On the other hand, the analysis of real-life times series require of consider the error propagation of input uncertainties. The observed data could be contaminated for different instrumental noise types as white noise or proportional to signal (the latter mainly arises from instrumental calibration). In modeling of chaotic times series, the prediction can be seen as erros-in-variable problem where the noisy is propagated into the prediction model. In the literature, the noisy impact on chaotic time series prediction have been scarce considered. We can found studies where the algorithms were tested from a theoretical point of view (for example, see [14, 15, 16, 17, 18]), and works where the implementation was applied of real-life time series (for example, see [19, 20, 15]). In the latter two references the authors proposed modification to the methods in order to improve the performance prediction in presence of noise.

In this work, we used a Mackey–Glass chaotic time series in order to study the short–term and long–term prediction with a hybrid ANN+PSO algorithm. Our performance prediction (of this *noiseless* cases) is compared with similar implementation available in the literature, in particular for the long–term forecast. Then, this *standard* hybrid ANN+PSO algorithm was complemented with a Gaussian stochastic procedure (called *stochastic* hybrid ANN+PSO) in order to obtain a new estimator of the predictions that also allowed us compute uncertainties of predictions for noisy Mackey-Glass chaotic time series. We study the impact of noise for three cases with a white noise level (σ_N) contribution of 0.01, 0.05 and 0.1.

2. Hybrid algorithm

A feed-forward neural network was used to represent non-linear relationships among variables [1]. This ANN was implemented replacing standard back-propagation algorithm with particle swarm optimization (PSO). PSO is a population-based optimization tool, whose operation is based on these two equations [21, 22]:

$$v_j^{k+1} = \omega v_j^k + c_1 r_1 \left(\psi_j^k - s_j^k\right) + c_2 r_2 \left(\psi_g^k - s_j^k\right) \tag{1}$$

$$s_j^{k+1} = s_j^k + v_j^{k+1} \tag{2}$$

where s and v denote a particle position and its corresponding velocity in a search space, respectively. In general, the value of each component in v can be clamped to the range $[-v_{\max}, +v_{\max}]$ control excessive roaming of particles outside the search space [21, 22].

On the other hand, the procedure to calculating the output values, using the input values, are described in detail in [23]. The net inputs (N) are calculated for the hidden neurons coming from the inputs neurons. In the case of a neuron in the hidden layer is had:

$$N_i^h = \sum_i^n w_{i,j}^h p_i + b_{i,j}^h \tag{3}$$

where p_i is the vector of the inputs of the training, $w_{i,j}^h$ is the weight of the connection among the input neurons with the hidden layer h, and the term $b_{i,j}^h$ corresponds to the bias of the neuron of the hidden layer h, reached in its activation. The PSO algorithm is very different then any of the traditional methods of training [21]. Each neuron contains a position and velocity. The position corresponds to the weight of a neuron $\left(s_i^k \rightarrow w_{i,j}^h\right)$. The velocity is used to update the weight $\left(v_i^{k+1} \rightarrow w_{i,j}^{\prime}\right)$. Starting from these inputs, the outputs (y_i) of the hidden neurons are calculated, using a transfer function f^h associated with the neurons of this layer. In the ANN, the hyperbolic tangent function (*tansig*) was used [24].

$$y_i = f^h \left(\sum_i^n w_{i,j}^h p_i + b_{i,j}^h\right) \tag{4}$$

Table 1. Adjustment parameters in the hybrid ANN+PSO algorithm.

ANN		PSO	
NN-type	feed-forward	Number of particles in swarm (N_{part})	50
Number of hidden layers	1	Number of iterations ($k_{\max}$)	1500
Transfer function	*tansig*	Cognitive component (c_1)	1.494
Number of iterations	1500	Social component (c_2)	1.494
Normalization range	[–1, 1]	Maximum velocity ($v_{\max}$)	12
Weight range	[–100, 100]	Minimum inertia weight ($\omega_{\min}$)	0.5
Bias range	[–10, 10]	Maximum inertia weight ($\omega_{\max}$)	0.7
Minimun error	1e–3	Objective function	RMSE

After finding the output values, the weights of all layers of the network are actualized $w_{i,j} \to w'_{i,j}$ by PSO, using equation 1 and 2 [22]. On each step, PSO compares each weight using the data set [21]. In this article, we used the mean square error (MSE) to determine network fitness for the entire training set. This process was repeated for the total number of patterns in the training set. For a successful process the objective of the algorithm is to modernize all the weights minimizing the total root mean squared error (RMSE):

$$\text{RMSE} = \sqrt{\frac{\sum_{i=1}^{n} \left(Y_i^{\text{true}} - Y_i^{\text{calc}}\right)^2}{n}}. \tag{5}$$

In PSO, the inertial weight ω, the constant c_1 and c_2, the number of particles N_{part} and the maximum speed of particle summary the parameters to syntonize for their application in a given problem. Table 1 shows the selected parameters for this hybrid algorithm.

3. Simulations of Mackey–Glass system

We computed the Mackey–Glass time series from the Mackey–Glass time-delay differential system [7, 25], which is described as follow:

$$\frac{dx}{dt} = \beta x(t) + \frac{\alpha x(t-\tau)}{1 + x(t-\tau)^{10}} \tag{6}$$

where x (unitless) is the series in the time t, and τ the time delay. Here, we assumed $\alpha = 0.2$, $\beta = 0.1$, $\tau = 17$ and $x(0) = 1.2$. Note that, if $\tau \geq 17$ the time series show a chaotic behaviour [4, 25]. Finally, the nominal Mackey–Glass chaotic time series was obtained from numerical integration via the method fourth order Runga Kutta. The series was computed with a time sampling of 1 second. Thus, $x(t)$ is derived for $0 \leq t \leq t_{\text{h}}$ with $x(t) = 0$ for $t < 0$, where t_{h} is the time horizon considered.

The algorithm implementation. We create as input a vector from d points of the time series spaced Δ apart, i.e. $\mathbf{x}(t) = [x(t), x(t+\Delta), \cdots, x(t+(d-1)\Delta)]$, in order to predict the future $x(t+T)$ value. In the standard Mackey–Glass analysis, we consider four non consecutive points in the time series in order to generate each input vector to predict the $x(t+6)$ and $x(t+84)$ as:

$$x(t+6) = F\left[x(t), x(t-6), x(t-12), x(t-18)\right] \tag{7}$$

$$x(t+84) = F\left[x(t), x(t-6), x(t-12), x(t-18)\right] \tag{8}$$

where this standard test assume $d = 4$ and $\Delta = T = 6$ [2, 4]. As reference, in general we use the 80% and 20% of data for the training and prediction, respectively.

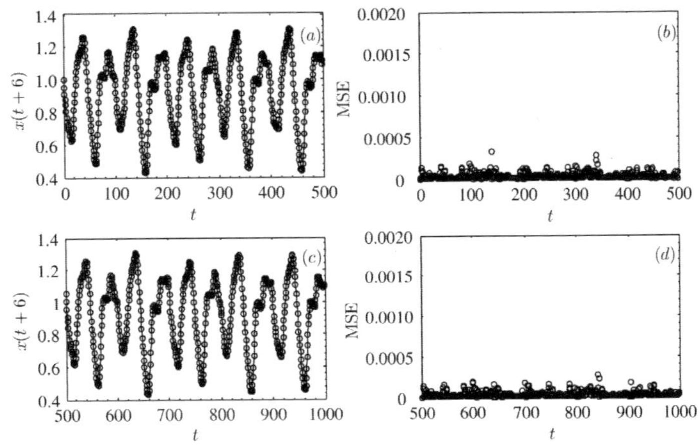

Figure 1. Training set and prediction set for the short-term $x(t+6)$.

Table 2. RRMSE for the computational methods found in the literature and our case studies.

Method	$\text{RMSE}_{x(t+6)}$	$\text{RMSE}_{x(t+84)}$
Linear model	0.5503	1.504
Cascade correlation NN	0.0624	0.170
Six order polynomial	0.0402	0.085
Back-propagation NN	0.0262	0.060
RBFNN	0.0114	0.097
This work *Case 1*	0.0053	0.038

4. Noiseless chaotic time series prediction

For the noiseless we generate one thousand data points ($t_{\text{h}} = 1000$), where the first five hundred were used for training while the others were used for testing.

Short-term prediction. The most basic architecture of a a neural network normally consisting of three or four layers [1, 26]. The input layer contains one neuron for each input parameter. The output layer has one node generating the scaled estimated value of the chaotic time series, $x(t+6)$. The optimum number of neurons was determined by adding neurons in systematic form and evaluating the MSE and RMSE of the sets during the learning process [21]. For this Case the optimum architecture was 4-6-1. In Figure 1 we present a comparison between recorded and predicted values of the standard configuration applied in the Mackey–Glass time series. Figure 1*a* shows the training set as a function of the time t. For this training set, the correlation coefficient R^2 was 0.99952. Figure 1*b* shows the MSE of training as a function of the time t, with a MSE_{max} of 0.00032. Figure 1*c* shows the prediction set. Note that for the prediction set, R^2 was 0.99953. Figure 1*d* shows the MSE obtained for the prediction set with MSE_{max} of 0.00027. Table 4 shows the RMSE of some computational methods [12, 6, 13, 5, 11] and the ANN+PSO method proposed in this work (for a more detailed table see [24]). The low errors found with the proposed method (MSE = 0.000028, and RMSE = 0.0053) indicate that it can estimate the Mackey–Glass time series $x(t+6)$ and $x(t+84)$ with better accuracy than other methods.

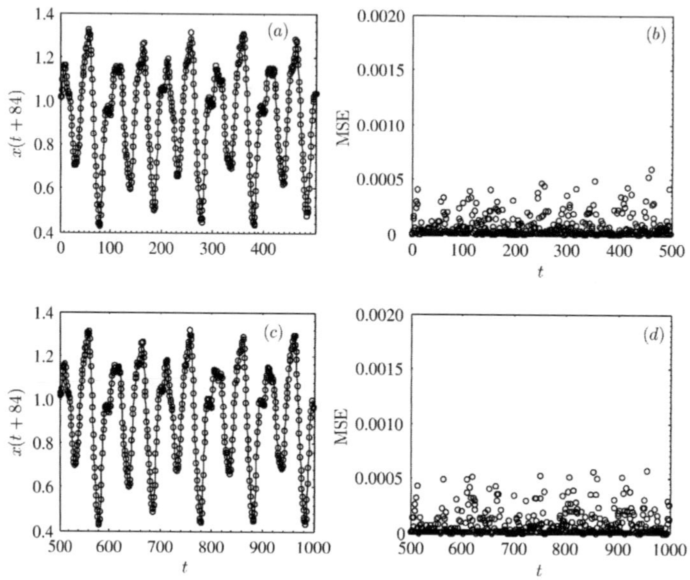

Figure 2. Training set and prediction set for the long-term $x(t+84)$.

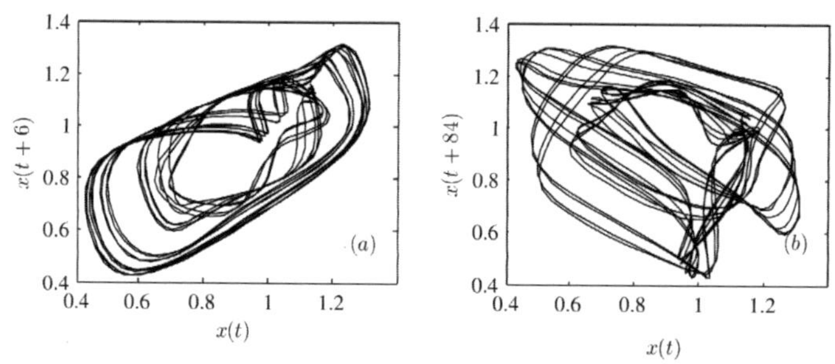

Figure 3. Chaotic attractors for the Mackey–Glass time series studies: (a) short-term representation; (b) long-term representation.

Long-term prediction. In the long-term prediction, the best architecture was 4-12-1. Figure 2 shows the recorded and predicted values of the long-term prediction. Figure 2*a* shows the training set as a function of the time t, with R^2 of 0.99877. Figure 2*b* shows the MSE for training set as a function of the time t, with MSE_{max} of 0.00059. Figure 2*c* shows the prediction set, with R^2 of 0.99853. Figure 2*d* shows the errors for the prediction set with MSE_{max} of 0.00057. Table 4 also shows a comparison between some computational methods found in the literature [27, 28, 29, 30] and our result for the long term prediction. This comparison was made using the normalized root mean squared error (NRMSE) [21]. The errors found with the proposed method (RMSE = 0.0080 and NRMSE = 0.0383) indicate that the method is efficient compared to others found in the literature [22].

Journal of Physics: Conference Series **720** (2016) 012002 doi:10.1088/1742-6596/720/1/012002

Finally, in the Figure 3 we present a representation of the chaotic attractor for the two cases studies. This Figure shows that with the above parameters, the system operates in a high-dimensional regime. The Mackey–Glass system is infinite dimensional (because it is a time-delay equation) and, thus, has an infinite number of Lyapunov exponents (λ_i) [25]. The Lyapunov exponents of dynamical systems are one of a number of invariants that characterize the attractors of the system in a fundamental way [31].

5. Noisy chaotic time series prediction

We consider the noisy times series as the contribution of a noise level on the nominal case (without noise). Here, we consider as nominal case ($x^{\mathrm{Noiseless}}$) the chaotic time series of Mackey-Glass with $\alpha = 0.2$, $\beta = 0.1$ and $\tau = 17$ (sampling of one second). We aim to the prediction of the short-term $x(t+6)$ with $x(t-0)$, $x(t-6)$, $x(t-12)$ and $x(t-18)$ terms. Also, we propose a complementary method that allow us compute the uncertainty of our forecast.

5.1. Simulations: Noisy chaotic time series

The Mackey–Glass noisy chaotic times series, $x_i \equiv x(t)$, is generated as:

$$x_{\mathrm{i}} = x_{\mathrm{i}}^{\mathrm{Noiseless}} + \eta_{\mathrm{i}} \tag{9}$$

where η_{i} is the particular contribution of noise of the i–element. It is estimated as $\eta_{\mathrm{i}} = GR(\sigma_{N,i})$, with $GR(\sigma_{N,i})$ a random number generator following a Gaussian distribution with mean zero and standard deviation equal to $\sigma^2_{N,i}$. Note that $\sigma^2_{N,i}$ corresponds to the noise level considered. Here, we assume that the original data are effected by a white noise, i. e., the noise level is the same in each i–element ($\sigma_{N,i} = \sigma_N$). We consider three white noise levels: $\sigma_N = 0.01$, $\sigma_N = 0.05$ and $\sigma_N = 0.1$. These values are nearly related with the 1%, 5% y 10% of the pick–to–pick amplitude ($I_{\mathrm{max}} - I_{\mathrm{min}} \sim max\{x^{Noiseless}\} - min\{x^{Noiseless}\}$) of nominal case.

Figure 4 shows the noisy chaotic time series of Mackey-Glass used in our work. As expected, the noisy time series with $\sigma_N = 0.01$ (blue line) is the closest to the nominal case. However, the case with $\sigma_N = 0.05$ and $\sigma_N = 0.1$ show a slightly more modified shape from the noiseless case, in particular with $\sigma_N = 0.1$ (red line).

5.2. The performance prediction

The *standard* hybrid ANN+PSO procedure applied in section 4 does not provides information about uncertainties of the predictions. Thus, the *standard* hybrid ANN+PSO algorithm was complemented with a Gaussian stochastic procedure (called *stochastic* hybrid ANN+PSO) in order to obtain a new estimator of predictions that also allow us compute uncertainties of predictions for noisy chaotic time series with a priori knowledge of the noise level.

- The *standard* ANN+PSO is the same procedure described in the section 2; however, here the root mean square error (RMSE, equation 5 in the learning process) is computed as:

$$RMSE = \sqrt{\frac{1}{\mathrm{n}} \sum_{\mathrm{i}=1}^{\mathrm{n}} \frac{\left(y_{\mathrm{i}}^{cal} - y_{\mathrm{i}}^{true}\right)^2}{\sigma^2_{\mathrm{N,i}}}} \tag{10}$$

where $\sigma_{\mathrm{N,i}}$ is the noise level of each i–element ($\sigma_{\mathrm{N,i}} = \sigma_{\mathrm{N}}$ for white noise). The prediction values obtained with this method are identified as y_i. Note that we continue use the RMSE (as defined in equation 5) for evaluation of the prediction performance and the comparison with literature. So that the noise level σ_N only is used in the learning phase (equation 10).

- The *stochastic* ANN+PSO is a procedure executed subsequent to the *standard* ANN+PSO. Once established the architecture (neuron number in hidden layer n, the transfer

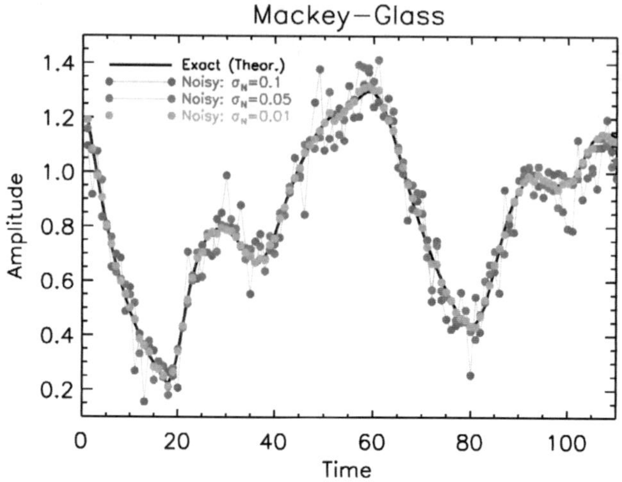

Figure 4. Noisy Mackey-Glass chaotic time series with $\alpha = 0.2$, $\beta = 0.1$ and $\tau = 17$. The black solid line is the noiseless case. The green, blue and red lines correspond to the series with a white noise level (σ_N) contribution of 0.01, 0.05 and 0.1, respectively.

functions f^h) and weights of each neuron ($w^h_{i,j}$ and $b^h_{i,j}$, equations 3 and 4) the neural network act as a function (called *function ANN*) whose output only depend on the input vector. The idea is generate simulations from the input data ($d_i \equiv d(t)$) via Gaussian random number generator in order to propagate the intrinsic data noise through of the *function ANN*. For each i–element of the input time series we generate k–simulations as:

$$d_{i,k} = d_i + GR_k(\sigma_{N,i}) \tag{11}$$

where the noise level $\sigma_{N,i}$ is known ($\sigma_{N,i} = \sigma_N$ for white noise). Note that we recover the data properties as $d_i =< d_{i,k} >$ and $\sigma_{N,i} =< d^2_{i,k} >^{1/2}$, where $< >$ is over the index k.

Now, and for the i-th element, each k input data $d_{i,k}$ provides an output $y_{i,k}$. These $y_{i,k}$ are used in the estimation of a new estimator of prediction ($\hat{y}_i$) and an error on the prediction ($\sigma_{\hat{y}}$) as follow:

$$\hat{y}_i =< y_{i,k} > \quad \text{and} \quad \sigma_{\hat{y}} =< y^2_{i,k} >^{1/2} . \tag{12}$$

In summary, and for each noisy time series, firstly the *standard* ANN+PSO is run in order to obtain the prediction y_i and characterize the *function ANN*. Then, we execute the *stochastic* ANN+PSO to compute $\hat{y}_i$ and $\sigma_{\hat{y}}$.

We applied the *standard* and *stochastic* ANN+PSO to our three noisy Mackay–Glass cases. In the *stochastic* ANN+PSO we run 1000 simulations ($k = 1000$). In left panel of Figure 5, we compare the y_i and $\hat{y}$ predictions with the original noisy MG-CTS for our more noisy case ($\sigma_N = 0.1$). As expected, even on this high noise level case, the y_i and $\hat{y}_i$ are in total agreement. One of a main goals of this section is estimate the uncertainty on the prediction, so that the measurement and the error bars obtained from the *stochastic* ANN+PSO for the noisy time series with $\sigma_N = 0.1$ are presented in the right panel of Figure 5. We confirm that our forecast and the input data, for the strong noise contribution, are in agreement at one-sigma (at 68.5% of confidential level) when the error bars are considered. Therefore, both picture of Figure 5 show that the *standard* and their complement (*stochastic* ANN+PSO) are a robust tool in the predictability of time series with a noise contribution.

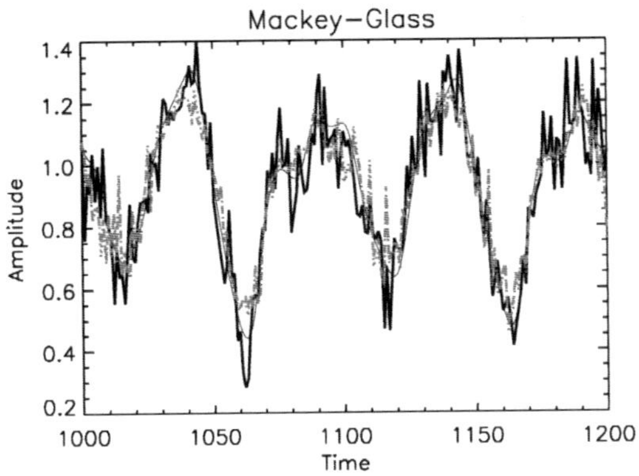

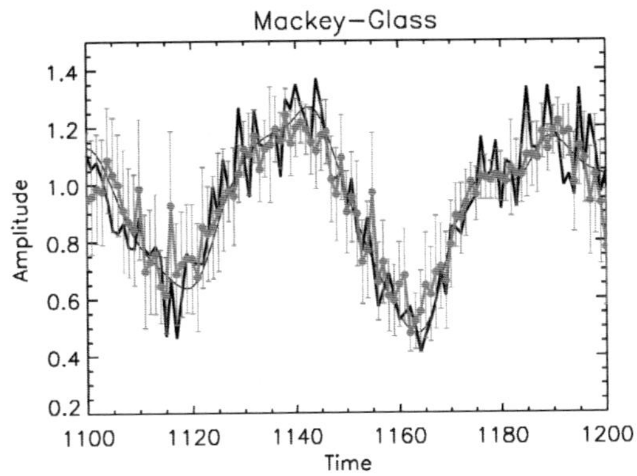

Figure 5. Predictions and uncertainties of noisy Mackey-Glass chaotic time series with a white noise of $\sigma_N = 0.1$. The black solid line correspond to the original noisy MG-CTS. *Left panel:* the dotted (red) and dashed (blue) lines are y_i (from *standard* ANN+PSO) and $\hat{y}_i$ (from *stochastic* ANN+PSO) predictions, respectively. *Right panel:* blue points correspond to the $\hat{y}_i$ prediction and the error bars are their uncertainties $\sigma_{\hat{y}}$.

In each case, and in order to evaluate the forecast performance, we used the y^{input} and the y^{output} to estimate the RMSE (following equation 9). The values computed are showed in Table 5.2. Note that both methods (*standard* and *stochastic*) provide very similar values of RMSE. As expected, the RMSE increases as growing the noise level from a 0.005 to 0.15 for noiseless and $\sigma_N = 0.1$ cases, respectively. Therefore, we confirm that a higher noise level of data leads to a poor estimation of the prediction estimator. We can use the ratio $r_{\text{RMSE}} = \text{RMSE}_{noisy}/\text{RMSE}_{noiseless}$ to study the impact of noise on the performance efficiency of our implementation (see values in Table 5.2). In the worst case, the performance efficiency (r_{RMSE}) is strongly affected by one order of magnitude with respect to noiseless case. Even so, the *standard* and *stochastic* ANN+PSO confirm be a powerful tools for making prediction of chaotic times series.

In the literature, we dont find a similar implementation (due to the ahead prediction, white noise level, etc.) that allow us a straightforward comparison of results. For example, we can contrast our results with those presented by Sheng et al. 2012 [20]. They applied the Echo State Network base on dual estimation (ESN) on a noisy Mackey-Glass time series ($\alpha = 0.2$, $\beta = 0.1$ and $\tau = 17$ with a sampling of 2 second) with a white noise level of $\sigma = 0.1$ (variance of 0.01). However, the ahead prediction was one, which is considerable lower than ours. Depend on the prediction performance they obtained RMSE of 0.05 for Generic ESN (hereafter GESN) and 0.04 for CKF/KF based ESN (henceforth CESN). These RMSE are almost two times better than our RMSE= 0.129 in a similar noise contribution ($\sigma_N = 0.1$). However, the impact of the noise noisy on the performance efficiency is lower in our implementation. In fact, we have a r_{RMSE} of 9.4 meanwhile we computed r_{RMSE} of 1161 and 33.5 for GESN and CESN, respectively.

6. Conclusions

In this work, a hybrid algorithm based on artificial neural network and particle swarm optimization (ANN+PSO) was used in the short-term and long-term prediction of Mackey-Glass chaotic time series. We also study the impact of noise on our hybrid method when a white

Table 3. Parameters used in the evaluation of the prediction performance of the *Standard* and *Stochastic* ANN+PSO approach.

	Noisy Mackey–Glass chaotic time series			
	Noiseless	$\sigma_N = 0.01$	$\sigma_N = 0.05$	$\sigma_N = 0.1$
RMSE	0.0138	0.016	0.067	0.129
r_{RMSE}	1	1.159	4.855	9.348

noise contribution affects the input data. The ANN+PSO was complemented with a procedure (*stochastic* ANN+PSO) in order to propagate the input uncertainties through of our method, and to obtain an estimation of error of the prediction.

Based on the results and discussion presented in this study, the following main conclusions are obtained: i) ANN+PSO method proposed in this work can be properly trained and used in the short-term prediction and long-term prediction of time series with acceptable accuracy; ii) the current value $x(t)$ and the past values used have influential effects on the good training and predicting capabilities of the chosen network; iii) simulation without noise contribution shows that this hybrid ANN+PSO algorithm is a very powerful tool for making prediction of chaotic time series, and the low deviations found with the proposed method shows a better accuracy than other methods available in the literature; iv) when the ANN+PSO is applied on simulated noisy times series our method carry out the prediction with acceptable accuracy (when the input noise level is considered); v) moreover, the performance efficiency is nearly related with the input noise level, for example, in the $\sigma_N = 0.1$ case the performance is worsened around to 10% ($r_{RMSE} = 9.4$); and vi) the hybrid ANN+PSO algorithm is a robust tool for making prediction of chaotic time series, including cases where the input data are contaminated by a noise contribution.

Acknowledgments

The authors thank the Direction of Research of the University of La Serena (DIULS) for the special support that made possible the preparation of this paper.

References

[1] Karunasinghe D S K and Liong S Y 2006 *J. of Hydrol.* **323** 92-105
[2] Chng E S, Chen S and Mulgrew B 1996 *IEEE Trans. on Neural Networks* **7** 190-4
[3] Zhang J S and Xiao X C 2000 *Chinese Phys. Lett.* **17** 88
[4] Mirzae H 2009 *Chaos Sol. and Frac.* **41** 2681-89
[5] Jang J S R, Sun C T and Mizutani E 1997 *Neuro-fuzzy and Soft Computing. A Computational Approach to Learning and Machine Intelligence.* (Upper Saddle River NJ: Prentice-Hall)
[6] Chen Y, Yang B and Dong J 2006 *Neurocomputing* **69** 449-65
[7] Mackey M C and Glass L 1977 *Science* **197** 287-9
[8] Ikeda K 1979 *Optics Communications* **30** 257-61
[9] Chua L O, Kocarev L, Eckert K and Itoh M 1992 *Int. J. of Bif. and Chaos* **2** 705-8
[10] Bezruchko B P, Karavaev A, Ponomarenko V and Prokhorov M 2001 *Phys. Rev. E* **64** 056216(1)-(6)
[11] Kim D and Kim D 1997 *IEEE Trans. on Fuzzy Systems* **5** 523-35
[12] Chen Y, Yang B and Dong J 2004 *Int. J. of Neural Systems* **14** 125-37
[13] Fu Y Y, Wu C J, Jeng J T and Ko C N 2010 *Exp. Sys. with Appl.* **37** 4441-51
[14] Girard A, Rasmussen C, Quinonero-Candela J and Murray-Smith R 2003 *Advances in Neural Information Processing Systems* (MIT Press) pp 529536.
[15] Sapankevych N and Sankar R 2009 *Comp. Intel. Mag. IEEE* **4** 2438
[16] Haykin S and Principe J 1998 *Sig. Proc. Mag. IEEE* **15** 66
[17] Li D, Han M and Wang J 2012 *IEEE Trans. on Neural Networks and Lear. Sys.* **23** 78799
[18] Han M and Wang X 2013 *Proc. The 2013 Int. Joint Conf. on Neural Networks (Dallas)* vol (no., 1-5

[19] Leung H, Lo T and Wang S 2001 *IEEE Trans. on Neural Networks* **13** 116372
[20] Sheng C, Zhao J, Liu Y and Wang W 2012 *Neurocomputing* **82** 186-95
[21] Lazzús J A 2010 *Fluid Phase Equilibria* **289** 176-84
[22] Lazzús J A 2010 *Comp. and Math. with Appl.* **60** 2260-69
[23] Lazzús J A 2009 *Therm. Acta* **489** 53
[24] Lazzús J A, Salfate I and Montecinos S 2014 *Neural Network World* **24** 601-17
[25] Farmer J D 1982 *Phys. D* **4** 366-93
[26] Han M and Wang Y 2009 *Exp. Sys. with Appl.* **36** 1280-90
[27] Awad M, Pomares H, Rojas I, Salameh O and Hamdon M 2009 *Int. Arab J. of Inf. Tech.* **6** 138-43
[28] Bersini H, Duchateau A and Bradshaw N 1997 *Proc. of the 6th IEEE Int. Conf. on Fuzzy Systems (Barcelona Spain)* vol (Barcelona: 21 IEEE Press) pp 1417-22
[29] Martinetz T M, Berkovich S G and Schulten K J 1993 *IEEE Trans. on Neural Networks* **4** 558-69
[30] Whitehead B A and Choate T D 1996 *IEEE Trans. on Neural Networks* **7** 869-80
[31] Brown R, Bryant P and Abarbanel H D I 1997 *Phys. Rev. A* **43** 2787-2806

Journal of Physics: Conference Series **720** (2016) 012003 doi:10.1088/1742-6596/720/1/012003

Temperature estimators in computer simulation

César Jara, Felipe González-Cataldo, Sergio Davis and Gonzalo Gutiérrez

Grupo de NanoMateriales, Departamento de Física, Facultad de Ciencias, Universidad de Chile, Casilla 653, Santiago, Chile.

E-mail: cesar.jara.uchile@gmail.com

Abstract. Temperature is a key physical quantity that is used to describe equilibrium between two bodies in thermal contact. In computer simulations, the temperature is usually estimated by means of the equipartition theorem, as an average over the kinetic energy. However, recent studies have shown that the temperature can be estimated using only the particles positions, which has been called configurational temperature. Through classical molecular dynamics simulations of 108-argon-atoms system, we compare the performance of four different temperature estimators: the usual kinetic temperature and three configurational temperatures. Our results show that the different estimators converge to the same value, but their fluctuations are different.

1. Introduction

The notion of temperature, as a statistical concept, relies on the ability to take averages over long periods of time and over a large number of particles. One of the questions that naturally arises is whether it is possible to extend the definition of temperature to small, finite systems. It would be convenient, then, to know how different temperatures estimators (or "thermometers") behave, in order to decide which one is more suitable for each system.

In computer simulations, such as molecular dynamics, the temperature is calculated as the time average of the kinetic energy of the system in equilibrium. This temperature is usually called kinetic temperature. However, this is not the only way to estimate temperature. Recently, it has been shown [1, 2, 3] that a family of estimators for the temperature exist, which can be expressed in terms of the velocity only, or positions only, or both.

In this work, we study the behavior of four different temperature estimators in a system of 108 argon atoms interacting via the Lennard-Jones potential in the microcanonical ensemble. The paper is structurated as follows. In section 2 we present the theoretical framework and the explicit forms of temperature estimators. Section 3 describes the computational details of the simulations, and section 4 show the results obtained from this simulations.

2. The Conjugate Variables Theorem (CVT) and configurational temperature

In the framework of statistical mechanics of hamiltonian systems, it has been recently proven, using continuous maximum entropy inference, a general theorem (named Conjugated Variables Theorem, CVT) connecting the Lagrange multipliers and the expectation values of a certain particularly constructed functions of the states of the system [3]. This theorem states that for a system of N particles governed by a hamiltonian

$$H = \sum_{j=1}^{N} \frac{\vec{p}_j^{\,2}}{2m_j} + V(\vec{r}_1, \ldots, r_N), \tag{1}$$

the following equality holds:

$$\frac{1}{k_B T} = \frac{\langle \nabla \cdot \vec{u} \rangle}{\langle \vec{u} \cdot \nabla H \rangle}, \tag{2}$$

where $\vec{u}$ is an arbitrary vector field, k_B the Boltzmann constant, and T the temperature of the system. This equation provides different estimators for the temperature, depending on the choice of the vector field $\vec{u}$. For instance, taking $\vec{u} = \nabla H / \|\nabla H\|^2$, we obtain

$$\frac{1}{k_B T} = \left\langle \nabla \cdot \frac{\nabla H}{\|\nabla H\|^2} \right\rangle. \tag{3}$$

This particular choice of $\vec{u}$ corresponds to the expression for the dynamical temperature obtained by Rugh [1], and shows that more general arbitrary definitions of temperature can be given. Note that equation (3) provides a way to obtain the temperature as an average of what we call the instantaneous temperature

$$\frac{1}{k_B \hat{T}} \equiv \nabla \cdot \frac{\nabla H}{\|\nabla H\|^2}. \tag{4}$$

This can be expressed in a more convenient way by taking the divergence of the quotient:

$$\frac{1}{k_B \hat{T}} = \nabla \cdot \frac{\nabla H}{\|\nabla H\|^2} = \frac{\nabla^2 H}{\|\nabla H\|^2} - \nabla H \cdot \frac{[\nabla(\nabla H \cdot \nabla H)]}{\|\nabla H\|^4}. \tag{5}$$

A remarkable fact is that the operator ∇ can be taken with respect to $\vec{p}$, or respect to $\vec{r}$, or both. In this way we have a family of estimators for the temperature, T.

We are interested in comparing the performance of the following three different temperature estimators:

1. *Kinetic Temperature*: This is the most common way in which the temperature is defined, and in this formalism is obtained by taking the gradient over the momenta ($\nabla \equiv \nabla_{\vec{p}}$) on equation (5). This leads to the expression

$$\frac{1}{k_B \hat{T}_1} = \frac{1}{K}\left(\frac{3N}{2} - 1\right), \tag{6}$$

where $K = \sum_{j=1}^{N} \frac{1}{2} m_j \vec{v}_j^{\,2}$ is the kinetic energy of the system.

2. *Configurational Temperature*: If the gradient goes over positions $\vec{r}$, we obtain

$$\frac{1}{k_B \hat{T}_2} = \frac{\nabla^2 V}{\|\nabla V\|^2} - \nabla V \cdot \frac{[\nabla(\nabla V \cdot \nabla V)]}{\|\nabla V\|^4}. \tag{7}$$

On this expression we can clearly see that the temperature depends only on the positions of the particles (or configurations of the system), not on their velocities. This is what Rugh called *dynamical temperature.*

3. *Truncated Configurational Temperature*: It is not difficult to prove that the second term in equation (7) goes as $1/N$, therefore the main contribution is on the first term. This means that
$$\frac{1}{k_B \hat{T}_3} = \frac{\nabla^2 V}{\|\nabla V\|^2} \tag{8}$$
is a good approximation for the configurational temperature when N is large.

In addition, we want to compare the average of these estimators, which corresponds to the inverse temperature $1/k_B T$, with the inverse temperature coming from the ratio between two averaged quantities:
$$\frac{1}{k_B T} = \frac{\langle \nabla^2 H \rangle}{\langle \|\nabla H\|^2 \rangle}. \tag{9}$$
This equation comes from the choice $\vec{u} = \nabla H$ in equation (2). Taking the gradient over the positions $\vec{r}$ leads to
$$\frac{1}{k_B T_4} = \frac{\langle \nabla^2 V \rangle}{\langle \|\nabla V\|^2 \rangle}. \tag{10}$$

3. Model and computational details

In order to study the performance of 4 different temperature estimators in finite system we use classical molecular dynamics simulations employing LPMD software [4]. The model consists of 108 Argon atoms interacting through the Lennard-Jones potential,
$$V(r) = 4\epsilon \left[\left(\frac{\sigma}{r}\right)^{12} - \left(\frac{\sigma}{r}\right)^{6} \right],$$
where r is the distance between atoms, whereas $\sigma = 3.41$ Å and $\epsilon = 0.0103$ eV are parameters fitted for the argon interaction [5]. The simulation is performed in the microcanonical ensemble, in three dimensions, without periodic boundary conditions.

Initially, we applied an energy corresponding approximately to 10 K, and then we run the system for 10000 steps (one time step is equal to 1 fs). The instantaneous temperature is estimated by three different expressions: $\hat{T}_1, \hat{T}_2$ and $\hat{T}_3$. The temperature is estimated as an average over the last 5000 steps, and we called them T_1, T_2, T_3 and T_4.

4. Results

Figure 1 displays the performance with respect to the time of three different temperature estimators, namely $\hat{T}_1$, $\hat{T}_2$ and $\hat{T}_3$. We can see that the value of the estimators are not very different and when $\hat{T}_2$ and $\hat{T}_3$ temperatures increases, $\hat{T}_1$ decreases, and vice versa. Also the fluctuations are similar.

For comparing the contribution of individual terms on the $\hat{T}_2$ estimator, we show in Figure 2 three important terms: $\nabla^2 V/\|\nabla V\|^2$, $\nabla V \cdot [\nabla(\nabla V \cdot \nabla V)]/\|\nabla V\|^4$ and $\nabla^2 V/\|\nabla V\|^2 - \nabla V \cdot [\nabla(\nabla V \cdot \nabla V)]/\|\nabla V\|^4$. We can see that the second term of $\hat{T}_2$ estimator is one hundred times smaller than the first term of that estimator. This is because the second term is of order $1/N$, where N is the number of atoms in the simulation. Also, this explains that the first term of $\hat{T}_2$ estimator and the complete term are similar.

Journal of Physics: Conference Series **720** (2016) 012003 doi:10.1088/1742-6596/720/1/012003

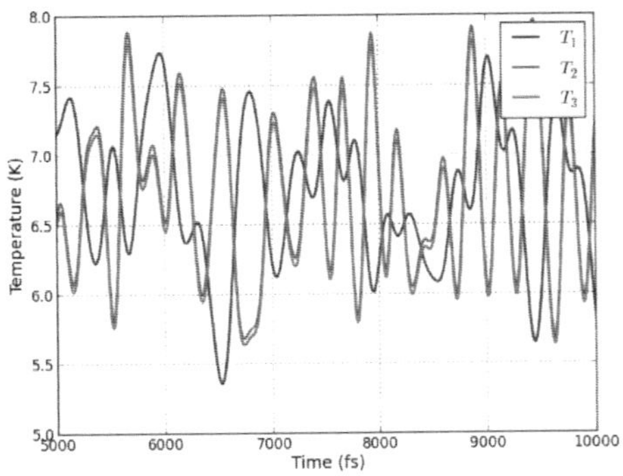

Figure 1. Instantaneous temperature $\hat{T}_1$, $\hat{T}_2$ and $\hat{T}_3$ for 108 argon atoms. Note that the average of $\hat{T}_2$ and $\hat{T}_3$ are similar.

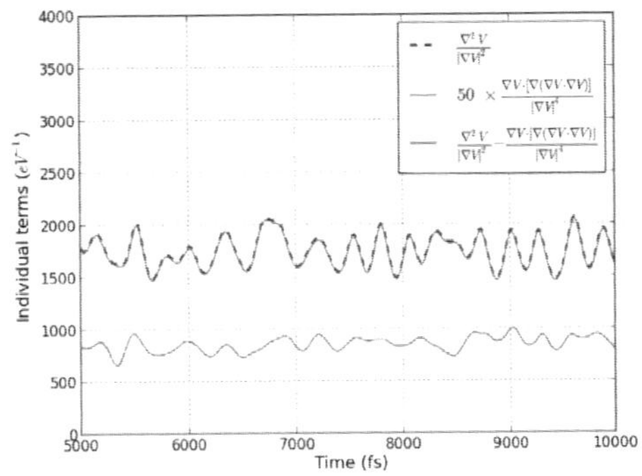

Figure 2. Contribution of the three terms that compose $\hat{T}_2$ estimator, for 108 argon atoms.

In order to study in more detail the behavior of the first term, $\nabla^2 V/\|\nabla V\|^2$, which is the main contributor to the estimator $\hat{T}_2$, we show in Figure 3 and in Figure 4 the numerator and denominator of that term. We can see that the average of $\nabla^2 V$ is two thousand times bigger than the average of $\|\nabla V\|^2$. On the other hand the fluctuations of $\nabla^2 V$ are smaller than fluctuations of $\|\nabla V\|^2$.

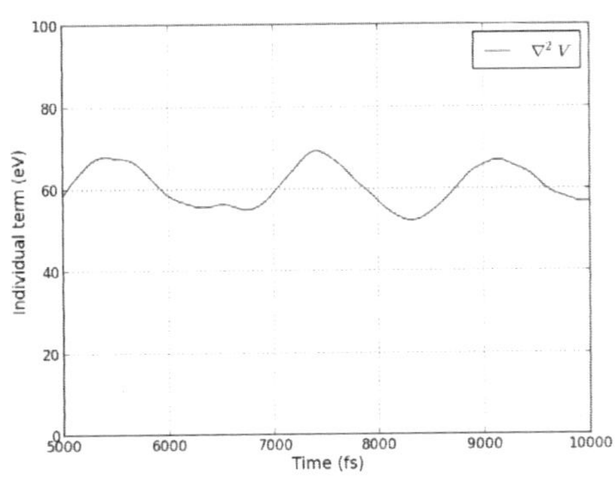

Figure 3. $\nabla^2 V$ for 108 argon atoms

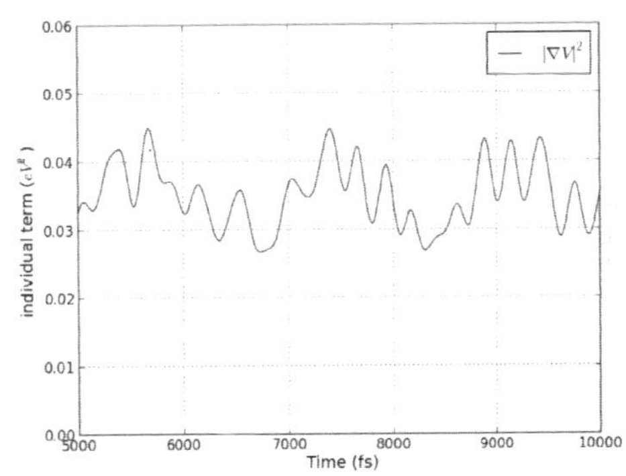

Figure 4. $\|\nabla V\|^2$ term for 108 argon atoms.

Finally, we compute the averages of different temperature estimators and display them in Table 1. Note that the value of T_1 and T_2, the kinetic and the configurational temperature, respectively, are almost the same, with a relative porcentual difference of 0.07%. Also, T_4 is close to T_1 and T_2. On the other hand, T_3, which is T_2 truncated, give the more different value, although, also very close to T_1 and T_2, with a relative porcentual difference of 0.98%

5. Concluding remarks

In conclusion, we have calculated the temperature of a small system in four different ways. The explicit functional forms of the temperature were derived from CVT theorem [3], and three of them are based on the positions only. We conclude that for long enough simulation time, the

Table 1. Temperatures calculated in the simulation.

T_1	T_2	T_3	T_4
6.681±0.042	6.686± 0.035	6.620±0.0345	6.679±0.034

different temperature estimators converge to the same value, while the fluctuations are not the same for each estimator.

Acknowledgments

We acknowledge to CONICYT-PIA Anillo ACT-1115 and Proyecto Bilateral Chile-Argentina Conicyt ACE-01/ANPCyT-PICT-2697.

References

[1] Rugh H H 1997 *Phys. Rev. Lett.* **78** 772–774
[2] Rickayzen G and Powles J G 2001 *The Journal of Chemical Physics* **114** 4333
[3] Davis S and Gutiérrez G 2012 *Physical Review E* **86** 051136
[4] Davis S, Loyola C, González F and Peralta J 2010 *Computer Physics Communications* **181** 2126–2139
[5] Kittel C 1986 *Introduction to Solid State Physics* 6th ed (John Wiley & Sons, Inc.)

Labyrinthine dissipative patterns

I Bordeu[1]

Departamento de Física, Facultad de Ciencias, Universidad de Chile, Casilla 653, Santiago, Chile.

E-mail: `ibordeu@imperial.ac.uk`

Abstract. A wide range of physical systems exhibit labyrinthine patterns as transient or steady states. One of the routes by which labyrinths emerge is through the destabilization of a localized structure through a curvature instability. In this work, we will study the emergence of labyrinths through this mechanism in different physical contexts, including nonlinear cavity optics, vegetation dynamics and chemical reactions. Furthermore, it is shown that the resulting labyrinths can be classified depending on their connectivities.

1. Introduction

Systems which are kept out of thermodynamic equilibrium can exhibit multistability of states. These states can be homogeneous or inhomogeneous. The latter are sustained by the injection and dissipation of energy matter and/or momentum, and correspond to a mechanism by which the systems self-organize [1, 2], minimizing–at least locally–some physical quantity, in variational systems this quantity is their energy, sometimes called nonequilibrium or Lyapunov functional [3]. It is in this framework that dissipative structures emerge [1]. This structures can be localized, this is, they are finite spatial domains (with a non diverging energy), or extended, corresponding to periodic or non periodic spatial structures which may possess a characteristic wavelength that defines them [2, 4, 5, 6].

It has been observed that a wide range of dissipative systems exhibit a state (or multiple states) of equilibrium in which labyrinth structures are formed, such as, experimental observations in magnetic fluids [7], cholesteric liquid crystals [8, 9], Langmuir monolayers [10], vegetation patterns [11], nonlinear cavity optics [12], among others. Such structures are characterized for having a well defined wavelength, but a small correlation length, this is, the pattern has no privileged direction, generating in this way an elaborated spatial structure [13]. This definition is purely statistical, thus, based on the average spatial distribution of the structure. In this work we study the route by which a localized spot destabilizes, and by an elliptical deformation it elongates, this elongation generates a stripe structure which suffers from a transversal instability propagating to all available space in the form of a labyrinthine structure, this *curvature mechanism* has been reported in [14]. Based on this result, we will test the universality of the mechanism presented, by showing its existence in a variety of physical systems. Namely, through a general competition-redistribution type of model and through the Gray-Scott reaction-diffusion model it will be shown that the transition from localized to

[1] Present address: Department of Mathematics, Imperial College London, 180 Queens Gate, London SW7 2BZ, United Kingdom.

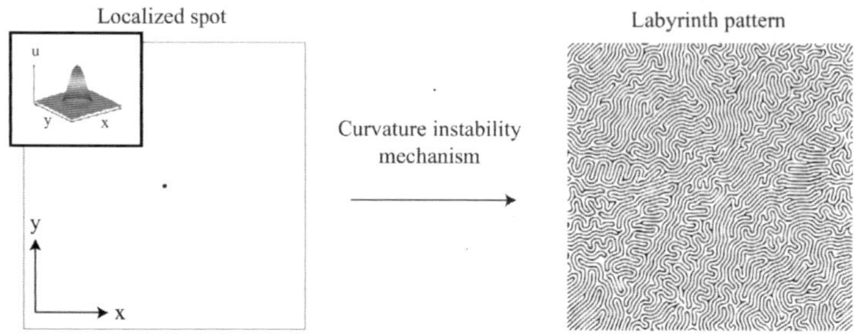

Figure 1. Curvature instability mechanism. Initially circular localized structure(left) and time evolved extended labyrinthine pattern final state (right) for Eq. (1). Parameters $\eta = -0.065$, $\epsilon = 2.45$, $\nu = 2.0$. Pseudo-spectral simulation (periodic boundary conditions), grid of 1024×1024 points and spacing $dx = 0.5$.

extended structure is also observed in the context of vegetation dynamics and chemical reactions, respectively. The different type of labyrinths observed will lead us to a simple classification of the labyrinth structures based on their connectivity.

2. From a localized to an extended labyrinthine pattern

2.1. Modified Swift-Hohenberg equation

The Swift-Hohenberg equation [15] is a real nonlinear partial differential equation (PDE), deduced in the context of fluid dynamic to describe the amplitude of the pattern formed by Rayleigh-Bénard convetion cells. By means of a generalized version of the Swift-Hohenberg equation it has been recently described how a localized structure loses stability, elongating, suffering a transversal instability to then generate an extended labyrinthine pattern [14]. The modified Swift-Hohenberg equation considered describes the evolution of a real scalar field $u(x, y, t)$, through an isotropic, symmetric under reflection, nonlinear PDE, which includes an constant term which breaks the field reflection symmetry. The modified Swift-Hohenberg equation reads

$$\frac{\partial u(x,y,t)}{\partial t} = \eta + \epsilon u(x,y,t) - u(x,y,t)^3 - \nu\nabla^2 u(x,y,t) - \nabla^4 u(x,y,t), \tag{1}$$

This equation was deduced originally as an approximation to the Maxwell-Bloch equations, in the context of nonlinear cavity optics to describe the nascent optical bistability with transversal effect [16, 17]. This equation predicted the existence and stability of circular [17] and rodlike [18] localized structures, and it also exhibits extended patterns, such as, labyrinths.

The field $u(x, y, t)$ accounts for the amplitude of the electric field, x and y are spatial coordinates and t is time. The bifurcation parameter ϵ measures the input field amplitude. The η parameter breaks the field reflection symmetry $u \to -u$. The parameter ν stands for the diffusion coefficient, positive values of ν lead to the formation of spatial patterns. The 2D Laplacian operator $\nabla^2 = \partial^2_{xx} + \partial^2_{yy}$ and the 2D bilaplacian operator ∇^4 act on the transverse plane (x, y), and they stand for the transport mechanisms or spatial coupling.

The modified Swift-Hohenberg equation (1) can be derived from a Lyapunov functional, i.e.

$$\frac{\partial u}{\partial t} = -\frac{1}{2}\frac{\delta F[u, \nabla u, \nabla^2 u]}{\delta u}, \tag{2}$$

where F is the Lyapunov functional or energy [19] and reads

$$F \equiv -\iint_{\mathbb{R}^2} dxdy \left[2\eta u + \epsilon u^2 - \frac{u^4}{2} + \nu(\nabla u)^2 - (\nabla^2 u)^2\right]. \tag{3}$$

It is clear that the η term accounts for the asymmetry between the homogeneous states. By using the solutions of Eq. (1), this functional satisfies

$$\frac{dF}{dt} = -\iint_{\mathbb{R}^2} dxdy\,(\partial_t u)^2 \leq 0. \tag{4}$$

This property guarantees that a system described by Eq. (1) evolves towards a state of minimal energy. The modified Swift-Hohenberg equation exhibits, for $\epsilon > 0$, coexistence between homogeneous and pattern states [20, 21], this allows the existence of stable localized structures.

This localized structures are azimuthally symmetric and occupy only a portion of the available space (see Fig. 1 left panel).

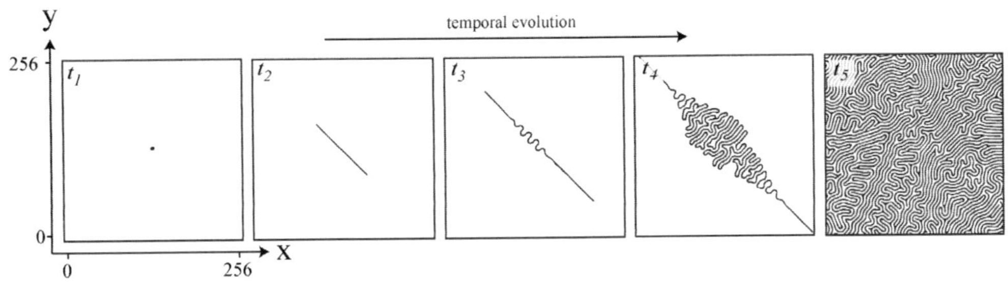

Figure 2. Curvature instability mechanism for the generation of a labyrinth from a localized structure as initial condition for the modified Swift-Hohenberg equation (1). Simulation made for parameters $\eta = -0.065$, $\epsilon = 2.45$, $\nu = 2.0$. Pseudo-spectral method was used with periodic boundary conditions, grid of 1024×1024 points and spacing $dx = 0.5$.

2.2. Curvature instability mechanism

As described in reference [14], Eq. (1) supports a curvature instability over localized spots. This instability deforms the spot into an elongated elliptical shape, producing a rod-like structure. This rod structure suffers from a transversal instability, undulations appear in its core. The non-saturating dynamics that follow take the system into an disordered extended structure that covers all the available space. Such structure shall be called a labyrinthine pattern or simply labyrinth, which is characterized for not having a global order or regularity. Due to the variational origin of Eq. (1) and given that the temporal dynamic moves towards a minimum of the Lyapunov functional (4), we know that the labyrinthine structure will reach a stationary state asymptomatically. In this case, the labyrinth is composed by a single interconnected structure or invaginated structure [14] (which protrudes from the more stable homogeneous state). Defects such as, dislocations, disinclination, and phase fronts contribute to the spatial disorder of the labyrinth. This behavior occurs far from any pattern forming instability and exists in a regime of bistability between homogeneous steady states.

3. Vegetation dynamics and chemical reaction models

The curvature instability mechanism for the formation of labyrinthine patterns from initially localized structures requires of few ingredients as the phenomenon is observed even in the simplest of the pattern forming equations as is Eq. (1).

3.1. Competition-redistribution model

In this section, a minimal but general model for vegetation pattern forming systems will be introduced. In this model the curvature instability mechanism also emerges in a regime of bistability of homogeneous steady states far from the pattern forming instability.

The emergence of spatial organization in extended landscapes has been an intriguing phenomenon for both botanists and nonlinear scientists community for decades and are still a matter of great discussion [22, 24, 25]. One of the transversal agreements among scientists is that competition for soil resources such as water and nutrients leads to self-organization phenomenon. Depending on the topography of the terrain, the climate conditions and the vegetation species, the process of self-organization leads to the formation of a wide variety of patterns, such as, tiger bush [22, 23], or fairy circles [24, 25].

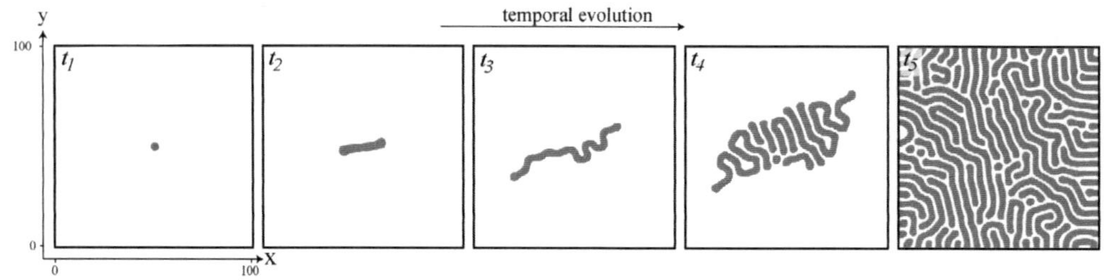

Figure 3. Curvature instability mechanism for the generation of a labyrinth from a localized patch as initial condition for the interaction-redistribution model (5). Simulation made for parameters $\eta = 0.085$, $\kappa = 0.6$, $\Delta = 0.005$, $\Gamma = 0.5$ and $\alpha = 0.125$. Finite differences method was used with periodic boundary conditions, grid of 200×200 points and spacing $dx = 0.5$.

Taking into account relation that exists between the structure of individual plants and the competition-facilitation interactions in a vegetation community is that the generic interaction-redistribution models emerge [22, 26]. These models have been successful in the prediction of formation of localized structures in arid and semi-arid environments, where the resource scarcity induces high competition between plants through their root networks. In this context, a localized structure emerges as a patch of vegetation in an unpopulated homogeneous state. By considering a logistic nonlinearity with nonlocal coupling between plants Tlidi et al. [27] where able via a weak gradient approximation, to deduce a mean-field nonlinear partial differential equation for the temporal evolution of the phytomass density $\rho(x, y, t)$:

$$\partial_t \rho = -\rho \left(\eta - \kappa\rho + \rho^2\right) + (\Delta - \Gamma\rho)\nabla^2\rho - \alpha\rho\nabla^4\rho, \tag{5}$$

where (x, y), and t are the spatial coordinates and time, respectively. This equation supposes an homogeneous and isotropic environment, and assumes that the interaction between plants does not change though their size. The parameters: η accounts for the decrease-to-growth rate ratio; κ is the facilitation-to-competition susceptibility ratio; Δ is proportional to the square root of the facilitation-to-competition range ratio, this three parameters are positive-defined. The parameters Γ and α are the nonlinear diffusion coefficients. The existence of a nonlinear hyperdiffusion is fundamental for Eq. (5) to satisfy the physical constrain for the phytomass density $\rho(x, y, t) \geq 0$. However, the presence of the nonlinear diffusion terms $u\nabla^2 u$ and $u\nabla^4 u$ render Eq. (5) nonvariational. Thus, Eq. 5 cannot be written as the variation of a Lyapunov functional.

Nonvariational systems can exhibit permanent complex spatiotemporal dynamics, such as chaos, intermittency, turbulence, among others.

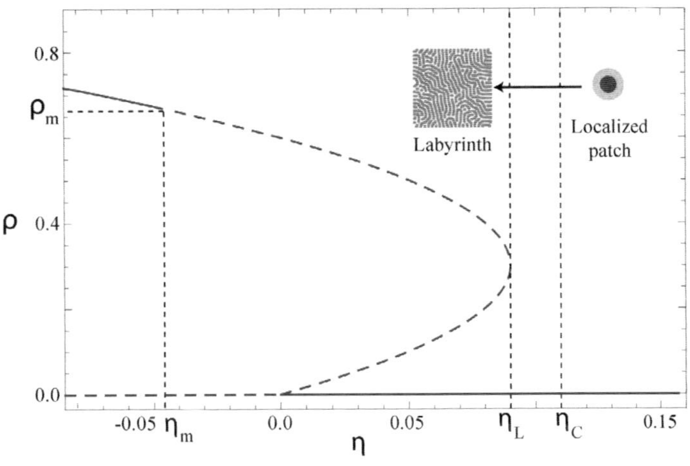

Figure 4. Bifurcation diagram for $\kappa = 0.6$, $\Delta = 0.005$, $\Gamma = 0.5$ and $\alpha = 0.125$, of the homogeneous plant population states (blue), dashed lines indicate unstable regime. Stable localized structures are observed for values of $\eta > \eta_C$. When the aridity η is decreased below η_L the curvature instability mechanism is observed.

The real order parameter equation (5) constitutes the simplest model of spatial dynamics in which competitive interactions between individuals occur locally.

The homogeneous solutions of Eq. (5) are (i) $\rho_s^0 = 0$, which correspond to a territory without vegetation, a bare state, (ii) and two non-zero states given by $\rho_s^{\pm} = (\kappa \pm \sqrt{\kappa^2 - 4\eta})/2$, a vegetated state. When $\kappa \leq 0$, only the homogeneous steady state ρ_s^+, is physically consistent, for $\eta < 0$. It decreases monotonously with μ and vanishes at $\eta = 0$. When $\kappa > 0$, the viable homogeneous solution extends up to the limit point $\rho_L = \kappa/2$ and $\eta_L = \kappa^2/4$. In the range $0 < \eta < \eta_L$, the biomass density exhibits a bistable behavior: the stable homogeneous branches of solutions ρ_s^0 and ρ_s^+ coexist with the intermediate unstable branch ρ_s^-. The former solution is always unstable.

The upper homogeneous state ρ_s^+ undergoes a Turing instability characterized by an intrinsic wavelength

$$\Lambda_m = 2\pi\sqrt{2\alpha}/\sqrt{\Gamma/\alpha - \Delta/\rho_m} \tag{6}$$

The threshold η_m, associated with this instability is solution of the following cubic equation:

$$(2\Gamma\rho_m - \Delta)^2 = 4\alpha\rho_m^2(2\rho_m - k) \tag{7}$$

We now focus on parameter regime where the uniform plant distribution exhibit bistability ($\kappa > 0$) as we know, from the previous section that bistability is a necessary ingredient for the curvature mechanism to exist. For parameters $\eta_m < \eta < \eta_L$ the homogeneous state becomes unstable with respect to the Turing instabilty. Thus, infinitesimal fluctuation around the uniform plant distribution ρ_s^+ will spontaneously evolution towards stationary, spatially periodic distributions of the biomass density which will invade the whole system. However, when a stable localized vegetation patch is considered (these exist for $\eta > \eta_C$) as an initial condition (high aridity), see Fig. 4, and the aridity is then varied in such quantity that the system reaches the coexistence of homogeneous states, then the localized structure suffers and elliptical deformation caused by the curvature instability and elongates into a rod structure (see Fig. 3 at t_2) in a process identical as previously observed for Eq. (1). Afterwards, the rod structure suffers from a transversal instability (see Fig. 3 at t_3), in this case the whole structure is compromised and not only the central portion as it was observed in the Swift-Hohenberg case (Fig. 1 for t_3.). As time further increases, the labyrinthine structure invades all the system.

Here, we have described the emergence of the curvature instability mechanism over a localized structure which by turn generates the emergence of a labyrinthine structure. This takes place when entering with an initially stable localized structure to the bistability parameter region.

3.2. Gray-Scott model

Pushing further the idea of universality of the curvature instability mechanism for the formation of labyrinths, we now enter the domain of chemical reactions. It is well-known that the reaction and diffusion of the components of a chemical reaction can produce an immense variety of spatial structures, such as lamelar structures ans self-replicating spots [28, 29], localized structures [30, 31] and Turing structure [28, 32, 33], and exhibit complex spatiotemporal dynamics, such as oscillatory dynamics, breathing localized structure and spatiotemporal chaos [34, 35].

In a series of seminal works [36, 37, 38] P. Gray and S.K. Scott introduced a variant of the autocatalytic model for glycolisis first proposed bt Sel'kov [34]. This prototype autocatalytic reaction (i.e. the product of the reaction is also the catalyst) is described by two irreversible processes

$$\begin{aligned} A + 2B &\rightarrow 3B, \\ B &\rightarrow C, \end{aligned} \tag{8}$$

where A, B and C are chemical species, C is a non-reactive product (inert) and the system is kept out-of-equilibrium by a permanent injection of reactive A, called feeding. In their papers, Gray and Scott show that this simple model presents multistability, hysteresis cycles, patterns, oscillatory dynamics and other exotic patterns.

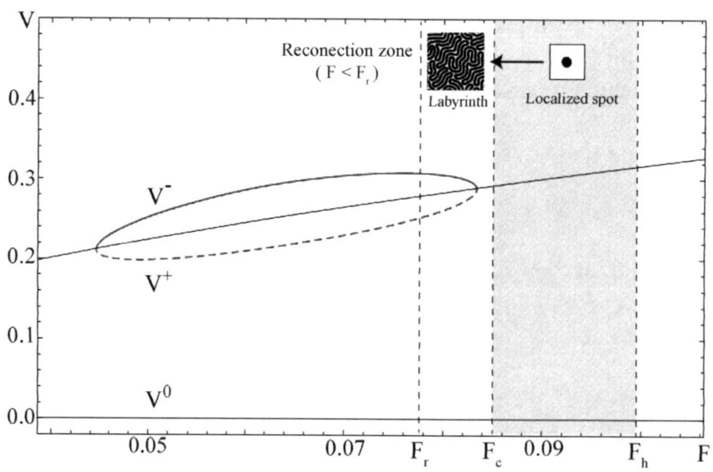

Figure 5. Bifurcation diagram the concentration V for $k = 0.061$, homogeneous states (blue), dashed lines indicate unstable regime. The $\sqrt{F}$-curve shows the position of the saddle-node bifurcation. For $F_c < F < F_h$ stable localized structures are observed (numerically). When changing the feeding parameter for a localized structure to values $F_r < F < F_c$, the curvature instability mechanism is observed. Reconnection zone indicates the zone where a radial instability is observed and structures can reconnect (see. Fig. 7). Here $F_h = 0.097$, $F_c = 0.085$ and $F_r = 0.078$.

Through a rescaling and including the diffusion phenomena in the reactions, a nonlinear partial differential equation can be deduced for the temporal evolution of the concentrations U and V of A and B, respectively [28]. This set of equations read

$$\begin{aligned} \frac{\partial U}{\partial t} &= D_u \nabla^2 U - UV^2 + F(1-U), \\ \frac{\partial V}{\partial t} &= D_v \nabla^2 V + UV^2 - (F+k)V, \end{aligned} \tag{9}$$

which are an example of reaction-diffusion equations, as they model the local evolution U and V, and the diffusion process of the chemical species. In Eq. (9) D_u and D_v correspond to the diffusion coefficients associated to U and V, respectively. F represents the feeding rate of the reactive A and dimensionless constant k (killing rate) accounts for the rate of the second reaction. Alike the interaction-redistribution model presented for vegetation dynamics, Eq. (5), the Gray-Scott model Eq. (9) does not have a variational structure, which is the reason behind its rich phenomenology which includes, from oscillatory dynamics, self-replication [28] and chaos [35] and others mentioned before. The existence and stability of patterns and localized spots (or single pulses) as homoclinic solutions of Eq. (9) has been well studied both in one and two dimensional cases [30, 39, 40, 41, 42]. This system supports a trivial homogeneous steady state $(U^0, V^0) = (1, 0)$ which is always stable. When the discriminator $d = 1 - 4(F+k)^2 > 0$ two additional homogeneous steady states $(U^\pm, V^\pm)$ appear, the first

$$(U^+, V^+) = \left(\frac{1}{2}(1+\sqrt{d}), \frac{F}{2(F+k)}(1-\sqrt{d})\right) \tag{10}$$

is always unstable. The second state

$$(U^-, V^-) = \left(\frac{1}{2}(1-\sqrt{d}), \frac{F}{2(F+k)}(1+\sqrt{d})\right) \tag{11}$$

is stable when $-V^2 + k < 0$ and $(F+k)(V^2 - F) > 0$. For $F > 1/4$ the system only exhibits the trivial steady state. For $F < 1/4$ and $k < 1/16$ the system exhibits three homogeneous steady states (one unstable). The bifurcation diagram for the homogeneous states is shown in Fig. (5). For a detailed study on the bifurcations and instabilities of Eq. 9 see Ref. [33] and references therein.

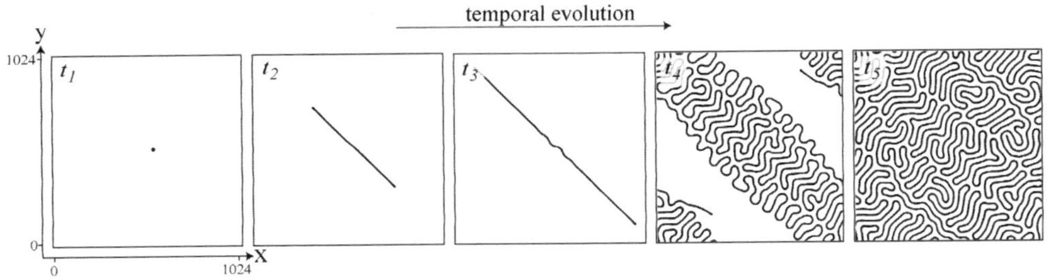

Figure 6. Curvature instability mechanism for the generation of a labyrinth from a localized concentration peak as initial condition for the Grey-Scott model, Eqs. 9. Simulation made for parameters $F = 0.080$, $k = 0.061$. Finite differences method was used with periodic boundary conditions, grid of 1024×1024 points and spacing $dx = 1$.

As indicated in Fig. 5, localized structures are observed in $F_c < F < F_h$, is F is increased further the system falls to the homogeneous state $V^0 = 0$. When the parameter F is reduced below F_c, perturbations become unstable and the structure suffers a curvature instability characterized by the elongation of the structure (see Fig. 6). As observed for the modified Swift-Hohenberg Eq. (1) and for the interaction-redistribution models, Eq. (5), the concentration field suffers then a transversal instability characterized by the wiggling of the central section of the structure. Non saturating evolution leads to the formation of a complex labyrinthine structure as seen in Fig. 6.

When the feeding parameter is decreased further ($F < F_r$), an initially circular localized structure suffers from a radial instability as seen in Fig. 7 at t_2, there two spots where given as initial conditions. As the radius of the structures grow, they suffer from a curvature instability. Non saturation causes the structure to propagate. Differently from what is observed from every labyrinthine structure showed previously, here the tips merge, this causes reconnection of the structures. In the transition from t_4 to t_5 one can realize that the two initially distinct structures reconnect into one single labyrinth.

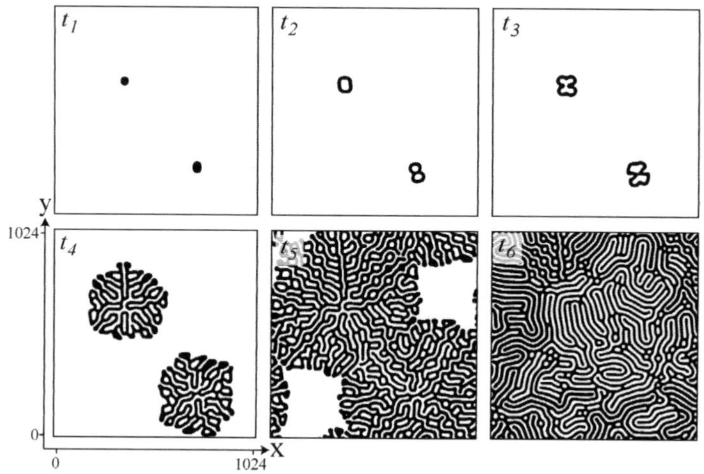

Figure 7. Alternative mechanism for the generation of a labyrinth from a localized concentration peak as initial condition for the Grey-Scott model 9. Simulation made for parameters $F = 0.070$, $k = 0.061$. In this parameter zone occurs reconnection between structures. Finite differences method was used with periodic boundary conditions, grid of 1024×1024 points and spacing $dx = 1$.

4. Labyrinth connectivities

The curvature instability of a localized structure was observed in the three systems presented previously. In the three cases (modified Swift-Hohenberg, interaction-redistribution and Grey-Scott models) an initially localized structure suffers from the curvature instability mechanism to finally fall into an extended labyrinthine structure. However each of these structures are different (see Figs. 2.t_5, 3.t_5, 6.t_5 and 7.t_6). Starting from a single localized structure the Swift-Hohenberg model Eq. (1), generates a single fully connected structure, this is, starting in the higher field on can reach every other higher value field without leaving it. The same happens in the fist labyrinth from the Gray-Scott model (cf. Fig. 6) only that in this case the structure shows no dislocations as the structure remains in a single line labyrinth. In the case of the interaction-redistribution model Eq. (3), starting from a single localized structure, the final labyrinthine state is disintegrated into a high number of small structures, thus, this labyrinth is highly disjoint. On contrary, when considering two initial structures in Fig. 7, the structures where able to reconnect increasing the connectivity from the initial state.

A classification of the labyrinth structures originated from localized initial states can be done based on the connectivity difference between the final and the initial states. If $\Delta = C_f - C_i$, where C_i and C_f are the initial and final number of disjoint structure, then, if $\Delta > 0$ we will say labyrinth is dissociative, if $\Delta = 0$ it is neutral and if $\Delta < 0$ the labyrinth is associative. We can conjecture after the previous observations that the type of labyrinth will depend on the capacity of a structure of preserving its integrity, here, surface tension, which keeps the structure together will play a fundamental role. In this sense, the type of labyrinths a system exhibits will give information of the surface properties of the system.

5. Conclusions

In this work the curvature destabilization of a localized structure had been show as a mechanism for the generation of extended labyrinthine patterns. Furthermore, the mechanism has been shown to exist in a wide range of physical systems including vegetation dynamics through a general interaction-redistribution model, and chemical reactions through the Gray-Scott reaction diffusion equations. It has been shown that depending not only in the context in which labyrinths emerge but also on the parameters considered for simulation/experimentation, different labyrinths emerge and even more, they can be classified as *associative*, *neutral* or *dissociative*, based on the difference between the initial and final connectivity of their structures. Future work will address quantitatively the effect of surface tension in the type of labyrinth formed.

Acknowledgments

I want to thank Professor Marcel G. Clerc for his support and valuable comments during the development of this work. I also thank Professor Mustapha Tlidi for his support. During the development of this work I was supported by CONICYT, Beca de Magister Nacional, number 22130947.

References

[1] Glansdorff P and Prigogine I (1971) *Thermodynamic Theory of Structures, Stability and Fluctuations* (Wiley, New York).
[2] Nicolis G and Prigogine I (1977) *Self-Organization in Non Equilibrium Systems* (J. Wiley & Sons, New York).
[3] Graham R and Tel T (1984) Phys. Rev. Lett. **52** 1.
[4] Pismen L M (2006) *Patterns and Interfaces in Dissipative Dynamics* (Springer, Berlin).
[5] Cross M C and Hohenberg P C (1993) Rev. Mod. Phys. **65** 851.
[6] Descalzi O, Clerc M G, Residori S, and Assanto G (2011) *Localized States in Physics: Solitons and Patterns* (Springer, New York).
[7] Dickstein A J, Erramilli S, Goldstein R E, Jackson D P, and Langer S A (1993) Science **261**.
[8] Ribiere P and Oswald P (1990) Journal de Physique **51** 16.
[9] Oswald P, Baudry J, and Pirkl S (2000) Physics Reports **337** 1.
[10] Heinig P, Helseth L E, and Fischer Th M (2004) New. J. Phys. **6** 189.
[11] von Hardenberg J, Meron E, Shachak M, and Zarmi Y (2001) Phyr. Rev. Lett. **87** 19.
[12] Taranenko V B, Staliunas K, and Weiss C O (1998) Phys. Rev. Lett. **81** 11.
[13] Le Berre M, Ressayre E, Tallet A, Pomeau Y, and Di Menza L (2002) Phys. Rev. E. **66** 026203.
[14] Bordeu I, Clerc M G, Lefever R, and Tlidi M (2015) Communications in Nonlinear Science and Numerical Simulation **29** 482.
[15] Swift J and Hohenberg P C (1977) Phys. Rev. A **15** 319.
[16] Tlidi M, Georgiou M, and Mandel P (1993) Phys. Rev. A **48** 4605.
[17] Tlidi M, Mandel P, and Lefever R (1994) Phys. Rev. Lett. **73** 640.
[18] Bordeu I and Clerc M G (2015) Phys. Rev. E **92** 042915.
[19] Vladimirov A G, Lefever R, and Tlidi M (2011) Phys. Rev. A **84** 043848.
[20] Coullet P, Riera C, and Tresser C (2000) Prog. Theor. Phys. Supp. **139**.
[21] del Campo F, Haudin F, Rojas R G, Bortolozzo U, Clerc M G, and Residori S (2012) Phys. Rev. E **86** 036201.
[22] Lefever R and Lejeune O (1997) Bulletin of Mathematical Biology **59** 263.

[23] Lejeune O and Tlidi M (1999) Journal of Vegetation Science **10** 201.
[24] Juergens N (2013) Science **339**.
[25] Fernandez-Oto C, Tlidi M, Escaff D, and Clerc M G (2014) Philosophical Transactions of the Royal Society A: Mathematical, Physical and Engineering Sciences **372**.
[26] Martínez-García R, Calabrese J M, Hernandez-Garcia E, and Lopez C (2013) Geophysical Research Letters **40** 6143.
[27] Lejeune O, Tlidi M, Couteron P (2002) Phys. Rev. E **66** 010901.
[28] Pearson J E (1993) Science **261** 189.
[29] Lee K J and Swinney H L (1995) Phys. Rev. E **51** 1899.
[30] Koga S, Kuramoto Y (1980) Progress of Theoretical Physics **63** 106.
[31] Vanag V K, Epstein I R (2007) Chaos: An Interdisciplinary Journal of Nonlinear Science **17** 037110.
[32] Lee K J, McCormick W D, Ouyang Q, and Swinney H L (1993) Science **261** 192.
[33] Mazin W, Rasmussen K E, Mosekilde E, Borckmans P, and Dewel G (1996) Mathematics and Computers in Simulation **40** 371.
[34] Sel'Kov E E (1968) European Journal of Biochemistry **4** 79.
[35] Nishiura Y and Ueyama D (2001) Physica D: Nonlinear Phenomena **150** 137.
[36] Gray P and Scott S K (1983) Chemical Engineering Science **38** 29.
[37] Gray P and Scott S K (1984) Chemical Engineering Science **39** 1087.
[38] Gray P and Scott S K (1985) The Journal of Physical Chemistry **89** 22.
[39] Doelman A, Gardner R A, and Kaper T J (1998) Physica D: Nonlinear Phenomena **122** 1.
[40] Hale J K, Peletier L A, and Troy W C (2000) SIAM Journal on Applied Mathematics **61** 102.
[41] Wei J (2001) Physica D: Nonlinear Phenomena **148** 20.
[42] Muratov C B and Osipov V V (2002) SIAM Journal on Applied Mathematics **62** 1463.

Riemannian geometry of fluctuation theory: An introduction

Luisberis Velazquez
Departamento de Física, Universidad Católica del Norte, Av. Angamos 0610, Antofagasta, Chile
E-mail: lvelazquez@ucn.cl

Abstract. Fluctuation geometry was recently proposed as a counterpart approach of Riemannian geometry of inference theory (*information geometry*), which describes the geometric features of the statistical manifold $\mathcal{M}$ of random events that are described by a family of continuous distributions $dp_{\xi}(x|\theta)$. This theory states a connection among geometry notions and statistical properties: separation distance as a measure of relative probabilities, curvature as a measure about the existence of irreducible statistical correlations, among others. In statistical mechanics, fluctuation geometry arises as the mathematical apparatus of a Riemannian extension of Einstein fluctuation theory, which is also closely related to Ruppeiner geometry of thermodynamics. Moreover, the curvature tensor allows to express some asymptotic formulae that account for the system fluctuating behavior beyond the gaussian approximation, while curvature scalar appears as a second-order correction of Legendre transformation between thermodynamic potentials.

1. Introduction

Riemannian geometries defined on statistical manifolds establish a direct correspondence among statistical properties of a parametric family of continuous distributions:

$$dp_{\xi}(x|\theta) = \rho_{\xi}(x|\theta)dx \tag{1}$$

and geometrical notions of certain statistical manifolds $\mathcal{M}$ and $\mathcal{P}$ associated to them. The advantage of these formalisms is that they enable a direct application of powerful tools of Riemannian geometry for statistical analysis. There exist two possible Riemannian geometries in the framework of continuous distribution (1). The first one is Riemannian geometry of inference theory, which is widely known as *information geometry* [1]. Distance notion of this geometry:

$$d\mathfrak{s}^2 = g_{\alpha\beta}(\theta)d\theta^{\alpha}d\theta^{\beta} \tag{2}$$

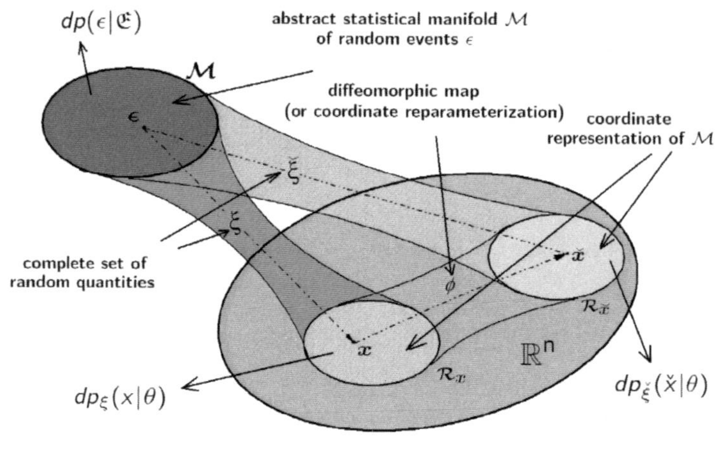

Figure 1. Continuous distributions $dp_\xi(x|\theta)$ and $dp_{\check{\xi}}(\check{x}|\theta)$ are diffeomorphic distributions, that is, a same abstract distribution $dp(\epsilon|\mathfrak{E})$ expressed into two different coordinate representations of the abstract statistical manifold $\mathcal{M}$.

establishes a *statistical separation* between two close distributions of parametric family (1), which characterizes distinguishing probability of these distribution during an statistical inference of control parameters $(\theta, \theta + d\theta)$. The second one is Riemannian geometry of fluctuation theory, or more briefly, *fluctuation geometry* [2, 3, 4]. Its distance notion:

$$ds^2 = g_{ij}(x|\theta)dx^i dx^j \tag{3}$$

establishes a *statistical separation* between two close values $(x, x + dx)$ of a random quantity ξ for a given member of parametric family (1).

A great advantage of differential geometry is the possibility to perform a *coordinate-free treatment*. An important concept here is the notion of *diffeomorphic distributions* [4]. There are those distributions whose random quantities ξ and ζ are related by a diffeomorphism $\phi : \xi \to \check{\xi}$, that is, a bijective and differentiable map that leaves invariant their respective probability distributions (see scheme in Fig.1):

$$\phi : dp_\xi(x|\theta) = dp_{\check{\xi}}(\check{x}|\theta) \Rightarrow \rho_{\check{\xi}}(\check{x}|\theta) = \rho_\xi(x|\theta)\left|\frac{\partial \check{x}}{\partial x}\right|^{-1}. \tag{4}$$

All these distributions are regarded as different representations of *a same abstract distribution* defined on the manifolds $\mathcal{M}$ y $\mathcal{P}$.

A simple example of transformation among random quantities is the one associated with *Box-Muller transformation* [5]:

$$\zeta_1 = \sqrt{-2\ln(\xi_1)}\cos(2\pi\xi_2) \text{ and } \zeta_2 = \sqrt{-2\ln(\xi_1)}\sin(2\pi\xi_2) \tag{5}$$

which is employed to generate Gaussian random numbers ζ_1 and ζ_2 from uniform random numbers ξ_1 and ξ_2. Continuous distribucions whose associated manifolds $\mathcal{M}$ are diffeomorphic to the real one-dimensional space $\mathbb{R}$ are always diffeomorphic distributions because of the only possible Riemannian geometry for these manifolds is the Euclidean one. In particular, *Gaussian distribution*:

$$dp_\xi(x|\mu,\sigma) = \frac{1}{\sqrt{2\pi}\sigma}\exp\left[-(x-\mu)^2/2\sigma^2\right]dx,\ -\infty < x < +\infty. \tag{6}$$

Cauchy distribution:

$$dp_{\breve{\xi}}(\breve{x}|\nu,\gamma) = \frac{1}{\pi}\frac{\gamma d\breve{x}}{\gamma^2+(\breve{x}-\nu)^2},\ -\infty < \breve{x} < +\infty. \tag{7}$$

Bimodal Gaussian distribution:

$$dp_{\tilde{\xi}}(\tilde{x}|\mu,\sigma) = \frac{1}{2\sqrt{2\pi}\sigma}\left\{\exp\left[-(\tilde{x}-\mu)^2/2\sigma^2\right]+\exp\left[-(\tilde{x}+\mu)^2/2\sigma^2\right]\right\}d\tilde{x}, -\infty < \tilde{x} < +\infty. \tag{8}$$

are fully equivalent from this geometric perspective, namely, all they can be regarded as different representations of a same abstract distribution. Of course, not all distributions can be regarded as diffeomorphic distributions. For random quantities ξ whose abstract statistical manifold $\mathcal{M}$ has a dimension $n \geq 2$ are possible the notions of curvature and statistical correlations. In particular, distributions family [4]:

$$dp_\xi(x,y|\theta) = \frac{1}{\mathcal{Z}(\theta)}\exp\left[-\frac{1}{2}(x^2+y^2)\right]\frac{\theta dxdy}{2\pi\sqrt{x^2+y^2+\theta^2}} \tag{9}$$

with normalization constant:

$$\mathcal{Z}(\theta) = \sqrt{\pi}e^{\frac{1}{2}\theta^2}\frac{\theta}{\sqrt{2}}\mathrm{erfc}\left(\frac{\theta}{\sqrt{2}}\right) \tag{10}$$

can be associated with curved geometry of surface of revolution represented in Fig.2. This family cannot be map to the product of two Gaussian distributions:

$$dp_\zeta(x,y|\sigma) = \frac{1}{2\pi\sigma}\exp\left[-\left(x^2+y^2\right)/2\sigma\right]dxdy \tag{11}$$

because of this last has Euclidean geometry of two-dimensional real space $\mathbb{R}^2$. Geometrical non-equivalence means that distributions (9) cannot be decomposed into the product of two independent distributions.

2. Fundamental equations and results of fluctuation geometry

For the sake of simplicity in notations, let us hereinafter omit the subindex of random quantity ξ in all mathematical expressions. Riemannian structure of the statistical manifold $\mathcal{M}$ allows us to introduce the *invariant volume element* $d\mu(x|\theta)$:

$$d\mu(x|\theta) = \sqrt{|g_{ij}(x|\theta)/2\pi|}dx, \tag{12}$$

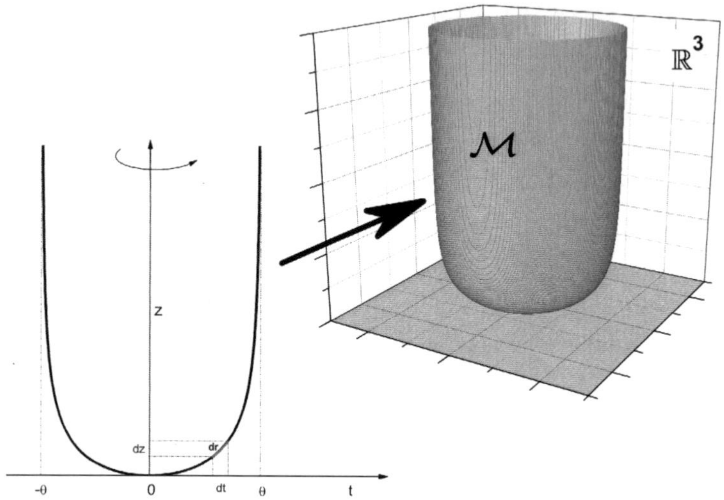

Figure 2. The geometry of the statistical manifold $\mathcal{M}$ associated with the distributions family (9) is fully equivalent to curved geometry defined on the revolution surface represented here.

which replaces the ordinary volume element dx (Lebesgue measure) that is employed in equation (1). The notation $|T_{ij}|$ represents the determinant of a given tensor T_{ij} of second-rank, while the factor 2π has been introduced for convenience. Additionally, one can define the *probabilistic weight* [3]:

$$\omega(x|\theta) = \rho(x|\theta)\sqrt{|2\pi g^{ij}(x|\theta)|}, \tag{13}$$

which is a scalar function that arises as *a local invariant measure of the probability.* Although the mathematical form of the probabilistic weight $\omega(x|\theta)$ depends on the coordinates representations of the statistical manifolds $\mathcal{M}$ and $\mathcal{P}$; the values of this function are the same in all coordinate representations. Using the above notions, the family of continuous distributions (1) can be rewritten as follows:

$$dp(x|\theta) = \omega(x|\theta)d\mu(x|\theta), \tag{14}$$

which is a form that explicitly exhibits the invariance of this family of distributions.

The notion of probability weight $\omega(x|\theta)$ can be employed to redefine the notion of information entropy for continuous distributions [6]:

$$\mathcal{S}_d\left[\omega|g,\mathcal{M}\right] = -\int_{\mathcal{M}} \omega(x|\theta)\log\omega(x|\theta)d\mu(x|\theta). \tag{15}$$

as a global invariant measure that depends on the metric tensor $g_{ij}(x|\theta)$ of the manifold $\mathcal{M}$. The quantity $\mathfrak{I}(x|\theta)$:

$$\mathfrak{I}(x|\theta) = -\log\omega(x|\theta) \tag{16}$$

Journal of Physics: Conference Series **720** (2016) 012005 doi:10.1088/1742-6596/720/1/012005

represents a *local invariant measurement of the information content*, where differential entropy (15) exhibits the same value for all diffeomorphic distributions. Introducing the *information potential* $\mathcal{S}(x|\theta)$ as the negative of the information content (16):

$$\mathcal{S}(x|\theta) = \log \omega(x|\theta) \equiv -\mathfrak{I}(x|\theta), \tag{17}$$

the metric tensor can be rewritten as follows [3]:

$$g_{ij}(x|\theta) = -D_i D_j \mathcal{S}(x|\theta) = -\frac{\partial^2 \mathcal{S}(x|\theta)}{\partial x^i \partial x^j} + \Lambda^k_{ij}(x|\theta)\frac{\partial \mathcal{S}(x|\theta)}{\partial x^k}, \tag{18}$$

where D_i is the *covariant derivative* associated with the Levi-Civita affine connections $\Lambda^k_{ij}(x|\theta)$ [10]:

$$\Lambda^k_{ij}(x|\theta) = g^{km}(x|\theta)\frac{1}{2}\left[\frac{\partial g_{im}(x|\theta)}{\partial x^j} + \frac{\partial g_{jm}(x|\theta)}{\partial x^i} - \frac{\partial g_{ij}(x|\theta)}{\partial x^m}\right]. \tag{19}$$

Covariant set of differential equations (18) can be rewritten into the alternative form:

$$g_{ij}(x|\theta) = -\frac{\partial^2 \log \rho(x|\theta)}{\partial x^i \partial x^j} + \Lambda^k_{ij}(x|\theta)\frac{\partial \log \rho(x|\theta)}{\partial x^k} + \frac{\partial \Lambda^k_{jk}(x|\theta)}{\partial x^i} - \Lambda^k_{ij}(x|\theta)\Lambda^l_{kl}(x|\theta) \tag{20}$$

in terms of probability density. According to expression (18), the metric tensor $g_{ij}(x|\theta)$ defines a positive definite distance notion (3), while the information potential $\mathcal{S}(x|\theta)$ is locally concave everywhere. This last behavior guarantees the uniqueness of the point $\bar{x}$ where the information potential reaches a global maximum, that is, the uniqueness of the point of global maximum $\bar{x}$ of the probabilistic weight $\omega(x|\theta)$.

The main consequence derived from equation (18) is the possibility to rewrite the distributions family (14) into the following *Riemannian gaussian representation* [2, 3]:

$$dp(x|\theta) = \frac{1}{\mathcal{Z}(\theta)}\exp\left[-\frac{1}{2}\ell^2_\theta(x,\bar{x})\right]d\mu(x|\theta), \tag{21}$$

where $\ell_\theta(x,\bar{x})$ denotes the *separation distance* between the arbitrary point x and the point $\bar{x}$ with maximum information potential $\mathcal{S}(x|\theta)$ (the arc-length Δs of the *geodesics* that connects these points). Moreover, the negative of the logarithm of gaussian partition function $\mathcal{Z}(\theta)$ defines the so-called *gaussian potential*:

$$\mathcal{P}(\theta) = -\log \mathcal{Z}(\theta), \tag{22}$$

which appears as the first integral of the problem (18):

$$\mathcal{P}(\theta) = \mathcal{S}(x|\theta) + \frac{1}{2}\psi^2(x|\theta). \tag{23}$$

Here, $\psi^2(x|\theta) = \psi^i(x|\theta)\psi_i(x|\theta) = g^{ij}(x|\theta)\psi_i(x|\theta)\psi_j(x|\theta)$ is the square norm of covariant vector field $\psi_i(x|\theta)$ defined by the gradient of the information potential $\mathcal{S}(x|\theta)$:

$$\psi_i(x|\theta) = -D_i\mathcal{S}(x|\theta) \equiv -\partial\mathcal{S}(x|\theta)/\partial x^i. \tag{24}$$

Journal of Physics: Conference Series **720** (2016) 012005 doi:10.1088/1742-6596/720/1/012005

The factor 2π of definition (12) guarantees that the gaussian partition function $\mathcal{Z}(\theta)$ drops the unity when the Riemannian structure of the manifold $\mathcal{M}$ is the same of Euclidean real space $\mathbb{R}^n$.

Riemannian gaussian representation (46) rephrases the distributions family (1) in term of geometric notions of the manifold $\mathcal{M}$. According to this result, the distance $\ell_\theta(x,\bar{x})$ is a measure of the *occurrence probability* of a deviation from the state $\bar{x}$ with maximum information potential. This result can be obtained combining equations (14) and (23) with the following the identity:

$$\psi^2(x|\theta) \equiv \ell^2_\theta(x,\bar{x}). \tag{25}$$

This last relation is a consequence of the *geodesic character* of the curves $x_g(s) \in \mathcal{M}$ derived from the following set of ordinary differential equations [3]:

$$\frac{dx^i_g(s)}{ds} = \upsilon^i \left[x_g(s)|\theta\right]. \tag{26}$$

Here, $\upsilon^i(x|\theta) = g^{ij}(x|\theta)\upsilon_j(x|\theta)$ is the contravariant form of the unitary vector field $\upsilon_i(x|\theta)$ associated with the vector field (24):

$$\upsilon_i(x|\theta) = \psi_i\left(x|\theta\right)/\psi\left(x|\theta\right), \tag{27}$$

while the parameter s is the arc-length of the curve $x_g(s)$. It is easy to check that this unitary vector field obeys the geodesic differential equation:

$$\upsilon^j(x|\theta)D_j\upsilon_i(x|\theta) = \upsilon^j(x|\theta)\left[g_{ij}(x|\theta) - \upsilon_i(x|\theta)\upsilon_j(x|\theta)\right] \equiv 0. \tag{28}$$

Identity (25) follows from the directional derivatives:

$$\frac{d\mathcal{S}\left(x_g(s)|\theta\right)}{ds} \equiv \psi(x_g(s)|\theta) \text{ and } \frac{d^2\mathcal{S}\left(x_g(s)|\theta\right)}{ds^2} \equiv -1, \tag{29}$$

which can be obtained from equation (26).

Let us now talk about the notion of curvature of fluctuation geometry. The affine connections $\Lambda^k_{ij} = \Lambda^k_{ij}(x|\theta)$ are employed to introduce of the *curvature tensor* $R^l_{ijk} = R^l_{ijk}(x|\theta)$ of the manifold $\mathcal{M}$:

$$R^l_{ijk} = \frac{\partial}{\partial X^i}\Lambda^l_{jk} - \frac{\partial}{\partial X^j}\Lambda^l_{ik} + \Lambda^l_{im}\Lambda^m_{jk} - \Lambda^l_{jm}\Lambda^m_{ik}. \tag{30}$$

Generally, the affine connections $\Lambda^k_{ij}(x|\theta)$ and the metric tensor $g_{ij}\left(x|\theta\right)$ are independent entities of Riemannian geometry. However, the knowledge of the metric tensor allows to introduce natural affine connections: the Levi-Civita connections (19). These affine connections are also referred to in the literature as the metric connections or the Christoffel symbols. The same ones follow from the consideration of a *torsion-free covariant differentiation* D_i that obeys the *condition of Levi-Civita parallelism* [10]:

$$D_k g_{ij}\left(x|\theta\right) = 0. \tag{31}$$

Journal of Physics: Conference Series **720** (2016) 012005 doi:10.1088/1742-6596/720/1/012005

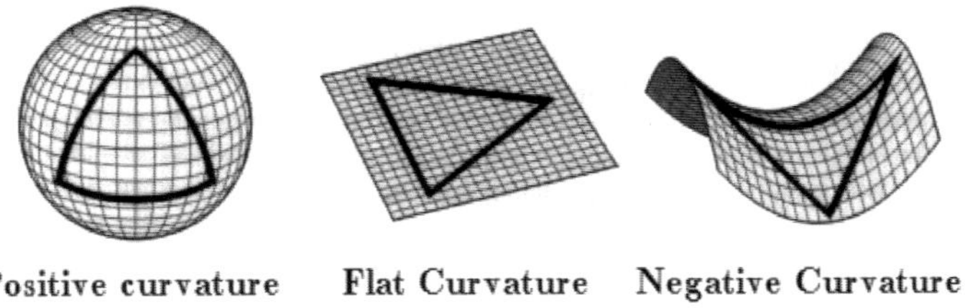

Figure 3. Curvature characterizes the deviation of local geometric properties of a manifold from the properties of the Euclidean geometry.

Using the Levi-Civita connections, the curvature tensor can be expressed in terms of the metric tensor $g_{ij}(x|\theta)$ and its first and second partial derivatives. Additionally, one can introduce the *Ricci curvature tensor* $R_{ij}(x|\theta)$:

$$R_{ij}(x|\theta) = R^{k}_{kij}(x|\theta) \tag{32}$$

as well as the *curvature scalar* $R(x|\theta)$:

$$R(x|\theta) = g^{ij}(x|\theta)R^{k}_{kij}(x|\theta) = g^{ij}(x|\theta)g^{kl}(x|\theta)R_{kijl}(x|\theta). \tag{33}$$

According to Riemannian geometry [10], the curvature scalar $R(x|\theta)$ is the only invariant derived from the first and second partial derivatives of the metric tensor $g_{ij}(x|\theta)$.

The curvature tensor characterizes the deviation of local geometric properties of a manifold $\mathcal{M}$ from the properties of the Euclidean geometry (see scheme of Fig.3). For example, the volume of a small sphere about a point x has smaller (larger) volume (area) than a sphere of the same radius defined on the n-dimensional real space $\mathbb{R}^n$ when the scalar curvature $R(x|\theta)$ is positive (negative) at that point. Quantitatively, this behavior is described by the following approximation formulae:

$$\frac{\mathrm{Vol}\left[\mathbb{S}^{(n-1)}(x|\ell)\subset\mathcal{M}\right]}{\mathrm{Vol}\left[\mathbb{S}^{(n-1)}(x|\ell)\subset\mathbb{R}^n\right]} = 1-\frac{R(x|\theta)}{6(n+2)}\ell^2+O(\ell^4), \tag{34}$$

$$\frac{\mathrm{Area}\left[\mathbb{S}^{(n-1)}(x|\ell)\subset\mathcal{M}\right]}{\mathrm{Area}\left[\mathbb{S}^{(n-1)}(x|\ell)\subset\mathbb{R}^n\right]} = 1-\frac{R(x|\theta)}{6n}\ell^2+O(\ell^4), \tag{35}$$

where the notation $\mathbb{S}^{(m)}(x|\ell)$ represents a m-dimensional sphere with small radius ℓ centered at the point x. Accordingly, the local effects associated with the curvature of the manifold $\mathcal{M}$ appears as second-order (and higher) corrections of the Euclidean formulae. The corresponding asymptotic formula for distribution (46) using spherical coordinates (ℓ, q) for radius ℓ sufficiently small:

$$dp(\ell,q|\theta) = \frac{1}{\mathcal{Z}(\theta)}\left[1-\frac{1}{24}\ell^2\mathcal{F}(q|\theta)+O(\ell^4)\right]dp_G(\ell,q|\theta). \tag{36}$$

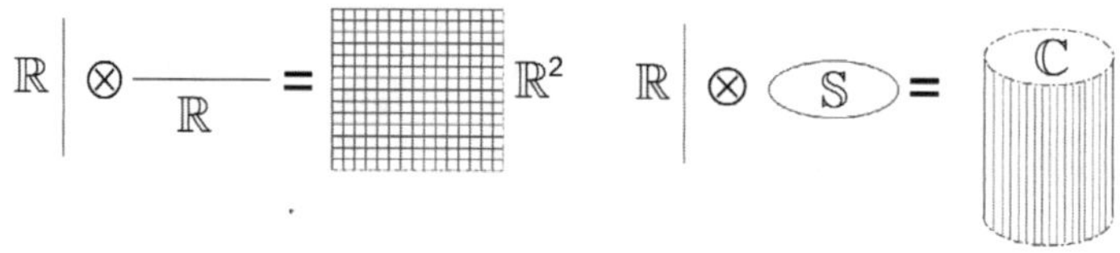

Figure 4. Schematic representation about the Cartesian product of manifolds.

Here, $dp_G(\ell, q|\theta)$ denotes the spherical coordinate representation of a gaussian distribution associated with the local Euclidean properties of the manifold $\mathcal{M}$ at the point $\bar{x}$ with maximum information potential:

$$dp_G(\ell, q|\theta) = \exp\left(-\frac{1}{2}\ell^2\right)\frac{\ell^{n-1}d\ell}{\sqrt{2\pi}}\sqrt{\left|\frac{\kappa_{\alpha\beta}(q)}{2\pi}\right|}dq. \tag{37}$$

where $\kappa_{\alpha\beta}(q) = \bar{g}_{ij}\xi^i_\alpha(q)\xi^j_\beta(q)$. The $(n-1)$ vector fields $\xi_\alpha(q) = \{\xi^i_\alpha(q)\}$ are obtained from the unitary vector field $\mathbf{e}(q) = \{e^i(q)\}$ associated with the spherical coordinates at the point $\bar{x}$ as follows:

$$\xi^i_\alpha(q) = \frac{\partial e^i(q)}{\partial q^\alpha}. \tag{38}$$

$\mathcal{F}(q|\theta)$ is a function on the spherical coordinates q defined as follows:

$$\mathcal{F}(q|\theta) = \bar{R}_{ijkl}\kappa^{\alpha\beta}(q)S^{ij}_\alpha(q)S^{kl}_\beta(q), \tag{39}$$

which is referred to as the *spherical function*. Moreover, $\bar{R}_{ijkl} = R_{ijkl}(\bar{x}|\theta)$ is the curvature tensor evaluated at the point $\bar{x}$, while the quantities $S^{ij}_\alpha(q)$ are defined as:

$$S^{ij}_\alpha(q) = e^i(q)\xi^j_\alpha(q) - e^j(q)\xi^i_\alpha(q). \tag{40}$$

Curvature of statistical manifold $\mathcal{M}$ is directly related to the notion of irreducible statistical correlations. Specifically, it is said that a continuous distribution $dp(x|\theta)$ exhibits a *reducible statistical dependence* if it possesses a diffeomorphic distribution $dp(\check{x}|\theta)$ that admits to be decomposed into independent distribution functions $dp^{(i)}(\check{x}^i|\theta)$ for each coordinate as follows:

$$dp(\check{x}|\theta) = \prod_{i=1}^{n} dp^{(i)}(\check{x}^i|\theta). \tag{41}$$

Otherwise, the distribution function $dp(x|\theta)$ exhibits an *irreducible statistical dependence*. The existence (or nonexistence) of a reducible statistical dependence for a given distributions family (1) is fully equivalent to the existence (or nonexistence) of a *Cartesian decomposition* of its associated statistical manifold $\mathcal{M}$ into two (or more) independent statistical manifolds $\left\{\mathcal{A}^{(i)}_\theta\right\}$:

$$\mathcal{M} = \mathcal{A}^{(1)} \otimes \mathcal{A}^{(2)} \ldots \otimes \mathcal{A}^{(l)}. \tag{42}$$

Journal of Physics: Conference Series **720** (2016) 012005 doi:10.1088/1742-6596/720/1/012005

A given manifold $\mathcal{A}$ is said to be an *irreducible manifold* when the same one does not admit the Cartesian decomposition (42). Moreover, a given Cartesian decomposition (42) is said to be an *irreducible Cartesian decomposition* if each independent manifold $\mathcal{A}^{(k)}$ is an irreducible manifold. In general, the question about the Cartesian decomposition of a Riemannian manifold into independent manifolds with arbitrary dimensions is better phrased and understood in the language of *holonomy groups*. The relation of holonomy of a connection with the curvature tensor is the main content of *Ambrose-Singer theorem*, while *de Rham theorem* states the conditions for a global Cartesian decomposition [10]. The flat character of the statistical manifold $\mathcal{M}$ implies the existence of a reducible statistical dependence for the family of distributions (1), while its curved character implies the existence of an irreducible statistical dependence.

3. Relevance in statistical mechanics

Redefining information potential (17) in units of Boltzmann constant k, the probability distribution (14) can be rewritten as follow:

$$dp(x|\theta) = \exp\left[\mathcal{S}(x|\theta)/k\right] d\mu(x|\theta). \tag{43}$$

Formally, this expression represents a sort of *covariant extension of Einstein postulate* of classical fluctuation theory [11], where the information potential $\mathcal{S}(x|\theta)$ is identified with the *thermodynamic entropy of closed system* (up to the precision of an additive constant). Hereinafter, the coordinates $x = (x^1, x^2, \ldots, x^n)$ are the relevant macroscopic observables of the closed system, e.g.: the internal energy U, the volume V, the total angular momentum $\mathbf{M}$, the magnetization $\mathcal{M}$, etc. Moreover, θ represents the set of control parameters of the given situation of thermodynamic equilibrium. The metric tensor $g_{ij}(x|\theta)$ of fluctuation geometry:

$$g_{ij}(x|\theta) = -D_i D_j \mathcal{S}(x|\theta) = -\partial_i \partial_i \mathcal{S}(x|\theta) + \Lambda^k_{ij}(x|\theta)\partial_k \mathcal{S}(x|\theta) \tag{44}$$

establishes a constraint between the entropy $\mathcal{S}(x|\theta)$ and the metric tensor $g_{ij}(x|\theta)$ of the statistical manifold $\mathcal{M}$ of macroscopic observables x. Expression (44) provides a generalization for the thermodynamic metric tensor of Ruppeiner geometry [7, 8]:

$$g_{ij}(\bar{x}) = -\frac{\partial^2 S(\bar{x}|\theta)}{\partial x^i \partial x^j}, \tag{45}$$

while the Riemannian Gaussian representation:

$$dp(x|\theta) = \frac{1}{\mathcal{Z}(\theta)} \exp\left[-\frac{1}{2k}\ell^2_\theta(x,\bar{x})\right] d\mu(x|\theta), \tag{46}$$

is an exact improvement of Gaussian approximation of classical fluctuation theory [11]:

$$dp(x|\theta) \simeq \exp\left[-g_{ij}(\bar{x})(x-\bar{x})^i(x-\bar{x})^j/2k\right] \sqrt{|g_{ij}(\bar{x}|\theta)/2\pi k|} d^n x. \tag{47}$$

According to the asymptotic formula (36), curvature characterizes deviation exact distribution beyond Gaussian approximation for thermodynamical fluctuations.

Applicability of Gaussian approximation is breakdown during the occurrence of phase transitions and critical phenomena, and therefore, curvature can be a useful notion to study these situations. It is noteworthy that similar connections have been established for curvature notion of information geometry [12, 13], as well as Ruppeiner geometry [9]. Direct application of the asymptotic formula (36) for the case of Boltzmann-Gibbs distributions [11]:

$$dp(x|\theta) = \frac{1}{Z(\theta)} \exp\left[-\theta_i x^i/k\right] \Omega(x) dx \tag{48}$$

can be employed to obtain the following result [4]:

$$P(\theta) \simeq \theta_i \bar{x}^i - s(\bar{x}|\theta) + k^2 R(\bar{x}|\theta)/6. \tag{49}$$

Here, $P(\theta)$ is the *Planck thermodynamic potential* [11]:

$$P(\theta) = -k \log Z(\theta), \tag{50}$$

while $s(x|\theta)$ is referred to as the *entropy of the open system*. This function is directly associated with the density of states $\Omega(x)$ via the metric tensor $g_{ij}(x|\theta)$:

$$\exp\left[s(x|\theta)/k\right] \sqrt{\left|\frac{g_{ij}(x|\theta)}{2\pi k}\right|} \equiv \Omega(x). \tag{51}$$

The entropy $s(x|\theta)$ is not an intrinsic property of the open system. Certainly, this entropy also depends on the metric tensor $g_{ij}(x|\theta)$, which accounts for the underlying environmental influence. Result (49) exhibits a very simple interpretation. Gaussian or zeroth-order approximation:

$$P(\theta) \simeq \bar{P}(\theta) = \theta_i \bar{x}^i - s(\bar{x}|\theta). \tag{52}$$

is just the known *Legendre transformation* that estimates the Planck thermodynamic potential $P(\theta)$ from the entropy of the open system $s(x|\theta)$. The curvature scalar $R(\bar{x}|\theta)$ introduces a correction of second-order of this transformation.

References

[1] Amari Sh 1990 *Differential-Geometrical Methods in Statistics: Lecture notes in Statistics* Vol. 28 (Berlin: Springer).
[2] Velazquez L 2011 *J. Stat. Mech.* **P11007**.
[3] Velazquez L 2012 *J. Phys. A: Math and Theo* **45** 175002.
[4] Velazquez L 2013 *J. Phys. A: Math and Theo* **46** 345003.
[5] Box G E P and Muller M E 1958 *Annals Math. Stat.* **29** 610.
[6] Jaynes E T 1963 *Information Theory and Statistical Mechanics*, in *Statistical Physics*, Ford K ed (New York: Benjamin).
[7] Ruppeiner G 1979 *Phys. Rev. A* **20** 1608.
[8] Ruppeiner G 1995 *Rev. Mod. Phys.* **67** 605.
[9] Ruppeiner G 2010 *Am. J. Phys.* **78** 1170.
[10] Berger A 2002 *A panoramic view of Riemannian geometry* (Berlin: Springer).
[11] Reichl L E 1980 *A modern course in Statistical Mechanics*, (Austin, TX: University of Texas Press).
[12] Janke W *et al* 2004 *Physica A* **336** 181-6.
[13] Brody D C and Hook D W 2009 *J. Phys. A: Math. Theor.* **42** 023001.

The Maximum Caliber principle applied to continuous systems

Diego González, Sergio Davis
Grupo de Nanomateriales, Departamento de Física, Facultad de Ciencias, Universidad de Chile
E-mail: dgonzalez@gnm.cl

Abstract. We give a brief presentation of the maximization of path entropy (Maximum Caliber) formalism as an approach to inference over trajectories. We develop its continuous-time version involving path integrals and present an identity between expectations over the most unbiased probability distribution of trajectories. This relation between expectations can be used to solve for the Lagrange multipliers in a Maximum Caliber problem without the need for the dynamical partition function.

1. Introduction

Although theoretically equilibrium is the fundamental concept in Thermodynamics, in practice most systems studied in a laboratory are systems out of equilibrium. This closeness to reality is what gives tremendous importance to the development of a formalism for describing these systems. Finding a formalism based on as few as possible principles and consistent with the principle of maximum entropy [1] in the case of equilibrium, which is still valid for a time-dependent treatment of statistical properties of a system, is the goal of much ongoing research. A proposal which has recently gained visibility in the statistical physics community is the principle of maximum path entropy or maximum *caliber* [2, 3].

In this framework, a probability functional associates a probability to every possible path the system can take when going from point A to point B. This probability functional can be directly used to predict expectations of functional properties and also instantaneous properties (through a *time-slicing* procedure), in what promises to become a useful tool for non-equilibrium systems. It is also expected that the non-equilibrium formalism by Evans *et al* [4] can be recovered from this single principle.

2. The Maximum Caliber principle

In the same way that the principle of Maximum Entropy (MaxEnt for short) produces the most unbiased probability distribution consistent with some constraint $\left\langle f(\vec{x})\right\rangle = F$ by maximizing the Gibbs-Shannon entropy functional

$$S[p] = -\int d\vec{x} p(\vec{x}) \ln p(\vec{x}) \tag{1}$$

under said constraints, the principle of Maximum Caliber (MaxCal) replaces the microstates $\vec{x}$ with micro-trajectories $x(t)$. Then, the functional corresponding to the Gibbs-Shannon entropy is called the *caliber*,

$$C[p] = -\int Dx()p[x()]\ln p[x()] \tag{2}$$

where the integral is now a functional integral over all trajectories (denoted by $Dx()$) and $p[x()]$ is a probability functional.

When the caliber in Eq. 2 is maximized under the constraints of normalization,

$$\int Dx()P[x()] = \Big\langle 1 \Big\rangle = 1, \tag{3}$$

and a known expectation of a functional

$$\Big\langle F[x()] \Big\rangle = \int Dx()P[x()]F[x()] = F_0, \tag{4}$$

the most unbiased distribution of trajectories is of the form

$$P[x()] = \frac{1}{Z}\exp(-\lambda F[x()]), \tag{5}$$

with Z a normalization constant, given by

$$Z = \int Dx()\exp(-\lambda F[x()]). \tag{6}$$

The Lagrange multiplier λ can be obtained, as in MaxEnt, from

$$-\frac{\partial}{\partial \lambda}\ln Z(\lambda) = F_0. \tag{7}$$

Equivalently, if an instantaneous property is known in expectation for each time t in an interval, $t \in [0, T]$,

$$\Big\langle G(x(t), \dot{x}(t), t) \Big\rangle = \int Dx()P[x()]G[x(); t] = g(t) \tag{8}$$

the probability functional is of the form

$$P[x()] = \frac{1}{\eta}\exp\left(-\int_0^T dt\lambda(t)G(x(t), \dot{x}(t); t)\right). \tag{9}$$

This is equivalent to the first kind of constraint (Eq. 4), because clearly if we integrate Eq. 8 in t from 0 to T we see that we are constraining the expectation of a *functional*

$$F[x()] = \int_0^T dt\lambda(t)G(x(t), \dot{x}(t); t), \tag{10}$$

in the sense of Eq. 4.

In the case of constraints as expressed in Eq. 8, the Lagrange multiplier function can be obtained, in an analogous way to Eq. 7, from

$$-\frac{\delta}{\delta\lambda(t)}\ln Z[\lambda()] = g(t). \tag{11}$$

If the instantaneous property G depends only on the instantaneous position $x(t)$ and velocity $\dot{x}(t)$ (and possible, explicitly on time), the functional F has the form of the classical action $A[x()]$ in mechanics

$$A[x()] = \int_0^T dt\mathcal{L}(x(t), \dot{x}(t); t), \tag{12}$$

and, in fact, if $\lambda > 0$ the most probable trajectory is the one which minimizes $A[x()]$, i.e., the principle of minimum action [5].

3. Conjugate variables theorem (CVT) for functionals

Once given the form of the probability functional (Eq. 5), we look for analytical tools that allows us to evaluate expectations of different properties. In MaxEnt there is such an identity, known as the conjugate variables theorem (CVT) [6], itself a direct consequence of the divergence theorem.

For an arbitrary distribution $p(\vec{x})$, we have

$$\Big\langle \nabla \cdot \vec{v}(\vec{x}) \Big\rangle + \Big\langle \vec{v}(\vec{x}) \cdot \nabla \ln p \Big\rangle = 0. \tag{13}$$

where the expectations are taken under the distribution p, and $\vec{v}(\vec{x})$ is an arbitrary (differentiable) vector field. We look for a generalization for MaxCal, i.e., when $\vec{x} \to x(t)$.

Consider the functional derivative of a product of functionals,

$$\frac{\delta}{\delta x(t)}\Big(P[x()]G[x()]\Big) = \frac{\delta P[x()]}{\delta x(t)}G[x()] + \frac{\delta G[x()]}{\delta x(t)}P[x()]. \tag{14}$$

If we rewrite the functional derivative of $P[x()]$ as

$$\frac{\delta P[x()]}{\delta x(t)} = P[x()]\frac{\delta}{\delta x(t)} \ln P[x()], \tag{15}$$

and replace in Eq. 14, we get

$$\frac{\delta}{\delta x(t)}\Big(P[x()]G[x()]\Big) = P[x()]\left[G[x()]\frac{\delta}{\delta x(t)} \ln P[x()] + \frac{\delta G[x()]}{\delta x(t)}\right]. \tag{16}$$

Now we integrate over all trajectories $x()$,

$$\int Dx()\frac{\delta}{\delta x(t)}\Big(P[x()]G[x()]\Big) = \int Dx()P[x()]G[x()]\frac{\delta}{\delta x(t)} \ln P[x()] + \int Dx()P[x()]\frac{\delta G[x()]}{\delta x(t)}, \tag{17}$$

and, as the probability functional P can be normalized, the "surface integral" on the left hand side vanishes. Therefore, we arrive at the following identity, generalization of Eq. 13 for functionals,

$$\Big\langle \frac{\delta G[x()]}{\delta x(t)} \Big\rangle + \Big\langle G[x()]\frac{\delta}{\delta x(t)} \ln P[x()] \Big\rangle = 0, \tag{18}$$

where $G[x()]$ corresponds to an arbitrary functional. For a Maximum Caliber probability functional (Eq. 5), this reduces to

$$\Big\langle \frac{\delta G[x()]}{\delta x(t)} \Big\rangle = \lambda \Big\langle G[x()]\frac{\delta}{\delta x(t)} F[x()] \Big\rangle. \tag{19}$$

This means we can avoid Eq. 7 to solve for λ, simply obtaining it from

$$\lambda = \frac{\Big\langle \frac{\delta}{\delta x(t)} G[x()] \Big\rangle}{\Big\langle G[x()]\frac{\delta}{\delta x(t)} F[x()] \Big\rangle}. \tag{20}$$

Using Eq. 19 with a constant functional, $G[x()] = 1$, we see that

$$\Big\langle \frac{\delta}{\delta x(t)} F[x()] \Big\rangle = \Big\langle \frac{\delta}{\delta x(t)} A[x()] \Big\rangle = 0, \tag{21}$$

that is, the principle of minimum action also holds in expectation. A particular case of this relation is the result in Ref. [5],

$$\Big\langle \dot{p}(t) \Big\rangle = -\Big\langle \Phi'(x(t)) \Big\rangle, \tag{22}$$

which is the expectation of Newton's second law.

4. Conclusions

We have shown that the Maximum Caliber principle is a powerful tool to study dynamical systems in a probabilistic framework. A generalization of the conjugate variables theorem (CVT) is presented for probability functionals, from which we prove that the minimum action principle holds also in expectation over the ensemble of trajectories. This "functional CVT" can also be used to find novel relations between expectations of functionals.

Acknowledgments

Both authors acknowledge funding from FONDECYT grant 1140514. DG also acknowledges funding from CONICYT PhD fellowship 21140914.

References

[1] Jaynes E T 1957 *Physical Review* **106** 620–630
[2] Jaynes E T 1980 *Ann. Rev. Phys. Chem.* **31** 579–601
[3] Pressé S, Ghosh K, Lee J and Dill K A 2013 *Reviews of Modern Physics* **85** 1115–1141
[4] Evans D J, Searles D J and Williams S R 2008 *J. Chem. Phys.* **128** 014504
[5] González D, Davis S and Gutiérrez G 2014 *Found. Phys.* **44** 923
[6] Davis S and Gutiérrez G 2012 *Phys. Rev. E* **86** 051136

Inverse temperature in Superstatistics

Humberto Loguercio, Sergio Davis
Grupo de Nanomateriales, Departamento de Física, Facultad de Ciencias, Universidad de Chile
E-mail: hloguercio@gmail.com

Abstract. In this work, it is shown that there are (at least) three alternative definitions of the inverse temperature for a non-canonical ensemble. These definitions coincide in expectation but, in general, not in their higher moments. We explore in detail the application to the recent formalism of Superstatistics (C. Beck, 2003), and, in particular, to the configurational probability distribution in the microcanonical ensemble.

1. Introduction

In Statistical Mechanics the canonical ensemble is the ubiquitous probabilistic model. This is the correct model for a system immersed in a thermal bath at temperature T, and it is given by

$$P(\vec{r},\vec{p}|\beta) = \frac{\exp(-\beta\mathcal{H}(\vec{r},\vec{p}))}{Z(\beta)} \tag{1}$$

with $\beta = 1/k_B T$ and $\mathcal{H}$ the Hamiltonian of the system. Despite its undisputed utility, there are systems (for instance, out of equilibrium systems and also in the so called *non-extensive* systems) for which predictions using the canonical ensemble are not verified. For these cases, the idea of Superstatistics [1] has been recently proposed. Superstatistics assumes the correct ensemble is not canonical, but a superposition of canonical ensembles at different (inverse) temperatures, weighted by a factor $f(\beta)$,

$$\exp(-\beta E) \to \int d\beta f(\beta)\exp(-\beta E)$$

In a Bayesian interpretation, Superstatistics is nothing but marginalization over the β parameter, thus

$$P(\vec{r},\vec{p}|\mathcal{I}) = \int_0^\infty d\beta P(\beta|\mathcal{I})P(\vec{r},\vec{p}|\beta) = \int_0^\infty d\beta P(\beta|\mathcal{I})\frac{\exp(-\beta\mathcal{H}(\vec{r},\vec{p}))}{Z(\beta)} \tag{2}$$

which means we have to make the identification

$$f(\beta) = \frac{P(\beta|\mathcal{I})}{Z(\beta)}. \tag{3}$$

Above, $\mathcal{I}$ represents our state of knowledge with respect to the system, which leads to the particular assignment of $f(\beta)$ against other possible distributions.

2. Temperature in non-canonical ensembles

For an arbitrary probabilistic model $P(\vec{x}|\mathcal{I})$ of continuous variables $\vec{x}$ based upon a state of knowledge $\mathcal{I}$, it can be shown [2], by using the divergence theorem that the following identity holds,

$$\Big\langle \nabla \cdot \vec{v}(\vec{x}) \Big\rangle_{\mathcal{I}} = -\Big\langle \vec{v}(\vec{x}) \cdot \nabla \ln P(\vec{x}|\mathcal{I}) \Big\rangle_{\mathcal{I}}, \tag{4}$$

where $\vec{v}(\vec{x})$ is an arbitrary (differentiable) vector field. Let us assume now a model whose dependence on $\vec{x}$ is given only through a function $F(\vec{x})$ by

$$P(\vec{x}|\mathcal{I}) = p(F(\vec{x})). \tag{5}$$

In this case, we have $\nabla P(\vec{x}|\mathcal{I}) = (dp/dF)\nabla F$ and therefore, the identity in Eq. 4 takes the form

$$\Big\langle \nabla \cdot \vec{v}(\vec{x}) \Big\rangle_{\mathcal{I}} = \Big\langle \hat{\lambda}(F(\vec{x}))\vec{v}(\vec{x}) \cdot \nabla F(\vec{x}) \Big\rangle_{\mathcal{I}}, \tag{6}$$

where we have defined the quantity

$$\hat{\lambda}(f) = -\frac{d}{df} \ln p(f). \tag{7}$$

In the case of the canonical ensemble, the function $p(f)$ is simply $p(f) = \exp(-\lambda f)/Z(\lambda)$ and we recover $\hat{\lambda}(f) = \lambda$, a constant. This suggests that $\hat{\lambda}$ is a generalization of the idea of inverse temperature in the case of a physical system with F corresponding to the Hamiltonian $\mathcal{H}$ and λ to the inverse temperature $\beta = 1/k_B T$. We will call $\hat{\lambda}(f)$ the **fundamental inverse temperature**, in order to distinguish it from similar functions we will present shortly.

This quantity has been previously used by Velázquez and Curilef [3] as an effective inverse temperature in the particular case of Superstatistics.

Let us take Eq. 6 and use $\vec{v} = \vec{\omega}/(\vec{\omega} \cdot \nabla F)$, this leads to

$$\Big\langle \nabla \cdot \left[\frac{\vec{\omega}(\vec{x})}{\vec{\omega}(\vec{x}) \cdot \nabla F(\vec{x})}\right] \Big\rangle_{\mathcal{I}} = \Big\langle \hat{\lambda}(F) \Big\rangle_{\mathcal{I}}. \tag{8}$$

where we recognize the quantity in the left hand side as **Rugh's inverse temperature** [4, 5],

$$\hat{\lambda}_R(\vec{x}) = \nabla \cdot \left[\frac{\vec{\omega}(\vec{x})}{\vec{\omega}(\vec{x}) \cdot \nabla F(\vec{x})}\right]. \tag{9}$$

For the kind of ensemble defined by the condition $P(\vec{x}|\mathcal{I}) = p(F(\vec{x}))$ we see that the probability of observing a particular value f of the function $F(\vec{x})$ is

$$P(F(\vec{x}) = f|\mathcal{I}) = \Big\langle \delta(F(\vec{x}) - f) \Big\rangle_{\mathcal{I}} = \int d\vec{x} \delta(F(\vec{x}) - f) p(F(\vec{x})) = p(f)\Omega(f) \tag{10}$$

where we have introduced the *density of states* of F,

$$\Omega(f) = \int d\vec{x} \delta(F(\vec{x}) - f). \tag{11}$$

The equivalent of the identity in Eq. 4 for the probability distribution of the values of F is

$$\Big\langle \frac{dg(f)}{df} \Big\rangle_{\mathcal{I}} = \Big\langle g(f) \left[\hat{\lambda}(f) - \hat{\lambda}_{\Omega}(f)\right] \Big\rangle_{\mathcal{I}}, \tag{12}$$

with

$$\hat{\lambda}_\Omega(f) = \frac{d}{df}\ln \Omega(f). \tag{13}$$

We will call this quantity the **microcanonical inverse temperature**, given that it coincides with the traditional definition of temperature in the microcanonical ensemble,

$$\frac{1}{T} = \frac{dS(E)}{dE} = \frac{d}{dE}k_B \ln \Omega(E). \tag{14}$$

If we choose $g(f) = 1$ in Eq. 12 we immediately see that, taking also into account Eq. 8,

$$\big\langle \hat{\lambda}_\Omega(f)\big\rangle_{\mathcal{I}} = \big\langle \hat{\lambda}(f)\big\rangle_{\mathcal{I}} = \big\langle \hat{\lambda}_R(\vec{x})\big\rangle_{\mathcal{I}}, \tag{15}$$

i.e., all three alternative definitions of temperature (fundamental, Rugh's and microcanonical) given by Eqs. 7, 9 and 13, are equivalent on expectation.

3. The particular case of Superstatistics

In the case of Superstatistics, $p(f)$ has the form

$$p(f) = \int d\lambda P(\lambda|\mathcal{I})\frac{\exp(-\lambda f)}{Z(\lambda)} \tag{16}$$

therefore the fundamental inverse temperature is given by

$$\hat{\lambda}(f) = -\frac{d}{df}\ln p(f) = \frac{\int d\lambda \cdot \lambda P(\lambda|\mathcal{I})\frac{\exp(-\lambda f)}{Z(\lambda)}}{p(f)}. \tag{17}$$

We can gain further understanding of this expression if we write $P(\lambda|\vec{x})$ using Bayes' theorem, as

$$P(\lambda|\vec{x}) = \frac{P(\lambda|\mathcal{I})P(\vec{x}|\lambda)}{P(\vec{x}|\mathcal{I})} = \frac{1}{p(F(\vec{x}))}P(\lambda|\mathcal{I})\frac{\exp(-\lambda F(\vec{x}))}{Z(\lambda)} \tag{18}$$

which allows us to write $\hat{\lambda}(f)$ as

$$\hat{\lambda}(F(\vec{x})) = \int d\lambda \cdot \lambda P(\lambda|\vec{x}) = \big\langle \lambda \big\rangle_{\vec{x}}. \tag{19}$$

This means the fundamental estimator $\hat{\lambda}(F(\vec{x}))$ in this case is the expectation of the "genuine" Superstatistical inverse temperature given that the system is in a known microstate $\vec{x}$. Taking expectation on both sides under a state of knowledge $\mathcal{I}$, we see that

$$\big\langle \hat{\lambda}\big\rangle_{\mathcal{I}} = \big\langle \lambda \big\rangle_{\mathcal{I}}, \tag{20}$$

i.e., the expectation of any of the three inverse temperatures is equal to the Superstatistical expectation of λ. In order to relate to the fluctuations of the superstatistical parameter λ, we differentiate $\big\langle \lambda \big\rangle_{\vec{x}}$ with respect to $F(\vec{x})$, obtaining

$$\frac{d}{dF}\big\langle \lambda \big\rangle_{\vec{x}} = -\big\langle \lambda^2 \big\rangle_{\vec{x}} + \big\langle \lambda \big\rangle_{\vec{x}}^2 \tag{21}$$

Taking expectation on both sides under $\mathcal{I}$ we see that

$$\big\langle \hat{\lambda}\big\rangle_{\mathcal{I}} = \big\langle \lambda \big\rangle_{\mathcal{I}}, \tag{22}$$

i.e., the expectation of any of the three inverse temperatures is equal to the expectation of the superstatistical parameter λ. In order to connect the fluctuations of the superstatistical parameter λ with the fluctuations of the estimators, we differentiate $\big\langle \lambda \big\rangle_{\vec{x}}$ with respect to $F(\vec{x})$, obtaining

$$\frac{d}{dF}\big\langle \lambda \big\rangle_{\vec{x}} = -\big\langle \lambda^2 \big\rangle_{\vec{x}} + \big\langle \lambda \big\rangle_{\vec{x}}^2 \tag{23}$$

Taking expectation on both sides under the state of knowledge $\mathcal{I}$ we see that

$$\Big\langle (\delta\lambda)^2 \Big\rangle_{\mathcal{I}} = \Big\langle (\delta\hat{\lambda})^2 \Big\rangle_{\mathcal{I}} - \Big\langle \frac{d\hat{\lambda}}{dF} \Big\rangle_{\mathcal{I}}, \tag{24}$$

which tells us that the fundamental inverse temperature does not necessarily have the same fluctuations of the parameter λ.

4. Application to the microcanonical coordinate distribution

For a physical system with Hamiltonian

$$\mathcal{H}(\vec{r},\vec{p}) = \frac{1}{2m}(\vec{p})^2 + \Phi(\vec{r}), \tag{25}$$

it can be shown [6] that the correct probability distribution for the coordinates at constant energy (microcanonical ensemble) is

$$P(\vec{r}|E) = \frac{1}{\eta(E)}\Theta(E-\Phi(\vec{r}))\sqrt{E-\Phi(\vec{r})}^{\,3N-2} \tag{26}$$

where $\eta(E)$ is proportional to the density of states $\Omega(E)$. For this distribution, the condition in Eq. 5 holds, with $F = \Phi$, and therefore

$$p(\phi) = \frac{1}{\eta(E)}\Theta(E-\phi)\sqrt{E-\phi}^{\,3N-2}. \tag{27}$$

The fundamental inverse temperature is then given by

$$\hat{\beta}(\phi) = -\frac{d}{d\phi}\ln p(\phi) = \frac{d}{d\phi}\ln\Theta(E-\phi) + \frac{3N-2}{2(E-\phi)}. \tag{28}$$

This leads to

$$\Big\langle \nabla\cdot\vec{v} \Big\rangle_E = \Big\langle \frac{3N-2}{2(E-\Phi)}\vec{v}\cdot\nabla\Phi \Big\rangle_E, \tag{29}$$

given that the first term on the right hand side of Eq. 28 vanishes in expectation. We can take the fundamental inverse temperature to be, without loss of generality,

$$\hat{\beta}(\phi) = \frac{3N-2}{2(E-\phi)}, \tag{30}$$

which, interestingly, coincides with the kinetic inverse temperature, as $E-\Phi = \frac{p^2}{2m}$ is the kinetic energy.

We have, finally, the equality of all three temperatures,

$$\Big\langle \nabla\cdot\Big[\frac{\vec{\omega}}{\vec{\omega}\cdot\nabla\Phi}\Big] \Big\rangle_E = \Big\langle \frac{3N-2}{2(E-\Phi)} \Big\rangle_E = \Big\langle \frac{d}{d\Phi}\ln\Omega(\Phi) \Big\rangle_E. \tag{31}$$

Journal of Physics: Conference Series **720** (2016) 012007 doi:10.1088/1742-6596/720/1/012007

5. Conclusions

We have shown that, for ensembles which are dependent of a Hamiltonian but are not canonical, at least three different definitions of inverse temperature are admissible. A particular case of this is Superstatistics, and consequently the q-exponential (Tsallis) distributions. All three estimators agree in expectation but their fluctuations are not necessarily equal. These results reveal a possible conceptual ambiguity in the definition of temperature for superstatistical systems, which must be resolved in order to construct an internally consistent theory of these systems.

Acknowledgments

Both authors acknowledge funding from FONDECYT grant 1140514.

References

[1] Beck C and Cohen E G D 2003 *Physica A* **322** 267–275
[2] Davis S and Gutiérrez G 2012 *Phys. Rev. E* **86** 051136
[3] Velazquez L and Curilef S 2009 *J. Phys. A: Math. Theor.* **42** 095006
[4] Rugh H H 1997 *Phys. Rev. Lett.* **78** 772–774
[5] Rickayzen G and Powles J G 2001 *J. Chem. Phys.* **114** 4333
[6] Ray J R 1991 *Phys. Rev. A* **44** 4061–4064

The Emergent Universe and the Anomalies in the Cosmic Microwave Background

Carlos Ríos, Pedro Labraña, Antonella Cid
Departamento de Física, Universidad del Bío-Bío, Casilla 5-C, Concepción, Chile.
E-mail: crios@ubiobio.cl

Abstract. The existence of anomalies in the Cosmic Microwave Background (CMB) on large angular scales (power deficit at low multipoles and hemispherical asymmetry) could be related to a pre-inflationary period, previous to standard inflation. Recent studies show that a period of super-inflation in the context of emergent universes (*EU*) generates a suppression of CMB anisotropies on large scales. We also study the implications of an *EU* model in the asymmetry of the average temperature in opposite hemispheres of the sky.

1. Introduction

The standard inflationary paradigm (SIP) is successful in solving the problems of the standard cosmological model (SCM) as horizon problem, flatness problem, monopole problem and the origin of structures [1, 2, 17]. However, recent measurements of Planck collaboration have reported a power deficit in the low-l CMB power spectrum at $l \lesssim 40$ [4] that can not be explained by the SIP.

In this work we use the Friedmann-Lemaître-Robertson-Walker (FLRW) metric:

$$ds^2 = dt^2 - a(t)^2 \left[\frac{dr^2}{1 - kr^2} + r^2 d\Omega^2 \right] \tag{1}$$

Where t is the cosmic time, $a(t)$ is the scale factor, k is a constant related to the curvature of the spatial sections (positive for closed models, negative for open ones, zero for a flat Universe). Throughout this work we adopt units such that $c = \hbar = 1$. In the case of a homogeneous and isotropic Universe, the field equations are give by:

$$H^2 + \frac{k}{a^2} = \frac{8}{3}\pi G\rho \tag{2}$$

$$\frac{\ddot{a}}{a} = -\frac{4\pi G}{3}(\rho + 3p) \tag{3}$$

where H is the Hubble parameter defined by $H(t) = \frac{\dot{a}(t)}{a(t)}$, ρ is the total energy density of the spacetime, p is the total pressure and G is the newtonian constant of gravitation. The Einstein equations also lead to the energy conservation equation:

$$\dot{\rho} + 3H(\rho + p) = 0 \tag{4}$$

We suppose that the early Universe is dominated by a canonical scalar field ϕ, called inflaton, whose energy density and pressure are given by:

$$\rho = \frac{1}{2}\dot{\phi}^2 + V(\phi) \tag{5}$$

$$p = \frac{1}{2}\dot{\phi}^2 - V(\phi) \tag{6}$$

where $V(\phi)$ is the scalar potential.

2. The Emergent Universe

The EU models suggest that the Universe had not origin but rather, the Universe has always existed [5, 6], initially in an Einstein static state which solves the problem of the initial singularity present in the SCM [24]. In an EU scenario, the evolution the scale factor can be modeled as follows:

$$a(t) \simeq a_0 + Ae^{H_0 t} \tag{7}$$

where a_0, A and H_0 are positive constants and t is the cosmic time. We note that this universe has a past that asymptotically tends to an Einstein Static State. To the future, the model approaches a de Sitter expansion phase.

With the scale factor described in (7), the normalized evolution of the Hubble parameter is shown in Figure 1.

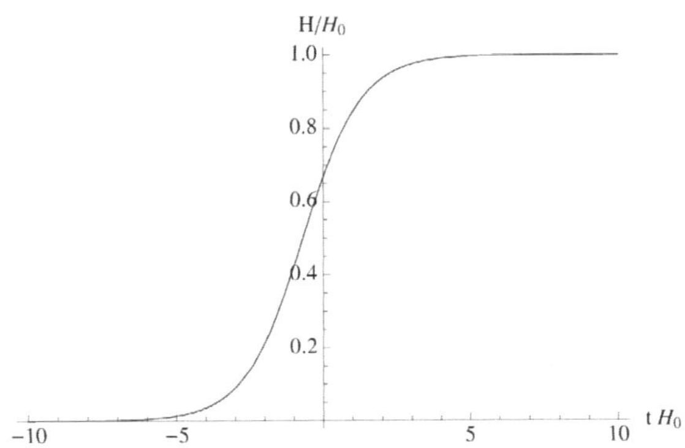

Figure 1. Evolution of the Hubble parameter corresponding to the scale factor (7). We distinguish three periods in the evolution, a past that asymptotically tends to an Einstein Static State ($H \to 0$), an era of super-inflation ($\dot{H} > 0$) and eventually a state of de Sitter expansion ($H \to Constant$).

The scalar field potential in a standard EU could be given by [6]:

$$V(\phi) = \frac{2}{\kappa a_0^2}(e^{B\phi} - 1)^2 \tag{8}$$

where $\kappa = 8\pi G$ and B is a constant, see Figure 2.

Journal of Physics: Conference Series **720** (2016) 012008 doi:10.1088/1742-6596/720/1/012008

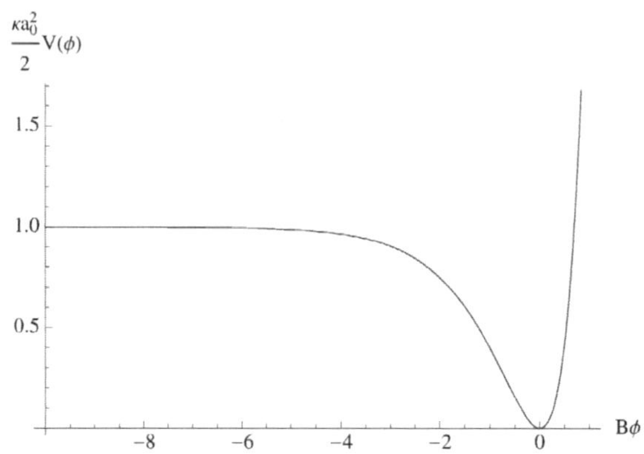

Figure 2. A schematic illustration of an EU potential normalized in units of $\frac{2}{\kappa a_0^2}$.

The evolution of the scalar field is determined by the Klein-Gordon equation:

$$\ddot{\phi} + 3H\dot{\phi} + V' = 0 \tag{9}$$

We are interested in the evolution of potential $\phi(t)$ in the flat region of the scalar field potential (see Figure 2). Then, if we consider that the potential is approximately constant $V \approx V_0$, we have

$$\ddot{\phi} + 3H\dot{\phi} \approx 0 \tag{10}$$

Using the Eqs. (2), (3), (5) and (6) we find that:

$$\dot{H} = \frac{k}{a^2} - \frac{\kappa}{2}\dot{\phi}^2 \tag{11}$$

If we make the change of variable $X = \frac{\kappa\dot{\phi}^2}{2}$ and using the Eq. (10), the Eq. (11) becomes in

$$\dot{X} + 6HX = 0 \tag{12}$$

On other hand, using the Eqs. (2), (5) and (11) we find that

$$\dot{H} = -\frac{2}{3}X - H^2 + \frac{\kappa V_0}{3} \tag{13}$$

We can write Eqs. (12) and (13) as an autonomous dynamical system using that $k = +1$ for an EU models [5, 6]. The directional field of the dynamical system is shown in Figure 2. We can note that the system of Eqs. (12) and (13) has an attractor for $H = \sqrt{\frac{\kappa V_0}{3}}$ and $\dot{\phi} = 0$, i.e. the system tends from the Einstein Static State to a de Sitter State. Then, we can model the Hubble parameter $H(t)$, giving one scale factor $a(t)$, getting well the form of $\dot{\phi}$ using Eq. (10). From here on, we will consider and approximately constant potential or $\dot{\phi} \propto a^{-3}$.

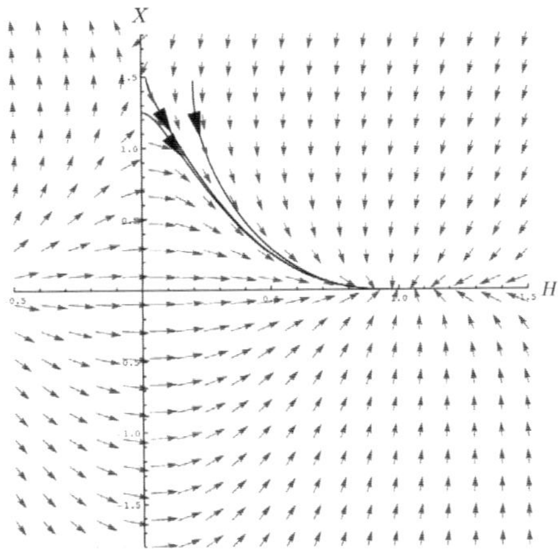

Figure 3. Directional field of the dynamical system and three numerical solutions. The instability point in this system is $(X = \frac{\kappa V_0}{2}, H = 0)$ and the attractor point is $(X = 0, H_0^2 = \frac{\kappa V_0}{3})$. In this graph $X = \frac{\kappa \dot{\phi}^2}{2}$ and $V_0 = \frac{3}{\kappa}$.

3. Primordial Perturbations

The CMB temperature is approximately the same in all directions in the sky. The first measurements of CMB by Penzias and Wilson in 1960s [26] showed no evidence for anisotropies in this temperature. The COBE satellite first measured anisotropies in the CMB temperature in 1992 [25]. Since then the satellites WMAP [3] and Planck [4] have measured there anisotropies with very high precision.

The scalar perturbations in the geometry of Friedmann-Lemaître-Robertson-Walker in the newtonian gauge, can be written as:

$$ds^2 = (1 + 2\Phi)dt^2 - (1 - 2\Phi)a(t)^2 d\vec{x}^2 \tag{14}$$

where Φ is the Newtonian gravitational potential [17]. We use a Mukhanov variables [7, 8] defined by

$$v_k = a\left(\delta\phi_k + \frac{\phi'}{h}\Phi_k\right) \tag{15}$$

where $\delta\phi_k$ are the perturbations of the canonical inflation field,$'$ denote derivate with respect to the conformal time η defined by $dt = a(\eta)d\eta$ and $h = \frac{a'}{a}$. If we consider that $\frac{z''}{z} \approx \frac{a''}{a}$ where $z = \frac{a\phi'}{h}$ and the scalar factor given in (7), the perturbed equation in the Fourier space is given by

$$v_k'' + \left(k^2 - \frac{z''}{z}\right)v_k = 0 \tag{16}$$

whose solution is [9]:

$$v_k = \frac{1}{\sqrt{2k}}\left[\frac{e^{-ik\eta}}{1 - e^{a_0 H_0 \eta}}\right] {}_2F_1(x_2 - 3, x_3 - 3; x_1; e^{a_0 H_0 \eta}) \tag{17}$$

Journal of Physics: Conference Series **720** (2016) 012008 doi:10.1088/1742-6596/720/1/012008

where ${}_2F_1$ is a hypergeometric function, and

$$x_1 = 1 - 2i\chi \tag{18}$$
$$x_2 = 2 - i\chi - \sqrt{1-\chi^2} \tag{19}$$
$$x_3 = 2 - i\chi + \sqrt{1-\chi^2} \tag{20}$$
$$\chi = \frac{k}{a_0 H_0} \tag{21}$$

It is important to note that this solution tends to the standard solution $v_k \approx \frac{e^{-ik\eta}}{\sqrt{2k}}$ in the limit of short wavelengths ($k >> a_0 H_0$). The spectrum of primordial perturbations is defined by the following expression [16]:

$$\mathcal{P}_{\delta\phi} = \frac{k^3}{2\pi^2}\left|\frac{v_k}{a}\right|^2 \tag{22}$$

In the context of the EU, we need to take the limit $\eta \to 0$, finally we get:

$$\mathcal{P}_{\delta\phi} = \frac{H_0^2}{\pi^2}\frac{\chi^2\Gamma[x_1]\Gamma[x_1^*]}{\Gamma[x_2]\Gamma[x_2^*]\Gamma[x_3]\Gamma[x_3^*]} \tag{23}$$

In the limit of small scales ($k >> a_0 H_0$) the standard result $\mathcal{P}_{\delta\phi} = \left(\frac{H_0}{2\pi}\right)^2$ is recovered from Eq. (23), see ref. [16].

4. Anomalies in the CMB

The WMAP and Planck satellites measured a lack of observed power in the lowest CMB multipoles of the power spectrum [3, 4]. This anomaly can be explained in the context of the EU, because this model naturally generates a suppression in the power spectrum of the primordial fluctuation at large scales (or low multipoles).

On the other hand, the data collected by the WMAP satellite [11, 14] and subsequently by the Planck satellite [15] indicated a difference between the CMB temperature fluctuations at two opposite hemispheres of the sky separated approximately in the ecliptic. A convenient way to model this asymmetry is by introducing a dipole in the sky using the following relation [19, 21]:

$$\mathcal{P}_{\mathcal{R}}^{1/2}(k,\vec{x}) = \left(1 + A(k)\frac{\hat{P}\cdot\vec{x}}{x_{ls}}\right)\mathcal{P}_{\mathcal{R}}^{1/2}(k) \tag{24}$$

where

$$\mathcal{P}_{\mathcal{R}}(k) \approx \left(\frac{H}{\dot{\phi}}\right)^2 \mathcal{P}_{\delta\phi}|_{k=aH} \tag{25}$$

is the spectrum of curvature perturbations evaluated in the horizon crossing condition $k = aH$ [17] , ϕ is the inflaton scalar field, $\hat{P}$ is the unit vector in the direction of dipolar modulation, $\vec{x}$ is a spatial vector, x_{ls} is the distance to the last scattering surface and $A(k)$ is the amplitude of modulation.

5. Effective Power Spectrum

In analogy with the standard inflationary paradigm, where $\mathcal{P}_{\mathcal{R}} \approx \frac{1}{\epsilon_{inf}}\mathcal{P}_{\delta\phi}$ and $\epsilon_{inf} < 0,008$ is the standard slow-roll parameter [13], we define a new parameter ϵ^*, such that:

$$\epsilon^*(\chi) = \frac{1}{2}\left[\frac{\dot{\phi}(\chi)}{H(\chi)}\right]^2 + \epsilon_{inf} \tag{26}$$

so that the spectrum of curvature perturbation is represented by:

$$\mathcal{P}_{\mathcal{R}} \approx \left(\frac{1}{\kappa a_0^2 H_0^2 \chi^2 (1+\chi)^4} + 0,008\right)^{-1} \mathcal{P}_{\delta\phi} \tag{27}$$

where we used the Klein-Gordon equation and we have considered that in the pre-inflationary phase $V \approx Constant$.

We have determined the power spectrum of curvature perturbations of our model generated during super-inflation period, but we must include the SIP. To get this, we follow [12] and [9]. A power spectrum that include both stages of evolution of the Universe is:

$$\mathcal{P}_{\mathcal{R}} = \frac{\bar{A}}{\epsilon^*}\left(\frac{k}{k_0}\right)^{n_s-1} \frac{\chi^2\Gamma[x_1]\Gamma[x_1^*]}{\Gamma[x_2]\Gamma[x_2^*]\Gamma[x_3]\Gamma[x_3^*]} \tag{28}$$

where k_0 is the pivot scale, $\bar{A}$ is a constant and n_s is the standard spectral index. Which can be written conveniently [9]:

$$\mathcal{P}_{\mathcal{R}} = \frac{A}{\epsilon^*}\left(\frac{k}{a_0H_0}\right)^2 \left(1+\frac{k}{a_0H_0}\right)^{-2} \left(\frac{k}{k_0}\right)^{n_s-1} \tag{29}$$

The temperature power spectrum is obtained using the $CLASS$ code [22, 23]. We use the following values for the constants $A = 2.05 * 10^{-9}$ and $a_0H_0 = 0.0002Mpc^{-1}$. The value of $k_0 = 0.002Mpc^{-1}$ is used by WMAP and $n_s = 0.967$ is reported by the same satellite [18].

The temperature power spectrum in terms of the C_ℓ [19] , in our model can be seen in Figure 4.

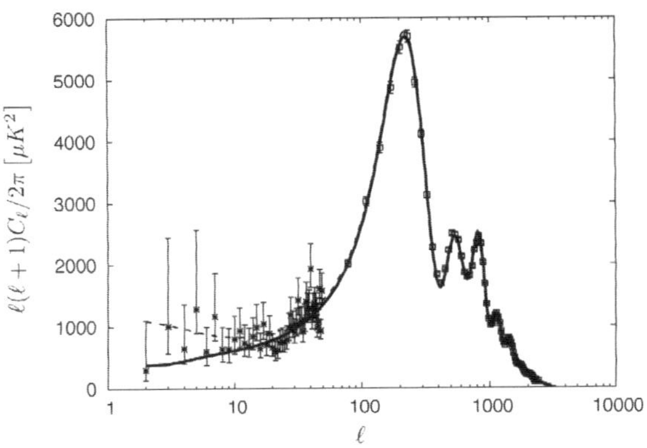

Figure 4. Temperature power spectrum in terms of the C_ℓ. The dashed line represent the SCM and the solid line is the EU scenario. The points show the Planck data.

We can also observe the evolution of the power spectrum $\mathcal{P}_{\mathcal{R}}(k)$ in terms of the scale in Figure 5.

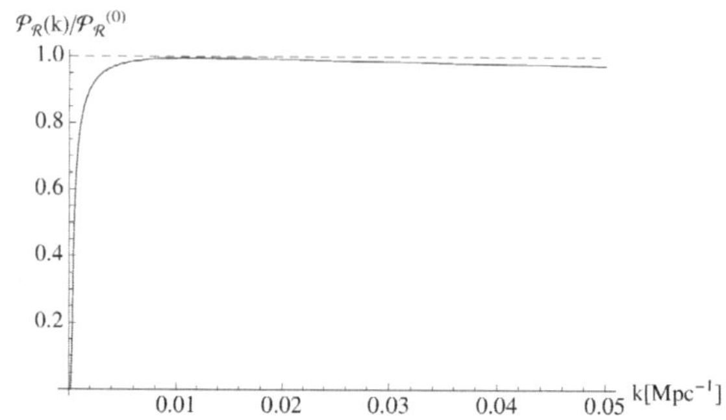

Figure 5. Normalized Power spectrum of curvature perturbations using EU (Solid line). It can be appreciated its dependence on the scale. The dashed line represent the scale invariant Power spectrum.

6. Hemispherical Asymmetry

Other anomaly that survives, even after the accurate observations of the Planck satellite [20] is a hemispherical asymmetry on the CMB power spectrum modeled by (24). Following [12] it can be shown that the modulation amplitude $A(k)$ satisfies:

$$|A(k)| \lesssim 0.02 \left| \frac{n_s(k) - 1}{2} \right| |1 - \epsilon(k)| \tag{30}$$

where ϵ is defined by

$$\epsilon \equiv -\frac{\dot{H}}{H^2} \tag{31}$$

and the spectral index $n_s(k)$ is obtained analytically by its definition:

$$n_s - 1 = \frac{d\ln\mathcal{P}_{\mathcal{R}}}{d\ln k} \tag{32}$$

the evolution of the modulation amplitude is shown in Figure 6.

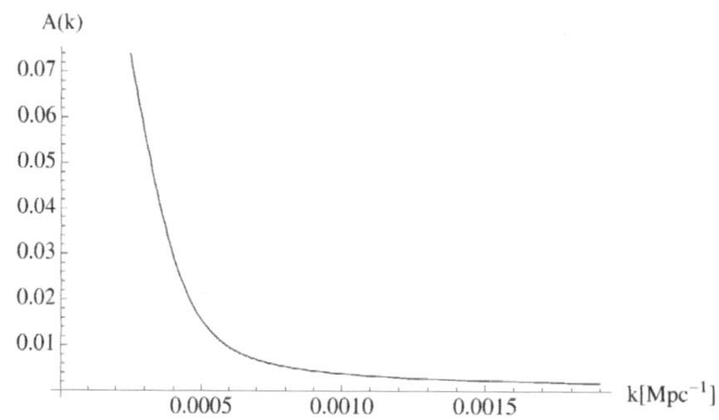

Figure 6. Evolution of the modulation amplitude as function of k. For small scales, Planck reports a value of $A = 0,07 \pm 0,02$.

Standard inflation predicts $n_s - 1 \approx 0$, then the modulation amplitude $A \approx 0$, but super-inflation in the context of EU indicates that the modulation amplitude depends on the scale (see Figure 6) then it is possible to explain that Planck reported $A = 0.07 \pm 0.02$ value $l \leq 40$ and $A \approx 0$ to $l \approx 500 - 600$ [10].

7. Conclusions

The Cosmic Microwave Background have several anomalies that survive even after the Plancks observations . One of this anomalies is the low power that has the power spectrum curvature of perturbation $\mathcal{P}_{\mathcal{R}}(k)$ in the lowest multipoles [4]. This anomaly can be explained in the context of the EU, because our model naturally generates a suppression in the power spectrum curvature of perturbation of the CMB at low multipoles. This suppression can be observed in Figure 5. On the other hand, Planck's observation show a Hemispherical Asymmetry in the CMB [20] unexplained when we use the SIP, because the standard inflation predicts $n_s - 1 \approx 0$ then n_s is constant and the modulation amplitude $A \approx 0$, but super-inflation in the context of EU indicates that the modulation amplitude depends on the scale then it is possible to explain that Planck reported $A = 0.07 \pm 0.02$ value $l \leq 40$ and $A \approx 0$ to $l \approx 500 - 600$ [10].

Acknowledgments
We thank to Dirección de Investigación of Universidad del Bío Bío for GI121407/VC (AC, PL, CR) and 141407 3/R (PL), to CONICYT for FONDECYT N° 11110507 (AC).

References

[1] Guth A., The inflationary universe: A possible solution to the horizon and flatness problems, 1981 Phys. Rev. D **23** 347.
[2] Linde A. D., A new inflationary universe scenario: A possible solution of the horizon, flatness, homogeneity, isotropy and primordial monopole problems, 1982 Phys. Lett. **108B** 389.
[3] E. Komatsu *et al.* [WMAP Collaboration], Astrophys. J. Suppl. **192**, 18 (2011). [arXiv:1001.4538 [astro-ph.CO]]. D. Larson, J. Dunkley, G. Hinshaw, E. Komatsu, M. R. Nolta, C. L. Bennett, B. Gold, M. Halpern *et al.*, Astrophys. J. Suppl. **192**, 16 (2011). [arXiv:1001.4635 [astro-ph.CO]]. N. Jarosik *et al.*, Astrophys. J. Suppl. **192**, 14 (2011). [arXiv:1001.4744 [astro-ph.CO]]. E. Komatsu *et al.* [WMAP Collaboration], Astrophys. J. Suppl. **180**, 330-376 (2009). [arXiv:0803.0547 [astro-ph]]. D. N. Spergel *et al.* [WMAP Collaboration], Astrophys. J. Suppl. **170**, 377 (2007). [astro-ph/0603449].
[4] P. A. R. Ade *et al.* [Planck Collaboration], Astron. Astrophys. **571**, A1 (2014) [arXiv:1303.5062 [astro-ph.CO]].
[5] Ellis G. F. R. and Maartens R., The emergent universe: Inflationary cosmology with no singularity, 2004 Class. Quant. Grav. **21** 223.
[6] Ellis G. F. R., Murugan J. and Tsagas C. G., The emergent universe: An explicit construction, 2004 Class. Quant. Grav. **21** 233.
[7] Mukhanov, V. F. and Chibisov, G. 1981, JETP Lett., 33, 532; Mukhanov, V. F. and Chibisov, G. 1982, Sov.Phys.JETP, 56, 258; Mukhanov, V. F. 1985, JETP Lett., 41, 493; Mukhanov, V. F., Feldman, H., and Brandenberger, R. H. 1992, Phys.Rept., 215, 203.
[8] J. Garriga and V. F. Mukhanov, Phys. Lett. B **458**, 219 (1999)
[9] P. Labrana, Phys. Rev. D **91**, no. 8, 083534 (2015) [arXiv:1312.6877 [astro-ph.CO]].
[10] P. A. R. Ade *et al.* [Planck Collaboration], Astron. Astrophys. **571**, A23 (2014) [arXiv:1303.5083 [astro-ph.CO]].
[11] H. K. Eriksen, A. J. Banday, K. M. Gorski, F. K. Hansen and P. B. Lilje, Astrophys. J. **660**, L81 (2007) [astro-ph/0701089].
[12] Z. G. Liu, Z. K. Guo and Y. S. Piao, Phys. Rev. D **88**, 063539 (2013) [arXiv:1304.6527 [astro-ph.CO]].
[13] P. A. R. Ade *et al.* [Planck Collaboration], Astron. Astrophys. **571**, A22 (2014) [arXiv:1303.5082 [astro-ph.CO]].
[14] J. Hoftuft, H. K. Eriksen, A. J. Banday, K. M. Gorski, F. K. Hansen and P. B. Lilje, Astrophys. J. **699**, 985 (2009) [arXiv:0903.1229 [astro-ph.CO]].

[15] Y. Akrami, Y. Fantaye, A. Shafieloo, H. K. Eriksen, F. K. Hansen, A. J. Banday and K. M. Grski, Astrophys. J. **784**, L42 (2014) [arXiv:1402.0870 [astro-ph.CO]].
[16] C. R. Contaldi, M. Peloso, L. Kofman and A. D. Linde, JCAP **0307**, 002 (2003) [astro-ph/0303636].
[17] S. Weinberg, *Cosmology*, Oxford (2008).
[18] C. L. Bennett *et al.* [WMAP Collaboration], Astrophys. J. Suppl. **208**, 20 (2013) [arXiv:1212.5225 [astro-ph.CO]].
[19] D. H. Lyth, JCAP **1308**, 007 (2013) [arXiv:1304.1270 [astro-ph.CO]].
[20] P. A. R. Ade *et al.* [Planck Collaboration], arXiv:1303.5083 [astro-ph.CO].
[21] M. H. Namjoo, A. A. Abolhasani, S. Baghram and H. Firouzjahi, JCAP **1408**, 002 (2014) [arXiv:1405.7317 [astro-ph.CO]].
[22] J. Lesgourgues, arXiv:1104.2932 [astro-ph.IM].
[23] D. Blas, J. Lesgourgues and T. Tram, JCAP **1107** (2011) 034 [arXiv:1104.2933 [astro-ph.CO]].
[24] Borde A. and Vilenkin A., Eternal inflation and the initial singularity, 1994 Phys. Rev. Lett. **72** 3305.
[25] G. F. Smoot et al., ApJ **396**, 1 (1992).
[26] A. A. Penzias and R. W. Wilson, Astrophys. J. **142**, 419 (1965).

Journal of Physics: Conference Series **720** (2016) 012009 doi:10.1088/1742-6596/720/1/012009

Can one hear the Riemann zeros in black hole ringing?

Rodrigo Aros[1], **Fabrizzio Bugini**[2] **and Danilo E Diaz**[1]

[1] Universidad Andres Bello, Departamento de Ciencias Fisicas, Republica 220, Santiago, Chile
[2] Universidad de Concepcion, Departamento de Fisica, Casilla 160-C, Concepcion, Chile

E-mail: raros@unab.cl, fbugini@udec.cl, danilodiaz@unab.cl

Abstract. We elaborate on an entry of the AdS/CFT dictionary relating functional determinants: the determinant of the one-loop contribution to the effective gravitational action by bulk scalars in an asymptotically locally AdS background X, and the determinant of the two-point function of the dual operator (a.k.a. scattering matrix) at the conformal boundary M. The formula originates from AdS/CFT heuristics that map a quantum contribution in the bulk gravitational partition function to a subleading large-N contribution in the boundary CFT partition function:

$$\frac{\det_-(\Delta_X + \lambda(\lambda - n))}{\det_+(\Delta_X + \lambda(\lambda - n))} = \det\, S_M(\lambda)\ .$$

The formula applies to quotients of AdS as well [1]. In the particular case of the BTZ black hole, a closed expression can be worked out in terms of an associated Patterson-Selberg zeta function $Z_{BTZ}(\lambda)$ [2]. The determinants can then be thought of as regularized products of either zeta zeros, scattering resonances or quasinormal frequencies [3]. In this sense, one could say that the zeros of $Z_{BTZ}(\lambda)$ *can be heard* in the spectrum of quasinormal modes of the BTZ black hole. The question we want to pose is whether a similar situation might exist for the celebrated zeros of the Riemann zeta function.

1. Introduction

The aim of this note is to describe a physical scenario that might shed light on the celebrated *Riemann hypothesis*. The question we pose is evidently inspired by the classical paper of Mark Kac (1966) *Can one hear the shape of a drum?*; however, in contrast to most of the physical approaches that look after the (as yet elusive) Hilbert-Pólya hamiltonian, our focus is the spectrum of black hole quasinormal modes or, for short, *black hole ringing.* We have an explicit example at hand, the BTZ black hole, where a connection can be established between the quasinormal spectrum and the zeros of an associated Patterson-Selberg zeta function

BTZ bh $\Rightarrow$ quasinormal freq's $\Rightarrow$ Patterson-Selberg zeros.

The natural, though perhaps naive, question that comes to mind is whether a similar construction exists, involving again a quotient of AdS space, where the Patterson-Selberg zeros are traded by the Riemann zeros

AdS/Γ ? $\Leftarrow$ quasinormal freq's ? $\Leftarrow$ Riemann zeros.

We first present the explicit AdS/CFT result for the BTZ black hole and then elaborate on the possibility to have the Riemann zeta, instead of the Patterson-Selberg zeta, and the restriction on the location of its nontrivial zeros.

2. The Rosetta stone of BTZ black hole

Ever since its appearance in the late 1990s in the form of Maldacena's conjecture, the AdS/CFT correspondence has evolved as a theoretical tool to address questions concerning strongly coupled physical systems within a dual holographic framework. It relates two seemingly different theories, gauge theory and gravity, via two deeply rooted ideas in physics, namely, the string emerging at the large-N limit of the gauge theory and the holographic principle. Many developments are, perhaps more appropriately, embraced under the name gauge/gravity dualities for they depart from the original canonical examples and the boundary theories capture the physics of condensed matter, liquids, superconductors, heavy-ion collisions, etc. These recent and exciting applications to physically relevant situations are more heuristic, and therefore mathematically exact results become rare; nonetheless, we will now exploit a *holographic formula*, one of the entries of the AdS/CFT dictionary that seems deeply rooted in spectral theory and conformal geometry.

2.1. Holographic formula

Our favorite entry of AdS/CFT dictionary is a relation (holographic formula) between the determinant of the one-loop contribution to the effective gravitational action by bulk scalars in an asymptotically locally AdS background X and the determinant of the two-point function of the dual operator (a.k.a. scattering matrix) at the conformal boundary M. The formula originates from AdS/CFT heuristics that map a quantum contribution in the bulk gravitational partition function to a subleading large-N contribution in the boundary CFT partition function:

$$\frac{\det_-(\Delta_X + \lambda(\lambda - n))}{\det_+(\Delta_X + \lambda(\lambda - n))} = \det\, S_M(\lambda)\ .$$

The bulk determinant $\det_+$ is computed in the usual way (via Green functions or heat kernel) and $\det_-$ is obtained by analytic continuation in the conformal dimension $\lambda_+ \to \lambda_- = n - \lambda$.

2.2. The case of BTZ black hole

Explicit results can be obtained for pure AdS and for quotients of the form AdS/Γ as well [1]. In particular, the determinants in the BTZ black hole background and the determinant on its boundary torus can be explicitly computed in terms of a Patterson-Selberg zeta function $Z_{BTZ}(\lambda)$ [2] defined by

$$Z_{BTZ}(\lambda) = \prod_{k_1,k_2\geq 0} \left[1 - \alpha_1^{k_1}\alpha_2^{k_2}e^{-(k_1+k_2+\lambda)l}\right]$$

where the parameters $\alpha_1 = e^{i\theta}$, $\alpha_2 = e^{-i\theta}$ and $l = 2\pi r_+$, $\theta = 2\pi|r_-|$ relate the horizon radii r_+, r_- of the black hole with the modular parameter of the boundary torus,

$$\frac{\det_-(\Delta_{BTZ} + \lambda(\lambda - 2))}{\det_+(\Delta_{BTZ} + \lambda(\lambda - 2))} = \det\, S_{T^2}(\lambda) = \frac{(Z_{BTZ}(2-\lambda))^2}{(Z_{BTZ}(\lambda))^2}\ .$$

Here we find a first connection: scattering poles (poles of the determinant S_{T^2}) are mapped to zeros of Z_{BTZ}. That is, the scattering poles are also the Patterson-Selberg zeros.

The second connection is provided by an independent result valid for the non-spinning black hole ($r_- = 0$): the bulk determinant $\det_+(\Delta_{BTZ}+\lambda(\lambda-2))$ for the massive scalar can be written as (regularized) product of quasinormal ω_{QN} and Matsubara frequencies[3]

$$Z_{BTZ} = \prod_{Matsub.,QN} (N + i\omega_{QN})\,.$$

In all, the Patterson-Selberg zeta attached to the BTZ black hole can then be written in three equivalent ways[1]:

$$Z_{BTZ}(\lambda) \sim \prod_{P-Szeros} (\lambda-\rho) \sim \prod_{scatt.poles} (\lambda-\xi) \sim \prod_{Matsub.,QN} (N + i\omega_{QN})$$

and, therefore, the Patterson-Selberg zeros can be 'heard' in the static BTZ black hole ringing.

3. From Patterson-Selberg to Riemann zeros

The challenge now is to trade the Patterson-Selberg zeros by the Riemann ones. We briefly report some pros and cons of this program.

The first thing to notice is that it is certainly possible to obtain the connection 'scattering poles = Riemann zeros'. This occurs for scattering in a *non-compact* quotient of H^2 (Euclidean AdS_2) of finite area with one cusp, as found in the early 70's by Faddeev and Pavlov; alternatively, a decade after, Gutzwiller found the same interesting result for scattering in the 'leaky torus'. The difficulty in establishing the connection with black hole ringing is that the quotient is taken by a parabolic element of the isometry group and this points to *extremal* black holes, whereas the quasi normal mode factorization only works for static situations.

Secondly, even if one succeeded in finding the Riemann zeros in a black hole quasinormal spectrum, there is no really a compelling reason that restricts their location implying the Riemann hypothesis. This was certainly the case of scattering in the modular domain, where Lax and Phillips were "able to contemplate the Riemann hypothesis peeking around the corner". Having said so, we still believe that a holographic approach via AdS/CFT may open a window to, at least, the known *semiclassical* prospects for the Hilbert-Pólya operator and provide a unifying view. These include random matrices and quantization of classical chaotic systems among many others (cf. [4] and [5] for recent progress). Beyond the semiclassical regime, it remains the hope that gauge/gravity dualities in a broader sense will bring in a physical reason that the Riemann hypothesis should be true.

Acknowledgments

We thank N.F. Svaiter and A. Hernández for useful conversations. Partial support by grant FONDECYT 11110430 is also acknowledged.

References

[1] Aros R and Diaz D E 2010 *Functional determinants, generalized BTZ geometries and Selberg zeta function* J. Phys. A: Math. Theor. 43 205402.

[2] Perry P and Williams F 2003 *Selberg zeta function and trace formula for the BTZ black hole* Int. J. Pure Appl. Math. 9 1.

[3] Denef F, Hartnoll S A and Sachdev S 2010 *Black hole determinants and quasinormal modes* Class. Quantum Grav. 27 125001.

[4] Pando-Zayas L and Reichmann D 2013 *A String Theory Explanation for Quantum Chaos in the Hadronic Spectrum* JHEP 1304 083.

[5] Dueñas J G, Svaiter N F and Menezes G 2014 *One-loop effective action and the Riemann Zeros* Int. J. Mod. Phys. A 29 31

[1] Hence Z_{BTZ} is our Rosetta stone (paraphrasing G. Sierra).

Journal of Physics: Conference Series **720** (2016) 012010 doi:10.1088/1742-6596/720/1/012010

Hybrid Expansion Law with Interacting Cosmic Fluids

Luis Avilés, Patricio Mella and Fabián Orellana

Departamento de Física, Universidad de Concepción, Casilla 160-C, Concepción, Chile.

E-mail: luisaviles@udec.cl; patriciomella@udec.cl; fabianorellana@udec.cl

Abstract. A theoretical model for hybrid expansion law in a spatially flat Friedmann-Robertson-Walker (FRW) model filled with two interacting perfect fluid is developed. We obtain the behaviour of the energy densities, pressures, and the energy transfer as functions of the cosmological time. This provides a new family of solutions for hybrid expansion law with interacting cosmic fluids.

1. Introduction

The observed late acceleration of the universe [1] could be explained by cosmological models with exchange of energy between two cosmic fluids. The energy transfer between cosmic fluids was first studied by Davidson [3]. Other examples of the physics producing the energy transfer were given by Sistero [4] and McIntosh [5]. For an exhaustive review of interacting fluids in cosmology see [6, 7, 8, 9, 10].

In this paper we analyze a flat FRW model with an interaction between cosmic fluids through a cosmic scale factor given by a hybrid expansion law [13], which mimics the power law and de Sitter cosmologies as special cases, providing a description of the transition from the early decelerating phase to the recent accelerating phase and generating a time dependent deceleration parameter.

Some time ago in Refs. [11, 12] the special case of hybrid expansion law for two fluids cosmological models with and without energy transfer was show, in this works the interaction term was seen as $Q(t) = 3H\sigma\rho_m$ and where ρ_m was considered as the energy density of matter. Notice that [13] study the hybrid expansion law whitout interaction and its observational constraints. On the other hand in [14] was studied a spatially homogeneous and anisotropic Bianchi type-V model without energy transfer.

Follow Ref.[2], the goal is to consider a hybrid expansion law and two barotropic perfect fluids without specify the nature of the fluid contents and find the interaction term $Q(t)$, the energy densities and pressures for the cosmic fluids in this configuration.

This paper is organized as follows. In Section 2, we review briefly the hybrid expansion law in a flat FRW model. In Section 3, we rewiew briefly the interaction for cosmic fluids. In Section 4 we presents the analysis of the interaction found. The conclusions are presented in Section 5.

2. Hybrid expansion law

We consider the cosmological scale factor as a hybrid expansion law [13]

$$a(t) = a_0 \left(\frac{t}{t_o}\right)^\alpha e^{\beta\left(\frac{t}{t_0}-1\right)} \tag{1}$$

where α and β are constants, here a_0 and t_0 denote the scale factor and the age of the Universe today, respectively. Setting $\beta = 0$ leads the power law cosmology and set $\alpha = 0$ the exponential-law cosmology. These cosmologies are the special cases of the Hybrid expansion law cosmology. For any other values of α and β will have new directions to explore cosmology in the context of the hybrid expansion law [13, 14].
From (1) we can obtain the Hubble parameter and deceleration parameter, these are respectively

$$H = \frac{\dot{a}(t)}{a(t)} = \frac{\alpha}{t} + \frac{\beta}{t_0} \tag{2}$$

$$q(t) = -\frac{a(t)\ddot{a}(t)}{\dot{a}(t)^2} = \frac{\alpha t_0^2}{(\beta t + \alpha t_0)^2} - 1 \tag{3}$$

if we choose α and β such that the power law-term dominates over the exponential term in the early universe [13], that is for $t \sim 0$, then the cosmological parameters approximate to the following:

$$a \sim a_0 \left(\frac{t}{t_0}\right)^\alpha, \qquad H \sim \frac{\alpha}{t}, \qquad q \sim \frac{1}{\alpha} - 1 \tag{4}$$

Similarly, if we choose α and β such that the exponential term dominates at late times [3] such that in the limit $t \to \infty$ we have

$$a \to a_0 e^{\beta(\frac{t}{t_0}-1)}, \qquad H \to \frac{\beta}{t_0}, \qquad q \to -1 \tag{5}$$

Notice when α and β both are non-zero, we have a variable deceleration parameter and we have a transition from deceleration to acceleration for $\beta > 0$ and $0 < \alpha < 1$ [13].

3. Interaction for cosmic fluids

We consider the flat FRW metric given by

$$ds^2 = -dt^2 + a(t)^2 \left(\frac{dr^2}{1 - kr^2} + r^2(d\theta^2 + \sin^2\theta d\phi^2)\right) \tag{6}$$

the Einstein's field equations for the metric (6) and the universe filled with two fluids ρ_1 and ρ_2 are

$$3H^2 = \kappa(\rho_1 + \rho_2) \tag{7}$$

$$-2\frac{\ddot{a}(t)}{a(t)} - H^2 = \kappa(P_1 + P_2) \tag{8}$$

where $\kappa = 8\pi G = 1$.
We assume that the perfect fluids interact through the interaction term Q according to

$$\dot{\rho}_1 + 3H(\rho_1 + P_1) = Q \tag{9}$$

$$\dot{\rho}_2 + 3H(\rho_2 + P_2) = -Q \tag{10}$$

Note that if $Q > 0$ we have there is a transfer of energy from the fluid ρ_2 to the fluid ρ_1. If $Q = 0$ we have two noninteracting fluids, and then each fluid satisfies the standard balance equation separately.
Now we consider barotropic equations of state given by $P_1 = \omega_1\rho_1$ and $P_2 = \omega_2\rho_2$, where $\omega_1 \neq \omega_2$. So, the balance equations (8) and (9) are

$$\dot{\rho}_1 + 3H\rho_1(1+\omega_1) = Q \tag{11}$$

$$\dot{\rho}_2 + 3H\rho_2(1+\omega_2) = -Q \tag{12}$$

4. The Interaction Q term

Taking into account $Q \neq 0$, the Friedmann equation (7), the equations (11) and (12) we have the solutions for the energy densities

$$\rho_1(t) = \frac{3\left[\frac{\alpha}{t} + \frac{\beta}{t_0}\right]^2(1+\omega_2) - 2\frac{\alpha}{t^2}}{\kappa(\omega_2 - \omega_1)} \tag{13}$$

and

$$\rho_2(t) = \frac{3\left[\frac{\alpha}{t} + \frac{\beta}{t_0}\right]^2(1+\omega_1) - 2\frac{\alpha}{t^2}}{\kappa(\omega_1 - \omega_2)} \tag{14}$$

where the interaction Q term takes the form

$$Q = \frac{4\frac{\alpha}{t^3} + 9\left[\frac{\alpha}{t} + \frac{\beta}{t_0}\right]^3(1+\omega_1)(1+\omega_2) - 6\frac{\alpha(t_o\alpha + t\beta)(2+\omega_1+\omega_2)}{t^3 t_0}}{\kappa(\omega_2 - \omega_1)} \tag{15}$$

Now we considerer the case $\beta = 0$, which give the results obtained by [2], that is, for the Hubble parameter and the deceleration parameter, we have

$$H = \frac{\alpha}{t} \tag{16}$$

and

$$q = \frac{1}{\alpha} - 1 \tag{17}$$

For the energy densities of the fluids ρ_1 and ρ_2

$$\rho_1(t) = \frac{3\left[\frac{\alpha}{t}\right]^2(1+\omega_2) - 2\frac{\alpha}{t^2}}{\kappa(\omega_2 - \omega_1)} \tag{18}$$

and

$$\rho_2(t) = \frac{3\left[\frac{\alpha}{t}\right]^2(1+\omega_1) - 2\frac{\alpha}{t^2}}{\kappa(\omega_1 - \omega_2)} \tag{19}$$

finally the interaction Q term

$$Q = \frac{4\frac{\alpha}{t^3} + 9\left[\frac{\alpha}{t}\right]^3(1+\omega_1)(1+\omega_2) - 6\frac{\alpha^2(2+\omega_1+\omega_2)}{t^3}}{\kappa(\omega_2 - \omega_1)} \tag{20}$$

5. Conclusions

We have shown that through of a hybrid expansion law with two interacting perfect fluid, we found the interaction term Q, which gives us the energy transfer between two cosmic fluids and without specify the nature of the fluid contents. We also show that in the special case $\beta = 0$, we recover the results obtained by [2].

Acknowledgments

This work was supported by Comisión Nacional de Ciencias y Tecnología through FONDECYT Grants 3130444 (PM).

References

[1] Perlmutter, S., et al. *Nature* **391**, 51 (1998); Perlmutter, S., et al. *Astrophys.J.***517**, 565 (1999); Riess, A.G., et al. *Astrophys. J.***116**, 1009 (1998); E.Komatsu *et al.* [WMAP Collaboration], *Astrophys.J.Suppl.* **192**, 18 (2011). [arXiv:1001.4538 [astro-ph.CO]]; D.Larson, J.Dunkley, G.Hinshaw, E.Komatsu, M.R.Nolta, C.L.Bennett, B.Gold, M.Halpern *et al.,Astrophys.J.Suppl.* **192**, 16 (2011). [arXiv:1001.4635 [astro-ph.CO]]; N.Jarosik *et al.*, *Astrophys.J.Suppl.* **192**, 14 (2011). [arXiv:1001.4744 [astro-ph.CO]]; P.A.R. Ade et al. [Planck Collaboration]: *Planck 2013 results. I. Overview of products and scientific results*, arXiv:1303.5062 [astro-ph.CO]; N.Suzuki, D.Rubin, C.Lidman, G.Aldering, R.Amanullah, K.Barbary, L.F.Barrientos and J.Botyanszki *et al.*, *Astrophys.J.* **746**, 85 (2012). [arXiv:1105.3470 [astro-ph.CO]].
[2] M.Cataldo, P.Mella, P.Minning, J.Saavedra, *Phys.Lett. B***662** (2008) 314-322.
[3] Davidson W., 1962, *MNRAS*, **124** , 79
[4] Sistero R.F. 1971, *Astrophys. and Space Science* **12** , 484
[5] McIntosh, C.B.G. 1967, *MNRAS* **138** , 423; McIntosh, C.B.G. 1968, *MNRAS* **140** , 461
[6] M.R.G.Maia , N.Pires, H.S.Gimenes, [arXiv:1503.00033 [gr-qc]].
[7] W.Zimdahl, *Int.J.Geom.Meth.Mod.Phys.* **11**, 1460014 (2014), [arXiv:1404.7334 [astro-ph.CO]].
[8] Y.L.Bolotin, A.Kostenko, O.A.Lemets and D.A.Yerokhin, *Int.J.Mod.Phys.D***24**, 1530007 (2015), [arXiv:1310.0085 [astro-ph.CO]].
[9] Y.L.Bolotin, V.A.Cherkaskiy, O.A.Lemets, I.V.Tanatarov and D.A.Yerokhin, [arXiv:1312.6556 [physics.ed-ph]].
[10] K.Bamba, S.Capozziello, S.Nojiri and S.D.Odintsov, *Astrophys.Space Sci.* **342**, 155 (2012), [arXiv:1205.3421 [gr-qc]].
[11] H.Amirhashchi, A.Pradhan and B.Saha, *Chin.Phys.Lett.* **28**, 039801 (2011), [arXiv:1011.3940 [gr-qc]].
[12] B.Saha, H Amirhashchi and A.Pradhan, *Astrophys.Space Sci.* **342**, 257 (2012), [arXiv:1108.2133 [gr-qc]].
[13] O. Akarsu, S. Kumar, R. Myrzakulov, M. Sami, Lixin Xu, *JCAP* **01** (2014) 022; B. Saha, H. Amirhashchi, A.Pradhan, *Astrophys.Space Sci.* **342** (2012) 257-267
[14] S.Kumar, *Grav.Cosmol.* **19**, 284 (2013), [arXiv:1010.1612 [physics.gen-ph]].

Journal of Physics: Conference Series **720** (2016) 012011 doi:10.1088/1742-6596/720/1/012011

Exact solutions for the motion of spinning massive particles in conformally flat spacetimes

Felipe A Asenjo[1] **and Sergio A Hojman**[1,2,3,4]

[1] Facultad de Ingeniería y Ciencias, Universidad Adolfo Ibáñez, Santiago, Chile
[2] Departamento de Ciencias, Facultad de Artes Liberales, Universidad Adolfo Ibáñez, Santiago, Chile
[3] Departamento de Física, Facultad de Ciencias, Universidad de Chile, Santiago, Chile
[4] Centro de Recursos Educativos Avanzados, CREA, Santiago, Chile

E-mail: `felipe.asenjo@uai.cl, sergio.hojman@uai.cl`

Abstract. The motion of massive spinning test particles (tops) on gravitational backgrounds is studied based on a Lagrangian formulation developed some forty years ago. In particular, the (non-geodesical) equations of motions on some conformally flat spacetimes are solved exactly. Since tops follow non-geodesical paths, velocity and momentum vectors are, in general, non-paralell. Some peculiar features such as superluminal motion are exhibited and discussed in some detail.

1. Introduction

It is largely known that the dynamics of spinning massive particles (tops) is described by non-geodesic equations. Massive spinning particles have a completely different behavior as compared to spinless massive particles. One may intuitively understand this behavior noticing that the equivalence principle (interpreted as stating that test particles in a gravitational field follow geodesics) is valid only for spinless point test particles. As a point particle cannot spin, the top is an extended particle. Therefore, tops experience tidal forces and, therefore, follow non-geodesic paths.

The equations governing the motion of tops in gravitational fields were first derived by Mathisson [1, 2] and later by Papapetrou [3] as the limiting case of rotating fluids moving in curved spacetimes. However, a correct derivation of the spherical top motion in gravitational fields, (described by the metric field $g_{\mu\nu}$) has been obtained at Lagrangian level [4, 5].

The complete development of this theory can be found in Refs. [4, 5], or more recently in Refs. [6, 7]. Here we just limit ourselves to mention that in this classical theory, the velocity u_μ and the canonical momentum P_μ vectors are, in general, not parallel. This is similar to the behavior of velocity and momentum of particles described by quantum theories, as for example, in the Dirac equation. The Lagrangian theory [4, 5] gives the evolution equation for the top dynamics

$$\frac{DP^\mu}{D\lambda} = -\frac{1}{2}R^\mu{}_{\nu\alpha\beta}u^\nu S^{\alpha\beta}\,, \qquad \frac{DS^{\mu\nu}}{D\lambda} = P^\mu u^\nu - u^\mu P^\nu\,, \tag{1}$$

where the covariant derivatives are defined through an arbitrary parameter λ, and $R^\mu{}_{\nu\alpha\beta}$ is the Riemann tensor. These results hold for an arbitrary Lagrangian (time-reparametrization invariant) function defined in terms of four scalars constructed in terms of the top's velocity vector and angular velocity (antisymmetric) tensor (for detail, see Refs. [4, 5]). It is worthwhile mentioning that the system is constrained to restrict the spin tensor to generate rotations only. From the Lagrangian theory one can derive the constraint $S^{\mu\nu}P_\nu = 0$, known as the Tulczyjew constraint.

The main features of this Lagrangian theory is that it can be proved that the square of the mass $m^2 \equiv P_\mu P^\mu > 0$ is conserved. Thus the momentum vector is timelike along the motion. However, as we will show, the velocity field may become spacelike [6, 7]. On the other hand, the scalar spin $J^2 \equiv (1/2)S_{\mu\nu}S^{\mu\nu}$ is also conserved on any curved background, where $S_{\mu\nu}$ is the antisymmetric spin tensor. Besides, in general, if ξ_μ is a vector de Killing, then it can be shown from Eqs. (1) that the quantity

$$C_\xi \equiv P^\mu \xi_\mu - \frac{1}{2} S^{\mu\nu} \xi_{\mu;\nu} \tag{2}$$

is also conserved along the top motion Refs. [4, 5].

The top dynamics (1) has been studied for several different metrics in Refs. [6, 7]. Interestingly enough, it has been shown that several exact solutions for the motion can be found. In all of cases, the velocity is not restricted to be timelike. Depending on the relations between the constants of motion of the particle (mass, spin, angular momentum and energy), the top velocity can become spacelike, implying superluminal motion.

In this work, we study the top dynamics in the particular case of a static spherically symmetric conformally flat spacetime. We explore different conformal factors for universes composed by radiation and a static perfect fluid. There exist exact solutions for universes filled with uniform electromagnetic radiation only, matter-dominated universes and radiation-dominated universes. These systems were studied in Ref. [7], and here we extend the analysis of our results.

2. Exact solution for conformally flat spacetimes

We plan to find an exact solution of Eqs. (1) for conformally flat spacetimes

$$ds^2 = \Omega^2 \eta_{\mu\nu} dx^\mu dx^\nu\,, \tag{3}$$

with the spherical symmetric conformal factor $\Omega \equiv \Omega(r)$ and where $\eta_{\mu\nu}$ is the spherical flat metric.

The main procedures to obtain the solutions are outlined in [7]. For extended calculations we refer the readers to the mentioned work. Along these lines we will just display the main results. The conserved energy for the top motion can be calculated in theory using the conserved quantities (2) derived from the equations of motion (1). For the Killing vector $\xi_\mu = (c^2\Omega^2, 0, 0, 0)$, we can find that the conserved energy is

$$E = P_t - c^2 \Omega \Omega' S^{tr}\,, \tag{4}$$

where P_t is the time-component of the momentum and similarly for the spin tensor S^{tr}. Here the symbol $'$ refers to the derivative with respect to r. Also we have added explicitly here and after the speed of light c in the expressions. On the other hand, using the Killing vector $\xi_\mu = (0, 0, 0, -r^2\Omega^2)$, we can find another conserved quantity in the equatorial plane $\theta = \pi/2$. This conserved quantity turns to be the component of the angular momentum orthogonal to the equatorial plane. Using (2) we find

$$j = -\Omega\left(\Omega + r\Omega'\right) r S^{\phi r} - P_\phi\,, \tag{5}$$

where P_ϕ is one of the angular component of the momentum.

Exact solutions for the momenta can be obtained using the equations of motion (1) and the conserved quantities (2) for the mass m and spin J. Following the calculations outlined in [7], we can find the momenta in terms of constants of motion

$$\begin{aligned} P_\phi(r) &= \frac{1}{1-\eta}\left(-j \mp \frac{EJ(\Omega + r\Omega')}{mc^2\Omega^2}\right), \\ P_t(r) &= \frac{1}{1-\eta}\left(E \pm \frac{jJ\Omega'}{mr\Omega^2}\right), \\ P_r(r) &= \pm\left[\frac{P_t^2}{c^2} - \frac{P_\phi^2}{r^2} - m^2c^2\Omega^2\right]^{1/2}, \end{aligned} \tag{6}$$

where $\eta(r) = \Omega'(\Omega + r\Omega')\, J^2/(r\Omega^4 m^2 c^2)$. The two signs correspond to the two possible spin orientations. Using these expressions in Eqs. (1) we can find the evolution of the top velocity in the equatorial plane [7]

$$\dot\phi = \frac{\zeta\,\gamma\, c^2}{r^2}\left(\frac{P_\phi}{P_t}\right), \quad \dot r = \zeta\, c^2\left(\frac{P_r}{P_t}\right), \tag{7}$$

where we have defined the auxiliary functions

$$\zeta = (\eta - 1)\left[\eta + 1 - \frac{J^2}{\Omega^3 m^2 c^2 r}\left(2\Omega' + r\Omega''\right)\right]^{-1}, \gamma = (1-\eta)^{-1}\left[1 + \frac{J^2}{\Omega^4 m^2 c^2}\left((\Omega')^2 - \Omega\Omega''\right)\right]. \tag{8}$$

At the same time, the trajectories for the top motion can be calculated directly from Eqs. (7). The trajectory in a parametric representation can be obtained by solving the integral

$$\phi(r) = \int dr \frac{\gamma\, P_\phi}{r^2\, P_r}. \tag{9}$$

These solutions allow us to evaluate the interval $ds^2 = g_{\mu\nu}dx^\mu dx^\nu = c^2\Omega^2 dt^2\left(1 - \dot r^2/c^2 - r^2\dot\phi^2/c^2\right)$ along the trajectory of the top. Notice that the interval is a measure of the velocity of the particle. Therefore, if $ds^2 > 0$, the velocity of the top is expected to be timelike moving the top at subluminal speed. Ordinary (spinless) particle always follow timelike trajectories, however if a particle could achieve luminal or superluminal velocities, the interval will behave as $ds^2 = 0$ or $ds^2 < 0$ respectively. The evaluation of the top interval in conformally flat spacetime can be found using the solutions (7). This turns out to be

$$\frac{ds^2}{c^2dt^2} = \Omega^2\left(1 - \zeta^2\right) + \frac{m^2c^4\zeta^2\,\Omega^4}{P_t^2}\left(1 + \frac{P_\phi^2\,(1-\gamma^2)}{\Omega^2 m^2 r^2 c^2}\right). \tag{10}$$

It is worth remarking that the previous interval is not always timelike. Depending on the constants of motion (specially spin J) and initial conditions, the interval of the top could be lightlike or spacelike. In particular, if the spin is neglected, then $ds^2/c^2dt^2 = \Omega^4(m^2c^4/E^2) > 0$, and the velocity of the top is always timelike. The main implication of considering the spin of particles is that the top could reach superluminal velocities without violating any law of General Relativity.

The previous exact solution requires an explicit expression for the conformal factor. Here we study a Universe composed by uniform electromagnetic radiation (through the electrostatic potential F_{0r}) and a static perfect fluid with energy density ϵ and pressure p. In this case an exact form for the conformal factor can be obtained from Einstein equations [8]. The conformal factor is

$$\Omega(r) = Q_1\left(r^{-2/(1+3\alpha)} - Q_2\beta\right)^{(1+3\alpha)/2}, \tag{11}$$

where Q_1 y Q_2 are constants, $\alpha = p/\epsilon$ and $\beta = (1+3\alpha)^{(3+3\alpha)/(1+3\alpha)}$. At the same time the energy density and the electromagnetic field could be obtained as

$$\begin{aligned} \epsilon &= \frac{3c^2\beta Q_2}{4\pi G Q_1^2(1+3\alpha)} r^{-\frac{4+6\alpha}{1+3\alpha}} \left(r^{-2/(1+3\alpha)} - Q_2\beta\right)^{-3-3\alpha}, \\ (F_{0r})^2 &= \frac{c^4\Omega^4 r^{\frac{-6-6\alpha}{1+3\alpha}}}{2GQ_1^2(1+3\alpha)} \left(r^{\frac{-2}{1+3\alpha}} - Q_2\beta\right)^{-3-3\alpha} \left[2+6\alpha - 4(2+3\alpha)\beta Q_2 r^{\frac{2}{1+3\alpha}}\right], \end{aligned} \tag{12}$$

where G is the gravitational constant and $Q_2 \geq 0$ is imposed in order to have positive semidefinitive energies. Notice that the pressure is obtained as $p = \alpha\epsilon$, which corresponds to an equation of state that depends on the type of Universe studied.

3. Tops in a Universe with radiation only

When $Q_2 = 0$, both the energy density and the pressure vanish. Thus, there is only radiation in the Universe. The conformal factor acquires the form $\Omega = Q/r$, which is generally called the Bertotti-Robinson solution [9, 10, 11]. In this simple case, the momenta (6) becomes $P_\phi = -j$, $P_t = E \pm jJ/(mQr)$. Using this, and the velocities expressions (7) we can calculate the interval for the top in this kind of Universe

$$\frac{ds^2}{c^2dt^2} = \frac{1}{r^4 P_t^2 m^4 c^2 Q^2}\left(2m^2c^2Q^2j^2J^2 + m^6c^6Q^6 - j^2J^4\right). \tag{13}$$

Notice that the interval depends on constants of motion only. There will be exist certain values of m, J, j and Q for which it may happen that $ds^2 \leq 0$ (lightlike or spacelike trajectories respectively). This implies that depending on the characteristics of the top, the massive particle could travel at superluminal speed for all moments of its motion.

In Fig. (1) we plot the region (in blue shadow) where the interval (13) becomes negative or zero. The plot is made in terms of the dimensionless quantities $J/(mcQ)$ and $j/(mcQ)$, as the expression (13) depends only on the relative values of those constants. Notice that due to the existence of the spin of the top, the region when $ds^2 \leq 0$ is larger than the one when $ds^2 > 0$.

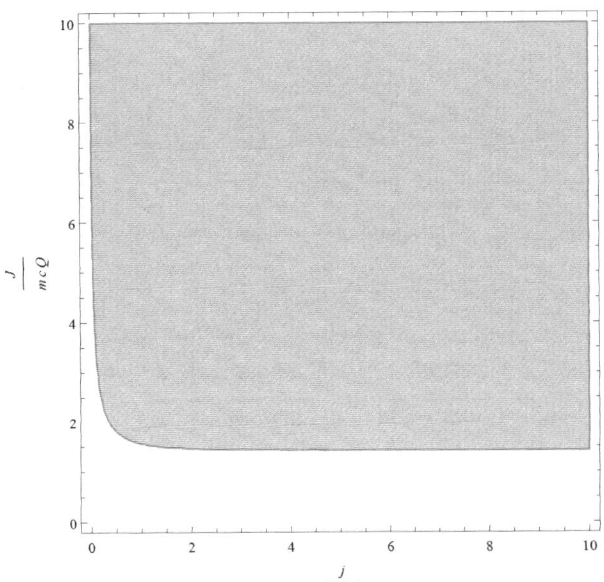

Figure 1. Behavior of the interval (13). The blue shadow region corresponds to the values of the constants of motion where $ds^2 \leq 0$.

4. Tops in a matter-dominated Universe

If the energy density of the fluid dominates over the pressure then the universe is matter-dominated (with non-relativistic matter). We can model this choosing $\alpha = 0$ [12]. In this case we have that the conformal factor is $\Omega = Q_1 \left(1/r^2 - Q_2\right)^{1/2}$, and the energy density and the strength of the electromagnetic field (12) are finite.

The solutions for the momenta are now $P_\phi = \left(-j \pm EJQ_1^2 Q_2/(mc^2\Omega^3)\right)/(1-\eta)$, and $P_t = \left(E \mp jJQ_1^2/(mr^4\Omega^3)\right)/(1-\eta)$, where now η is reduced to $\eta = J^2Q_1^4Q_2/(r^4m^2c^2\Omega^6) > 0$. Similarly to the previous case, we can use the solutions to evaluate the interval, which for this case becomes

$$\frac{ds^2}{c^2dt^2} = \frac{\Omega^2}{(\eta+1-2\eta^2)^2{P_t}^2}\left[3\eta(2+\eta)\left(1-\eta\right)^2{P_t}^2 + m^2c^4\Omega^2(1-\eta)^4 - \left(1-\eta\right)^2{P_\phi}^2\left(\frac{8\eta(1+\eta)c^2}{r^2} + \left(\frac{J^2Q_1^4}{\Omega^6 c\, m^2 r^7}\right)^2 - \frac{(2+6\eta)J^2Q_1^4}{\Omega^6 m^2 r^8}\right)\right]. \quad (14)$$

Again, it is possible for some combination of the different top's parameters to have $ds^2 \leq 0$ along the trajectories. The behavior of the interval depends on the relative strength of the following five dimensionless parameter Q_1/r, $Q_1^2Q_2$, $J\sqrt{Q_2}/(mc)$, $j\sqrt{Q_2}/(mc)$, and $E/(mc^2)$. On the contrary to the previous case (where the Universe is composed only by uniform radiation), now the behavior of ds^2 explicitly depends on the position r of the top.

In Fig. (2) we plot several examples of regions where ds^2 given in (14) could be zero or negative. As it is not possible to have a graphical representation of ds^2 using the five parameters, we have fixed two of them arbitrarily, $Q_1^2Q_2$ and $J\sqrt{Q_2}/(mc)$. Notice that these choices do not fix neither Q_1, Q_2 nor m. In Figs. (2a), (2b) and (2c) we have fixed the values to be $Q_1^2Q_2 = 1$ and $J\sqrt{Q_2}/(mc) = 6$, $Q_1^2Q_2 = 5$ and $J\sqrt{Q_2}/(mc) = 10$, and $Q_1^2Q_2 = 7$ and $J\sqrt{Q_2}/(mc) = 16$, respectively. We can notice that as both values increase, the size of the region where $ds^2 \leq 0$ diminishes. On the other hand, in Fig. (2d), we plot the region for the values $Q_1^2Q_2 = 15$ and $J\sqrt{Q_2}/(mc) = 3$, showing that for large values of $Q_1^2Q_2$, the region where $ds^2 \leq 0$ is bounded for large values of r (small values of Q_1/r).

5. Tops in a radiation-dominated Universe

Another type of Universe can be studied choosing the equation of state appropriately. If the composition of the Universe is such that radiation dominates over the fluid (for example in the very-early Universe), then $\alpha = 1/3$ [12]. In this case, the conformal factor reduces to $\Omega = Q_1\left(1/r - 4Q_2\right)$, while the momenta become $P_\phi = \left(-j \pm 4EJQ_1Q_2/(mc^2\Omega^2)\right)/(1-\eta)$, and $P_t = \left(E \mp jJQ_1/(mr^3\Omega^2)\right)/(1-\eta)$, where $\eta = 4J^2Q_1^2Q_2/(r^3m^2c^2\Omega^4) > 0$.

With the solutions, the evaluation of the interval produces

$$\frac{ds^2}{c^2dt^2} = \frac{\Omega^2}{(1-\eta^2)^2{P_t}^2}\left[4\eta\left(1-\eta\right)^2{P_t}^2 + m^2c^4\Omega^2(1-\eta)^4 - \left(1-\eta\right)^2{P_\phi}^2\left(\frac{3\eta(2+\eta)c^2}{r^2} + \frac{J^4Q_1^4}{\Omega^8 m^4 c^2 r^{10}} - \frac{(2+4\eta)J^2Q_1^2}{\Omega^4 m^2 r^6}\right)\right], \quad (15)$$

from where again we can see that ds^2 may become timelike, lightlike or spacelike depending on the top's properties and its position.

In Fig. (3) we plot different regions where (15) could be non-positive. In this case, the values of the interval (15) will depend on the five dimensionless parameter Q_1/r, Q_1Q_2, $J/(mcQ_1)$, $j/(mcQ_1)$, and $E/(mc^2)$. For the graphical representation (3) we have arbitrarily fixed the parameters Q_1Q_2 and $J/(mcQ_1)$. Notice that for the radiation-dominated Universe, the region

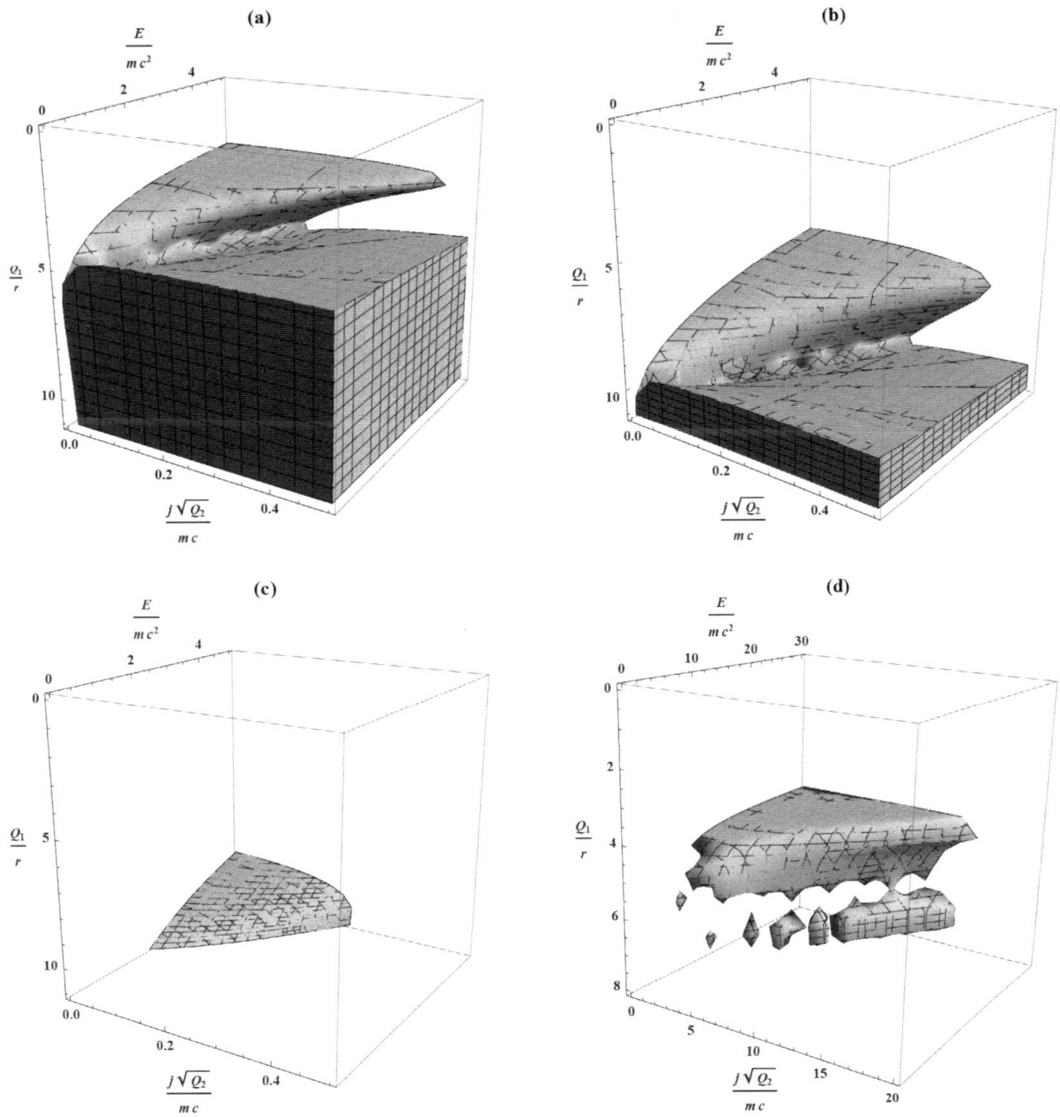

Figure 2. Behavior of the interval (14). The region shows when $ds^2 \leq 0$ along the top motion. The region depends on the position r of the top through the parameter Q_1/r. (a) Interval (14) with chosen values of $Q_1^2 Q_2 = 1$ and $J\sqrt{Q_2}/(mc) = 6$. (b) Interval (14) for $Q_1^2 Q_2 = 5$ and $J\sqrt{Q_2}/(mc) = 10$. (c) Interval (14) for $Q_1^2 Q_2 = 7$ and $J\sqrt{Q_2}/(mc) = 16$. (d) Interval (14) for $Q_1^2 Q_2 = 15$ and $J\sqrt{Q_2}/(mc) = 3$.

for superluminal motion is restricted to some values of r. In Figs. (3a), (3b) and (3c) we have fixed the parameter values to be $Q_1 Q_2 = 1$ and $J/(mcQ_1) = 3$, $Q_1 Q_2 = 5$ and $J/(mcQ_1) = 10$, and $Q_1 Q_2 = 9$ and $J/(mcQ_1) = 19$, respectively. One sees that as both parameter values increase, the region where $ds^2 \leq 0$ increases for smaller values of r (larger values of Q_1/r). In Fig. (3d), we have plotted the region $ds^2 \leq 0$ for the values $Q_1 Q_2 = 18$ and $J/(mcQ_1) = 4$, which shows that superluminal motion can be achieved even for smaller values of r.

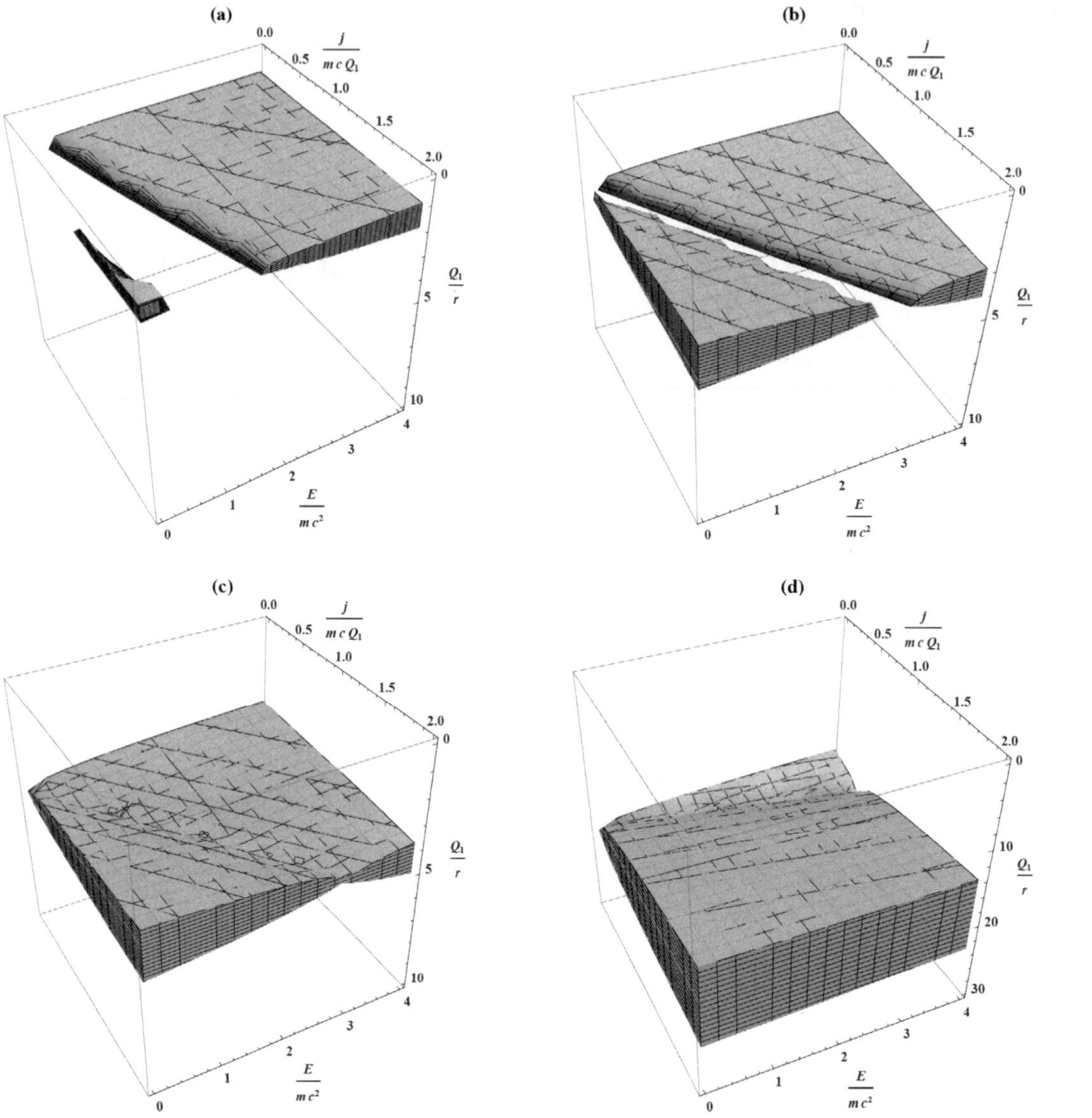

Figure 3. Behavior of the interval (15). The region shows when $ds^2 \leq 0$ in the top motion. Again it depends on the position r. (a) Interval (15) with chosen values of $Q_1^2 Q_2 = 1$ and $J\sqrt{Q_2}/(mc) = 3$. (b) Interval (15) for $Q_1^2 Q_2 = 5$ and $J\sqrt{Q_2}/(mc) = 10$. (c) Interval (15) for $Q_1^2 Q_2 = 9$ and $J\sqrt{Q_2}/(mc) = 19$. (d) Interval (15) for $Q_1^2 Q_2 = 18$ and $J\sqrt{Q_2}/(mc) = 4$.

6. Conclusions

Along this work we have studied the exact solutions for the motion of the top in spherical conformally flat spacetimes. We have shown that the spin of the massive particles plays an extremely important role changing the dynamics as compared to a spinless particle case.

The inclusion of the spin implies that the massive particle follows non-geodesic paths. This has as a consequence that the top trajectories could be lightlike or spacelike as well as timelike. This behavior depends on the values of the constants of motion of the top and on the spacetime metric.

We review three kind of metrics for a Universe filled with uniform radiation only, a matter-

Journal of Physics: Conference Series **720** (2016) 012011 doi:10.1088/1742-6596/720/1/012011

dominated Universe, and a radiation-dominated Universe. In all three Universes, superluminal motion of the top can be achieved. This seems to be a robust characteristic of Eqs. (1) which has already been studied for different spacetimes [6, 7].

For the case of a Universe filled with uniform radiation only, the lightlike or spacelike trajectories will depend only on the features of the massive spinning particle, and not on its position. This means that the nature of the top (whether it is a bradyon or a tachyon) is completely determined by the values of the constant of motion of the particle. In the matter-dominated and radiation-dominated Universes, the type of trajectories depends on the position of the top. In the plots (2) and (3), we depict that there are a huge number of possibilities for superluminal behavior of the top relying on the different values of its constants of motion an on its position. In particular we have shown that in a radiation-dominated Universe it is possible to have superluminal motion for small values of r, which may be relevant for the study of the early Universe.

Acknowledgments

FAA thanks to CONICyT-Chile for Funding no. 79130002.

References

[1] Mathisson M 1937 Acta Phys. Pol. **6** 163
[2] Mathisson M 1937 Acta Phys. Pol. **6** 218
[3] Papapetrou A 1951 Proc. R. Soc Lond. **209A** 248
[4] Hojman S 1975 Electromagnetic and gravitational interactions of a spherical relativistic top *PhD Thesis* Princeton University (unpublished)
[5] Hojman S 1978 *Phys. Rev. D* **18** 2741
[6] Hojman S A and Asenjo F A 2013 *Class. Quantum Grav.* **30** 025008
[7] Zalaquett N, Hojman S A and Asenjo F A 2014 *Class. Quantum Grav.* **31** 085011
[8] Banerjee A and Santos N O 1981 *J. Math. Phys.* **22** 824
[9] Bertotti B 1959 *Phys. Rev.* **116** 1331
[10] Robinson I 1959 Bull. Acad. Polon. Sci. **7** 351
[11] Stefani H, Kramer D, MacCallumM, Hoenselaers C and Herlt E 2003 *Exact Solutions of Einsteins Field Equations* (Cambridge: Cambridge University Press)
[12] Ryden B 2003 *Introduction to Cosmology* (Reading, MA: Addison-Wesley)

Numerical simulations of Modified Newtonian Dynamics

G N Candlish[1], R Smith[2,1] and M Fellhauer[1]

[1]Departamento de Astronomía, Universidad de Concepción, Casila 160-C, Concepción, Chile
[2]Yonsei University, Graduate School of Earth System Sciences-Astronomy-Atmospheric Sciences, Yonsei-ro 50, Seoul 120-749, Republic of Korea

E-mail: gcandlish@astro-udec.cl

Abstract. The ΛCDM standard cosmological model is strongly supported by multiple lines of evidence, particularly from observations at large scales such as the CMB and large scale structure. There are some indications, however, of problems at smaller scales. An alternative to the CDM approach is to modify the gravitational force, as exemplified by the MOdified Newtonian Dynamics (MOND) idea. While evidence suggests MOND cannot account for dynamics at all scales without dark matter, it has been successful at galactic scales. Due to the complexity of the theory, however, most tests of MOND have extended no further than using a simple scaling relation to determine rotation curves or velocity dispersions. Therefore, to test the concept more thoroughly we require numerical simulations. We discuss the development and testing of a new N-body solver, using two distinct formulations of MOND, that is incorporated into the RAMSES code. The theory of MOND as a modification of Newtonian gravity is briefly summarised. We then show how it is implemented in the code, providing an example of an idealised test case and future applications.

1. Introduction

According to the standard model of cosmology a cold dark matter component comprises around 27% of the energy budget of the Universe. Observational support for this picture includes CMB observations [1], large-scale structure, gravitational lensing and cluster dynamics (see e.g. [2] or [3] for a more recent review and references). At small scales, however, there remains several unresolved problems, such as the missing satellites problem, the cusp/core issue and the predicted existence of satellite halos that are "too big to fail" (see e.g. [4] for a recent review of these issues). Recent observations [5, 6] have also pointed to the existence of large thin rotating disks of satellite galaxies around both the Milky Way and Andromeda, which may be difficult to incorporate into the heirarchical formation mechanism of the standard ΛCDM cosmological model.

An alternative approach, that may alleviate these problems, is to treat the dark matter phenomenology as indicative of a modification of gravity in the very weak field regime. One well-known example of this idea is the MOdified Newtonian Dynamics (MOND) first proposed in [7], in which the standard Newtonian gravitational force is enhanced for accelerations below an empirically determined value of $a_0 \sim 1.2 \times 10^{-10}$ m/s^2. Applied to galaxy rotation curves this simple idea has been remarkably effective [8]. The proposal has been developed into a full relativistic theory, which is necessary for the construction of a cosmological model, in [9, 10, 11] among others (see [12] and references therein), although so far none of these theories is known to pass all cosmological tests. Furthermore, it is known that additional unseen matter is required in MOND at cluster scales and beyond [13].

Nevertheless, the success of the MOND approach at galactic scales suggests it is worthwhile to investigate the possible consequences to galaxy dynamics of such a modification of gravity. To investigate such questions we require numerical simulations using a MOND gravitational solver. Several numerical studies of MOND have been undertaken, in particular the pioneering work of [14]. In [15] a MOND N-body code was used to investigate the stability of MOND disks, as well as the behaviour of galaxy mergers. A modification of the cosmological N-body code AMIGA along similar lines is developed in [16]. A publically available code known as NMODY [17] has been available for some time, and has been used to investigate various aspects of stellar dynamics in MOND, such as galaxy mergers and dynamical friction [18, 19]. This code works on a fixed spherical grid, however, making it cumbersome to perform simulations of systems that exhibit little symmetry.

As we wish to study systems covering a wide range of length scales (as necessary to consider satellite galaxies orbiting around hosts, or galaxies falling into clusters, for example) and to include gas physics, we have chosen to modify the powerful N-body/hydrodynamics adaptive mesh refinement code RAMSES [20]. Here we will briefly describe the code itself, which we call "RAyMOND," and we will demonstrate the performance of the code using an idealised test. In the course of developing our code a similar code was developed in [21], although our code uses two formulations of MOND, rather than just one, and a somewhat different numerical treatment. A more thorough description of the code and testing may be found in [22].

2. MOND formulations

Several theories that reproduce the MOND phenomenology in the weak field limit have been proposed over the years. Here we focus on two specific examples, referred to as the AQUAL formulation [23] and the QUMOND formulation [24]. The modified Poisson equation of the AQUAL formulation arises from an aquadratic gravitational Lagrangian of the form

$$\mathcal{L}_G = -\frac{1}{8\pi G} a_0^2 \mathcal{F}\left(\frac{|\nabla\phi|^2}{a_0^2}\right) \tag{1}$$

for which the associated modified Poisson equation is

$$\nabla \cdot \left[\mu\left(\frac{|\nabla\phi|}{a_0}\right)\nabla\phi\right] = 4\pi G\rho, \tag{2}$$

where $\mu(x) = d\mathcal{F}(z)/dz$, $z = x^2$, ρ is the density distribution, ϕ is the gravitational potential and a_0 is the MOND acceleration scale. The function $\mu(x)$ is the interpolation function, common to all current MOND formulations, that interpolates between the Newtonian regime in the presence of large accelerations, to the deep MOND regime when the accelerations are small. Therefore the form of $\mu(x)$ is only constrained at the limits:

$$\begin{aligned} \mu(x) &\to 1, \quad x \gg 1 \\ \mu(x) &\to x, \quad x \ll 1. \end{aligned} \tag{3}$$

Eq. 2 is clearly highly non-linear, requiring the application of particular numerical algorithms, as will be discussed below.

An alternative formulation, more amenable to a numerical treatment, is the quasi-linear QUMOND formulation. In this case the gravitational Lagrangian involves an auxiliary potential and may be written as

$$\mathcal{L}_G = -\frac{1}{8\pi G}\left[2\nabla\phi\cdot\nabla\phi^N - a_0^2\mathcal{Q}\left(\frac{|\nabla\phi^N|^2}{a_0^2}\right)\right]. \tag{4}$$

A variation of the action involving this Lagrangian leads to the following equations for the ϕ and ϕ^N fields:

$$\nabla^2\phi^N = 4\pi G\rho \qquad \nabla^2\phi = \nabla\cdot\left[\nu\left(\frac{|\nabla\phi^N|}{a_0}\right)\nabla\phi^N\right] \tag{5}$$

where $\nu(y) = d\mathcal{Q}(z)/dz$, $y = z^2$, and the field ϕ^N satisfies the standard Newtonian Poisson equation. The function $\nu(y)$ again performs an interpolation between Newtonian and MONDian regimes, depending on the accelerations in the system. The limiting behaviour in this formulation is

$$\begin{aligned} &\nu(y) \to 1, \quad y \gg 1 \\ &\nu(y) \to y^{-1/2}, \quad y \ll 1. \end{aligned} \tag{6}$$

This formulation is more straightforward numerically: one need only solve two standard linear Poisson equations, albeit with an unusual source term for the second equation. This source term, as well as including the contribution from standard matter, also contains a component of so-called "phantom dark matter," which is responsible for the MOND phenomenology in this formulation. Unlike the non-baryonic cold dark matter of the standard cosmology, however, the distribution of this component is fully determined by the normal "baryonic" matter of the galaxy.

2.1. Relativistic extensions

To be a complete theory of modified gravity MOND clearly requires a relativistic completion. Although we will not use the relativistic extensions of MOND in our work, it is nevertheless worth pointing out that several proposals for such a theory now exist, most of them reducing in the weak field limit to the AQUAL formulation discussed above. Almost all of these theories (except the recently proposed metric-only theory of [25]) propose to extend the degrees of freedom in the gravitational sector beyond those of a standard massless spin-2 gravitational field. Perhaps the most well-studied example is the TeVeS model of [9], which involves a metric, a vector field and a scalar field. Many more details regarding this theory may be found in the review of modified gravity theories of [26] and references therein. TeVeS appears to have dfficulty reproducing certain cosmological observables, particularly CMB data.

There is also a relativistic completion of the QUMOND formulation which adds additional degrees of freedom in the form of a second metric, and so lies in the class of bimetric theories of gravity [11]. While somewhat less well-studied than TeVeS, early indications seem promising, assuming that the theory does not exhibit the pathologies that are known to afflict many other multi-metric theories, such as inconsistencies and ghosts.

3. Modifications to RAMSES for the MOND calculation

The technical details of the RAMSES code may be found in [20] and the details of the modified code RAyMOND may be found in [22]. Here we will summarise the changes necessary to implement both the AQUAL and QUMOND formulations.

The non-linearity of the AQUAL modified Poisson equation limits the possible numerical schemes that may be practically utilised in an N-body solver. In particular, we cannot use direct force summation methods such as those used in tree codes. Nor can we use the fast Fourier transform method that is used to solve the Poisson equation in Fourier space. Instead

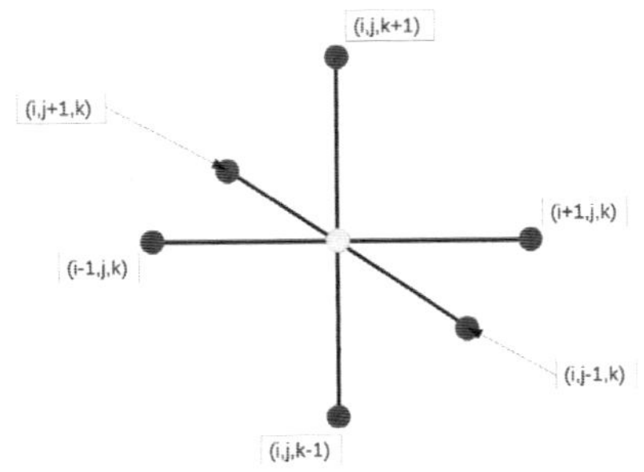

Figure 1. 6-point stencil used in the Gauss-Siedel iterative solver of standard RAMSES.

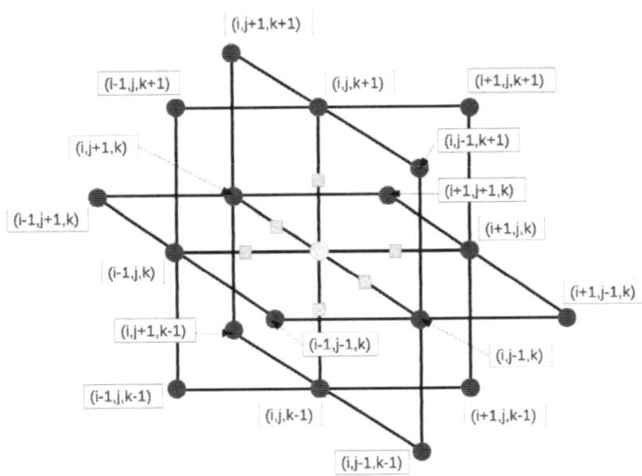

Figure 2. 18-point stencil used for the MOND solver. The squares indicate the locations where the μ function is evaluated.

we must use an iterative solver, which is readily adaptable to non-linear equations such as Eq. 2 above.

With this in mind, RAMSES is a good candidate for modification to MOND gravity, as it uses an iterative Gauss-Siedel solver accelerated with a multigrid scheme (see e.g. [27] for details of multigrid techniques) to solve the Newtonian Poisson equation. Both the numerical stencil for the iterative solver and the multigrid acceleration algorithm must be modified for the non-linear AQUAL formulation. We give a very brief description of these changes here, more details may be found in [22].

3.1. The modified Poisson solver

3.1.1. AQUAL The Gauss-Siedel solver in RAMSES uses a standard 7-point stencil to update the value of ϕ at the central grid point by the average of ϕ on the six neighbouring grid points. This stencil is shown for comparison in Fig. 1.

We will follow the procedure of [14], [15] and [16] by modifying the Gauss-Siedel solver used in RAMSES to use an extended stencil of points around each grid point in the calculation of ϕ, as shown in Fig 2. The updated value of ϕ at each grid point is now given by

$$\begin{aligned}\phi_{i,j,k} = \big(&\mu_{i+1/2,j,k}\phi_{i+1,j,k} + \mu_{i-1/2,j,k}\phi_{i-1,j,k} \\ &+\mu_{i,j-1/2,k}\phi_{i,j-1,k} + \mu_{i,j+1/2,k}\phi_{i,j+1,k} \\ &+\mu_{i,j,k-1/2}\phi_{i,j,k-1} + \mu_{i,j,k+1/2}\phi_{i,j,k+1} \\ &-4\pi dx^2\rho_{i,j,k}\big) \,/\, \sum_N \mu_N .\end{aligned} \tag{7}$$

The subscripts on the μ interpolation function denote that this is to be evaluated midway between the grid points, as indicated by the squares in Fig. 2, and the denominator is the sum over μ at all 6 neighbouring midway points. The evaluation of μ requires a finite-difference calculation of the gradient of ϕ at the neighbouring grid points, and this requires the extended stencil shown in Fig. 2. As an example, the gradients of ϕ at the $i+1/2, j, k$ grid point (used in the evaluation of μ at that location) are calculated as follows:

$$\begin{aligned}(\nabla\phi)^x_{i+1/2,j,k} &= \frac{\phi_{i+1,j,k} - \phi_{i,j,k}}{dx} \\ (\nabla\phi)^y_{i+1/2,j,k} &= \frac{\phi_{i,j+1,k} + \phi_{i+1,j+1,k} - \phi_{i,j-1,k} - \phi_{i-1,j-1,k}}{4dx} \\ (\nabla\phi)^z_{i+1/2,j,k} &= \frac{\phi_{i,j,k+1} + \phi_{i+1,j,k+1} - \phi_{i,j,k-1} - \phi_{i-1,j,k-1}}{4dx} .\end{aligned} \tag{8}$$

Similar expressions apply for the 5 other midway points at which μ is evaluated. Clearly the dependence of μ on ϕ leads to a ϕ dependence in the coefficients used in the Gauss-Siedel solver and thus a non-linear behaviour. We can easily see from Eq. 7 that we recover the Newtonian version of the Gauss-Siedel solver when $\mu = 1$, as is the case whenever all accelerations in the system are well above the MOND scale.

The extended stencil requires further modification of the RAMSES code as we now need ϕ values at *diagonal* neighbours, such as $\phi_{i+1,j+1,k}$. The data structure of RAMSES that allows access to the grid values only stores the six nearest neighbour points shown in Fig. 1. Rather than substantially revise this data structure to incorporate the diagonal grid points as "neighbours," we choose to use the existing linked list architecture to extract the points we require by finding the neighbours of neighbours.

Finally, the non-linear AQUAL formulation of MOND requires that we also modify the multigrid algorithm, in addition to modifying the Gauss-Siedel solver. Several approaches are possible: we choose the Full Approximation Storage scheme. In this scheme the full non-linear equation is solved on all coarse levels, rather than simply the correction to the fine grid solution, as in the standard linear algorithm. More details regarding this scheme may be found in [27].

3.1.2. QUMOND The QUMOND version of the code does not require any modifications to the Gauss-Siedel solver, as this formulation involves solving the standard Poisson equation, albeit twice, and with a modified density field the second time. The source term of Eq. 5 may be rewritten in terms of an additional density component, as discussed earlier:

$$\nabla^2\phi = 4\pi G(\rho + \rho') \tag{9}$$

where ρ is the physical density of the system and ρ' is given by

$$\rho' = \frac{1}{4\pi G}\nabla\cdot\left(\tilde{\nu}\left(\frac{|\nabla\phi^N|}{a_0}\right)\nabla\phi^N\right). \tag{10}$$

The calculation of ρ' makes use of the extended stencil described earlier and the Newtonian potential ϕ_N is determined from a standard pass through the RAMSES multigrid algorithm. Note that the interpolation function used here is related to $\nu(y)$ by $\nu = \tilde{\nu} + 1$.

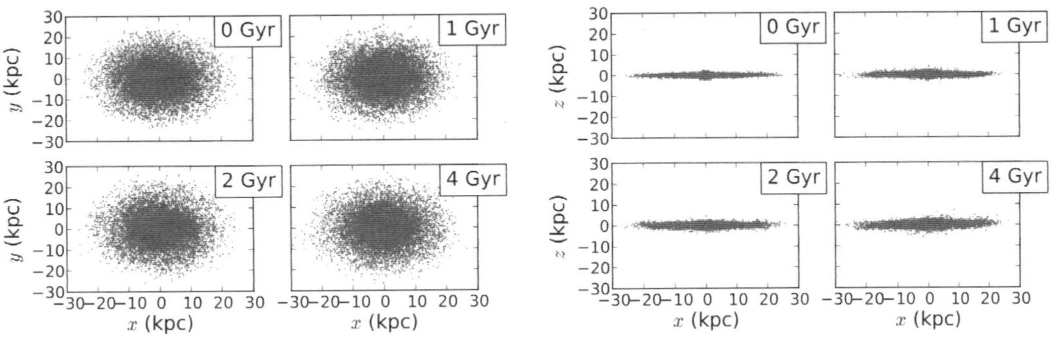

Figure 3. Newtonian disk and bulge (dark matter halo is not plotted). These are subsets of 20000 particles.

The principal changes necessary for this scheme are additional arrays to store the standard and phantom density components, and the Newtonian and MONDian potentials. The execution of the code through a timestep is also modified in order to solve two Poisson equations rather than just one.

4. Code test - spiral galaxy model

We show an example test of the RAyMOND code using a galaxy model comprising a disk and a bulge component. We use the initial conditions code *GalactICs* [28] to generate a Newtonian disk and bulge inside a dark matter halo. We run a simulation of this model as a control run using the normal Newtonian RAMSES code. For the MOND runs, we simply remove the dark matter halo particles from the initial conditions and run the disk and bulge particles only using RAyMOND. Fortuitously, this seems to result in a disk with a very similar rotation curve: it appears that in this case the Newtonian dark matter halo corresponds closely with the MONDian "phantom dark matter" halo determined by the stellar disk and bulge, but this may not generally be true.

The parameters in the disk, bulge and halo are those of the MW-A model in [28]. The disk is modelled as an exponential disk, with mass $4.2 \times 10^{10} M_{\odot}$, scale radius 4.5 kpc, and scale height 0.43 kpc. The bulge is a King model, with mass $2.1 \times 10^{10} M_{\odot}$ and the halo uses a lowered Evans model with mass $2.7 \times 10^{11} M_{\odot}$ (see [28] for the full set of parameters used to initialise the model). We use 60000 particles in the disk, 30000 in the halo and 10000 in the bulge.

Snapshots of the Newtonian run viewed face-on and edge-on are shown in Fig. 3. Although not clear in the plots, the disk forms weak spiral structure soon after the simulation starts, but is clearly globally stable, as expected due to the presence of the dark matter halo. There is some disk thickening over time, as is standard in numerical disk simulations. The AQUAL run snapshots are shown in Fig. 4. The disk forms more pronounced spiral structure than in the Newtonian case, but again exhibits no global instability. A bar instability does develop after around 3 Gyr of evolution. The disk thickening is considerably less pronounced in this case, although the bar becomes clearly visible by 4 Gyr.

The average rotational velocities as a function of radius in the disk are shown in Fig. 5. The rotation curve at the beginning of the simulation (identical in both the Newtonian and AQUAL runs) is shown as a black line, the Newtonian rotation curve after 4 Gyr is shown as a dashed blue line, while the AQUAL rotation curve is shown as a solid green line. In the Newtonian case, the flat rotation curve is evident due to the presence of the dark matter halo, and it remains stable throughout the evolution. The AQUAL rotation curve remains effectively flat at large radii, entirely due to the MOND enhancement of the gravitational accelerations in the disk, as

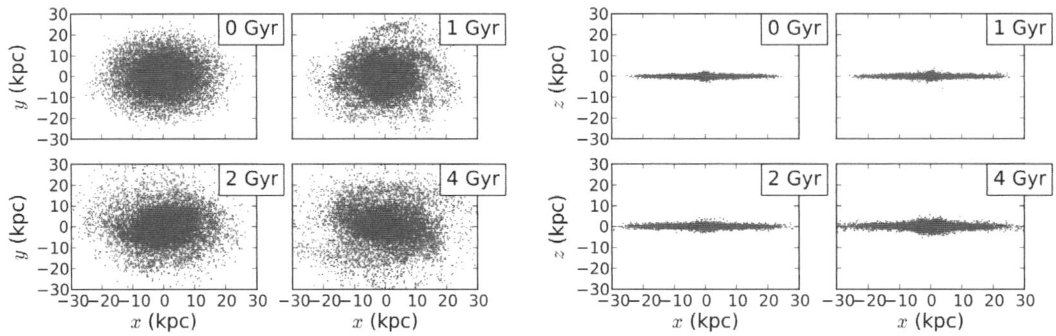

Figure 4. AQUAL disk and bulge (no dark matter halo). These are subsets of 20000 particles.

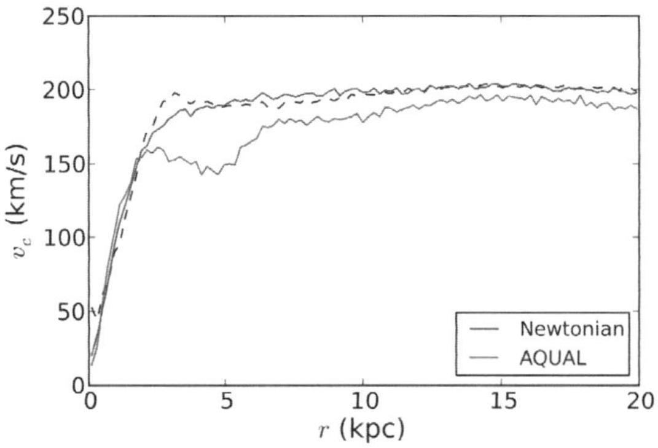

Figure 5. Mean rotational (azimuthal) velocity of the Newtonian disk and bulge and the AQUAL disk and bulge at the beginning and end of their respective runs, in r-bins of width 200 pc. The black line denotes the initial conditions of the simulations, the blue dashed line shows the Newtonian rotation curve after approximately 4 Gyr, while the green solid line shows the AQUAL rotation curve after the same time.

there is no dark matter halo in this simulation. The MOND curve is slightly lower than that of the Newtonian system after 4 Gyr, and the existence of the bar is evident in the dip in the rotation curve at small radii.

This test demonstrates that our code can evolve a rotationally-supported system over a significant timescale, and that our code may be used in the study of secular processes in MOND disks.

5. Case study - collisional ring galaxies

We will now discuss an explicit example of a study for which the RAyMOND code is being used: a collisional ring galaxy system. In these systems a spiral galaxy suffers a collision through the disk by a smaller, elliptical, companion. The resulting perturbation of the disk causes a density wave to propagate through the galaxy, producing a striking ring structure in the disk. The famous Cartwheel Galaxy is the paradigmatic example of such an object (for a review of collisional ring galaxies see [29]). Numerical simulations, such as that of [30], allow us to gain considerable insight into the formation and evolution of these systems. An example simulation in MOND gravity is shown in Figs. 6 and 7. We can see the ring structure resulting from the

Journal of Physics: Conference Series **720** (2016) 012012 doi:10.1088/1742-6596/720/1/012012

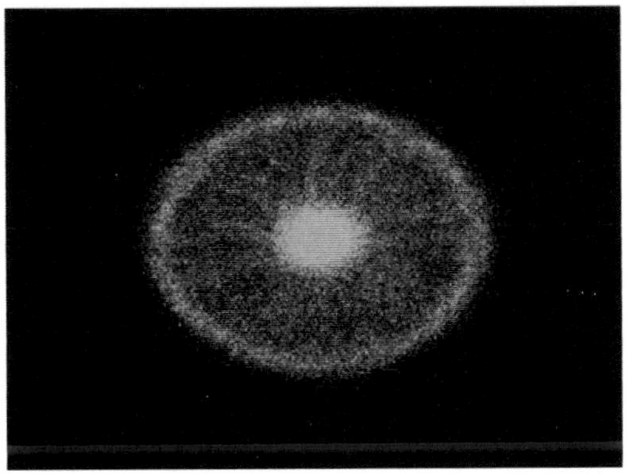

Figure 6. Collisional ring galaxy, face-on

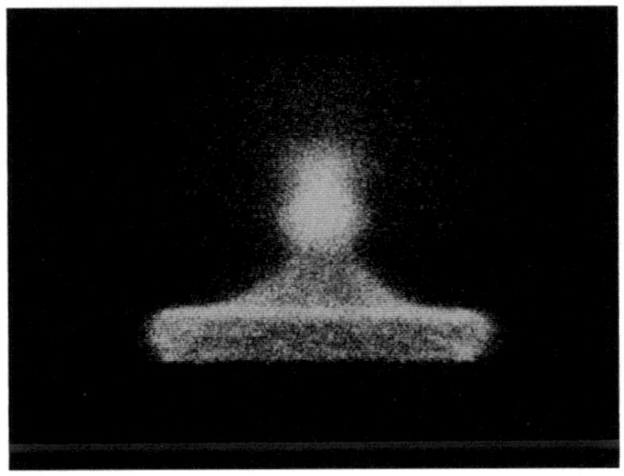

Figure 7. Collisional ring galaxy, edge-on

collision clearly in the stellar particles comprising the disk, as well as the dragging of stars out of the disk by the elliptical collider. Similar structures are visible in observed collisional ring systems in the gaseous material as well, which may also be simulated with the RAyMOND code.

These systems provide an opportunity to examine one key difference between MONDian dynamics and Newtonian dynamics in the presence of dark matter: the differing amounts of dynamical friction. Normally, the presence of large dark matter haloes leads to dynamical friction, where the gravitational interaction between a galaxy (and its associated dark matter halo) and the particles comprising a larger host halo gives rise to a friction effect, slowing the motion of the galaxy [31]. In MOND, the effect of dynamical friction is only due to the baryonic material (i.e. the stars and gas) when the two galaxies undergo a collision. While the enhanced gravity of MOND will *increase* the effectiveness of dynamical friction during the collision, there is essentially no effect after the collision. In the ΛCDM context the presence of the dark matter haloes leads to longer dynamical friction timescales, and potentially more slowing of the galactic motion.

Thus galaxy collisions proceed differently in MOND and ΛCDM. In a collisional ring galaxy, observations of the ring properties can provide constraints on the timescales of the collision, and thus we may find significant differences in the modelling of these systems.

6. Conclusions

We have developed a modification of the N-body/hydrodynamics code RAMSES that enables high resolution simulations using MOND gravity. The code utilises two formulations of MOND, known as AQUAL and QUMOND, allowing a direct comparison between them and the investigation of possible observational consequences from the different manifestations of MONDian behaviour. The execution time of the code has also been demonstrated to be as fast as may be expected, considering the complexity of the MOND calculation. Furthermore, using idealised tests we have confirmed that the RAyMOND code is able to evolve dispersion-supported and rotation-supported stellar systems in a stable manner in a MOND gravitational potential.

There are, of course, many possible applications of our code at galactic scales, including galaxy interactions, the behaviour of dwarf satellites, star cluster dynamics in the Milky-Way and so on. A particular application of the code to the study of collisional ring galaxy systems was discussed. Further studies of the effects of MOND gravity on galaxy dynamics in various environments will be the subject of future studies.

Acknowledgments
The authors wish to thank Romain Teyssier for providing very helpful information about RAMSES via the RAMSES users mailing list. They would also like to thank Francoise Combes for suggesting the use of RAMSES for this project, and Marcel Pawlowski for additional comments. GC, RS and MF acknowledge the support of FONDECYT grants 3130480, 3120135 and 1130521 respectively.

References
[1] Planck collaboration 2013 *Preprint* 1303.5076
[2] Weinberg 2008 *Cosmology* (Oxford: Oxford University Press)
[3] Peebles P J E 2012 *Preprint* 1203.6334
[4] Weinberg D H, Bullock J S, Governato F, Kuzio de Naray R, Peter A H G 2013 *Preprint* 1306.0913
[5] Pawlowski M S, Pflamm-Altenburg J and Kroupa P 2012 *MNRAS* **423** 1109
[6] Ibata R A, Lewis G F, Conn A R, Irwin M J, McConnachie A W, Chapman S C, Collins M L, Fardal M, Ferguson A M N, Ibata N G, Mackey A D, Martin N F, Navarro J, Rich R M, Valls-Gabaud D, Widrow L M 2013 *Nature* **493** 62
[7] Milgrom M 1983 *ApJ* **270** 365
[8] Sanders R H and Noordermeer E 2007 *MNRAS* **379** 702
[9] Bekenstein J D 2004 *Phys. Rev.* D **70** 083509
[10] Skordis C 2008 *Phys. Rev.* D **77** 123502
[11] Milgrom M 2009 *Phys. Rev.* D **80** 123536
[12] Famaey B and McGaugh S S 2012 *Living Rev. Relativity* **15**
[13] Sanders R H 2003 *MNRAS* **342** 901
[14] Brada R and Milgrom M 1999 *ApJ* **519** 590
[15] Tiret O and Combes F 2007 *A&A* **464** 517
[16] Llinares C, Knebe A, Zhao H S 2008 *MNRAS* **391** 1778
[17] Londrillo P and Nipoti C 2009 *Memorie della Società Astronomica Italiana Supplement* **13** 89
[18] Nipoti C, Londrillo P and Ciotti L 2007 *MNRAS* **381** L104
[19] Nipoti C, Ciotti L, Binney J and Londrillo P 2008 *MNRAS* **386** 2194
[20] Teyssier R 2002 *A&A* **385** 337
[21] Lüghausen F, Famaey B and Kroupa P 2015 *Can. J. Phys.* **93** 232
[22] Candlish G, Smith R and Fellhauer M 2015 *MNRAS* **446** 1060
[23] Bekenstein J and Milgrom M 1984 *ApJ* **286** 7
[24] Milgrom M 2010 *MNRAS* **403** 886
[25] Deffayet C, Esposito-Farèse G and Woodard R P 2011 *Phys. Rev.* D **84** 124054
[26] Clifton T, Ferreira P G, Padilla A and Skordis C 2011 *Phys. Rep.* **513** 1
[27] Trottenberg U, Oosterlee C and Schüller A 2001 *Multigrid* (London: Academic Press)
[28] Kuijken K and Dubinski J 1995 *MNRAS* **277** 1341
[29] Appleton P N and Struck-Marcell C 1996 *Fundamentals of Cosmic Physics* **16** 111
[30] Smith R, Lane R R, Conn B C and Fellhauer M 2012 *MNRAS* **423** 543
[31] Binney J and Tremaine S 2008 *Galactic Dynamics* (Princeton: Princeton University Press)

Journal of Physics: Conference Series **720** (2016) 012013 doi:10.1088/1742-6596/720/1/012013

Truncated γ-exponential models for tidal stellar systems

Y. J. Gomez-Leyton and L. Velazquez

Departamento de Física, Universidad Católica del Norte, Av. Angamos 0610, Antofagasta, Chile.

E-mail: yuvineza.gomez@gmail.com; lvelazquez@ucn.cl

Abstract. We introduce a parametric family of models to characterize the properties of astrophysical systems in a quasi-stationary evolution under the incidence evaporation. We start from an one-particle distribution $f_\gamma(\mathbf{q}, \mathbf{p}|\beta, \varepsilon_s)$ that considers an appropriate deformation of Maxwell-Boltzmann form with inverse temperature β, in particular, a power-law truncation at the scape energy ε_s with exponent $\gamma > 0$. This deformation is implemented using a generalized γ-exponential function obtained from the *fractional integration* of ordinary exponential. As shown in this work, this proposal generalizes models of tidal stellar systems that predict particles distributions with *isothermal cores and polytropic haloes*, e.g.: Michie-King models. We perform the analysis of thermodynamic features of these models and their associated distribution pro les. A nontrivial consequence of this study is that pro les with isothermal cores and polytropic haloes are only obtained for low energies whenever deformation parameter $\gamma < \gamma_c \simeq 2.13$. This study is a rst approximation to characterize a self-gravitating system, so we consider equal to all the particles that constitute the system.

1. Introduction

Gravitation is an example of attractive interaction that is unable to confine particles. It is always possible that some particles acquire sufficiently energy and escape from gravitational influence of the system. Evaporation is an unavoidable process that drives dynamical evolution of astrophysical systems in ature. eedless to say that it is a very important ingredient to explain structure and evolution of the so-called tidal stellar systems, such as globular clusters [1, 2]. In general, the effects of evaporation crucially depend on the relaxation mechanisms that are present in the microscopic dynamics, which depend on concrete conditions of a given particular system [3, 4, 5].

e propose in this work a parametric family of astrophysical models that accounts for the influence of evaporation under different relaxation regimes: the *truncated γ-exponential models* [6]. These phenomenological models constitutes a suitable generalization of some models of tidal stellar systems available in the literature, such as ooley & Dickens truncated isothermal model [7], as well as ing s models of globular clusters [8, 9] and ilson models [10]. This proposal enable us to analyze the influence of the evaporation on the thermodynamical behavior and distribution profiles. Moreover, it provides a unification framework of the known isothermal and polytropic profiles considered in hydrodynamic models of stellar systems [11]. Remarkably, these models are sufficiently amenable to allow an analytical derivation of most of their associated hydrodynamic quantities.

The paper is organized into sections as follows. Second section is devoted to introduce the parametric family of quasi-stationary distributions, the derivation of their associated hydrodynamic quantities, as well as some details about the integration of the associated nonlinear Poisson equation. Afterwards, third section is devoted to discuss numerical results obtained from this model, both information concerning to thermodynamical description as well as distribution profiles. Some conclusions are drawn in forth section.

2. A arametric family of quasi-stationary one-body distributions

Observational evidences and theoretical analysis suggest that distribution profiles of many stellar systems can be explained considering a quasi-stationary one-particle distribution with a mathematical form close to Maxwell-Boltzmann distribution [12]:

$$f_{MB}(\mathbf{q},\ ) \sim \exp\left[-\beta\varepsilon(\mathbf{q},\ )\right], \tag{1}$$

with $\varepsilon(\mathbf{q},\ ) = \ ^2/2m + m\phi(\mathbf{q})$ being the mechanical energy of a given particle with momentum located at the position $\mathbf{q}$ with a mean field gravitational potential $\phi(\mathbf{q})$. The incidence of evaporation introduces a deformation of the above distribution, specifically, a truncation for energies above a certain *escape energy* ε_s. Some examples of distributions that introduce this kind of deformation are truncated isothermal models of ooley and Dickens [7], Michie- ing models [8, 9] and ilson models [10]. Quasi-stationary one-body distributions of all these models can be expressed using certain truncated exponential function $E_t(x)$ as follows:

$$f_{qs}(\mathbf{q},\ |\beta,\varepsilon_s) = AE_t(x), \tag{2}$$

where $x = \beta\left[\varepsilon_s - \varepsilon(\mathbf{q},\ )\right]$. The function $E_t(x)$ adopts the following forms:

$$\begin{aligned} E_{WD}(x) = \exp(x)H(x),\ E_{MK}(x) = \left[\exp(x) - 1\right]H(x) \\ E_W(x) = \left[\exp(x) - 1 - x\right]H(x) \end{aligned} \tag{3}$$

for each ones of these models, respectively, with $H(x)$ being the Heaviside step function.

Our interest is to propose a generalization of evaporation truncation scheme considered in the above astrophysical models. Firstly, let us express them into a unifying form as follows:

$$f_\gamma(\mathbf{q},\ |\beta,\varepsilon_s) = AE_\gamma(x), \tag{4}$$

where $E_\gamma(x)$ is a certain function obtained from the *truncation of exponential function* $\exp(x)$, with γ being a deformation parameter. At first glance, the truncation schemes

Journal of Physics: Conference Series **720** (2016) 012013 doi:10.1088/1742-6596/720/1/012013

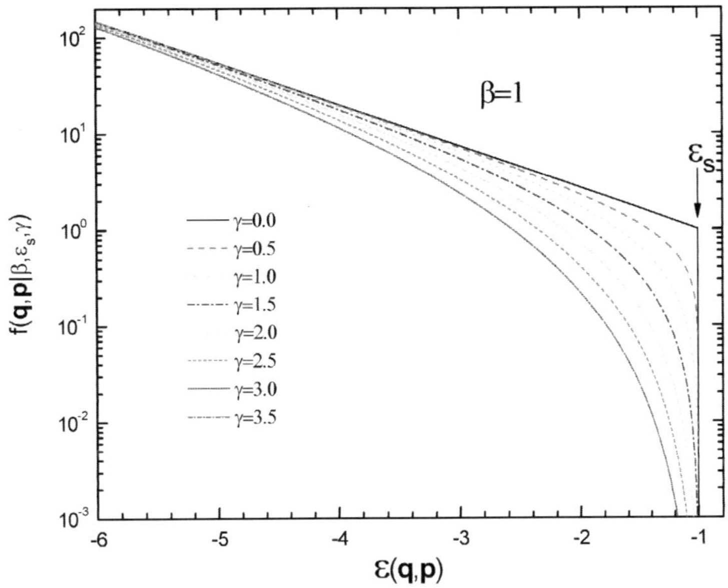

Figure 1. One-particle distributions *versus* mechanical energy $\varepsilon(\mathbf{q},\ )$ corresponding to truncated γ-exponential model (4) for some values of deformation parameter γ. e have employed prefixed values $\varepsilon_s = -1$ and $\beta = 1$ for the scape energy and the inverse temperature, respectively.

(3) could be generalized for any nonnegative integer $\gamma = n$ subtracting the first n terms of power-expansion of exponential function:

$$E_n(x) = \begin{cases} \sum_{k=0}^{+\infty} x^{n+k}/(n+k)!, & \text{for } x > 0, \\ 0, & \text{otherwise.} \end{cases} \quad (5)$$

Precisely, definitions (3) correspond to the particular cases $\gamma = 0$, $\gamma = 1$ and $\gamma = 2$. A more general definition for any nonnegative real number γ is the following:

$$E_\gamma(x) \equiv E(x;\gamma) = \sum_{k=0}^{\infty} \frac{x^{\gamma+k}}{\Lambda(\gamma+1+k)} \equiv \frac{1}{\Lambda(\gamma)} \int_0^x \eta^{\gamma-1} \exp(x-\eta) d\eta, \quad (6)$$

which is obtained replacing factorial function $(n+k)!$ by the Gamma function $\Lambda(\gamma+1+k)$. Hereinafter, this function will be referred to as *γ-exponential function*, and the same one will be equivalently denoted as $E_\gamma(x)$ or $E(x;\gamma)$. Strictly speaking, this function represents fractional integration of ordinary exponential function [13].

The proposed family of quasi-stationary distributions describe the influence of evaporation with two independent parameters: the escape energy ε_s and the deformation parameter γ. This family of distributions is schematically shown in figure 1. The same ones exhibit a Maxwell-Boltzmann profile (1) for energies $\varepsilon(\mathbf{q},\ )$ that are far enough from below the escape energy ε_s. Besides, they show a *power-law truncation* at ε_s

Journal of Physics: Conference Series **720** (2016) 012013 doi:10.1088/1742-6596/720/1/012013

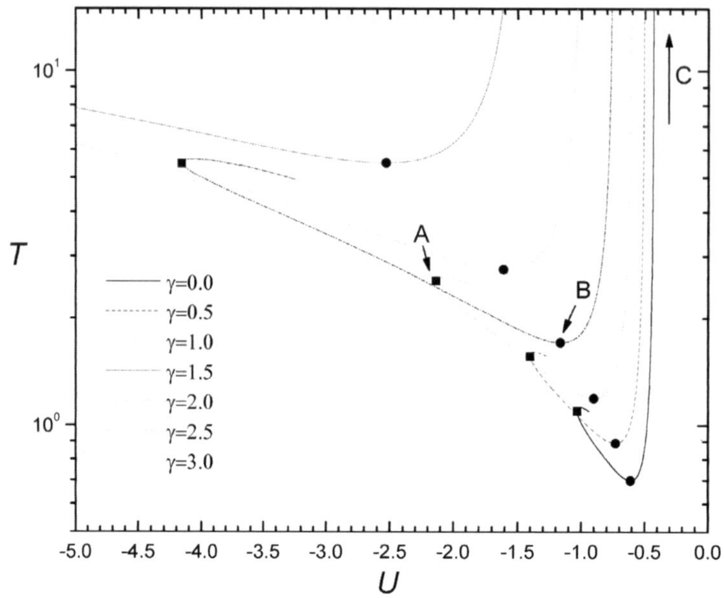

Figure 2. Behavior of the caloric curves T *versus* U. Additionally, we have included the following notable points: (squares) critical point of gravothermal collapse U_A; (circles) critical point of isothermal collapse U_B; (triangles) critical point of evaporation disruption U_C.

with exponent γ. The increasing of deformation parameter γ characterizes a larger deviation from Maxwell-Boltzmann profile (1). For a phenomenological viewpoint, a larger value of the deformation parameter γ describes a stronger influence of evaporation, or equivalently, the incidence of a weaker relaxation mechanism.

3. tructure equation and observables

Truncation due to evaporation leads to spherical-symmetric distributions that are located inside a region of finite radius R_t, the so-called *tidal radius* [1]. Rewriting the escape energy $\varepsilon_s = m\phi_s$ using the surface gravitational potential ϕ_s, one can express R_t as follows:

$$R_t = -GM/\phi_s, \tag{7}$$

where M is the system total mass. Introducing the dimensionless potential $\;=\;(\mathbf{q})$:

$$(\mathbf{q}) = \beta m\left[\phi_s - \phi\left(\mathbf{q}\right)\right], \tag{8}$$

one can express the mass density ρ in characteristic units $\rho_c = M/R_t^3$ as follows:

$$\rho = \frac{\kappa}{\eta}E\left(\;;\gamma + \frac{3}{2}\right), \tag{9}$$

Journal of Physics: Conference Series **720** (2016) 012013 doi:10.1088/1742-6596/720/1/012013

where η is the dimensionless inverse temperature:

$$\eta = \beta\frac{GMm}{R_t} \tag{10}$$

and κ a certain integration parameters defined below. Mathematical form of dimensionless potential is obtained from integration of the following nonlinear Poisson problem:

$$\frac{1}{\xi^2}\frac{d}{d\xi}\left[\xi^2\frac{d}{d\xi}\ \ (\xi)\right] = -4\pi E\left[\ \ (\xi);\gamma+\frac{3}{2}\right]. \tag{11}$$

e have considered here the dimensionless radius:

$$\xi = \sqrt{\kappa}r/R_t, \tag{12}$$

where r is the physical radius, $\kappa = Gm^2\beta CR^2$ is an auxiliary constant, R_t is the tidal radius. Problem (11) can be solved in a numerical way by demanding the following conditions at the origin:

$$\ (0) = \ {}_0,\quad \ '(0) = 0, \tag{13}$$

which is integrated till the vanishing of dimensionless potential at the surface of the system with dimensionless radius $\xi_c = \sqrt{\kappa}$:

$$\ (\xi_c) = 0,\quad \ '(\xi_c) = -\frac{\eta}{\xi_c}. \tag{14}$$

The total energy $U = K + V$ can be obtained from the total kinetic K and total potential energy V:

$$\begin{aligned} V &= -\frac{1}{2} - \frac{1}{2\eta^2\xi_c}\int_0^{\xi_c} E\left[\ \ (\xi);\gamma+\frac{3}{2}\right]\ \ (\xi)\ 4\pi\xi^2 d\xi, \\ K &= \frac{3}{2\eta^2\xi_c}\int_0^{\xi_c} E\left[\ \ (\xi);\gamma+\frac{5}{2}\right] 4\pi\xi^2 d\xi, \end{aligned} \tag{15}$$

which are expressed in characteristic units $U_c = GM^2/R_t$.

4. Results and discussions

According the asymptotic behavior of truncated γ-exponential function, the particles density (9) describes a power-law dependence for small with exponent $n = \gamma + 3/2$, while an exponential dependence for large enough:

$$\rho(\ ;\gamma) = \begin{cases} \propto\ \ ^n, & <<1, \\ \propto \exp(\ \ ), & >>1. \end{cases} \tag{16}$$

Consequently, this family of models unifies the known isothermal and polytropic profiles of hydrodynamic astrophysical models [1, 11]. Since the deformation parameter γ is nonnegative and the polytropic profiles are unstable for $n \geq 5$ [11], the admissible values of the polytropic exponent n are restricted to the interval $3/2 \leq n < 5$, which correspond

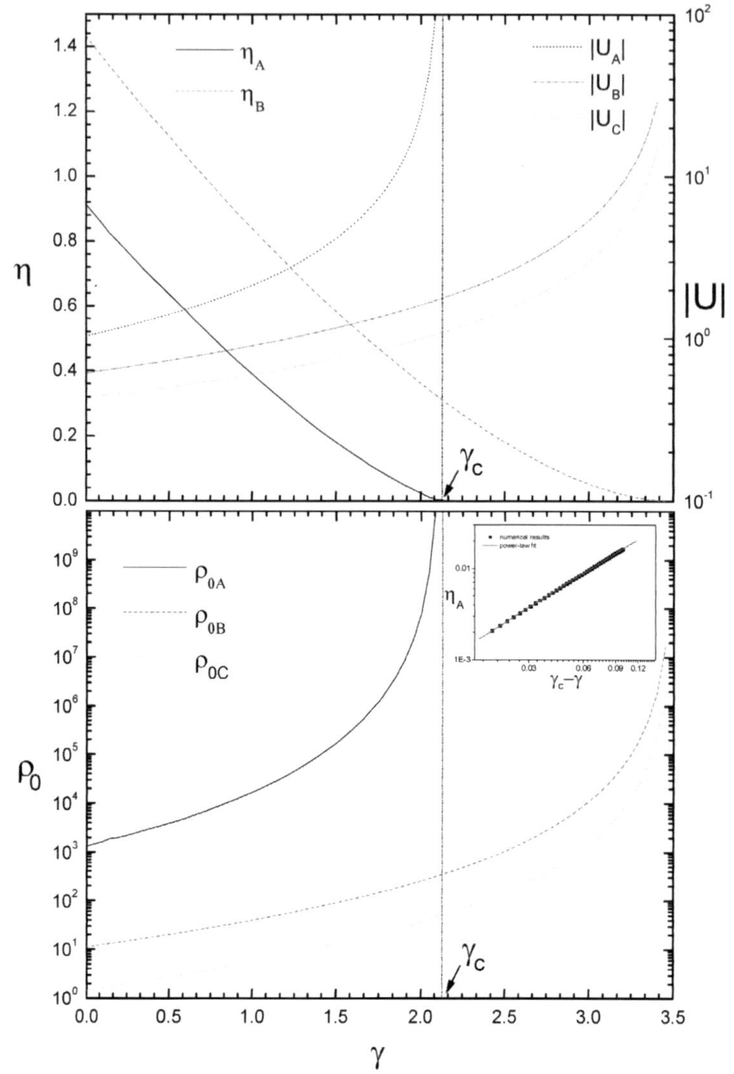

Figure 3. Thermodynamic quantities associated with the notable points *versus* deformation parameter γ: Panel a) the inverse temperatures (η_A, η_B) and the modulus of energies (U_A, U_B, U_C); Panel b) the central particles densities $(\rho_{0A}, \rho_{0B}, \rho_{0C})$. Inset: A power-law fit $\eta_A(\gamma) = A\,|\gamma_c - \gamma|^{\alpha}$ that provides a more precise estimation for the notable value $\gamma_c = 2.1310.003$

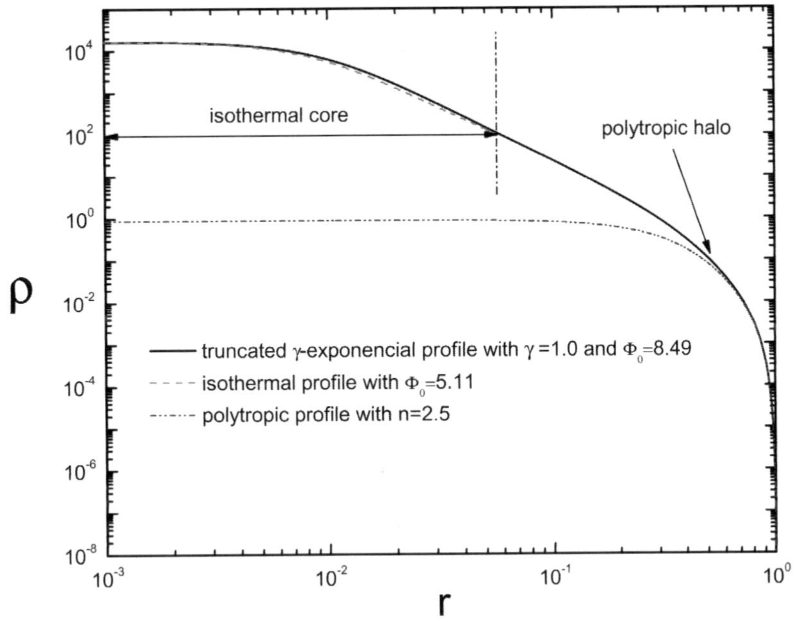

Figure 4. Truncated γ-exponential profile with $\gamma = 1$ (Michie- ing profile) corresponding to the point of gravothermal collapse and its comparison with isothermal and polytropic profiles using *log-log scales*. Accordingly, the proposed family of models can describe distribution profiles that exhibit isothermal cores and polytropic haloes.

to the interval $0 \leq \gamma < 7/2$ for the deformation parameter γ. This restriction will be manifested in the thermodynamic quantities and distribution profiles.

Dependencies of the temperature T *versus* the dimensionless energy U are shown in figure 2 for different values of deformation parameter γ. All quasi-stationary configurations obtained from these models have always negative values for the energy U. Moreover, one can recognize the existence of three notable points: the critical point of gravothermal collapse U_A, the critical point of isothermal collapse U_B and the critical point of evaporation disruption U_C [4, 5]. All quasi-stationary configurations are located inside the energy region $U_A \leq U \leq U_C$. Stable quasi-stationary configurations inside the energy range $U_A \leq U \leq U_B$ exhibit *negative heat capacities* $C < 0$, which is a distinguished feature of thermodynamics of astrophysical systems [14, 15]. For the sake of a better understanding about the influence of deformation parameter γ, we have calculated dependence of some thermodynamic quantities at the notable points. Such results are shown in figure 3. The increasing of the deformation parameter γ provokes a systematic decreasing of inverse temperatures (η_A, η_B), and the increasing of absolute values of energies (U_A, U_B, U_C) and their associated central particles densities $(\rho_{0A}, \rho_{0B}, \rho_{0C})$. Interestingly, the inverse temperature η_A at the notable point of gravothermal collapse vanishes when $\gamma \geq \gamma_c \simeq 2.13$. This means that both the total energy U_A and the temperature T_A diverge at this point as a consequence of divergence of the central density ρ_A. hen γ is below the point γ_c the system develops a gravothermal collapse at the finite energy U_A and when the deformation parameter γ is above the point γ_c, the system should release an infinite amount of energy. Additionally, one observes a divergence in the other thermodynamic variables when the deformation parameter γ approaches its maximum admissible value $\gamma_m = 7/2$, which is related to instability of polytropic profiles for $n \geq 5$.

e show in figure 4 a distribution profile with deformation parameter $\gamma = 1$ at the point of gravothermal collapse, which corresponds to a Michie- ing profile with lowest energy. As clearly shown in this figure, the inner region, the core, can be well-fitted with an isothermal profile [14], while the outer one, the halo, is well-fitted with a polytropic profile [11]. Isothermal cores disappears for energies sufficiently large, so that, distribution profiles are practically polytropic. For a better illustration, dependence of distribution profiles on the deformation parameter γ and the internal energy U is shown in figure 5, specifically, twelve profiles corresponding to three notable points and forth different values of deformation parameter γ. Particles concentration in the inner regions decreases with the increasing of the internal energy U. The increasing of the deformation parameter γ produces distribution profiles with more diluted haloes, and consequently, more dense cores.

e have shown in figure 6 what is considered as the phase diagram of truncated γ-exponential models in the plane $\gamma -\ {}_0$. For each value of deformation parameter γ, the admissible values of the central dimensionless potential ${}_0$ are located inside the interval $0 \leq\ {}_0 \leq\ {}_A(\gamma)$, where ${}_A(\gamma)$ corresponds to the critical point of gravothermal collapse U_A. Central values of dimensionless potential ${}_0$ above dependence ${}_A(\gamma)$ correspond to nonphysical on unstable configurations (white region). Additionally, we have included dependence ${}_B(\gamma)$ associated with the point of isothermal collapse U_B. Configuration between dependencies ${}_B(\gamma) \leq\ {}_0 \leq\ {}_A(\gamma)$ exhibit negative heat capacities (dark gray region), while those one with central dimensionless potential ${}_0$ belonging to interval $0 <\ {}_0 <\ {}_B(\gamma)$ exhibit positive heat capacities (light gray region). It is remarkable that dependence ${}_A(\gamma)$ is weakly modified by a change in the deformation parameter γ for values below the notable point γ_c. However, this function experiences an sudden change above this notable value. As expected, this behavior accounts for a sudden change in behavior of distribution profiles: the proposed models can exhibit profiles with isothermal cores for $\gamma < \gamma_c$, while they only exhibit profiles without isothermal cores for $\gamma \geq \gamma_c$.

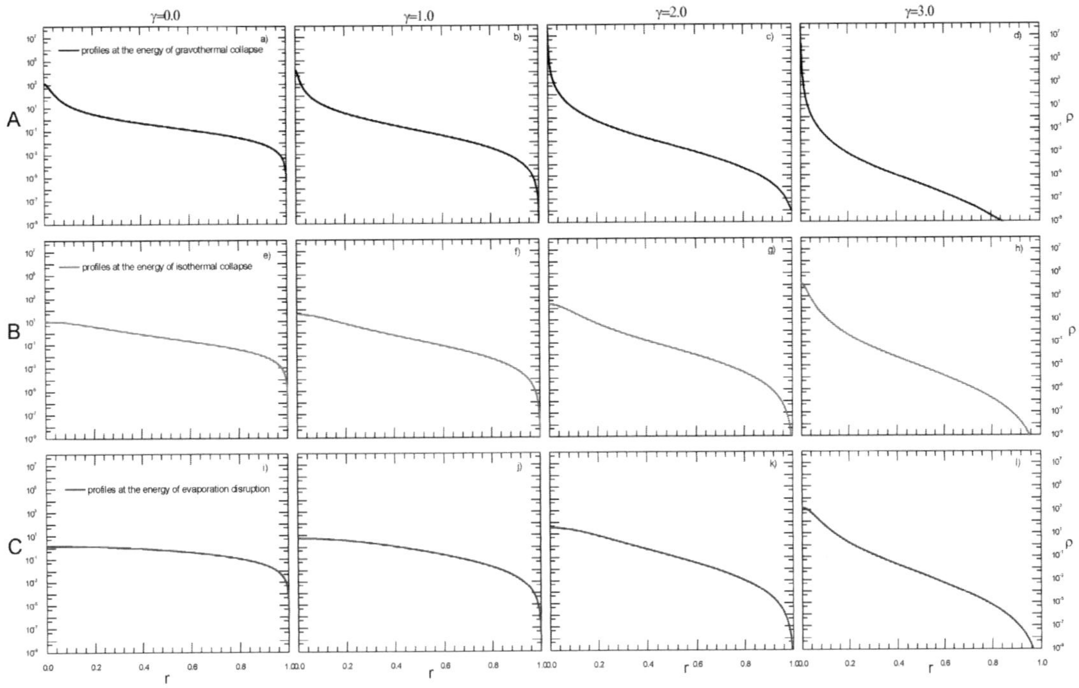

Figure . Dependence of distribution profiles on the deformation parameter γ for three notable values of the internal energy U. Panels a)-c) Distribution profiles at the energy of gravothermal collapse U_A for three values of deformation parameter with $\gamma < \gamma_c$. All them exhibit isothermal cores and polytropic haloes. Panel d) A distribution profile at gravothermal collapse with deformation parameter $\gamma > \gamma_c$. ote that this profile does not exhibit an isothermal core. Panels i)-l) Distribution profiles very near the energy of evaporation disruption U_C are everywhere polytropic. Panels e)-h) Transitional profiles at the energy of isothermal collapse U_B. These profiles hardly differ from polytropic profiles i)-l) because of they exhibit more dense cores.

Acknowledgments
Velazquez thanks the financial support of CO ICyT Programa Bicentenario de Ciencia y Tecnolog a PSD . He also thanks partial financial support from VRIDT-UC research programme. Gomez-Leyton thanks financial support from CO IC T Beca Mag ster acional **2214003** (2014). She also thanks Organizers of XIX Chilean Symposium on Physics for partial financial support.

References

[1] Binney J and Tremaine S 1994 *Galactic Dynamics*, (New Jersey: Princenton University Press).
[2] Spitzer L Jr. and Harm R 1958 *Astrophys. J.* **127** 544.
[3] Cipriani P and Pettini M 2003 *Ap & SS* **283** 347.
[4] Velazquez L and Guzman F 2003 *Phys. Rev. E* **68** 066116.

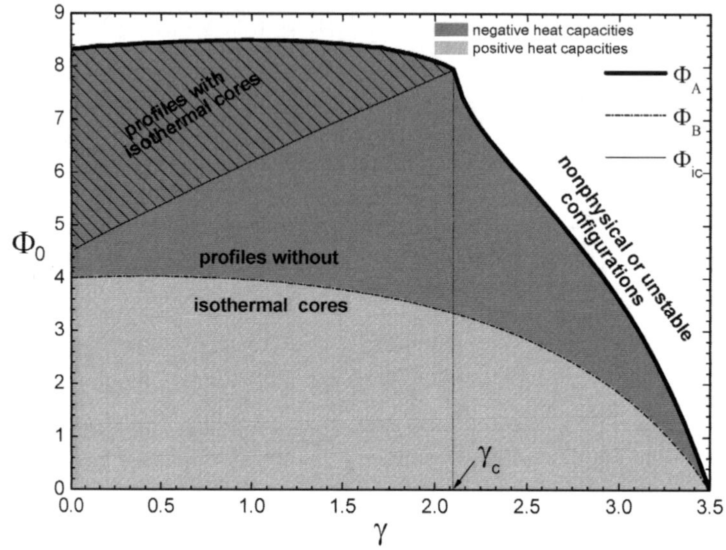

Figure . Phase diagram of truncated γ-exponential models in the plane of integration parameters $\gamma -\ _0$.

[5] Velazquez L, Gómez-García S and Guzman F 2009 *Phys. Rev. E* **79** 011120.
[6] Y J Gomez-Leyton and L Velazquez, *J. Stat. Mech. Theo. Exp.* **041-1113** (2014).
[7] Wooley R.v.d.R and Dickens R J 1961 *Royal Obs. Bull.* No. **42**.
[8] Michie R W 1963 *Mon. Not. R. Astron. Soc.* **125** 127 ibid **126** 331.
[9] King I A 1962 Astron. J. **67** 471 ibid 1965 **70** 376 1965 **71** 64 1966 **71** 276.
[10] Wilson Ch P, Astron. J. **80**, 1430 (1975)
[11] Chandrasekhar S 1960 *Principles of Stellar Dynamics* (New York: Dover Publications Inc.).
[12] Hjorth J and Madsen J 1993 *Mon. Not. R. Astron. Soc.* **265** 237.
[13] Miller K S and Ross B 1993 *An Introduction to the Fractional Calculus and Fractional Differential Equations* (New York: John Wiley).
[14] Antonov V A 1962, *Vest. Leningrad Univ.* **7** 135 1995 Translation IAU Symposium **113** 525.
[15] Lynden-Bell D 1967 Mon. Not. R. Astr. Soc. 136 101.

Journal of Physics: Conference Series **720** (2016) 012014 doi:10.1088/1742-6596/720/1/012014

Generalized equations of state and regular universes

F Contreras[1], N Cruz[2] y E González[2]

[1] Departamento de Matemática y Ciencia de la Computación, Universidad de Santiago de Chile, Las Sophoras 173, Santiago, Chile

[2] Departamento de Física, Universidad de Santiago de Chile, Avenida Ecuador 3493, Santiago, Chile

E-mail: norman.cruz@usach.cl

Abstract. We found non singular solutions for universes filled with a fluid which obey a Generalized Equation of State of the form $P(\rho) = -A\rho + \gamma\rho^{\lambda}$. An emergent universe is obtained if $A = 1$ and $\lambda = 1/2$. If the matter source is reinterpret as that of a scalar matter field with some potential, the corresponding potential is derived. For a closed universe, an exact bounce solution is found for $A = 1/3$ and the same λ. We also explore how the composition of theses universes can be interpreted in terms of known fluids. It is of interest to note that accelerated solutions previously found for the late time evolution also represent regular solutions at early times.

1. Introduction

The study of Generalized Equations of State (EoS) for the main fluid component of the universe has a long time, inspired, as the best of our knowledge, in the particular behavior of Friedmann models during inflationary scenarios. In order to extend the range of known inflationary behaviors, Barrow [1] assumed the matter stress has pressure P and density ρ, related by the following model equation of state

$$P(\rho) = -\rho + \gamma\rho^{\lambda}, \tag{1}$$

where γ and λ are both constants and $\gamma \neq 0$. The standard EoS of a perfect fluid, $P = (\gamma - 1)\rho$ is recovered when $\lambda = 1$. An emergent flat universe solution was found for the case $\lambda = 1/2$ and $\gamma > 0$ although not discussed in this context. In this solution $a(t \to \infty) \to \infty$ and $a(t \to -\infty) \to 1$. The Hubble parameters and its derivatives were not investigated in detail. It is interesting to mention that the doubled exponential behavior of this solution was previously found for a bulk viscous source in the presence of an effective cosmological constant [2]. This is a consequence of the inclusion of bulk viscosity in the Eckart's theory, which leads to and viscous pressure Π of the form $-3\xi H$, where ξ is assumed usually in the form $\xi = \xi_0\rho^{\lambda}$.

A variation of Eq.(1) was considered by Mukherjee *et al* [3] assuming the form

$$P(\rho) = -A\rho - \gamma\rho^{1/2}, \tag{2}$$

where for $A < 1$ and $\gamma > 0$ it is possible to obtain a solution describing an emergent flat universe, by with a scale factor of the form

$$a(t) = a_0(\beta + e^{\alpha t})^{\omega}, \tag{3}$$

where a_0 and β are constants and α, ω are given in terms of the parameters A, γ. For same values of A the energy density of this fluid can be interpreted as the sum of three known fluids such radiation, string and a cosmological constant. The case with $A = 1$ and $\lambda = 1/2$ was considered in [4], [5]. In both works the cosmological solutions of dark energy models with this fluid was analyzed, focusing in the future expansion of the universe. A late time behavior of a universe filled with a dark energy component with an EoS given by Eq.(1) has been investigated in [6], [7], where the allowed values of the parameters A and γ were constrained using H(z)-z data, a model independent BAO peak parameter an cosmic parameter (WMAP7 data).

The research of late time evolution of the universe has also assumed EoS of the type given by Eq.(1) motivated by the fact that the constraints from the observational data implies $\omega \approx -1$ for the EoS of the dark energy component if it assumed ruled by a barotropic EoS, but the values $\omega < -1$, corresponding to a phantom fluid, or $\omega > -1$, corresponding to quintessence can not be discarded. Also theoretical studies like the so called running vacuum energy in QFT (see [8]) gives rise to a cosmological constant with a dynamical evolution during the cosmic time, so EoS of the type given in Eq.(1) could represent effectively the results of this approach under some specific assumptions.

The aim of the present work is to explore non singular solutions for universes filled with a fluid which obey and EoS given by Eq.(1). In particular, in the next section we show that previous solutions found for a late time solutions represents also an emergent universe. We evaluate also the field potential $V(\phi)$ if the matter source is reinterpreted as that of a scalar matter.

In section 3 we found an exact solution for a bouncing universe with positive curvature. We show that for this case the behavior of the density energy and pressure is like that tree known fluids. In section 4 we discuss our results.

2. Exact solution for an emergent universe

As it was mentioned above an EoS of the form given in Eq.(1) leads to an emergent flat universe solution for the case $\lambda = 1/2$, i. e., for an EOS given by

$$P(\rho) = -\rho - \gamma\rho^{1/2}. \tag{4}$$

The Einstein equations for a flat FRW metric without cosmological constant can be write in the following form

$$\rho = 3\left(\frac{\dot{a}}{a}\right)^2, \tag{5}$$

$$P = -2\left(\frac{\ddot{a}}{a}\right) - \left(\frac{\dot{a}}{a}\right)^2. \tag{6}$$

From the equations (4) to (6) and assuming $\dot{a}, a \neq 0$, we obtain

$$\frac{\dot{a}}{a} = K\exp\left(\frac{\gamma\sqrt{3}}{2}t\right), \tag{7}$$

so the solution for the scale factor as a function of the cosmic time is given by

$$a(t) = \sigma\exp\left[\frac{2K}{\gamma\sqrt{3}}\exp\left(\frac{\gamma\sqrt{3}}{2}t\right)\right], \tag{8}$$

where K y σ are positive integration constants. This is the solution already found by Barrow [1]. Note that for $\gamma < 0$, $a(t \to -\infty) \to 0$ and we have zero scale factor in the infinity past. An emergent universe solution is obtained for $\gamma > 0$, since $a(t \to -\infty) \to \sigma$ so the scale factor tends to a finite value different from zero in the infinity past. If we choose the initial conditions $a(t_0) = a_0$ and $\rho(t_0) = \rho_0$ the integration constants K y σ are given by

$$K = \frac{\rho_0^{1/2}}{\sqrt{3}}\exp\left(-\frac{\gamma\sqrt{3}}{2}t_0\right), \quad \sigma = a_0\exp\left(-\frac{2\rho_0^{1/2}}{3\gamma}\right). \tag{9}$$

Then the solution for the scale factor with these initial conditions becomes

$$a(t) = a_0 \exp\left(\frac{2\rho_0^{1/2}}{3\gamma}\left[\exp\left(\frac{\gamma\sqrt{3}}{2}(t-t_0)\right)-1\right]\right), \tag{10}$$

and the Hubble parameter and its derivatives are given by

$$H(t) = \frac{\rho_0^{1/2}}{\sqrt{3}}\exp\left[\frac{\gamma\sqrt{3}}{2}(t-t_0)\right], \tag{11}$$

$$\dot{H}(t) = \frac{\rho_0^{1/2}\gamma}{2}\exp\left[\frac{\gamma\sqrt{3}}{2}(t-t_0)\right], \tag{12}$$

$$\ddot{H}(t) = \frac{\rho_0^{1/2}\gamma^2\sqrt{3}}{4}\exp\left[\frac{\gamma\sqrt{3}}{2}(t-t_0)\right]. \tag{13}$$

The above expressions indicates that H, $\dot{H}$ and $\ddot{H}$ are positive. Thus H, $\dot{H}$ and $\ddot{H}$ tends to zero for $t \to -\infty$ and to infinity when $t \to \infty$.

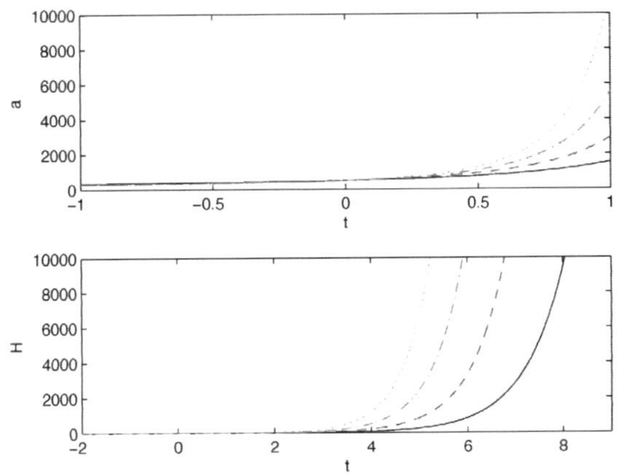

Figure 1. Graphics scale factor (upper) and the Hubble parameter (lower) as a function of time for the emergent universe with $a_0 = 500$ and $t_0 = 0$. The black line (——) is for $\gamma = 1.4$ and $\rho_0 = 1$, blue line (— — —) for $\gamma = 1.6$ and $\rho_0 = 2$, red line (— · —) for $\gamma = 1.8$ and $\rho_0 = 3$ and magenta line (· · · · · ·) for $\gamma = 2.0$ and $\rho_0 = 4$.

2.1. Composition of the emergent universe

In the following we shall study the energy density as a function of the scale factor in order to looking for a possible equivalence between our fluid in terms of other knows fluids. Using the energy conservation equation

$$\dot{\rho} + 3\frac{\dot{a}}{a}(\rho + P) = 0, \tag{14}$$

and taken the EoS given by Eq.(4) we obtain

$$\rho(a) = \left[\frac{3\gamma}{2}\ln\left(\frac{a}{a_0}\right) + \rho_0^{1/2}\right]^2. \tag{15}$$

Using Eq. (10) in Eq. (15) we obtain the energy as a function of time

$$\rho(t) = \rho_0 \exp\left(\gamma\sqrt{3}(t-t_0)\right), \tag{16}$$

which indicates that ρ tends to zero for $t \to -\infty$. Introducing Eq.(15) in Eq.(4) we obtain the fluid pressure as a function of the scale factor, which is given by

$$P(a) = -\left[\frac{3\gamma}{2}\ln\left(\frac{a}{a_0}\right) + \rho_0^{1/2}\right]^2 - \gamma\left[\frac{3\gamma}{2}\ln\left(\frac{a}{a_0}\right) + \rho_0^{1/2}\right]. \tag{17}$$

Journal of Physics: Conference Series **720** (2016) 012014 doi:10.1088/1742-6596/720/1/012014

In order to obtain now the fluid pressure as a function of time we use Eq.(16) in Eq.(4)

$$P(t) = -\rho_0 \exp\left(\gamma\sqrt{3}(t-t_0)\right) - \gamma\rho_0^{1/2} \exp\left(\frac{\gamma\sqrt{3}}{2}(t-t_0)\right). \tag{18}$$

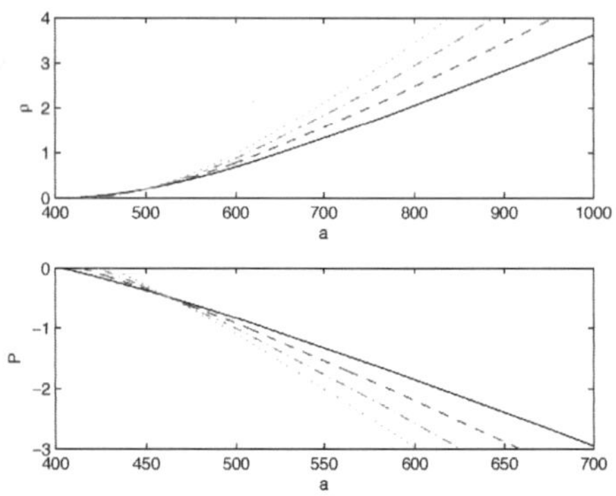

Figure 2. Graphic energy density (upper) and pressure (lower) as a function of scale factor for the emergent universe with $a_0 = 500$, $t_0 = 0$ and $\rho_0 = 0.2$. The black line (——) is for $\gamma = 1.4$, blue line (— — —) for $\gamma = 1.6$, red line (— · —) for $\gamma = 1.8$ and magenta line (······) for $\gamma = 2.0$.

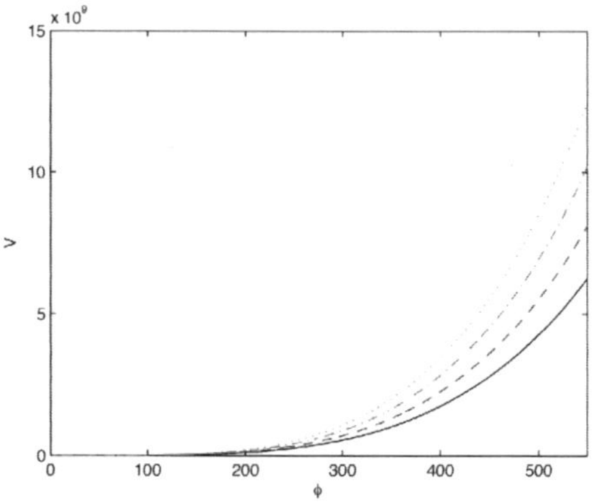

Figure 3. Graphic of the scalar field potential for the emergent universe with $t_0 = 0$. The black line (——) is for $\gamma = 1.4$, $\phi_0 = 1$ and $\rho_0 = 0.2$, blue line (— — —) for $\gamma = 1.6$, $\phi_0 = 2$ and $\rho_0 = 0.4$, red line (— · —) for $\gamma = 1.8$, $\phi_0 = 3$ and $\rho_0 = 0.6$ and magenta line (······) for $\gamma = 2.0$, $\phi_0 = 4$ and $\rho_0 = 0.8$.

For the EoS given by Eq.(4) we can reinterpret the matter source as that of a scalar matter field with some potential $V(\phi)$. Since a $\omega = P/\rho$, for the EoS considered we obtain

$$\omega = -1 - \frac{\gamma}{\rho^{1/2}}, \tag{19}$$

so we always have a phantom behavior for $\gamma > 0$. Therefore, Friedmann equations can be solved taken in the energy density and in the pressure a negative kinetic term

$$\rho = -\frac{1}{2}\dot{\phi}^2 + V(\phi), \tag{20}$$

$$P = -\frac{1}{2}\dot{\phi}^2 - V(\phi). \tag{21}$$

Using Eqs. (4), (5) and (20), (21) yields

$$V(\phi) = \frac{3\gamma^2}{256}[3(\phi-\phi_0)^4 + 8(\phi-\phi_0)^2], \tag{22}$$

for the field potential. In [2] a similar but negative expression was found for the potential, but without the inclusion of phantom matter. The behavior of field ϕ and the potential as a function of time is easily obtained

$$\phi(t) = \frac{4\rho_0^{1/4}}{\sqrt{3}\gamma} \exp\left(\frac{\gamma\sqrt{3}}{4}(t-t_0)\right) + \phi_0, \tag{23}$$

$$V(t) = \rho_0 \exp\left(\gamma\sqrt{3}(t-t_0)\right) + \frac{\gamma\rho_0^{1/2}}{2}\exp\left(\frac{\gamma\sqrt{3}}{2}(t-t_0)\right). \tag{24}$$

Since the energy density decrease as t goes to $-\infty$, the ω parameter becomes smaller than -1 and $\omega(t \to -\infty) = -\infty$ (see the upper plot in Fig. 6). The above emergent solution was found in the context of phantom cosmologies without big rip singularity, so discussed for late times [9]

3. Exact solution for a bouncing universe

For a closed universe exist a family of bouncing solutions if the parameter A of Eq.(2) is taken close to the value 1/3. These solutions can be obtained solving numerically the equation of motion.

For a universe with positive curvature ($k = 1$), Eq. (5) becomes

$$\rho = 3\left(\frac{\dot{a}}{a}\right)^2 + \frac{3}{a^2}, \tag{25}$$

and taken $A = 1/3$ in Eq.(2) yields

$$P(\rho) = -\frac{1}{3}\rho - \gamma\rho^{1/2}. \tag{26}$$

In this case the following exact solution is found

$$a(t) = \frac{2}{\gamma\sqrt{3}}\cosh\left(\frac{\gamma\sqrt{3}}{2}t + \alpha\right) + \beta, \tag{27}$$

where α y β are integration constants. The bouncing solution is obtained when $\gamma > 0$. This solution represents a universe expanding exponentially for $t = -\infty, \infty$. The scale factor takes a minimum value $a(t = t_{min}) = \frac{2}{\gamma\sqrt{3}} + \beta$, where $t_{min} = -\frac{2\alpha}{\gamma\sqrt{3}}$. The positivity of the scale factor constraints β to be in the following range $(-\frac{2}{\gamma\sqrt{3}}, \infty)$.

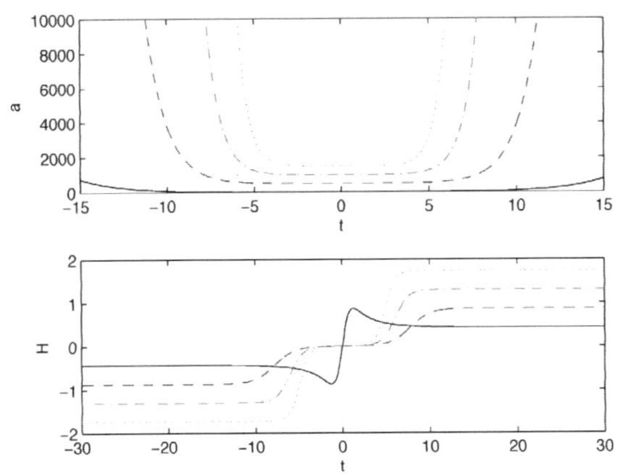

Figure 4. Graphics scale factor (upper) and the Hubble parameter (lower) as a function of time for the bouncing universe with $\alpha = 0$. The black line (——) is for $\gamma = 0.5$ and $\beta = -2$, blue line (— — —) for $\gamma = 1.0$ and $\beta = 500$, red line (— · —) for $\gamma = 1.5$ and $\beta = 1000$ and magenta line (· · · · · ·) for $\gamma = 2.0$ and $\beta = 1500$.

Using the above solution we can evaluate H, $\dot{H}$ and $\ddot{H}$, which are given by the following expressions

$$H(x) = \frac{\sinh(x)}{\frac{2}{\gamma\sqrt{3}}\cosh(x) + \beta}, \tag{28}$$

Journal of Physics: Conference Series **720** (2016) 012014 doi:10.1088/1742-6596/720/1/012014

$$\dot{H}(x) = \frac{1 + \frac{\gamma\sqrt{3}}{2}\beta\cosh(x)}{\left[\frac{2}{\gamma\sqrt{3}}\cosh(x) + \beta\right]^2}, \tag{29}$$

$$\ddot{H}(x) = \frac{\sinh(x)\left(\frac{3\gamma^2\beta^2}{2} - 4\right) - \frac{\gamma\sqrt{3}\beta}{2}\sinh(2x)}{2\left[\frac{2}{\gamma\sqrt{3}}\cosh(x) + \beta\right]^3}, \tag{30}$$

where $x = \frac{\gamma\sqrt{3}}{2}t + \alpha$. For $\beta > 0$ the Hubble parameter is a strictly increasing function, so there are no critical points and we have $H(t \to -\infty) = -\frac{\gamma\sqrt{3}}{2}$ and $H(t \to \infty) = \frac{\gamma\sqrt{3}}{2}$. So for the late times this solution behaves as a de Sitter phase.

3.1. Composition of the bouncing universe

We can obtain the energy density as a function of the scale factor by introducing the EoS given by Eq.(26) in the continuity equation (14)

$$\rho(a) = \left(\frac{3\gamma}{2} + \frac{\delta}{2a}\right)^2, \tag{31}$$

where $\delta = -3\gamma\beta$. Note that if $\beta = 0$ then ρ is constant.

Introducing Eq.(31)in Eq.(26) we obtain the fluid pressure as a function of the scale factor

$$P(a) = -\frac{1}{3}\left(\frac{3\gamma}{2} + \frac{\delta}{2a}\right)^2 - \gamma\left(\frac{3\gamma}{2} - \frac{\delta}{2a}\right). \tag{32}$$

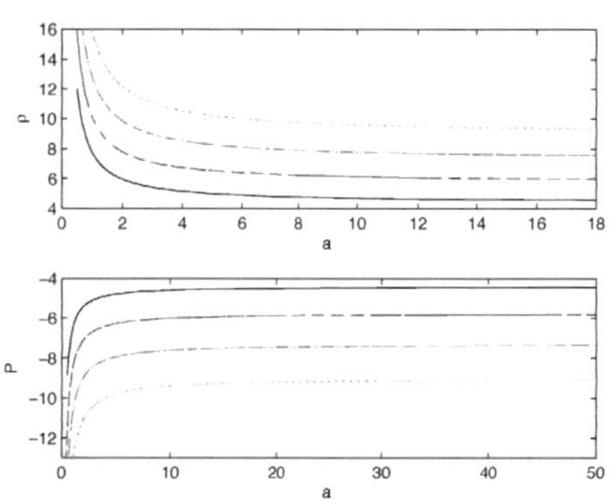

Figure 5. Graphic energy density (upper) and pressure (lower) as a function of scale factor for the bouncing universe with $\alpha = 0$ and $\beta = -0.33$. The black line (——) is for $\gamma = 1.4$, blue line (— — —) for $\gamma = 1.6$, red line (— · —) for $\gamma = 1.8$ and magenta line (······) for $\gamma = 2.0$.

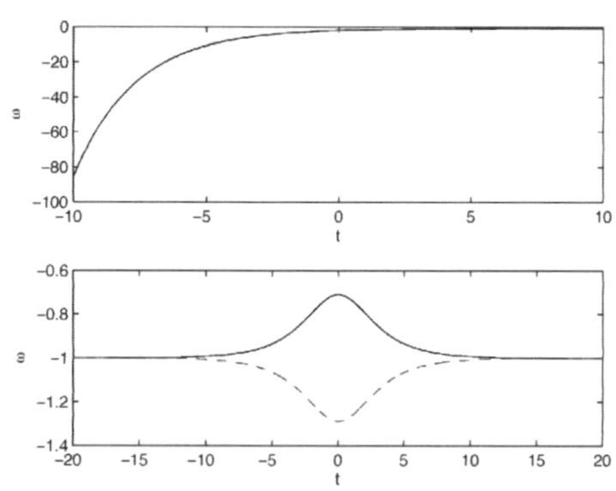

Figure 6. Graphic of the ω parameter for emergent universe (upper) with $\gamma = 0.5$, $\rho_0 = 0.2$, $a_0 = 500$ and $t_0 = 0$ and bouncing universe (lower) with $\gamma = 0.5$ and $\alpha = 0$ as a function of time. In the case of the bouncing universe, the black line (——) is for $\beta = -1$ and blue line (— — —) for $\beta = 1$.

Expanding the terms of the above both expressions yields

$$\rho(a) = \frac{9\gamma^2}{4} + \frac{3\gamma\delta}{2a} + \frac{\delta^2}{4a^2} = \rho_1 + \rho_2 + \rho_3, \tag{33}$$

$$P(a) = -\frac{9\gamma^2}{4} - \frac{\gamma\delta}{a} - \frac{\delta^2}{12a^2} = P_1 + P_2 + P_3. \tag{34}$$

Comparing each terms of the expansions our fluid can be seen as the sum of three fluids with the EoS given by $\omega_1 = P_1/\rho_1 = -1$, $\omega_2 = P_2/\rho_2 = -2/3$ and $\omega_3 = P_3/\rho_3 = -1/3$, respectively. So the first fluid correspond to a cosmological constant, the second is a quintessence and the third correspond to the presence of positive curvature. In this case the used EoS has a $\omega = P/\rho$ given by

$$\omega = -\frac{1}{3} - \frac{\gamma}{\rho^{1/2}}. \tag{35}$$

Introducing Eq.(27) in Eq.(31) we can obtain an expression for the energy density as a function of the cosmic time. Using this in Eq.(19) the behavior for the ω parameter can be obtained as a function as time. The lower plot in Fig. 6 depicted this behavior. It can seen that in the case with $\beta = -1$ the fluid ruled by the EoS given in Eq.(35) behaves like a cosmological constant for $t \to +\infty$ and $t \to -\infty$ and like quintessence for the lapse associated at the time of bouncing.

4. Discussion

We have show that cosmological solutions for a flat universe filled with a generalized EoS which represent accelerated expansion for late time evolution, also represent emergent universes for $t \to -\infty$. The behavior of the Hubble parameter indicates that these solutions are geodesically complete. For a scalar matter field with the EoS given by Eq.(4), the corresponding potential takes the form $V(\phi) \sim \alpha\phi^4 + \beta\phi^2$.

For a closed universe and taken $A = 1/3$ an exact solution was found and the energy density and pressure evaluated as a function of the scale factor. These expressions allows us to see that the generalized EoS behaves in this case as the sum of a cosmological constant, quintessence and the equivalent fluid due to curvature. For the emergent solution the fluid always violated the NEC condition, except for $t \to +\infty$ when the evolution ends in a de sitter phase.

In the bouncing solution, we have the case where the NEC condition is satisfied during the entire cosmic evolution (case $\beta < 0$), which is possible due the presence of positive curvature.

Acknowledgements

This work was supported by CONICYT through Grant FONDECYT N° 1140238 (NC), and by Proyecto Basal, Hacia una cultura de indicadores en la Educación Superior, USA 1298 (EG).

References

[1] J. D. Barrow, Phys. Lett. B **235**, 40 (1990).
[2] J. D Barrow, Nucl. Phys. B **310**, 743 (1988).
[3] S.Mukherjee, B. C. Paul, N. K. Dadhich, S. D. Maharaj and A. Beesham A., Class. Quantum Grav. **23** 6927-6933 (2006).
[4] Nojiri S., Odintsov S. D., and Tsujikawa S., Phys. Rev. D71 (2005) 063004; Nojiri S., Odintsov S. D., Phys. Rev. D72 (2005)023003.
[5] Stefancic, Phys. Rev. D71 (2005) 084024.
[6] Paul B. C., Thakur P., Ghose S., Mont. Not. R. Astron. Soc., **407**, 415 (2010).
[7] Paul B. C., Ghose S., Thakur P., Mont. Not. R. Astron. Soc., **413**, 686 (2011).
[8] Shapiro I. L. and Solà J., JHEP, 02, 006 (2002); Shapiro I. L. and Solà J., Nucl. Phys. Proc. Suppl., 127, 71 (2004); Basilakos, S. and Solà J., Mont. Not. R. Astron. Soc., **437**, 3331 (2014).
[9] A. Astashenrok, S. Nojiri, S. Odinstsov and A. V. Yurov, Phys. Lett. B **709**, 396 (2012).

5D EChS Cosmology with Perfect Fluid

Luis Avilés, Patricio Mella and Patricio Salgado

Departamento de Física, Universidad de Concepción, Casilla 160-C, Concepción, Chile.

E-mail: luisaviles@udec.cl; patriciomella@udec.cl; pasalgad@udec.cl

Abstract. We consider a Chern-Simons gravity theory in the cosmological context for a flat Friedmann-Robertson-Walker metric in five dimensions, chosen barotropic perfect fluids filled the Universe, that is, an equation of state with constant state parameter for the fluid representing the h field of Einstein-Chern-Simons cosmology and an equation of state with variable state parameter for the fluid representing the matter field.

1. Introduction

The Einstein-Chern-Simons (EChS) cosmology was introduced by [1] in order to have an dynamical description to a five dimensional Friedmann-Robertson-Walker (FRW) cosmology from Chern-Simons gravity theory for a certain Lie algebra B [2], which was obtained from the AdS algebra and a particular semigroup S by means of the S-expansion procedure development by [3, 4].

Using the extended Cartan's homotopy formula as in Ref.[5], and integrating by parts, the five dimensional EChS Lagrangian for the B algebra can be written as

$$L_{EChS}^{(5)} = \alpha_1 l^2 \varepsilon_{abcde} R^{ab} R^{cd} e^e + \alpha_3 \epsilon_{abcde} \left(\frac{2}{3} R^{ab} e^c e^d e^e + 2l^2 k^{ab} R^{cd} T^e + l^2 R^{ab} R^{cd} h^e \right) \qquad (1)$$

where α_1, α_3 are parameters of the theory, l is a coupling constant, $T^a = \mathrm{d}e^a + \omega^a{}_b e^b$ is the torsion, $R^{ab} = \mathrm{d}\omega^{ab} + \omega^a{}_c \omega^{cb}$ corresponds to the curvature 2-form in the first-order formalism related to the 1-form spin connection [6], [7], [8], and e^a, h^a and k^{ab} are others gauge fields presents in the theory [2].

We consider now the fields equations for the lagrangian given by

$$L = L_{\mathrm{ChS}}^{(5)} + \kappa L_M \qquad (2)$$

where $L_{\mathrm{ChS}}^{(5)}$ is the five-dimensional Chern-Simons Lagrangian given by (1), $L_M = L_M(e^a, h^a, \omega^{ab})$ is the matter Lagrangian and κ is a coupling constant related to the effective Newton's constant. If $T^a = 0$ and $k^{ab} = 0$. The variation of the lagrangian (2), leads to the following field equations

$$\varepsilon_{abcde} \left(2\alpha_3 R^{ab} e^c e^d + \alpha_1 l^2 R^{ab} R^{cd} \right) = \kappa \frac{\delta L_M}{\delta e^e}, \qquad (3)$$

$$\alpha_3 l^2 \varepsilon_{abcde} R^{ab} R^{cd} = \kappa \frac{\delta L_M}{\delta h^e}, \qquad (4)$$

$$\frac{\delta L_M}{\delta k^{ab}} = 0, \quad (5)$$
$$2\alpha_3 l^2 \varepsilon_{abcde} R^{cd} D_\omega h^e = \kappa \frac{\delta L_M}{\delta \omega^{ab}}, \quad (6)$$

consider here $\frac{\delta L_M}{\delta \omega^{ab}} = 0$, imposed for consistency with $T^a = 0$.

Notice that from this fields equations we recover the odd-dimensional Einstein gravity theory when the curvature R^{ab} takes values not excessively large and the parameter l takes small values ($l \longrightarrow 0$) [2] and the constant α takes values not excessively large, namely

$$\varepsilon_{abcde} R^{ab} e^c e^d \approx 4\kappa_5 \frac{\delta L_M}{\delta e^e}, \quad (7)$$

where $\kappa_5 = \kappa / 8\alpha_3$.

If R^{ab} is not large then $\delta L_M / \delta e^a$ is also not large. This means that General Relativity can be seen as a low energy limit of Einstein-Chern-Simons gravity. So that, in the range of validity of the General Relativity, the equations (3-6) are given by

$$\varepsilon_{abcde} R^{ab} e^c e^d = 4\kappa_5 \frac{\delta L_M}{\delta e^e}, \quad (8)$$
$$\varepsilon_{abcde} R^{cd} D_\omega h^e = 0. \quad (9)$$

On the another hand, if R^{ab} is large enough, so that when it is multiplied by l^2 (which is very small) will have a non-negligible results, then we will find that $\delta L_M / \delta h^a$ is not negligible. This means that, in this case, we must consider the entire system of equations (3-6).

Next we consider a flat five dimentional FRW metric and given a equation of state for the h field of the EChS theory like a barotropic perfect fluid with constant state parameter, this generates a variable state parameter in the equation of state chosen for the matter field. We think that is important because this no was explored in [1] and this open a new family of exact solutions to the EChS cosmology.

2. 5D-FRW-CHS field equations: Case k=0.

In order to solve the EChS field equations, we consider the FRW metric of the universe in five dimensions given by

$$ds^2 = -dt^2 + a^2(t)\left(\frac{dr^2}{1 - kr^2} + r^2(d\theta^2 + sin^2\theta(d\phi^2 + sin^2\phi d\psi^2))\right), \quad (10)$$

with $k = 0$ and the energy-momentum tensors

$$T_{\mu\nu} = (\rho + P) u_\mu u_\nu + P g_{\mu\nu}, \quad (11)$$
$$T^{(h)}_{\mu\nu} = (\rho^{(h)} + P^{(h)}) u_\mu u_\nu + P^{(h)} g_{\mu\nu}, \quad (12)$$

where ρ and P are the energy density and pressure of matter, and $\rho^{(h)}$ and $P^{(h)}$ are the energy density and pressure for h field, so the field equations are [1]

$$6\, H^2 + \alpha l^2 H^4 = \kappa_1 \rho, \quad (13)$$
$$\dot{\rho} + 4H(\rho + P) = 0, \quad (14)$$
$$l^2 H^4 = \kappa_2 \rho^{(h)}, \quad (15)$$
$$\dot{\rho^{(h)}} + 4H(\rho^{(h)} + P^{(h)}) = 0, \quad (16)$$
$$(g - f)H + \dot{g} = 0, \quad (17)$$

where $H = \frac{\dot{a}}{a}$.

2.1. Solutions to the barotropic equation of state

We further assume that the perfect fluid for the field h obeys the barotropic equation of state

$$P^{(h)} = \omega^{(h)}\rho^{(h)}, \tag{18}$$

where $\omega^{(h)}$ is a constant.

In order to obtain solutions for the scale factor, energy density of the field h, field matter matter, we analyze the case when $\omega^{(h)} \neq -1$ and when $\omega^{(h)} = -1$ separately.

2.1.1. ***Case*** $\omega^{(h)} \neq -1$. Writing the balance equation for the field h (16) in terms of the Hubble parameter H and using the equation of state (18), we obtain

$$H(t) = \frac{1}{\left(\omega^{(h)}+1\right)(t-t_0)}, \tag{19}$$

from this result we can find the scalar factor

$$a(t) = a_0\left[\left(\omega^{(h)}+1\right)(t-t_0)\right]^{\frac{1}{\omega^{(h)}+1}}. \tag{20}$$

Now introducing (20) in (15), we obtain

$$\rho^{(h)}(t) = \frac{l^2}{\kappa_2\left[\left(\omega^{(h)}+1\right)(t-t_0)\right]^4} \tag{21}$$

and introducing (20) in (13), we obtain

$$\rho(t) = \frac{6\left[\left(\omega^{(h)}+1\right)(t-t_0)\right]^2 + \alpha l^2}{k_1\left[\left(\omega^{(h)}+1\right)(t-t_0)\right]^4}. \tag{22}$$

If we now propose an equation of state with variable state parameter $P = \omega(t)\rho$ for the field matter, we can find from (14) that is

$$\omega(t) = \frac{\alpha l^2\omega^{(h)} + 3(\omega^{(h)}-1)(\omega^{(h)}+1)^2(t-t_0)^2}{\alpha l^2 + 6(\omega^{(h)}+1)^2(t-t_0)^2} \tag{23}$$

this result is new and not reported on Ref.[1].

2.1.2. ***Case*** $\omega^{(h)} = -1$. Writing the balance equation for the field h (16) in terms of the Hubble parameter H and using the equation of state (18), we obtain

$$H(t) = H_0 \tag{24}$$

from this result we can find the scalar factor

$$a(t) = a_0 e^{tH_0}. \tag{25}$$

Now introducing (25) in (15), we obtain

$$\rho^{(h)}(t) = \frac{l^2 H_0^4}{\kappa_2} \tag{26}$$

and introducing (25) in (13), we obtain

$$\rho(t) = H_0^2\left[\frac{6 + l^2\alpha H_0^2}{\kappa_1}\right]. \tag{27}$$

Journal of Physics: Conference Series **720** (2016) 012015 doi:10.1088/1742-6596/720/1/012015

If we now propose an equation of state with variable state parameter $P = \omega(t)\rho$ for the field matter, we can find from (14) that

$$\omega = -1 \tag{28}$$

This result agrees with that found in Ref.[1].

3. Conclusions

In this paper we have considered a Chern-Simons gravity theory in the cosmological context for a flat FRW metric in five dimensions, considering an equation of state with $\omega^{(h)}$ constant for the fluid that represents the h field of EchS theory and an equation of state with $\omega(t)$ variable for the fluid representing the matter field. With this we have obtained an exact solution for our cosmological model, finding the Hubble parameter, the scalar factor, pressures and energy densities for the matter field and h field, and the variable parameter of state for the matter field, considering the cases $\omega^{(h)} \neq -1$ and $\omega^{(h)} = -1$.

Acknowledgments

This work was supported by Comisión Nacional de Ciencias y Tecnología through FONDECYT Grants 3130444 (PM).

References

[1] Gomez, F., Minning, P., Salgado, P.: Phys. Rev. **D**84, 063506 (2011)
[2] F. Izaurieta, P. Minning, A. Perez, E. Rodriguez, P. Salgado, Phys. Lett. **B** 678 (2009) 213
[3] F. Izaurieta, E. Rodriguez, P. Salgado, Jour. Math. Phys. 47 (2006) 123512.
[4] F. Izaurieta, A. Perez, E. Rodriguez, P. Salgado, Jour. Math. Phys. 50 (2009) 073511.
[5] F. Izaurieta, E. Rodriguez, P. Salgado, Lett. Math. Phys. 80 (2007) 127
[6] J. Zanelli, "Lecture notes on Chern-Simons (super)gravities. Second edition (February 2008), 2005.
[7] A. H. Chamseddine, Phys. Lett. B 233 (1989) 291.
[8] A. H. Chamseddine, Nucl. Phys. B 346 (1990) 213.
[9] K. Andrew, B. Bolen, and C. Middleton, Gen. Relativ. Gravit. 39, 2061 (2007).
[10] N. Mohammedi, Phys. Rev. D 65, 104018 (2002).

Journal of Physics: Conference Series **720** (2016) 012016 doi:10.1088/1742-6596/720/1/012016

Emergent Universe Scenario and the Low CMB Multipoles

Pedro Labraña

Departamento de Física, Universidad del Bío-Bío, Casilla 5-C, Concepción, Chile.

E-mail: plabrana@ubiobio.cl

Abstract. In this paper, based on Phys. Rev. D **91**, no. 8, 083534 (2015), we study super inflation in the context of the Emergent Universe (EU) scenario. The existence of a super inflating phase before the onset of slow roll inflation arises in any Emergent Universe model. We found that the super inflationary period in the EU scenario produces a suppression of the CMB anisotropies at large scale which could be responsible of the observed lack of power at large angular scales of the CMB.

1. Introduction

Cosmological inflation has become an integral part of the standard model of the Universe. Apart from being capable of removing the shortcomings of the standard cosmology, it gives important clues for large scale structure formation [2, 3, 4, 5] (see [6] for a review).

The standard slow-roll inflation predicts a slightly red-tilted power spectrum of the primordial perturbation. This means that large scale (small k) perturbation has more power than small scale perturbation. This red tilt arises from the fact that the Hubble parameter slowly decreases with time as the scalar field rolls downwards to its potential.

The recent cosmological observations are entirely consistent with the simplest slow-roll inflationary models [8, 9, 10]. Probing that the observed universe is almost flat, the bispectral non-Gaussianity parameter f_{NL} measured is consistent with zero. The scalar perturbation spectral index n_s is less than one, which is a measure of the tilt discussed above ($n_s < 1$ red tilt, $n_s > 1$ blue tilt and $n_s = 1$ scale invariant spectrum). All these results are predicted by the simplest slow-roll inflationary models.

However, there are intriguing observations on cosmic microwave background radiation, suggesting a lack of power at large angular scales (very low multipoles, $l < 40$). These results were first obtained by COBE [11] and WMAP [8] and now are confirmed by Planck [9]. Although these results are well within our cosmic variance and statistically their significance is still low, the power deficit is not insignificant. The Planck Collaboration reported a power deficit in the low multipoles CMB power spectrum of order $5\% - 10\%$ (with respect to the Planck best-fit ΛCDM model [12, 13]) with statistical significance $2.5 \sim 3\,\sigma$.

There are different approaches developed in order to explain this problem. One possibility is to introduce a cutoff in the primordial power spectrum [15, 16, 17, 18, 19, 20, 21, 22, 23, 24]. This cutoff is normally introduced by hand but linked to the spatial curvature scale [25], string physics [20, 22, 23], the bouncing universe scheme [23, 24], or a fast-rolling stage in the evolution of the inflaton field [17]. This approach is interesting since it has been claimed that from the

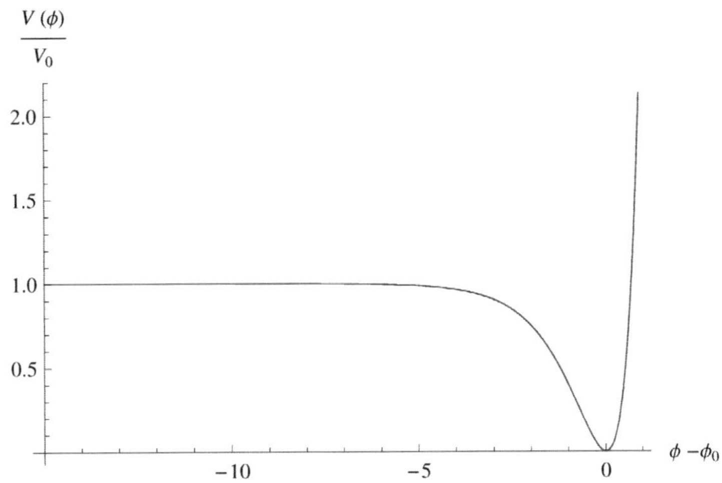

Figure 1. Schematic representation of a potential for a standard emergent universe scenario.

observed angular power spectrum, it is possible to deconvolve the primordial power spectrum for a given set of cosmological parameters. The most prominent feature of the recovered primordial power spectrum is a sharp, infrared cutoff on the horizon scale [26, 27].

In this respect, it has been suggested that the low-l power could be related to a period of superinflation, previous to the standard slow-roll inflationary regimen [29, 28, 30], where the superinflationary period is characterized by the condition $\dot{H} > 0$. In particular, in Ref. [28], this possibility was studied in the context of bouncing universes and was discussed regarding the emergent universe scheme.

The emergent universe (EU) refers to models in which the universe emerges from an Einstein static state, inflates, and is then submitted into a hot big bang era. Such models are appealing since they provide specific examples of nonsingular (geodesically complete) inflationary universes. These models have been studied in different contexts during recent years; see [31, 32, 33, 34, 35, 36, 37, 38, 39, 40, 41, 42, 43, 44, 46, 45, 47, 48, 49, 50, 51, 52].

In this work, we study the period of superinflation in the context of the EU scenario. The existence of a superinflating phase before the onset of slow-roll inflation arises in any EU scenario [28] and has not been studied thus far. As an arena to explore the superinflation phase, we consider the EU model developed in Ref. [31, 32]. This EU model is based on general relativity and considered a closed universe dominated by a scalar field minimally coupled to gravity. We found that the superinflationary period in the EU scenario produces a suppression of the CMB anisotropies at large scale which could be responsible of the observed lack of power at large angular scales of the CMB [11, 8, 9].

2. The Emergent Universe scenario

In the emergent universe scenario, the Universe is initially in a past-eternal classical Einstein static state which eventually evolves into a subsequent inflationary phase; see [31, 32, 33, 34, 35, 36, 37, 38]. During the past-eternal static regime, it is assumed that the scalar field is rolling on the asymptotically flat part of the scalar potential with a constant velocity, providing the conditions for a static universe. But once the scalar field exceeds some value, the scalar potential slowly drops from its original value. The overall effect of this is to distort the equilibrium behavior breaking the static solution. If the potential has a suitable form in this region, slow-roll inflation will occur.

Journal of Physics: Conference Series **720** (2016) 012016 doi:10.1088/1742-6596/720/1/012016

In the EU scenario, the evolution of the scale factor could be modeled by the following expression (see[31, 32]):

$$a(t) \simeq a_0 + A\, e^{H_0\, t} \tag{1}$$

where a_0, A, H_0 are positive constants. This universe is past asymptotic to an Einstein static state, since $a(t) \to a_0$ as $t \to -\infty$. Thus, a_0 is identified with the radius of the Einstein static universe. At late times, on the other hand, $a(t) \to A\, e^{H_0\, t}$ and the model approach a de Sitter expansion phase.

We can note that a generic characteristic of the EU scenario is the existence of a superinflation phase where $\dot{H} > 0$ before slow-roll inflation. In this model the evolution described in Eq. (1) corresponds precisely to the superinflationary phase of the evolution of the EU, which asymptotically approaches the de Sitter expansion phase.

The mechanism which generates this superinflationary period depends on the particular EU model under consideration, but it is a generic characteristic of the EU scenario. For example, in the models in [31, 32], it is considered a closed universe. Then, in this case, the spatial curvature is responsible for the superinflationary period.

3. The primordial perturbation in EU

The scalar perturbations to the FRW geometry, in the longitudinal gauge, can be written as follows:

$$ds^2 = (1 + 2\Phi)\, dt^2 - (1 - 2\Phi)\, a(t)^2\, d\vec{x}^2, \tag{2}$$

where Φ is the Newtonian gravitational potential.

Normally in the emergent universe scenario, the Universe is positively curved; see for example [31, 32]. In this first approach to the problem, we have neglected the contributions of the space curvature to the primordial perturbation, but we have a first approach to this point in Ref. [1], see also [55].

The equation for the perturbations in momentum space is given by

$$v_k'' + k^2\, v_k - \frac{z''}{z}\, v_k = 0\,. \tag{3}$$

where we have used the Mukhanov variable [7, 59],

$$v_k = a\left(\delta\phi_k + \frac{\phi'}{h}\,\Phi_k\right), \tag{4}$$

where $\delta\phi_k$ are the perturbations in the inflaton field and $'$ is derivative with respect to the conformal time $\eta = \int dt/a$ and we have used units $M_p = (8\pi G)^{-1/2} = 1$. Also, we have defined

$$z = \frac{a\phi'}{h}, \tag{5}$$

$$h = \frac{a'}{a}. \tag{6}$$

In order to solve Eq. (3), we consider the evolution of the scale factor $a(t)$ given in Eq. (1) written in the conformal time,

$$a(\eta) = \frac{a_0}{1 - e^{a_0 H_0 \eta}}, \quad \eta < 0\ . \tag{7}$$

From Eq. (3) and using Eq. (7), we obtain

$$v_k'' - (a_0\, H_0)^2 e^{a_0 H_0 \eta} \left(\frac{(1 + e^{a_0 H_0 \eta})}{(-1 + e^{a_0 H_0 \eta})^2} \right) v_k + k^2 v_k = 0\,, \tag{8}$$

where we have considered $z''/z \approx a''/a$. This equation is solved by

$$v_k(\eta) = \frac{1}{\sqrt{2k}} \left[\frac{e^{-ik\,\eta}}{1 - e^{a_0 H_0 \eta}} \right] \quad \times \tag{9}$$

$$_2F_1\Big(-1 - \frac{ik}{a_0 H_0} - \sqrt{1 - \left(\frac{k}{a_0 H_0}\right)^2}, \quad - \quad 1 - \frac{ik}{a_0 H_0} + \sqrt{1 - \left(\frac{k}{a_0 H_0}\right)^2}; 1 - 2\frac{ik}{a_0 H_0}; e^{a_0 H_0 \eta} \Big)\,,$$

where $2F_1$ is the hypergeometric function.

In the solution (9) we have considered (and appropriately normalized) the solution of Eq. (8) such that in the short wavelength limit, the normalized positive frequency modes correspond to the minimal quantum fluctuations,

$$v_k \approx \frac{e^{-ik\eta}}{\sqrt{2k}}\,, \quad a\,H << k\ . \tag{10}$$

Following [17], we consider the spectrum of $Q \equiv v/a$ which becomes constant at late time,

$$P_Q = \frac{k^3}{2\pi^2} |Q|^2 \longrightarrow \frac{H_0^2}{\pi^2} \frac{\chi^2\, \Gamma[x_1]\, \Gamma[x_1^*]}{\Gamma[x_2]\, \Gamma[x_2^*]\, \Gamma[x_3]\, \Gamma[x_3^*]}\,, \tag{11}$$

where we have defined

$$x_1 = 1 - 2\,i\chi \tag{12}$$

$$x_2 = 2 - i\chi - \sqrt{1 - \chi^2} \tag{13}$$

$$x_3 = 2 - i\chi + \sqrt{1 - \chi^2} \tag{14}$$

$$\chi = \frac{k}{a_0\, H_0} \tag{15}$$

We can note that in the short wavelength limit($k >> H_0$), we recovered the standard result of a nearly scale-invariant spectrum,

$$P_Q \to \left(\frac{H_0}{2\pi} \right)^2 . \tag{16}$$

At superhorizon scales, the two modes Q_k and Φ_k are related by a k-independent rescaling so that the spectrum given by Eq. (11) directly translates into the spectrum of Φ. In Fig. (2) we have plotted the spectrum of P_Q as a function of χ obtained from the analytical calculation Eq. (11), solid line. We can note that there is a suppression of the long wave modes as we expected given the superinflationary regimen.

4. The effective power spectrum

In the last section we study the spectrum generated during the superinflationary regimen. From this result and by following Ref. [24] we can estimate an effective power spectrum generated by the EU model which takes into account the early period of superinflation and the subsequent period of standard slow roll. In this respect, from the result of previous section, we can note that the scale invariance of the spectrum is the result of inflationary evolution after the

Journal of Physics: Conference Series **720** (2016) 012016 doi:10.1088/1742-6596/720/1/012016

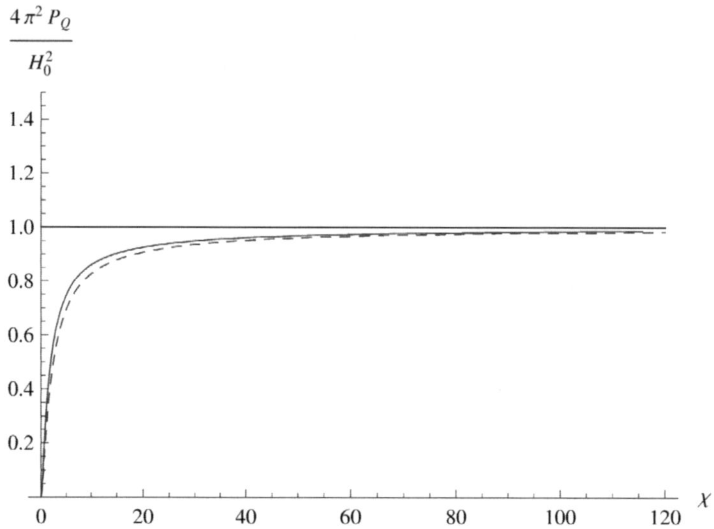

Figure 2. Power spectrum for Q. The analytical computation Eq. (11), solid line. The approximate spectrum, dashed line Eq. (19).

superinflationary stage. Then, if the inflationary period is generated by the standard slow-roll conditions, we can expect the usual slight red-tilted spectrum during this period. From Eq. (11) and by following Ref. [24], we could model this situation by considering the following scalar spectrum:

$$\mathcal{P}_\Phi = A\, k^{n_s-1} \frac{\chi^2\,\Gamma[x_1]\,\Gamma[x_1^*]}{\Gamma[x_2]\,\Gamma[x_2^*]\,\Gamma[x_3]\,\Gamma[x_3^*]}\,. \tag{17}$$

For small k, this spectrum reproduces the behavior of spectrum Eq. (11), i.e., suppression of the long wave modes. On the other hand, when $k \gtrsim H_0\, a_0$, the spectrum reproduces the usual slightly red spectrum of slow-roll inflation.

Another possibility is to consider a cutoff in the spectrum. This cutoff is related to the transition from the superinflationary regimen to slow-roll inflation which occurs when the mode k_{max} exits the horizon, after which we would get the usual red-tilted power law spectrum. In this case, we can write

$$\mathcal{P}_\Phi = \begin{cases} \bar{A}\,\frac{\chi^2\,\Gamma[x_1]\,\Gamma[x_1^*]}{\Gamma[x_2]\,\Gamma[x_2^*]\,\Gamma[x_3]\,\Gamma[x_3^*]}\,, & k < k_{max} \\ A\left(\frac{k}{k_{max}}\right)^{n_s-1}\,, & k > k_{max} \end{cases} \tag{18}$$

The constants A and $\bar{A}$ are chosen in order to match the superinflationary power spectrum Eq. (11) with the slow-roll inflationary power spectrum at $k = k_{max}$.

5. CMB Anisotropies

We compute the theoretical CMB power spectrum of this model by using CLASS code [53, 54]. In this first approximation, in the context of the EU scenario, for the study of the consequences of the superinflationary phase to the CMB power spectrum, we have simplified the code for computing the C_l by considering an approximation of the spectrums Eqs. (17) and (18). In

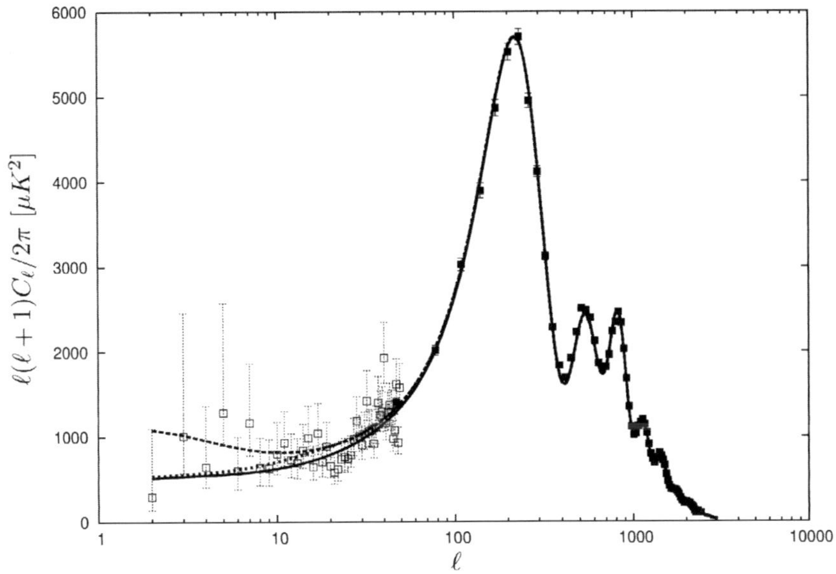

Figure 3. Temperature power spectrum for the pure power law (dashed line), EU scenario (solid line), and EU scenario with a cutoff (points line). The points show the Planck data.

particular, we approximate the spectrum of the superinflation phase Eq. (11) with the following expression (see for example [28]),

$$P_Q \sim \frac{H_0^2}{4\pi^2}\frac{\chi^2}{(1+\chi)^2}, \tag{19}$$

which is obtained by considering the evolution of the scale factor given by Eq.(1) and the Hubble crossing condition,

$$k = a\,H \quad \Rightarrow\ k = H_0\,A\,e^{H_0 t}\,. \tag{20}$$

In Fig. 2 we have plotted the spectrum Eq. (19), where we can compare it with the analytic calculation Eq. (11). We can note that the approximated spectrum reproduces very well the spectrum obtained by the analytic calculation.

In order to show the suppression at large scales coming from the EU scenario, we have plotted in Fig. 3 the temperature power spectrum obtained from the pure power law (dashed line), from the EU scenario (solid line), and from the EU scenario with a cutoff (dotted line). The points show the Planck data. We can note that compared to the standard power-law model, the C_l spectrum in the emergent universe scenario is suppressed at large scales.

In these examples, we have considered the following values for the parameters in the case of, the EU scenario in Eqs. (17) and (19): $A = 2.07\times10^{-9}$, $n_s = 0.9603$, and $a_0\,H_0 = 0.0002\,\mathrm{Mpc}^{-1}$. In the case of the EU scenario with the cutoff Eqs. (18) and (19), we consider $A = 2.42\times10^{-9}$, $n_s = 0.967$, $a_0\,H_0 = 0.0003\,\mathrm{Mpc}^{-1}$, and $k_{max} = 0.0015\,\mathrm{Mpc}^{-1}$. At this moment we are not doing a best-fit calculation of theses parameters, just showing two particular cases and how they produce a suppression of the spectrum at large scales.

6. Conclusions

In recent cosmological observations, there are intriguing results on the cosmic microwave background radiation suggesting a lack of power at large angular scales. This situation is interesting because it may give us clues towards physics at the very early Universe, perhaps before slow-roll inflation. In this context, the EU models are an interesting arena to explore preinflationary physics and its possible implications for the CMB anomalies.

In this work, based on Ref. [1], we study the primordial perturbations in the context of the EU scenario. In particular, we focus on the primordial perturbations generated during the superinflationary phase. We find that the superinflationary period in the EU scenario produces a suppression of the primordial perturbations at large scale which could be responsible for the observed lack of power at large angular scales of the CMB.

7. Acknowledgments

P.L. was supported by Dirección de Investigación de la Universidad del Bío-Bío through grants GI121407/VBC and 141407 3/R. All figures in this review are taken from Phys. Rev. D **91**, no. 8, 083534 (2015). Copyright (2015) by The American Physical Society.

References

[1] P. Labrana, Phys. Rev. D **91**, no. 8, 083534 (2015) [arXiv:1312.6877 [astro-ph.CO]].
[2] Guth A., The inflationary universe: A possible solution to the horizon and flatness problems, 1981 Phys. Rev. D **23** 347.
[3] Albrecht A. and Steinhardt P. J., Cosmology for grand unified theories with radiatively induced symmetry breaking, 1982 Phys. Rev. Lett. **48** 1220.
[4] Linde A. D., A new inflationary universe scenario: A possible solution of the horizon, flatness, homogeneity, isotropy and primordial monopole problems, 1982 Phys. Lett. **108B** 389.
[5] Linde A. D., Chaotic inflation, 1983 Phys. Lett. **129B** 177.
[6] Linde A. D., *Particle physics and inflationary cosmology*, arXiv:hep-th/0503203 (Harwood Academic Publishers, 1990).
[7] Mukhanov, V. F. and Chibisov, G. 1981, JETP Lett., 33, 532; Mukhanov, V. F. and Chibisov, G. 1982, Sov.Phys.JETP, 56, 258; Mukhanov, V. F. 1985, JETP Lett., 41, 493; Mukhanov, V. F., Feldman, H., and Brandenberger, R. H. 1992, Phys.Rept., 215, 203.
[8] E. Komatsu *et al.* [WMAP Collaboration], Astrophys. J. Suppl. **192**, 18 (2011). [arXiv:1001.4538 [astro-ph.CO]]. D. Larson, J. Dunkley, G. Hinshaw, E. Komatsu, M. R. Nolta, C. L. Bennett, B. Gold, M. Halpern *et al.*, Astrophys. J. Suppl. **192**, 16 (2011). [arXiv:1001.4635 [astro-ph.CO]]. N. Jarosik *et al.*, Astrophys. J. Suppl. **192**, 14 (2011). [arXiv:1001.4744 [astro-ph.CO]]. E. Komatsu *et al.* [WMAP Collaboration], Astrophys. J. Suppl. **180**, 330-376 (2009). [arXiv:0803.0547 [astro-ph]]. D. N. Spergel *et al.* [WMAP Collaboration], Astrophys. J. Suppl. **170**, 377 (2007). [astro-ph/0603449].
[9] P. A. R. Ade et al. [Planck Collaboration]: *Planck 2013 results. I. Overview of products and scientific results*, Astron. Astrophys. **571**, A1 (2014) [arXiv:1303.5062 [astro-ph.CO]].
[10] P. A. R. Ade *et al.* [Planck Collaboration], Astron. Astrophys. **571**, A22 (2014) [arXiv:1303.5082 [astro-ph.CO]].
[11] G. F. Smoot et al., ApJ **396**, 1 (1992).
[12] P. A. R. Ade *et al.* [Planck Collaboration], Astron. Astrophys. **571**, A15 (2014) [arXiv:1303.5075 [astro-ph.CO]].
[13] P. A. R. Ade *et al.* [Planck Collaboration], Astron. Astrophys. **571**, A23 (2014) [arXiv:1303.5083 [astro-ph.CO]].
[14] P. A. R. Ade *et al.* [Planck Collaboration], arXiv:1303.5086 [astro-ph.CO].
[15] S. L. Bridle et al., Mon.Not.Roy.Astron.Soc. 342 L72 (2003). astro-ph/0302306.
[16] J. P. Uzan et al., astro-ph/0302597 (2003).
[17] C. R. Contaldi, M. Peloso, L. Kofman and A. D. Linde, JCAP **0307**, 002 (2003) [astro-ph/0303636].
[18] J. M. Cline, P. Crotty, and J. Lesgourgues, JCAP 0309 010 (2003). astro-ph/ 0304558 (2003).
[19] B. Feng, and X. Zhang, Phys.Lett. B570 145-150 (2003). astro-ph/0305020 (2003).
[20] M. Bastero-Gil, K. Freese, and L. Mersini-Houghton, Phys.Rev. D68 123514 (2003). hep-ph/0306289 (2003).
[21] E. Ramirez and D. J. Schwarz, Phys. Rev. D **85**, 103516 (2012) [arXiv:1111.7131 [astro-ph.CO]].
[22] E. Dudas, N. Kitazawa, S. P. Patil and A. Sagnotti, JCAP **1205**, 012 (2012) [arXiv:1202.6630 [hep-th]].
[23] Y. -S. Piao, B. Feng and X. -m. Zhang, Phys. Rev. D **69**, 103520 (2004) [hep-th/0310206].
[24] Z. -G. Liu, Z. -K. Guo and Y. -S. Piao, Phys. Rev. D **88**, 063539 (2013) [arXiv:1304.6527 [astro-ph.CO]].
[25] G. Efstathiou, astro-ph/0303623 (2003).
[26] A. Shafieloo and T. Souradeep, Phys. Rev. D **70** (2004) 043523 [astro-ph/0312174].

Journal of Physics: Conference Series **720** (2016) 012016 doi:10.1088/1742-6596/720/1/012016

[27] D. K. Hazra, A. Shafieloo and T. Souradeep, arXiv:1303.5336 [astro-ph.CO].
[28] T. Biswas and A. Mazumdar, Class. Quant. Grav. **31**, 025019 (2014) arXiv:1304.3648 [hep-th].
[29] D. -H. Zhang and C. -Y. Sun, Chin. Phys. Lett. **21**, 1865 (2004) [astro-ph/0310127].
[30] Z. -G. Liu, Z. -K. Guo and Y. -S. Piao, arXiv:1311.1599 [astro-ph.CO].
[31] Ellis G. F. R. and Maartens R., The emergent universe: Inflationary cosmology with no singularity, 2004 Class. Quant. Grav. **21** 223.
[32] Ellis G. F. R., Murugan J. and Tsagas C. G., The emergent universe: An explicit construction, 2004 Class. Quant. Grav. **21** 233.
[33] Mulryne D. J., Tavakol R., Lidsey J. E. and Ellis G. F. R., An emergent universe from a loop, 2005 Phys. Rev. D **71** 123512.
[34] Mukherjee S., Paul B. C., Maharaj S. D. and Beesham A., Emergent universe in Starobinsky model, arXiv:gr-qc/0505103.
[35] Mukherjee S., Paul B. C., Dadhich N. K., Maharaj S. D. and Beesham A. , Emergent universe with exotic matter, 2006 Class. Quant. Grav. **23** 6927.
[36] A. Banerjee, T. Bandyopadhyay and S. Chakraborty, Emergent universe in brane world scenario, Grav. Cosmol. **13**, 290 (2007).
[37] Nunes N. J., Inflation: A graceful entrance from loop quantum cosmology, 2005 Phys. Rev. D **72** 103510.
[38] Lidsey J. E. and Mulryne D. J., A graceful entrance to braneworld inflation, 2006 Phys. Rev. D **73** 083508.
[39] A. Beesham, S. V. Chervon and S. D. Maharaj, Class. Quant. Grav. **26**, 075017 (2009) [arXiv:0904.0773 [gr-qc]].
[40] B. C. Paul and S. Ghose, arXiv:0809.4131 [hep-th].
[41] U. Debnath, Class. Quant. Grav. **25**, 205019 (2008) [arXiv:0808.2379 [gr-qc]].
[42] A. Banerjee, T. Bandyopadhyay and S. Chakraborty, Gen. Rel. Grav. **40**, 1603 (2008) [arXiv:0711.4188 [gr-qc]].
[43] S. del Campo, R. Herrera and P. Labraña, Emergent universe in a Jordan-Brans-Dicke theory, JCAP **0711**, 030 (2007).
[44] S. del Campo, R. Herrera and P. Labrana, JCAP **0907**, 006 (2009) [arXiv:0905.0614 [gr-qc]].
[45] S. del Campo, E. I. Guendelman, A. B. Kaganovich, R. Herrera and P. Labrana, Phys. Lett. B **699**, 211 (2011) [arXiv:1105.0651 [astro-ph.CO]].
[46] S. del Campo, E. I. Guendelman, R. Herrera and P. Labrana, JCAP **1006**, 026 (2010) [arXiv:1006.5734 [astro-ph.CO]].
[47] P. Labrana, Phys. Rev. D **86**, 083524 (2012) [arXiv:1111.5360 [gr-qc]].
[48] Y. -F. Cai, M. Li and X. Zhang, Phys. Lett. B **718**, 248 (2012) [arXiv:1209.3437 [hep-th]].
[49] P. Rudra, Mod. Phys. Lett. A **27**, 1250189 (2012) [arXiv:1211.2047 [gr-qc]].
[50] Z. -G. Liu and Y. -S. Piao, Phys. Lett. B **718**, 734 (2013) [arXiv:1207.2568 [gr-qc]].
[51] E. Guendelman and P. Labrana, IJMP D Vol. 22, **No 9** 1330018 (2013). arXiv:1303.7267 [astro-ph.CO].
[52] Y. -F. Cai, Y. Wan and X. Zhang, Phys. Lett. B **731**, 217 (2014) [arXiv:1312.0740 [hep-th]].
[53] J. Lesgourgues, arXiv:1104.2932 [astro-ph.IM].
[54] D. Blas, J. Lesgourgues and T. Tram, JCAP **1107** (2011) 034 [arXiv:1104.2933 [astro-ph.CO]].
[55] P. Labrana, A cid and C. Rios, under preparation.
[56] M. Cicoli, S. Downes, B. Dutta, F. G. Pedro and A. Westphal, JCAP **1412**, no. 12, 030 (2014) [arXiv:1407.1048 [hep-th]].
[57] D. J. Schwarz and E. Ramirez, arXiv:0912.4348 [hep-ph].
[58] E. Ramirez, Phys. Rev. D **85** (2012) 103517 [arXiv:1202.0698 [astro-ph.CO]].
[59] J. Garriga and V. F. Mukhanov, Phys. Lett. B **458**, 219 (1999)
[60] L. F. Abbott and R. K. Schaefer, Astrophys. J. **308**, 546 (1986).

Integrability properties of a coupled KdV system and its supersymmetric extension

Adrián Sotomayor[1], **Alvaro Restuccia**[2]

[1] Departament of Matemathics , Antofagasta University, Antofagasta, Chile
[2] Physics Departament, Antofagasta University, Antofagasta, Chile

E-mail: adrian.sotomayor@uantof.cl,alvaro.restu@uantof.cl

Abstract. We discuss several integrability properties of a coupled KdV system. We obtain a new generalization of the already known static solutions for the system. We then consider the supersymmetric extension of the coupled KdV system, it is a new integrable system. We show that for particular Grassmann algebras the system is the limit of a Clifford algebra valued system with nice stability properties. We briefly discuss the hamiltonian structures of this supersymmetric integrable system.

1. Introduction

The Korteweg-de Vries (KdV) equation has relevant applications in low energy physics as well as in high energy physics. Several extensions of KdV have been proposed. Some of them are the coupled KdV systems, the supersymmetric KdV equation and operatorial extensions.

Recently [1] we introduced a parametric coupled KdV system which includes the complex version of KdV equation for a particular value of the parameter as well as one of the Hirota-Satsuma coupled systems. We found a Bäcklund transformation, the Gardner transformation and we showed the permutability theorem for the system. In particular we found an infinite family of multisolitonic and periodic solutions for the system. On the other side a supersymmetric extension of KdV equation has been considered in the literature. Although it has infinite conserved quantities as the KdV equation, only one local hamiltonian structure has been found for it in distinction to the KdV case which is bihamiltonian.

In this work we present some new interesting properties about the solutions to the parametric coupled KdV system and we show also that for particular Grassmann algebras, which are the basic algebraic structures to define the supersymmetric extensions of the KdV equation, the supersymmetric system is bihamiltonian, in the sense we will explain in this work. In particular, one of the Hirota-Satsuma coupled systems is contained in a supersymmetric system. Consequently, this supersymmetric system has all the integrability properties shown in [1].

2. The parametric coupled KdV system and some of its relevant properties

The parametric coupled KdV system is given by

$$u_t + uu_x + u_{xxx} + \lambda v v_x = 0 \tag{1}$$
$$v_t + u_x v + v_x u + v_{xxx} = 0, \tag{2}$$

where λ is a real parameter and u, v are rapidly decreasing real valued functions depending on the temporal and spatial variables t and x respectively. The system contains, for $\lambda = -1$, the complex version of KdV and for $\lambda = 0$ includes the ninth Hirota-Satsuma system given in [2] (after a suitable change of the variables). For the case $\lambda = 1$ the system decouples in to two KdV equations, in terms of $u \pm v$ respectively; for the other values of the parameter λ ($\lambda \neq 0$) a simple change in the variable v reduces it to $\lambda = \pm 1$.

The system defined by (1) and (2) has applications as a physics model in a two-layer liquid, see [3], and also has solitonic solutions, obtained using the Hirota approach [4].

In [1] it was proven the following

Lemma 1 *Let $r, s \in C_{\downarrow}^{\infty}$ be a solution of the following ϵ-parameter partial differential equations (called the Gardner system)*

$$r_t + r_{xxx} + r r_x + \lambda s s_x - \frac{1}{6}\varepsilon^2 \left[\left(r^2 + \lambda s^2\right) r_x + 2\lambda r s s_x \right] = 0$$
$$s_t + s_{xxx} + r s_x + s r_x - \frac{1}{6}\varepsilon^2 \left[\left(r^2 + \lambda s^2\right) s_x + 2 r s r_x \right] = 0.$$

Then $u, v \in C_{\downarrow}^{\infty}$ defined through the relations (called the Gardner transformation)

$$u = r + \varepsilon r_x - \frac{1}{6}\varepsilon^2 \left(r^2 + \lambda s^2\right)$$
$$v = s + \varepsilon s_x - \frac{1}{3}\varepsilon^2 r s$$

are solutions of the system (1),(2).

As a direct consequence of lemma 1 we have that the system (1),(2) has infinite polynomial conserved quantities in the fields u, v and its spatial derivatives (making the system integrable in this sense for every value of λ) and they are explicitly obtained from the first two conserved quantities of the Gardner system. The first few of them are

$$\int_{-\infty}^{+\infty} u dx \quad , \quad \int_{-\infty}^{+\infty} v dx,$$
$$\int_{-\infty}^{+\infty} \left(u^2 + \lambda v^2\right) dx \quad , \quad \int_{-\infty}^{+\infty} u v dx,$$
$$\int_{-\infty}^{+\infty} \left(\frac{1}{3}u^3 + \lambda u v^2 - \lambda (v_x)^2 - (u_x)^2\right) dx \quad , \quad \int_{-\infty}^{+\infty} \left(\frac{1}{2}u^2 v - u_x v_x + \frac{1}{6}\lambda v^3\right) dx.$$

We proposed in [1] for the system (1), (2) the following Bäcklund transformation. If (w, y) and (w', y'), where $w_x = u, w'_x = u'$ and $y_x = v, y'_x = v'$ satisfy the following equations (Bäcklund transformation)

$$w_x + w'_x = 2\eta - \frac{1}{12}(w - w')^2 - \frac{\lambda}{12}(y - y')^2, \tag{3}$$
$$\begin{aligned} w_t + w'_t = {} & \frac{1}{6}(w - w')(w_{xx} - w'_{xx}) + \frac{\lambda}{6}(y - y')(y_{xx} - y'_{xx}) - \frac{1}{3}w_x^2 - \\ & - \frac{1}{3}{w'_x}^2 - \frac{1}{3}w_x w'_x - \frac{\lambda}{3}y_x^2 - \frac{\lambda}{3}{y'_x}^2 - \frac{\lambda}{3}y_x y'_x, \end{aligned} \tag{4}$$
$$y_x + y'_x = 2\mu - \frac{1}{6}(w - w')(y - y'), \tag{5}$$
$$\begin{aligned} y_t + y'_t = {} & \frac{1}{6}(w - w')(y - y')_{xx} + \frac{1}{6}(w - w')_{xx}(y - y') - \\ & - \left(\frac{2}{3}w_x y_x + \frac{2}{3}w'_x y'_x + \frac{1}{3}w_x y'_x + \frac{1}{3}w'_x y_x\right) \end{aligned} \tag{6}$$

on an open set $\Omega \subset \mathbb{R}^2$ and

$$(w - w')^2 - \lambda(y - y')^2 \neq 0$$

on Ω, then (w, y) and (w', y') are two (different) solutions of the system

$$Q_1(w, y) \equiv w_t + \frac{1}{2}(w_x)^2 + w_{xxx} + \frac{1}{2}\lambda(y_x)^2 = 0 \tag{7}$$

$$Q_2(w, y) \equiv y_t + w_x y_x + y_{xxx} = 0, \tag{8}$$

and consequently $w_x = u, y_x = v, w'_x = u', y'_x = v'$ are solutions of the original system (1), (2) over the corresponding set in the plane. We observe that (7), (8) is essentially our original system.

It was also proven in [1] the following

Theorem 1 *(Permutability theorem) Let w_{12}, y_{12} be the solution of (7),(8) obtained from the Bäcklund transformation following the sequence*

$$(w_0, y_0) \rightarrow_{\eta_1} (w_1, y_1) \rightarrow_{\eta_2} (w_{12}, y_{12})$$

and (w_{21}, y_{21}) the solution following the sequence

$$(w_0, y_0) \rightarrow_{\eta_2} (w_2, y_2) \rightarrow_{\eta_1} (w_{21}, y_{21}).$$

Then $w_{12} = w_{21}, y_{12} = y_{21}$ and

$$w_{12} - w_0 = \frac{24(\eta_1 - \eta_2)(w_1 - w_2)}{(w_1 - w_2)^2 - \lambda(y_1 - y_2)^2}$$

$$y_{12} - y_0 = \frac{-24(\eta_1 - \eta_2)(y_1 - y_2)}{(w_1 - w_2)^2 - \lambda(y_1 - y_2)^2}.$$

We notice that the denominator in the above formulas is the same expression which appears in the assumptions given for the Bäcklund transformation. This condition is necessary in order to have regular solutions. In the case $\lambda = -1$, the denominator becomes

$$(w_1 - w_2)^2 + (y_1 - y_2)^2. \tag{9}$$

3. On the explicit solutions from the Bäcklund transformation

In [1] the following set of solutions for (1), (2) were found, using the Bäcklund transformation : for any $\eta > 0$ and $\lambda = -1$

$$u(x, t) = 4\eta \frac{\left[1 - \frac{3C}{\eta A}\cosh(ax + b)\right]}{\left[\cosh(ax + b) - \frac{3}{\eta}\frac{C}{A}\right]^2}, \quad v(x, t) = -\frac{\rho}{A} a \frac{\sinh(ax + b)}{\left[\cosh(ax + b) - \frac{3}{\eta}\frac{C}{A}\right]^2}, \tag{10}$$

for any $\eta < 0$ and $\lambda = -1$

$$u = \frac{-4|\eta|\left(1 + \frac{3C\epsilon}{|\eta|\hat{A}}\cos(ax + b)\right)}{\left(\epsilon\cos(ax + b) + \frac{3C}{|\eta|\hat{A}}\right)^2}, \quad v = \frac{\rho a}{\hat{A}} \frac{\epsilon\sin(ax + b)}{\left(\epsilon\cos(ax + b) + \frac{3C}{|\eta|\hat{A}}\right)^2} \tag{11}$$

and finally for $\eta = 0$ and $\lambda = -1$ (and for any value of the parameters $\mathcal{C}, H$ and $\rho \neq 0$)

$$u = w_x = \frac{C^2 \frac{\rho^2}{12} - 12C^4(x+H)^2}{\left[\left(\frac{\rho}{12}\right)^2 + C^2(x+H)^2\right]^2}, \quad v = y_x = \frac{-2C^3\rho\,(x+H)}{\left[\left(\frac{\rho}{12}\right)^2 + C^2(x+H)^2\right]^2}. \tag{12}$$

The solutions are solitonic, periodic and static (independent of t) respectively. For the details about the parameters involved in the solutions see [1].

We can also use the expressions given for the permutability theorem and the explicit formulas for w_{12}, y_{12} to generate new solutions from old ones with respect our original system (1), (2), in particular new multisolitonic solutions.

In distinction with what occurs with the Bäcklund transformation for the KdV equation, we can use directly w_{12} and y_{12} in terms of the regular solutions presented before to find new regular solutions of the coupled KdV system. In fact, the denominator in the expressions for w_{12}, y_{12} is manifestly positive and $\mathcal{C}^\infty$. Moreover, we are going to show that if we choose adequately the parameters of the given solitonic solutions the denominator in the expressions for w_{12}, y_{12} is strictly positive and the new solutions are regular solutions.

An interesting result about the regularity of the solutions, in the case of solitonic ones, is the following (see also [1])

Theorem 2 *For any value of the parameters* $\eta_1 > 0, \rho_1 \neq 0, \mathcal{C}_1$ *and* $\eta_2 > 0, \rho_2 \neq 0, \mathcal{C}_2$ *such that* $\eta_1 \neq \eta_2$ *and*

$$\frac{\mathcal{C}_1}{\eta_1\rho_1} = \frac{\mathcal{C}_2}{\eta_2\rho_2}$$

the solutions for the coupled KdV system with $\lambda = -1$ *obtained from the permutability formulas are regular.*

4. About the static solutions for the parametric coupled KdV system

In the most simple case of (3),(4),(5) and (6) the Bäcklund transformation reduces to

$$w_x = 2\eta - \frac{1}{12}w^2 \tag{13}$$

$$w_t = \frac{1}{6}ww_{xx} - \frac{1}{3}{w_x}^2. \tag{14}$$

Clearly, we have started with the trivial solution and we have assumed that the field y is identically zero. We have thus the Bäcklund transformation for the scalar KdV equation given by

$$u_t + uu_x + u_{xxx} = 0. \tag{15}$$

For $\eta = 0$ and assuming $w_t = 0$, that is, w is only a function of the spatial variable x we get from (13) $w_{xx} = -\frac{1}{6}ww_x$ and substituting this in (14) we get $-\frac{w_x}{3}\left(\frac{w^2}{12} + w_x\right) = 0$. The conclusion is that for this special type of solutions equation (14) gives only known information about equation (13).

We also have that if w is a solution of (13) (for $\eta = 0$) then $w_x = u$ satisfies $\frac{u^2}{2} + u_{xx} = 0$ and this automatically implies that $u_t = 0$, because we have

$$u_t + \left(\frac{u^2}{2} + u_{xx}\right)_x = 0.$$

For example, a static solution for the equation (15) in the case of $\eta = 0$ can be obtained directly integrating the equation $w_x = -\frac{1}{12}w^2$, giving

$$w = -\frac{12}{x+K},$$

where K is a constant of integration. We note however that this solution is a singular solution, making a substantial difference with the solutions given by (12) (for the case $\lambda = -1$ and $\eta = 0$), which are regular solutions.

This argument can be generalized to the Bäcklund transformation (3),(4),(5),(6). In fact, the result about the static solution can be extended as follows

Lemma 2 *Let w, y a solution of the parametric equations*

$$w_x = -\frac{1}{12}\left(w^2 + \lambda y^2\right) \quad (16)$$

$$y_x = -\frac{1}{6}wy. \quad (17)$$

Then $u = w_x, v = y_x$ is a static solution of equations (1),(2).

Proof of Lemma 1 *From (16),*

$$u_x = w_{xx} = -\frac{1}{12}\left(2ww_x + 2\lambda y y_x\right) = \frac{w^3}{72} + \frac{3}{72}\lambda y^2 w$$

$$u_{xx} = \frac{3}{72}w^2 w_x + \frac{3}{72}\lambda\left(2yy_x w + y^2 w_x\right) = -\frac{1}{288}\left(w^2+\lambda y^2\right)^2 - \frac{\lambda}{72}y^2 w^2.$$

Then (1) becomes

$$u_t + \left(\frac{u^2}{2} + u_{xx} + \lambda\frac{v^2}{2}\right)_x = 0, \quad (18)$$

and when replacing the previous results we obtain

$$\frac{u^2}{2} + u_{xx} + \lambda\frac{v^2}{2} = 0.$$

Consequently any solution of the system (16),(17) is a static solution of (1). In the same way we can check that equation (2) is satisfied.

Remark: This result generalizes the formulas given by (12) for static solutions of the system (1),(2).

5. The supersymmetric extension

The parametric coupled KdV system (1),(2) has a supersymmetic extension. It can be constructed from the $N = 1$ supersymmetric KdV equation [5] by considering the fields valued on a Z^λ algebra [6] generated by e_1, e_2:

$$e_1 \circ e_1 = e_1, e_1 \circ e_2 = e_2, e_2 \circ e_2 = \lambda e_1. \quad (19)$$

If we express the super KdV fields $U = ue_1 + ve_2, \xi = \mu e_1 + \nu e_2$ then the new supersymmetric transformations becomes

$$\begin{aligned} \delta u &= \epsilon_1 \mu' + \lambda \epsilon_2 \nu' \,,\; \delta v = \epsilon_1 \nu' + \epsilon_2 \mu' \\ \delta \mu &= \epsilon_1 u + \lambda \epsilon_2 v \,,\; \delta \nu = \epsilon_1 v + \epsilon_2 u, \end{aligned} \quad (20)$$

where ϵ_1 and ϵ_1 are two independent odd Grassmann supersymmetric parameters.

The supersymmetric extension is

$$\begin{aligned} u_t &= -u''' + 6uu' + 6\lambda v'v - 3\mu\mu'' - 3\lambda\nu\nu'' \\ v_t &= -v''' + 6(vu)' - 3\mu\nu'' - 3\nu\mu'' \\ \mu_t &= -\mu''' + 3(u\mu + \lambda v\nu)' \\ \nu_t &= -\nu''' + 3(u\nu + v\mu)'. \end{aligned} \tag{21}$$

This system is invariant under the supersymmetric transformation (20).

In particular, if $\lambda = 0$ the system (1),(2) corresponds to one of the Hirota-Satsuma integrable equations [2, 7]. The system (21) for $\lambda = 0$ becomes its supersymmetric extension.

The system (21), for any value of λ, has an infinite sequence of local conserved quantities. It is a new integrable system. In particular if we consider the Grassmann algebra to have only one generator e, then the system (21), for $\lambda = 0$, reduces to

$$\begin{aligned} u_t &= -u''' + 6uu' \\ \mu_t &= -\mu''' + 3(u\mu)' \\ v_t &= -v''' + 6(vu)' \\ \nu_t &= -\nu''' + 3(u\nu + v\mu)'. \end{aligned} \tag{22}$$

If in addition, we take $v = \nu = 0$, the system becomes

$$\begin{aligned} u_t &= -u''' + 6uu' \\ \mu_t &= -\mu''' + 3(u\mu)'. \end{aligned} \tag{23}$$

In [8] we introduced the system

$$\begin{aligned} u_t &= -u''' - uu' - \tfrac{1}{4}(\mathcal{P}\,(\xi\bar{\xi}))' \\ \xi_t &= -\xi''' - \tfrac{1}{2}(\xi u)'. \end{aligned} \tag{24}$$

It arises from a supersymmetric breaking of the $N = 1$ super KdV system. If we perform a scale transformation $\xi \to \alpha\hat{\xi}$, assuming that the Clifford algebra has only one generator, and take the limit $\alpha \to 0$ leaving $\hat{\xi}$ and μ finite we obtain the system (23). The interesting point is that (23) is a supersymmetric system, it has a hamiltonian structure and an infinite sequence of conserved quantities, consequently system (24) in the above limit has those properties. For $\alpha \neq 0$ (24) has only a finite number of conserved quantities. Hence in the $\alpha \to 0$ limit we obtain an enhancement of the symmetries of the system.

Consequently the system (23) has two hamiltonian structures, although one of it arises after a limit process. The hamiltonian structures are not compatible. This is a new aspect to be investigated since it is well known that supersymmetric KdV equations only have one local hamiltonian structure.

6. Conclusions

We discussed the integrability properties of a coupled KdV system [1]. We obtained a new generalization of the static solutions to the system. We also obtained new properties of the supersymmetric extension of the coupled KdV system, which is a new integrable system. For the particular case of a Grassmann algebra with only one odd generator the supersymmetric system is the limit of a Clifford valued system with nice stability properties. Hence we expect that the same stability properties will be valid for the supersymmetric system. It has two non-compatible hamiltonian structures one arising from its supersymmetric structure and the other one arising from the Clifford system in the limit procedure. This interesting aspect will be analyzed elsewhere.

Acknowledgments
A. S. and A. R. are partially supported by Project Fondecyt 1121103, Chile.

References
[1] L. Cortés Vega , A. Restuccia and A. Sotomayor 2014 *Preprint* arXiv: math/ph 1407.7743 v3
[2] S. Yu Sakovich 1999 *J. Nonlin. Math. Phys.* **6** N3 255-262
[3] S. Y. Lou, B. Tong, H. C. Hu and X. Y. Tang 2006 *J. Phys. A: Math. Gen.* **39** 513-527
[4] J. R. Yang and J. J. Mao 2008 *Commun. Theor. Phys.* **49** 22-26
[5] P. Mathieu 1988 *J. Math. Phys.* **29** 2499
[6] D. Zuo 2014 *J. Geom. Phys.* **86** 203-210
[7] P. Casati and G. Ortenzi 2006 *J. Geom. and Phys.* **56** 418-449
[8] A. Restuccia and A. Sotomayor 2013 *Boundary Value Problems* **2013:224**

Journal of Physics: Conference Series **720** (2016) 012018 doi:10.1088/1742-6596/720/1/012018

Yang-Mills connections valued on the octonionic algebra

A Restuccia[1,2] **and J P Veiro**[3]

[1]Universidad de Antofagasta, Departamento de Física, Antofagasta, Chile
[2]Universidad Simón Bolívar, Departamento de Física, Apartado 89000, Caracas 1080-A, Venezuela
[3]Universidad Simón Bolívar, Departamento de Matemáticas Puras y Aplicadas, Apartado 89000, Caracas 1080-A, Venezuela

E-mail: `arestu@usb.ve` and `jpveiro@usb.ve`

Abstract. We consider a formulation of Yang-Mills theory where the gauge field is valued on an octonionic algebra and the gauge transformation is the group of automorphisms of it. We show, under mild assumptions, that the only possible gauge formulations are the usual $\mathfrak{su}(2)$ or $\mathfrak{u}(1)$ Yang-Mills theories.

1. Introduction

Yang-Mills theory with appropriate structure group is the basic ingredient in the description of the fundamental forces in nature. A notorious extension of Yang-Mills theory is Super Yang-Mills in 3, 4, 6, and 10 spacetime dimensions. It is well known the relation between the division algebras: the reals $\mathbb{R}$, complex $\mathbb{C}$, quaternion $\mathbb{H}$, the octonions $\mathbb{O}$, and Super Yang-Mills [1, 2, 3, 4, 5, 6, 7, 8, 9]. In particular octonions are relevant since the algebra $\mathfrak{sl}(2, \mathbb{O})$ is isomorphic to the $\mathfrak{so}(9, 1)$ Lorentz algebra [10, 11, 12, 15, 3, 13, 14], the relativistic symmetry in string theory. It can also be used to describe representations of the Lorentz group in $D = 11$ dimensional space-time where M-theory is formulated. The octonionic algebra was relevant to obtain Yang-Mills instanton solutions in seven and eight dimensions [16, 17]. Recently the relation between Yang-Mills, division algebras, and triality has been emphasized [9, 18], in particular an octonionic formulation of the $N = 1$ Supersymmetry algebra in $D = 11$ space-time dimensions has been obtained [19].

In all these interesting formulations the gauge field is always valued on a Lie algebra. We will consider in this work the construction of Yang-Mills theory where the gauge field is valued on the algebra of the octonions. In order to do so we have to define the corresponding gauge transformations and consider its action on the curvature of the connection one-form. The formulation we will consider is based on a proposition introduced by Okubo [20].

The main result of this work is that under mild assumptions: (5), (6), and (7) in Section 2, the only possible gauge formulation, with the gauge field valued on the octonionic algebra, of a Yang-Mills theory are the ones on which the connection is valued on an associative subalgebra of the octonions.

Journal of Physics: Conference Series **720** (2016) 012018 doi:10.1088/1742-6596/720/1/012018

2. The Yang-Mills formulation on a non-associative algebra

Given a Lie algebra L, let us denote by Λ_p, for $p > 0$, the space of all L-valued p-forms and $\Lambda_0 = L$. The boundary operator, $\mathrm{d} : \Lambda_p \to \Lambda_{p+1}$, satisfies the usual properties:

(i) $\mathrm{dd} = 0$,

(ii) $\mathrm{d}(\omega_p \wedge \omega_q) = \mathrm{d}\omega_p \wedge \omega_q + (-1)^p \omega_p \wedge \mathrm{d}\omega_q$ where $\omega_p \in \Lambda_p$ and $\omega_q \in \Lambda_q$ are L-valued p-forms and q-forms respectively.

Consider ω to be an L-valued one-form which we shall write

$$\omega = \sum_{\mu=1}^{m} A_\mu \mathrm{d}x^\mu,$$

where the $\mathrm{d}x^\mu$ are anti-commutative one-forms, and the curvature is then given by the L-valued two-form $R = \mathrm{d}\omega + \omega \wedge \omega$ which can be expressed in terms of $F_{\mu\nu}$ via

$$R = \sum_{1 \leq \mu < \nu \leq m} F_{\mu\nu} \mathrm{d}x^\mu \wedge \mathrm{d}x^\nu = \frac{1}{2} \sum_{\mu,\nu=1}^{m} F_{\mu\nu} \mathrm{d}x^\mu \wedge \mathrm{d}x^\nu.$$

The gauge transformations are given by

$$\omega \longrightarrow U^{-1}\omega U + U^{-1}\mathrm{d}U \tag{1}$$

and they are such that the curvature transforms covariantly by

$$R \longrightarrow U^{-1}RU. \tag{2}$$

Given $A \in L$ write $g_U(A) = U^{-1}AU$, they are automorphisms of the Lie algebra. It is clear that $g_U \in GL(L)$ and it can also be extended to Λ_p in a natural way for which it can simply be considered $g_U \in GL(\Lambda_p)$. If we write $\xi = U^{-1}\mathrm{d}U$ then ξ satisfies the Maurer-Cartan relation

$$\mathrm{d}\xi + \xi \wedge \xi = 0.$$

In this manner, Equations (1) and (2) can be expressed as

$$\begin{aligned} \omega &\longrightarrow g_U(\omega) + \xi, \\ R &\longrightarrow g_U(R). \end{aligned}$$

Consider now the octonion algebra $\mathbb{O}$. In [20] Okubo proposed a generalization of Yang-Mills theory to non-associative algebras. Okubo's main idea for constructing a gauge theory based on a non-associative algebra K is to use the group of automorphisms over K as the group of gauge transformations. Let ω now be a one-form with coefficients valued over the non-associative algebra $\mathbb{O}$. To be precise,

$$\omega = \sum_{\mu=1}^{m} \omega_\mu \mathrm{d}x^\mu$$

with $\omega_\mu \in \mathbb{O}$. Analogously, the curvature two-form valued over the non-associative algebra $\mathbb{O}$ is given by

$$R = \mathrm{d}\omega + \omega \wedge \omega.$$

For some $g \in GL(\mathbb{O})$, and some $\xi \in \Lambda_1(\mathbb{O})$, consider the gauge transformation

$$\omega \longrightarrow \omega' = g(\omega) + \xi. \tag{3}$$

In order that R transforms covariantly as

$$R \longrightarrow g(R) = \mathrm{d}\omega' + \omega' \wedge \omega' \tag{4}$$

it is sufficient for $g \in GL(\Lambda_1)$ and $\xi \in \Lambda_1(\mathbb{O})$ to satisfy the following conditions [20]:

$$g(\omega \wedge \omega) = g(\omega) \wedge g(\omega), \tag{5}$$
$$\mathrm{d}\xi + \xi \wedge \xi = 0, \tag{6}$$
$$\mathrm{d}(g(\omega)) = g(\mathrm{d}\omega) - \xi \wedge g(\omega) - g(\omega) \wedge \xi. \tag{7}$$

We notice that these conditions are satisfied for a Lie algebra Yang-Mills theory, however it is not known if they can be satisfied for an octonionic valued gauged field theory. The condition shown in Equation (5) is satisfied whenever g is an automorphism over $\mathbb{O}$, that is $g \in \mathrm{Aut}(\mathbb{O})$ which is the group G_2. In order to deal with Equations (6) and (7) it is convenient to translate them from the Lie group $\mathrm{Aut}(\mathbb{O})$ to the Lie algebra $\mathfrak{Der}(\mathbb{O})$. Given an infinitesimal parameter ϵ and a derivation $D \in \mathfrak{Der}(\mathbb{O})$, we can always write

$$g = \exp(\epsilon D) = \mathrm{id} + \epsilon D + O(\epsilon^2)$$

where id is the identity transformation and

$$\xi = \epsilon \mathrm{d}k + O(\epsilon^2)$$

for some $k \in \mathbb{O}$. Since the boundary operator is such that $dd = 0$, Equation (6) holds automatically and Equation (7) is equivalent to

$$\mathrm{d}(D(\omega)) = D(\mathrm{d}\omega) - \mathrm{d}k \wedge \omega - \omega \wedge \mathrm{d}k. \tag{8}$$

Following Okubo's assumptions, (5), (6), and (7), it suffices to find pairs (D, k) of elements in a subalgebra of $\mathfrak{Der}(\mathbb{O}) \times \mathbb{O}$ that satisfy Equation (8) for all $\omega \in \Lambda_1(\mathbb{O})$ so that the gauge transformation (3) induces the covariant transformation (4) as desired. This is the first step in the construction of the Yang-Mills theory. One should finally provide the action for the theory. Although (5), (6), and (7) are mild assumptions, the non-associative context imposes very restrictive conditions on the gauge theory.

3. The derivation algebra

We shall consider a basis for the octonions where the multiplication of the imaginary elements is represented in the Fano plane shown in Figure 1. The identity element e_0 is left outside because it commutes and associates with any other element. The diagram is read in the following way: the result of the product of any two elements is the only other element laying on the line that passes through the first two elements, and the sign is determined by the arrows (for example, $e_5e_2 = -e_4$).

Recalling Cartan's work [21] on the classification of simple Lie groups, G_2, the smallest exceptional Lie group, is the group of automorphisms over the octonions. Hence $G_2 = \mathrm{Aut}(\mathbb{O})$. The tangent space to a group of automotphisms is an algebra of derivations. Therefore, the Lie algebra $\mathfrak{g}_2$ of the Lie group G_2 is $\mathfrak{Der}(\mathbb{O})$. Given two octonions, a and b, the map

$$\begin{aligned} D_{a,b}(x) &= \big([L_a, L_b] + [L_a, R_b] + [R_a, R_b]\big)(x) \\ &= \frac{1}{2}\big([[a,x],b] + [a,[b,x]] + [[a,b],x]\big) \\ &= [[a,b],x] - 3[a,b,x], \end{aligned}$$

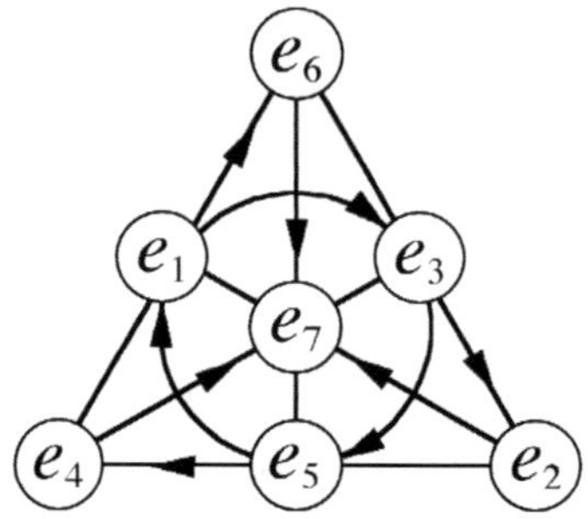

Figure 1. Fano plane representing the multiplication table for the octonions.

where $L_a(x) = ax$, $R_a(x) = xa$, the bracket with two entries is the commutator $[a,b] = ab - ba$, and the bracket with three entries is the associator $[a,b,x] = (ab)x - a(bx)$ acting over an octonion x, is a derivation. They clearly satisfy the Leibniz rule

$$D_{a,b}(xy) = (D_{a,b}(x))y + x(D_{a,b}(y)),$$

the Generalized Jacobi Identity

$$D_{a,b}\left(D_{c,d}\right) = D_{D_{a,b}(c),d} + D_{c,D_{a,b}(d)} + D_{c,d}\left(D_{a,b}\right),$$

and they are anti-symmetric in the sense that $D_{a,b} = -D_{b,a}$. For each index $p \in \{1,\ldots,7\}$, consider the three pairs of indices (i,j), (r,s), and (u,v) such that $e_p = e_ie_j = e_re_s = e_ue_v$, then

$$D_{e_i,e_j} + D_{e_r,e_s} + D_{e_u,e_v} = 0.$$

A basis for $\mathfrak{g}_2$ can be given with 14 of these transformations grouped in seven pairs, each of them constituted by D_{e_i,e_j} and D_{e_r,e_s} where $e_ie_j = e_re_s = e_p$ for a different p in every one.

4. The analysis of the Okubo restrictions

We are interested in solving Equation (8) for the non-associative algebra of the octonions. It suffices to find pairs of elements (D,k) in a subalgebra of $\mathfrak{Der}(\mathbb{O}) \times \mathbb{O}$ that satisfy Equation (8) for all $\omega \in \Lambda_1(\mathbb{O})$. In what follows we will show that Okubo's formulation, for the case of the octonions being the non-associative algebra over which the geometric objects shall be valued, fails to extend the usual Yang-Mills theory formulated over Lie algebras.

Given two one-forms valued over the octonions, ρ and ω, they may be interpreted as one-forms with octonionic coefficients

$$\rho = \rho_1 \mathrm{d}x^1 + \cdots + \rho_n \mathrm{d}x^n = \left(\sum_{i=0}^{7} \rho_1^i e_i\right) \mathrm{d}x^1 + \cdots + \left(\sum_{i=0}^{7} \rho_n^i e_i\right) \mathrm{d}x^n$$

and also as octonions with coefficients that are one-forms

$$\rho = \rho^0 e_0 + \cdots + \rho^7 e_7 = \left(\sum_{i=1}^{n} \rho_i^0 \mathrm{d}x^i\right) e_0 + \cdots + \left(\sum_{i=1}^{n} \rho_i^7 \mathrm{d}x^i\right) e_7$$

Journal of Physics: Conference Series **720** (2016) 012018 doi:10.1088/1742-6596/720/1/012018

simply by interchanging the summation order in the general expression. The product of one-forms valued over the octonions is given by

$$\rho \wedge \omega = \sum_{1\leq i<j\leq n} \big(\rho_i\omega_j - \rho_j\omega_i\big)\mathrm{d}x^i \wedge \mathrm{d}x^j$$

which is a two-form valued over the octonions. Seen as an octonion, its real part is not necessarily equal to zero. Nonetheless, both $\omega \wedge \omega$ and $\rho \wedge \omega + \omega \wedge \rho$ do possess real part equal to zero. To be precise,

$$\omega \wedge \omega = \sum_{1\leq i<j\leq n} [\omega_i, \omega_j]\,\mathrm{d}x^i \wedge \mathrm{d}x^j = \frac{1}{2}\sum_{i,j=1}^{n} [\omega_i, \omega_j]\,\mathrm{d}x^i \wedge \mathrm{d}x^j$$

and

$$\begin{aligned}
\rho \wedge \omega + \omega \wedge \rho &= \sum_{1\leq i<j\leq n} \Big([\rho_i, \omega_j] + [\omega_i, \rho_j]\Big)\mathrm{d}x^i \wedge \mathrm{d}x^j \\
&= \frac{1}{2}\sum_{i,j=1}^{n} \Big([\rho_i, \omega_j] + [\omega_i, \rho_j]\Big)\mathrm{d}x^i \wedge \mathrm{d}x^j \\
&= \sum_{i,j=1}^{n} [\rho_i, \omega_j]\,\mathrm{d}x^i \wedge \mathrm{d}x^j.
\end{aligned}$$

On the other hand, considering $\mathcal{G} \in \mathfrak{g}_2$ and using the fact that $\mathrm{d}^2 = 0$, we are able to calculate

$$\begin{aligned}
\mathrm{d}\big(\mathcal{G}(\omega)\big) - \mathcal{G}(\mathrm{d}\omega) &= \sum_{i=1}^{n}\Big(\mathrm{d}\big(\mathcal{G}(\omega_i)\big) \wedge \mathrm{d}x^i\Big) - \mathcal{G}\left(\sum_{i=1}^{n} \mathrm{d}\omega_i \wedge \mathrm{d}x^i\right) \\
&= \sum_{i=1}^{n}\left(\mathrm{d}\left(\sum_{j=1}^{7} \omega_i^j \mathcal{G}(e_j)\right) \wedge \mathrm{d}x^i - \left(\sum_{j=1}^{7} \mathrm{d}\omega_i^j \mathcal{G}(e_j)\right) \wedge \mathrm{d}x^i\right) \\
&= \sum_{i=1}^{n}\left(\sum_{j=1}^{7} \omega_i^j \mathrm{d}\mathcal{G}(e_j)\right) \wedge \mathrm{d}x_i
\end{aligned}$$

noticing that $\mathcal{G}(e_0) = 0$. Since $\mathcal{G}$ is a derivation over the octonions, it can be written as a linear combination over a basis of $\mathfrak{g}_2$. The transformation $\mathrm{d}\mathcal{G}$ consists on applying the exterior derivative to the coefficients that determine $\mathcal{G}$. For example, if we write $\mathcal{G} = \sum_{i,j=1}^{7} \lambda_{ij} D_{e_i,e_j}$ then

$$\begin{aligned}
\mathrm{d}\mathcal{G} &= \mathrm{d}\left(\sum_{i,j=1}^{7} \lambda_{ij} D_{e_i,e_j}\right) = \sum_{i,j=1}^{7} \mathrm{d}\lambda_{ij} D_{e_i,e_j} = \sum_{i,j=1}^{7}\left(\sum_{k=1}^{n} \partial_k \lambda_{ij} \mathrm{d}x^k\right) D_{e_i,e_j} \\
&= \sum_{k=1}^{n} \partial_k \mathcal{G} \mathrm{d}x^k
\end{aligned}$$

where $\partial_k \mathcal{G} = \sum_{i,j=1}^{7} \partial_k \lambda_{ij} D_{e_i,e_j}$. Therefore,

$$\begin{aligned} \mathrm{d}\big(\mathcal{G}(\omega)\big) - \mathcal{G}(\mathrm{d}\omega) &= \sum_{i,j=1}^{n} \Big(\omega_j^1 \partial_i \mathcal{G}(e_1) + \cdots + \omega_j^7 \partial_i \mathcal{G}(e_7)\Big) \mathrm{d}x^i \wedge \mathrm{d}x^j \\ &= \sum_{i,j=1}^{n} \Big(\big(\partial_i \mathcal{G}\big)(\omega_j)\Big) \mathrm{d}x^i \wedge \mathrm{d}x^j. \end{aligned}$$

Finally, in order to solve the equation

$$\mathrm{d}\big(\mathcal{G}(\omega)\big) - \mathcal{G}(\mathrm{d}\omega) = \rho \wedge \omega + \omega \wedge \rho$$

where $\mathcal{G} \in \mathfrak{g}_2$ and $\rho = -\mathrm{d}k$ is a one-form valued over the octonions, it is necessary that $\mathcal{G}$ and ρ satisfy the following relation, for any $\omega \in \Lambda_1(\mathbb{O})$,

$$\big(\partial_i \mathcal{G}\big)(\omega_j) = [\rho_i, \omega_j] \tag{9}$$

for all indices $i, j \in \{1, \ldots, n\}$ such that $i \neq j$. Nonetheless, Equation (9) only possesses non-trivial solutions, for the case of the octonions, when the expression is considered inside an associative subalgebra. To understand this situation, notice that $\partial_i \mathcal{G}$ is still a vector in $\mathfrak{g}_2$, thus a derivation and must satisfy the Leibniz rule. Therefore, if a and b are two octonions then

$$\partial_i \mathcal{G}(ab) = \big(\partial_i \mathcal{G}(a)\big)b + a\big(\partial_i \mathcal{G}(b)\big) \tag{10}$$

and on the other hand

$$[\rho_i, ab] = [\rho_i, a]b + a[\rho_i, b] + 3[\rho_i, a, b] \tag{11}$$

by straightforward calculation. If it were possible that $\big(\partial_i \mathcal{G}\big)(\,\cdot\,) = [\rho_i, \,\cdot\,]$ then subtracting Equations (10) and (11) yields $[\rho_i, a, b] = 0$. Since the octonion ω_i can always be written as a product of two octonions, Equation (9) does not hold in general.

Then, finding solutions to Equation (9) is equivalent to answer the following question: When is it true that the transformation induced by the commutator bracket represents a transformation in $\mathfrak{g}_2$? The answer to this question is that it happens only when all the octonions into consideration belong to an associative subspace; that is, when the octonions involved belong to a subalgebra isomorphic to the quaternions. Artin's Theorem establishes that any subalgebra of the octonions generated by only two elements is associative; this statement is formulated in a more general context of alternative algebras [22, p. 29]. Returning to the question of wheter the commutator bracket is a derivation, given three octonions, a, b, and c, then

$$[c, ab] = [c, a]b + a[c, b] + 3[c, a, b]$$

showing that $[c, ab]$ is equal to $[c, a]b + a[c, b]$ if and only if $[c, a, b] = 0$. That is, the bracket on the right hand side of (9) is a derivation only for an associative subalgebra of the octonions. This explains why it is impossible to find non-trivial solutions for Equation (8) when the non-associative algebra into consideration is the octonions. In this sense, the only formulation for a Yang-Mills theory on an octonionic algebra, under assumptions (5), (6), and (7), is the usual Yang-Mills theory for the algebras $\mathfrak{su}(2)$ and $\mathfrak{u}(1)$, since both the imaginary part of the quaternions and their algebra of derivations can be identified with $\mathfrak{su}(2)$ and it is in that context where the terms $\mathrm{d}(D(\omega)) - D(\mathrm{d}\omega)$ and $-\mathrm{d}k \wedge \omega - \omega \wedge \mathrm{d}k$ are compatible.

5. Conclusions

It was known since the work of Okubo that the conditions (5), (6), and (7) for a non-associative algebra imply that the curvature two-form transforms in a natural way as an automorphism of the non-associative algebra in analogy with the Lie algebra case (in this case the three conditions are satisfied). We consider the construction of a Yang-Mills theory where the connection one-form is valued on the octonion algebra, in distinction to the construction where the structure group is a Lie group. We proved that the conditions (5), (6), and (7) are not simultaneously satisfied for the octonion algebra. However they are satisfied for the associative subalgebras of the octonions: the quaternions and the complex numbers. They are associated to the $\mathfrak{su}(2)$ and $\mathfrak{u}(1)$ Yang-Mills theories respectively.

Acknowledgments

A.R. is partially supported by Project Fondecyt 1121103.

References

[1] Kugo T and Townsend P 1983 Supersymmetry and the division algebras *Nucl. Phys.* B **221**(2) 357-80
[2] Duff M J 1988 Supermembranes - the 1st 15 weeks *Classical Quantum Gravity* **5** 189-205
[3] Manogue C A and Schray J 1993 Finite Lorentz transformations, automorphisms, and division algebras *J. Math. Phys.* **34**(8) 3746-67
[4] Baez J and Huerta J 2010 Division algebras and supersymmetry I *Superstrings, Geometry, Topology, and C^*-Algebras* Proc. of Symp. in Pure Mathematics vol 81 ed Doran R *et al* (Texas: AMS) pp 65-80
[5] Baez J and Huerta J 2011 Division algebras and supersymmetry II *Adv. Theor. Math. Phys.* **15**(5) 1373-410
[6] Evans J M 1988 Supersymmetric Yang-Mills theories and division algebras *Nucl. Phys.* B **298**(1) 92-108
[7] Evans J M 1995 Auxiliary fields for super Yang-Mills from division algebras *Lecture Notes in Phys.* **447** 218-23
[8] Borsten L, Dahanayake D, Duff M J, Ebrahim H and Rubens W 2009 Black holes, qubits and octonions *Phys. Rep.* **471**(3-4) 113-219
[9] Anastasiou A, Borsten L, Duff M J, Hughes L J and Nagy S 2014 Super Yang-Mills, division algebras and triality *J. High Energy Phys.* 2014:80
[10] Sudbery A 1983 Octonionic description of exceptional Lie superalgebras *J. Math. Phys.* **24**(8) 1986-8
[11] Sudbery A 1984 Division algebras, (pseudo)orthogonal groups and spinors *J. Phys.* A **17**(5) 939-55
[12] Barton C H and Sudbery A 2003 Magic squares and matrix models of Lie algebras *Adv. Math.* **180**(2) 596-647
[13] Manogue C A and Dray T 1999 Octonionic möbius transformations *Modern Phys. Lett.* A **14**(19) 1243-55
[14] Dray T and C. A. Manogue C A 2010 Octonionic cayley spinors and E_6 *Comment. Math. Univ. Carolin.* **51**(2) 193-207
[15] Veiro J P 2014 Octonionic presentation for the Lie group $SL(2, \mathbb{O})$ *J. Algebra Appl.* **13**(6) 1450017
[16] Günaydin M and Nicolai H 1995 Seven-dimensional octonionic Yang-Mills instanton and its extension to an heterotic string soliton *Phys. Lett.* B **351**(1-3) 169-72
[17] Fubini S and Nicolai H 1985 The octonionic instanton *Phys. Lett.* B **155**(5-6) 369-72
[18] Borsten L, Duff M J, Hughes L J and Nagy S 2014 Magic square from Yang-Mills squared *Phys. Rev. Lett.* **112**(13) 131601
[19] Anastasiou A, Borsten L, Duff M J, Hughes L J and Nagy S 2014 An octonionic formulation of the M-theory algebra *J. High Energy Phys.* 2014:22
[20] Okubo S 1995 *Introduction to Octonion and Other Non-Associative Algebras in Physics* (Cambridge University Press)
[21] Cartan E 1914 Les groupes réels simples, finis et continus *Ann. Sci. Éc. Norm. Supér.* **31** 263-355
[22] Schafer R D 1966 *An introduction to nonassociative algebras* (New York: Academic Press)

Arguments towards the construction of a matrix model groundstate

L Boulton[1], M P Garcia del Moral[2] and A Restuccia[2]

E-mail: L.Boulton@hw.ac.uk; maria.garciadelmoral@uantof.cl; alvaro.restuccia@uantof.cl

Abstract. We discuss the existence and uniqueness of wavefunctions for inhomogenoeus boundary value problems associated to x^2y^2-type matrix model on a bounded domain of $\mathbb{R}^2$. Both properties involve a combination of the Cauchy-Kovalewski Theorem and a explicit calculations.

1. Introduction

In this note we sketch a technique for determining the existence and uniqueness of groundstates of supersymmetric matrix models with global symmetry, subject to Dirichlet boundary conditions. We consider one of the simplest possible benchmark models available. This allows us to illustrate the actual procedure on a concrete setting. The hamiltonian, given by (1), is the canonical toy model which is often used to test spectral properties of the $D = 11$ supermembrane. In the following we call it the x^2y^2-model.

We believe the approach we present below is new in the context of matrix models and its scope of applicability includes a wide range of physical settings. Wavefunctions of matrix models have been investigated by means of different methods in the past. They provide a better understanding of M-theory, either from the point of view of Supermembrane Theory [1, 2, 3] or from the point of view of other matrix models [4]. They also play a role in Supersymmetric Yang-Mills theories in the slow-mode regime [5, 6, 7, 8].

2. The x^2y^2-model

The x^2y^2-model was first introduced in [9]. In [10] it was shown that zero is not an eigenvalue. Numerical tests carried out in [11] suggest the existence of more general Weyl-type wavefunctions. See also [12]. We now show analytically that the inhomogeneous boundary value problem associated to a zero eigenvalue for Dirchlet boundary conditions on a compact domain Ω has a solution which is unique. See (3)-(2). Without loss of generality we assume that Ω is a two-dimensional ball of finite radius $R > 0$.

The supersymmetric hamiltonian to be considered is

$$H = p_x^2 + p_y^2 + x^2y^2 + x\sigma_3 + y\sigma_1 \tag{1}$$

where σ_i are the Pauli matrices. The supersymmetric charges in this case are given by the expression

$$Q = Q^\dagger = \begin{pmatrix} -xy & i\partial_x - \partial_y \\ i\partial_x - \partial_y & xy \end{pmatrix}.$$

The wavefunctions are $\Psi = \begin{pmatrix} \Psi_1 \\ \Psi_2 \end{pmatrix} \in H^2(\Omega) \cap H_0^1(\Omega)$. Note they are subject to zero boundary conditions,

$$\Psi = 0 \quad \text{on} \quad \partial\Omega. \tag{2}$$

3. Existence and uniqueness of wavefunctions at zero

The resolvent of H is compact, so the inhomogeneous system

$$H\Psi = \Phi \tag{3}$$

for regular enough right hand side Φ will have a unique solution if and only if zero is not an eigenvalue of H. We show that this is indeed the case.

Assume that

$$H\Psi = 0. \tag{4}$$

Then

$$Q\Psi = Q^\dagger\Psi = 0 \quad \text{in} \quad \Omega. \tag{5}$$

According to the regularity properties of the elliptic operators, H in this case, this condition holds true pointwise up to the boundary of Ω.

Let $\mathbf{x} \in \partial\Omega$ and denote by $(\mathbf{n}_1, \mathbf{n}_2)$ the components of the normal to $\partial\Omega$ at $\mathbf{x}$. The tangent to $\partial\Omega$ at $\mathbf{x}$ is then $(\mathbf{n}_2, -\mathbf{n}_1)$ and we must have

$$(\mathbf{n}_2\partial_x - \mathbf{n}_1\partial_y)\Psi(\mathbf{x}) = 0.$$

The solution of problem (4) is regular, so we can extend it continuously up to the boundary. Then (5) yields

$$(i\partial_x + \partial_y)\Psi_2(\mathbf{x}) = (i\partial_x - \partial_y)\Psi_1(\mathbf{x}) = 0$$

pointwise. Since $(\mathbf{n}_1, \mathbf{n}_2) \neq 0$, if $\mathbf{n}_2 \neq 0$,

$$\left(1 + i\frac{\mathbf{n}_1}{\mathbf{n}_2}\right)\partial_y\Psi_2(\mathbf{x}) = 0 \quad \text{implies} \quad \partial_y\Psi_2(\mathbf{x}) = 0 \text{ and } \partial_x\Psi_2(\mathbf{x}) = 0,$$

and

$$\left(-1 + i\frac{\mathbf{n}_1}{\mathbf{n}_2}\right)\partial_y\Psi_1(\mathbf{x}) = 0 \quad \text{implies} \quad \partial_y\Psi_1(\mathbf{x}) = 0 \text{ and } \partial_x\Psi_1(\mathbf{x}) = 0.$$

A similar conclusion is obtained for $\mathbf{n}_1 \neq 0$. Hence Ψ and $\partial_{\mathbf{n}}\Psi$ must vanish on the boundary $\partial\Omega$.

Now, by virtue of the Cauchy-Kovalewski Theorem, cf. [13], and from the fact that the potential is analytic, we conclude that $\Psi = 0$ pointwise on the whole of $\overline{\Omega}$. That is, the only solution to (4) is the zero wavefunction.

In a paper soon to be completed [14], we will examine the existence and uniqueness of the solution of boundary value problems associated to the $SU(N)$ regularized model of the $D = 11$ supermembrane on a bounded domain. This theory was introduced in [1] for an unbounded domain. We expect to being able to examine the existence of massless groundstates in supersymmetric matrix models with global and gauge symmetries.

4. Discussion

The x^2y^2-model [9] is a well-known benchmark for testing spectral properties of the $D = 11$ supermembrane. The existence of a groundstate was analyzed in [10] by considering the whole space $\mathbb{R}^2$. In the latter case the model does not admit a solution [10]. We have considered this system confined to a ball of finite radius. In this case zero is also not an eigenvalue for the space of wavefunctions which are zero on the boundary. Additionally we show that the inhomogeneous boundary value problem associated to zero eigenvalue has always a solution which is unique. We consider a viewpoint to determine the existence of groundstate wavefunctions which we believe is new in the context of matrix models. The conditions of supersymmetry and ellipticity of the hamiltonian operator are essential to this analysis. The results announced in this note can be further generalized to more interesting settings. A rigorous general framework in this respect is currently being consider [14].

References

[1] B. de Wit, J. Hoppe, H. Nicolai, Nucl. Phys. **B305** (1988) 545.
[2] Jens Hoppe, *On The Construction of Zero Energy States in Supersymmetric Matrix Models III* arXiv:hep-th/9711033.
[3] J. Froehlich, J. Hoppe, *On Zero-Mass Ground States in Super-Membrane Matrix Models* arXiv:hep-th/9701119.
[4] T. Banks, W. Fischler, S.H. Shenker and L. Susskind, Phys.Rev.**D55** (1997) 5112-5128.
[5] M. Claudson, M. B. Halpern, Nucl.Phys. B **250** (1985) no. 4, 689.
[6] P. Yi, *Nucl. Phys. B* (1997) **505**, 307.
[7] M. Porrati, A. Rozenberg, *Nucl. Phys. B* (1998) **515**, 184-202.
[8] S. Sethi, M. Stern, *Commun. Math. Phys.* (1998) **194**, 675.
[9] B. de Wit, M. Luscher, H. Nicolai, Nucl.Phys. **B320** (1989) 135
[10] G.M. Graf, D. Hasler, J. Hoppe, *No zero energy states for the supersymmetric x^2y^2 potential*, arXiv:math-ph/0109032.
[11] L. Boulton, M.P. Garcia del Moral, Alvaro Restuccia, *Nucl.Phys. B* (2012) **856**, 716-747.
[12] P. Korcyl, *Phys.Rev.D* (2006) **74**,115012.
[13] G. Folland, Introduction to Partial Differential Equations. Second Edition. Princeton University Press, Princeton, 1995.
[14] L. Boulton, M.P. Garcia del Moral, A. Restuccia, In preparation.

Setting the Renormalization Scale in QFT

Carlos Contreras[†], **Benjamin Koch**[*] **and Paola Rioseco**[*]

[†] Departamento de Física, Universidad Técnica Federico Santa María
Casilla 110-V, Valparaíso, Chile
[*]Instituto de Física,Pontificia Universidad Católica de Chile
Av. Vicuña Mackenna 4860, Santiago, Chile

E-mail: carlos.contreras@usm.cl,bkoch@fis.puc.cl

Abstract. In this work a setting for the "scale-field" is proposed at the level of effective action, which is consistent with the conservation of the stress-energy tensor. The mechanism and its potential is exemplified for scalar ϕ^4 theory and for Einstein-Hilbert-Maxwell theory.

1. Introduction

These proceedings on "Setting the Renormalization Scale in QFT" are based on a more complete article on the same topic [1], where also complementary discussions can be found. The effective action approach [2] can be seen as an elegant way of defining a generating functional for one-particle-irreducible Green's functions. Following Wilson's idea [3] one can study the effect of integrated quantum degrees of freedom at different scales k. The scale dependent effective action Γ_k is to be understood as interpolation between the ultraviolet (UV) bare action Γ_∞ and the fully integrated action in the IR Γ_0. Γ_k contains scale dependent couplings g_k^a which are obtained from a suitable flow equation $k\partial_k\Gamma_k$. The space of solutions g_k^a is called the "coupling flow" [4]. A specific trajectory is selected out of this flow by imposing conditions for the couplings at an initial scale k_0. The evaluation of the effective action Γ_k is typically hampered by various technical difficulties such as singularities, anomalies, and non-localities. However, in many cases those difficulties can be overcome by the "regularization - renormalization" technique, where infinities are absorbed in the initial conditions at a scale k_0.

Minimizing a given effective action with respect to variations of its (average) field content ϕ_a gives the equations of motion of the effective action

$$\frac{\delta\Gamma_k}{\delta\phi_a} = 0 \quad . \tag{1}$$

Those equations are typically non-linear and sometimes non-local differential equations and are frequently referred to as "gap equations" [5]. Therefore, finding solutions for the "gap equations" is highly relevant for defining a self-consistent background in quantum field theory.

However, even if it is technically possible to solve the "gap equations", the physical interpretation of such a solution is still biased by the way the scale k is related to the quantities x_i, $Q_i, \ldots$ (for example positions and charges) that are used to describe the physical system. Choosing a relation $k = k(x_i, Q_i, \ldots)$ is called "scale setting". The main focus of this article is on the role of scale setting in the quest of finding self-consistent solutions of (1). The method

Journal of Physics: Conference Series **720** (2016) 012020 doi:10.1088/1742-6596/720/1/012020

of improving solutions has been successfully applied in many different contexts (see refs.[6] to [22]).

2. Scale-field setting at the level of effective action

In this section a general scale setting at the level of effective action will be proposed. Lets assume that within the quantum field theoretical model it was possible, to evaluate the corresponding coupling flow and to select a particular trajectory due to the choice of initial conditions ($g_i(k_0) = g_{i,0}$). Thus, one can start with the effective quantum action [2]

$$\Gamma_k(\phi_a(x), \partial\phi_a(x), g_k^a) = \int d^4x\sqrt{-g}\mathcal{L}(\phi_a(x), \partial\phi_a(x), g_k^a) \quad , \tag{2}$$

where ϕ_a are actually the expectation values of the quantum fields and g_k^a are the scale dependent couplings, including the coupling multiplying the kinetic term.

Note that doing this, one frequently has to truncate higher order and nonlocal couplings [23, 24, 25] from the model, that might appear due to the quantum integration procedure. Then, it will be assumed that all relevant couplings are taken into account and a local expansion are consider.

Now, one can derive the equations of motion for the average quantum fields ϕ_a from

$$\frac{\delta\Gamma_k}{\delta\phi_a} = 0 \quad . \tag{3}$$

As mentioned in the introduction, the solutions $\bar{\phi}_a(x,k)$ of those "gap equations" will also be functions of the arbitrary scale k. From a physical point of view this is however not yet satisfactory since no possible observable can be a function of an a priory arbitrary scale. In order to obtain a physical quantity one has to define some kind of scale setting procedure, that establishes a relation between the physical quantities (charges Q_i and positions x_j) of a given problem and the scale k. When doing this one can borrow an idea from the calculation of observables $\langle T\phi(x_i)\ldots\rangle_k$ in standard quantum field theory. Also there, the observables turn out to be scale "k" dependent quantities[1]. Subsequently, the scale setting for those observables in terms of initial conditions and kinematical variables $k = k(x_i;\, Q_i\ldots)$ is chosen such that any k dependence of the time ordered correlation function for observable is minimized

$$\left.\frac{d}{dk}\langle T\phi_1(x_1)\phi_2(x_2)\ldots\rangle_k\right|_{k=k_{opt}} \equiv 0 \quad . \tag{4}$$

This is the key philosophy that is used when deriving the "Callan-Symanzik" equations [26, 27], the "principal of minimal sensitivity [28], or the "principle of maximal conformality" [29, 30].

It is proposed to implement an analogous philosophy at the level of the effective action Γ_k. This means that one should choose an optimal scale setting prescription for which a variation of k has a minimal impact on the self-consistent background $\bar{\phi}_i$. This principle can be implemented by promoting the a priory arbitrary scale to a physical scale-field in the effective quantum action

$$\Gamma_k(\phi_a(x), \partial\phi_a(x), g_k^a) \to \Gamma(\phi_a(x), \partial\phi_a(x), k(x), g_k^a) \quad . \tag{5}$$

This leads to the coupled equations of motion

$$\frac{\delta\Gamma}{\delta\phi_a} = 0 \quad , \quad \left.\frac{d}{dk}\mathcal{L}(\phi_a(x), \partial\phi_a(x), k(x), g_k^a)\right|_{k=k_{opt}} = 0 \quad . \tag{6}$$

[1] It is, argued that this k dependence is an artifact of the truncation in the loop expansion

Clearly it is not guaranteed that a solution for (6) can be found, but such a prescription is not limited to be a variation of a classical solution or to a saddle point approximation. The procedure (6) has already been applied for some particular gravitational actions [31, 32, 33, 34] but in this work it is discussed in a broader context. A nice feature of such a procedure is that any solution of the equations (6) is automatically independent of k, which is actually the fundamental precondition for a physical observable in the language of the renormalization group approach.

Promoting the scale k to a scale-field $k(x)$ raises the question whether this new field only appears in the couplings g_k^a, or whether it has to be equipped with other additional couplings, for instance a proper kinetic term. A standard procedure when introducing new fields into a Lagrangian is to incorporate actually all couplings that are in agreement with the symmetry of the Lagrangian. This abundant freedom is then restricted by imposing some other additional conditions such as renormalizability, simplicity, and/or agreement with experimental constraints. However, in the presented approach the philosophy is different. The scale-field k is understood to have its origin in the process of renormalization and throughout this process, no such extra couplings are taken into account. In particular, the beta functions of the couplings g_k^a are calculated without any additional couplings. Therefore, the presented version of scale-field setting is chosen in a sense "minimal", since it contemplates the appearance of $k(x)$ only as dictated by the running couplings g_k^a.

Finally, there is one other consistency condition one would like to impose, the conservation of the stress-energy tensor, even at the quantum-improved level:

$$\nabla^\mu T_{\mu\nu} \equiv 0 \quad . \tag{7}$$

Therefore, the idea will be studied for some examples, where self-consistency of the approach can be shown explicitly.

3. Scale-field setting for scalar ϕ^4 theory

As most simple example without any further complications due to gauge symmetry lets study the scale-field setting procedure for scalar ϕ^4 theory. There are various ways of writing the effective action for ϕ^4 theory. One of them is in terms of a scale dependent wave function renormalization Z_k, running mass m_k, and running quartic coupling g_k. The other way of writing this action is terms of separate couplings for every term appearing in the Lagrangian, which are a coupling for the kinetic term α_k, a coupling for the ϕ^2 term $\tilde{m}_k^2$, and a coupling for the quartic term $\tilde{g}_k$. As long as the scale k is assumed to be fixed, the formalism for both is exactly equivalent. However, in the context of scale-field setting $k \to k(x)$, derivatives do not necessarily commute with $k(x)$ and both formulations could to be treated differently. This subtlety will be exemplified in the following subsection, before applying the scale setting to ϕ^4 theory at the one loop level.

3.1. Consistency in scalar ϕ^4 theory

Let us consider the effective action

$$\Gamma_k = \int d^4x \left(\frac{\alpha_k}{2}(\partial\phi)^2 - \frac{\tilde{m}_k^2}{2}\phi^2 - \frac{\tilde{g}_k}{4!}\phi^4 \right) \tag{8}$$

with two fields ϕ and k. The couplings α_k, $\tilde{m}_k^2$, and $\tilde{g}_k$ are functions of the field k. This implies an equation of motion for $\delta\phi$:

$$\partial_\mu(\alpha_k \partial^\mu \phi) + \tilde{m}_k^2 \phi + \frac{\tilde{g}_k}{6}\phi^3 = 0 \quad , \tag{9}$$

an other equation of motion for k:

$$\alpha_k'(\partial\phi)^2 - (\tilde{m}_k^2)'\phi^2 - \frac{1}{12}\tilde{g}_k'\phi^4 = 0 \quad , \tag{10}$$

where $\alpha' = \partial_k \alpha = (\partial_x \alpha)dx/dk$, and we will consider the conserved energy momentum tensor, which is obtained as variation with respect to the metric tensor

$$T_{\mu\nu} = \alpha_k(\partial_\mu\phi)(\partial_\nu\phi) - g_{\mu\nu}\left(\frac{\alpha_k}{2}(\partial\phi)^2 - \frac{\tilde{m}_k^2}{2}\phi^2 - \frac{\tilde{g}_k}{4!}\phi^4\right) \quad . \tag{11}$$

and the corresponding conservation law reads

$$0 = \partial^\mu T_{\mu\nu} = -\frac{1}{2}\left(\alpha_k'(\partial\phi)^2 - (\tilde{m}_k^2)'\phi^2 - \frac{1}{12}\tilde{g}_k'\phi^4\right)\cdot\partial_\nu k \quad , \tag{12}$$

and the equation (12) is identical to the equation of motion for k (10), which shows that the approach at the level of effective action (6) is on the one hand implementing the idea of minimal scale dependence and on the other hand maintaining the validity of improved equations of motion and the fundamental conservation law.

Instead of writing a coupling for the kinetic term one frequently works with wave function renormalization where the bare field is $\phi_B = \sqrt{Z_k}\phi$. In this case one could simply identify $\alpha_k = Z_k$, $\tilde{m}_k^2 = Z_k m_k^2$, and $\tilde{g}_k = g_k Z_k^2$ and observe that the corresponding effective action is completely equivalent to (8). However, if one allows for field valued scales $k = k(x)$, this identification is not the only possibility, since the derivatives of the kinetic term, acting on the scale field might contribute to the action. Still, even in this case it can be shown that the scale setting procedure is consistent with the conservation law, just like in (12).

In order to demonstrate the functionality of the procedure, an approximated self-consistent scale setting for spherically symmetric backgrounds in ϕ^4 theory is calculated at the one loop level. The loop expansion of ϕ^4 theory has been calculated up to high order in perturbation theory [35], and the one-loop beta functions γ_Z, β_g and β_{m^2} can be fund in [36]. Integrating those flow equations with initial conditions at $k = k_0$, for $Z_{k_0} \equiv 1$, $g_{k_0} = g_0$ and $m_{k_0}^2 = m_0^2$, one can find the flow trajectory g_k and m_k^2.

Considering the equation of motion for ϕ, the scale setting equation (10), the flow trajectories, and the specific case of static spherical symmetry the only allowed coordinate dependence of k is with respect to the radial distance $k = k(r)$. In this case one can find a scale-field setting which actually reproduces the standard $k \sim 1/r$ behavior for very small radii, but it has more complex behavior.

4. Scale-field setting for Einstein Hilbert Maxwell action

In order to show the consistency of the proposed scale setting, with conservation laws in a less-trivial example, one can study the approach for gravity coupled to a $U(1)$ gauge field and to a cosmological constant. Gravity is exemplary for a non-trivial field theory that is notoriously perturbatively not renormalizable and the situation becomes even less favorable when it is coupled to matter. Using the non-perturbative methods: ERG and the Asymptotic Safety approach one can calculate effective actions and scale dependent couplings for this (see e.g. refs. [37] to [53]). Therefore, it is reasonable to investigate the proposed scale setting procedure in the context of a gravitational action coupled to matter. Let us consider to the discussion the Einstein-Hilbert-Maxwell action:

$$\Gamma_k[g_{\mu\nu}, A_\alpha] = \int_M d^4x\sqrt{-g}\left(\frac{R - 2\Lambda_k}{16\pi G_k} - \frac{1}{4e_k^2}F_{\mu\nu}F^{\mu\nu}\right) \quad , \tag{13}$$

Journal of Physics: Conference Series **720** (2016) 012020 doi:10.1088/1742-6596/720/1/012020

where R is the Ricci scalar and $F_{\mu\nu} = D_\mu A_\nu - D_\nu A_\mu$ is the electromagnetic field strength tensor and the scale dependent couplings are: the Newtons coupling G_k, the cosmological coupling Λ_k, and the electromagnetic coupling e_k. The flow of those couplings has been derived non-perturbatively in [51]. Let us start our study considering the equations of motion for the metric field in (13) which are

$$G_{\mu\nu} = -g_{\mu\nu}\Lambda_k - \Delta t_{\mu\nu} + 8\frac{\pi G_k}{e_k^2}T_{\mu\nu} \quad , \tag{14}$$

where $\Delta t_{\mu\nu} = G_k\left(g_{\mu\nu} - \nabla_\mu\nabla_\nu\right)\frac{1}{G_k}$ is an additional contribution to the stress-energy tensor [42] coming from the $k-$dependence of G_k and $T_{\mu\nu} = F_\nu{}^\alpha F_{\mu\alpha} - \frac{1}{4}g_{\mu\nu}F_{\mu\nu}F^{\mu\nu}$. The equations of motion (Maxwell equations) for this $U(1)$ gauge field are

$$D_\mu\left(\frac{1}{e_k^2}F^{\mu\nu}\right) = 0 \quad , \tag{15}$$

and finally the equations of motion for the scale-field k are

$$\left[R\nabla_\mu\left(\frac{1}{G_k}\right) - 2\nabla_\mu\left(\frac{\Lambda_k}{G_k}\right) - \nabla_\mu\left(\frac{4\pi}{e_k^2}\right)F_{\alpha\beta}F^{\alpha\beta}\right]\cdot(\partial^\mu k) = 0 \quad . \tag{16}$$

The above equations of motion are complemented by the relations corresponding to gauge invariance of the system: For the diffeomorphism invariance one has $\nabla^\mu G_{\mu\nu} = 0$ and for the internal $U(1)$ symmetry $\nabla_{[\mu}F_{\alpha\beta]} = 0$.

The new ingredient due to the scale-field is the equation (16), therefore it is important to check whether this equation is actually non-trivial and consistent with the equations (14 and 15). The consistency can be shown by explicitly deriving (16) from (14) and (15) and further imposing the gauge symmetries conditions. This confirms that, the approach at the level of effective action (6) is on the one hand an elegant way of minimizing scale setting ambiguities and on the other hand maintaining the validity of improved equations of motion and the fundamental conservation laws of the effective action.

4.1. Scale-field setting for Einstein Hilbert Maxwell action in UV - Asymptotic Safety

Lets exemplify our proposed method with the Einstein Hilbert Maxwell action (13) at the fixed points. In the deep UV limit ($k \to \infty$) there is strong evidence [38, 39, 41] for the existence of a non-Gaussian fixed point for the two gravitational couplings

$$G_k \approx \frac{g^*}{k^2} \quad , \quad \Lambda_k \approx \lambda^* k^2 \quad , \tag{17}$$

and there exists further evidence for one UV fixed point for the electromagnetic couplings [51]

$$\lim_{k\to\infty}\frac{1}{e_{k,2}^2} \approx \frac{1}{e_2^{*2}} \quad . \tag{18}$$

Since this fixed point in the electromagnetic coupling is not an attractor it is only approached by particular trajectories in the corresponding flow. Other trajectories either run into a Landau pole type of divergence at finite k, or they run to vanishing values of $e_{k,1}$ at infinite k [51]

$$\lim_{k\to\infty}\frac{1}{e_{k,1}^2} \approx \frac{1}{e_1^{*2}}\cdot(k^2)^B \quad , \tag{19}$$

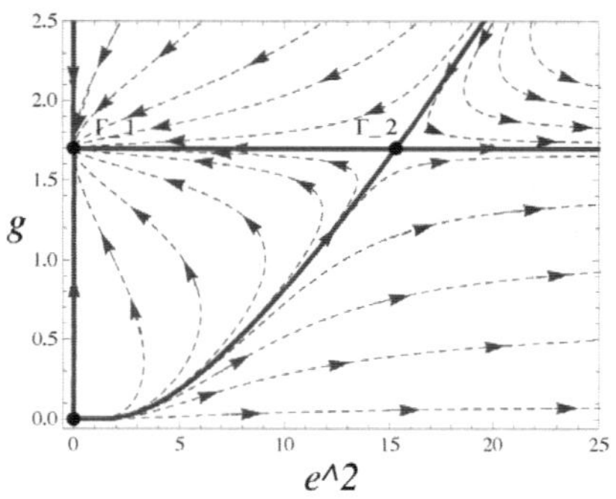

Figure 1. Coupling flow projected to the dimensionless coupling g, e^2 where $B \equiv 1$ was selected. One observes three fixed points: the Gaussian fixed point at vanishing couplings, one fixed point with finite g^* but vanishing e^2 (labeled Γ_1), which corresponds to the limit (19), and an other with finite g^* and e^2 (labeled Γ_2), which corresponds to the limit (18). The arrows point in direction of increasing k.

where the value of B depends on the method of calculation and ranges from 0.8 to 1.6. The behavior of this ER flow using the functions of [51] is shown in figure 1.

In order to integrate out the scale-field from the effective action (13) one has to solve the corresponding equation of motion (16) for k^2. In the UV limit (17) one finds for the fixed point in (18)

$$k_2^2|_{UV} = \frac{R}{4\lambda^*} \tag{20}$$

and for the asymptotic behavior (19) one finds in the same limit

$$k_1^2|_{UV} = \frac{R - \frac{4\pi g^*}{e_1^{*2}} F^2}{4\lambda^*} \quad . \tag{21}$$

Those field-scale settings relate the UV scale k^2 proportional to the curvature scalar R, in agreement with the literature [54, 55, 56, 57, 58], but they go beyond and determine the constant proportionality factor and modifications due to the electromagnetic field strength.

After the "integrating out" is applied, the effective actions valid in the deep UV and at the fixed point in (18) and with the corresponding scale setting (20) is described by:

$$\tilde{\Gamma}_{UV,2} = \int d^4x\sqrt{-g}\left[\frac{R^2}{128\,\pi g^* \lambda^*} - \frac{F^2}{4e_2^{*2}}\right] \quad , \tag{22}$$

and for the behavior (19) and thescale setting (21) the UV effective action results to be

$$\tilde{\Gamma}_{UV,1} = \int d^4x\sqrt{-g}\left[\frac{\left(R - \frac{4\pi g^*}{e_1^{*2}} F^2\right)^2}{128\,\pi g^* \lambda^*}\right] \quad . \tag{23}$$

For the case $F^2 \sim 0$, the R^2 dependence of the UV effective action in Asymptotic Safe gravity is indeed renormalizable [59], in agreement with the literature [55, 56, 58]. Those UV results confirm that the fixed points (18) and (19) correspond to different physical systems with different effective equations of motion for the background.

5. Summary and conclusion

In this proceeding we show a procedure for the scale setting, by promoting the scale k to a scale-field $k(x)$ at the level of effective action Γ_k. We have showed that one might define a scale setting that keeps consistency at the level of improved equations of motion and that the conservation of the stress-energy tensor can be guaranteed throughout the improving solutions procedure.

In order to show the functionality of the procedure, we discussed initially the approach for scalar ϕ^4 theory with spherically symmetric backgrounds at the one loop level.

Finally, this scale-field method is used to study the scale setting prescription in gravity coupled to an electromagnetic stress-energy tensor, represented by Einstein-Hilbert-Maxwell theory. It is explicitly shown that also in this example the conservation of a generalized stress-energy tensor is guaranteed by the scale-field setting. As application, the UV scale-field setting of Asymptotically Safe gravity coupled to an electromagnetic field strength is calculated and the scale independent effective action valid in the UV of this theory is derived by integrating the scale-field out.

Acknowledgements

The work of B.K. was supported by Fondecyt 1120360 and anillo Atlas Andino 10201. C.C was supported by Fondecyt 1120360 and DGIP-USM grant 11.15.41. The work of P.R. was supported by Beca de Postgrado Conicyt.

References

[1] Koch B, Rioseco P and Contreras C 2015 *Phys. Rev.* D **91** 025009
[2] Coleman S R and Weinberg E 1973 *Phys. Rev.* D **7** 1888
[3] Wilson K G 1975 *Rev. Mod. Phys.* **47** 773
[4] Gell-Mann M and Low F 1954 *Phys. Rev.* **95** 1300
[5] Avan J and de Vega H 1984 *Phys. Rev.* D **29** 2891
[6] Bonanno A and Reuter M 1999 *Phys. Rev.* D **60** 084011, 2000 *Phys. Rev.* D **62** 04300, 2006 *Phys. Rev.* D **73** 083005
[7] Ward B 2006 *Acta Phys. Polon.* B **37**, 1967
[8] Falls K, Litim D and Raghuraman A 2012 *Int. J. Mod. Phys.* A **27** 1250019
[9] Basu S and Mattingly D 2010 *Phys. Rev.* D **82** 124017
[10] Reuter M and Tuiran E 2011 *Phys. Rev.* D **83** 044041
[11] Cai Y and Easson D 2010 *JCAP* **1009** 002
[12] Frolov A and Wardlaw D 2012 *European Physical Journal* B **85** 348
[13] Falls K and Litim D 2014 *Phys. Rev.* D **89** 084002
[14] Hewett J and Rizzo T 2007 *JHEP* **0712** 009
[15] Litim D and Plehn T 2008 *Phys. Rev. Lett.* **100** 131301
[16] Koch B 2008 *Phys. Lett.* B **663** 334
[17] Burschil T and Koch B 2010 *Zh. Eksp. Teor. Fiz.* **92** 219
[18] Contillo A, Hindmarsh M and Rahmede C 2012 *Phys. Rev.* D **85** 043501
[19] Cai Y and Easson D 2011 *Phys. Rev.* D **84** 103502
[20] Hindmarsh M, Litim D and Rahmede C 2011 *JCAP* **1107** 019
[21] Koch B and Saueressig F 2014 *Class. Quant. Grav.* **31** 015006
[22] Litim D and Nikolakopoulos K 2014 *JHEP* **1404** 021
[23] Reuter M and Saueressig F 2002 *Phys. Rev.* D **66** 125001
[24] Machado P and Saueressig F 2008 *Phys. Rev.* D **77** 124045
[25] Codello A 2010 *Annals Phys.* **325** 1727
[26] Callan C 1970 *Phys. Rev.* D **2** 1541
[27] Symanzik K 1970 *Commun. Math. Phys.* **18** 227
[28] Stevenson P 1981 *Phys. Rev.* D **23** 2916
[29] Brodsky S, Lepage G and Mackenzie P 1983 *Phys. Rev.* D **28** 228
[30] Brodsky S and Di Giustino L 2012 *Phys. Rev.* D **86** 085026
[31] Koch B and Ramirez I 2011 *Class. Quant. Grav.* **28** 055008
[32] Domazet S and Stefancic H 2012 *Class. Quant. Grav.* **29** 235005
[33] Contreras C, Koch B and Rioseco P 2013 *Class. Quant. Grav.* **30** 175009
[34] Koch B, Contreras C, Rioseco P and Saueressig F 2013 *Preprint* hep-th/1311.1121
[35] Kleinert H, Neu J, Schulte-Frohlinde V, Chetyrkin K and Larin S 1991 *Phys. Lett.* B **272** 39
[36] Peskin M and Schroeder D 1995 *An Introduction to Quantum Field Theory* (Perseus Books)
[37] Dou D and Percacci R 1998 *Class. Quant. Grav.* **15** 3449
[38] Souma W 1999 *Prog. Theor. Phys.* **102** 181
[39] Reuter M and Saueressig F 2002 *Phys. Rev.* D **65** 065016

[40] Lauscher O and Reuter M 2002 *Class. Quant. Grav.* **19** 483
[41] Litim D 2001 *Phys. Rev.* D **64** 105007
[42] Reuter M and Weyer H 2004 *Phys. Rev.* D **69** 104022, 2004 *JCAP* **0412** 001
[43] Codello A, Percacci R and Rahmede C 2008 *Int. J. Mod. Phys.* A **23** 143
[44] Benedetti D, Machado P and Saueressig F 2010 *Nucl. Phys.* B **824** 168
[45] Niedermaier M 2009 *Phys. Rev. Lett.* **103** 101303
[46] Benedetti D and Caravelli F 2012 *JHEP* **1206** 017
[47] Dietz J and Morris T 2013 *JHEP* **1301** 108
[48] Nagy S, Krizsan J and Sailer K 2012 *JHEP* **1207** 102
[49] Christiansen N, Litim D, Pawlowski J and Rodigast A 2012 *Prepint* hep-th/1209.4038
[50] Demmel D, Saueressig F and Zanusso O 2012 *JHEP* **1211** 131
[51] Harst U and Reuter M 2011 *JHEP* **1105** 119
[52] Eichhorn A, Gies H and Scherer M 2009 *Phys. Rev.* D **80** 104003
Groh K and Saueressig F 2010 *J. Phys.* A **43** 365403
Eichhorn A and Gies H 2010 *Phys. Rev.* D **81** 104010
Eichhorn A 2013 *Phys. Rev.* D**87** 124016
[53] Christiansen N, Knorr B, Pawlowski J and Rodigast A 2014 *Prepint* hep-th/1403.1232
[54] Frolov A and Guo J 2011 *Prepint* astro-ph/1101.4995
[55] Bonanno A 2012 *Phys. Rev.* D **85** 081503
[56] Hindmarsh M and Saltas I 2012 *Phys. Rev.* D **86** 064029
[57] Copeland E, Rahmede C and Saltas I 2013 *Prepint* gr-qc/1311.0881
[58] Koch B and Saueressig F 2014 *Int. J. Mod. Phys.* A **29** 1430011
[59] Stelle K 1977 *Phys. Rev.* D **16** 953

Journal of Physics: Conference Series **720** (2016) 012021 doi:10.1088/1742-6596/720/1/012021

Structure and cooling of neutron stars: nuclear pairing and superfluid effects

F Isaule[1] **and H F Arellano**[2]
Departamento de Física-FCFM, Universidad de Chile, Avenida Blanco Encalada 2008, Santiago, Chile
E-mail: [1]`fisaule@ing.uchile.cl`, [2]`arellano@dfi.uchile.cl`

Abstract. We investigate structure and cooling properties of neutron stars making use of microscopic descriptions of nuclear matter based on realistic nucleon-nucleon interactions. The equation of state is obtained from solutions of the Brueckner-Hartree-Fock equation for interacting nucleons at zero temperature. The structure aspects are described by the Tolman-Oppenheimer-Volkoff equation for hydrostatic equilibrium of non-rotating stars. Furthermore we have also solved Bardeen-Cooper-Schrieffer (BCS) pairing gap equations to identify superfluid states in neutronic matter. We obtain mass-radius behavior and cooling curves consistent with current understanding.

1. Introduction

Neutron stars are extremely dense and compact objects with mean mass density ranging between $2\rho_0$ and $3\rho_0$, where $\rho_0 = 2.8 \times 10^{14}\,\mathrm{g\,cm^{-3}}$ is the nuclear saturation density. Their masses are of the order of the solar mass while their radii are about 10 km. The inner layers of neutron star are governed by the equation of state (EoS) of dense matter, which is poorly known. For this reason neutron stars become a natural laboratory for nuclear physics. Observations of neutron star properties, like radius and surface temperature, provide valuable information to constrain the equation of state of nuclear matter [1].

New observations in the last few years have attracted much attention in the field. Observation of high-mass neutron stars may have ruled out soft equations of state for dense matter [2, 3]. Furthermore, the cooling of the neutron star in Cassiopeia A provides strong evidence for superfluidity in the star core [4], particularly in the $^3\mathrm{PF}_2$ channel [5].

Here we report results for structure and cooling properties based on microscopic descriptions of nuclear matter. The equation of state of nuclear matter is based on solutions of the Brueckner-Hartree-Fock (BHF) equations for interacting nucleons in symmetric and neutronic matter at zero temperature [6], a reasonable starting point for ab-initio calculations. We use the Argonne v_{18} bare nucleon-nucleon (NN) interaction [7]. We calculate a equation of state for neutron star matter and obtain a mass-radius relation for non-rotating stars solving the Tolman-Oppenheimer-Volkoff (TOV) equations [8]. Furthermore, cooling curves for neutron stars using these results are calculated, including superfluid effects [9]. Pairing gaps in the $^1S_0$ and $^3PF_2$ channels are obtained solving the BCS gap equation [10].

Journal of Physics: Conference Series **720** (2016) 012021 doi:10.1088/1742-6596/720/1/012021

2. Nuclear equation of state and neutron star structure

Infinite nuclear matter at zero temperature can be studied within the framework of the BHF approximation considering homogeneous matter in normal state. The effective interaction between pairs is described by the g matrix, which satisfies

$$g(\omega) = v + v\frac{Q}{\omega + i\eta - \hat{h}_1 - \hat{h}_2}g, \tag{1}$$

where v is the bare interaction between nucleons, $\hat{h}_i$ the single-particle (sp) energy of nucleon i and Q the Pauli blocking operator. The solution of this equation allows the evaluation of the mass operator

$$M(k;E) = \sum_{|p|\leq k_f} \langle\frac{1}{2}(k-p)|g_K(E+e_p)|\frac{1}{2}(k-p)\rangle, \tag{2}$$

where $K = k + p$, is the total momentum, and

$$e_p = \frac{p^2}{2m} + U(p), \tag{3}$$

the sp energy. In the BHF approximation the sp potential is

$$U(k) = ReM(k;e_k). \tag{4}$$

Eqs. (1-4) are solved self-consistent following the *continuous choice* for $U(k)$ [11]. $^1S_0$ and $^3SD_1$ channels are treated following Ref. [12]. These considerations have led to coexisting solutions in symmetric matter which in this study appears to have negligible effects.

Self-consistent solutions for $U(k)$ at zero temperature were obtained using Argonne v_{18} potential. Solutions have been obtained for both symmetric and neutronic matter. In the case of symmetric nuclear matter (SMN) the Fermi momentum (k_F) ranges between 0.25 fm^{-1} and 2.2 fm^{-1}. In the case of pure neutronic matter (PNM) $k_f \leq 3$ fm^{-1}. In Fig. 1 we show the binding energy per nucleon, B/A, as a function of density. It is observed the gas-like behavior of neutronic matter. In the case of SNM the saturation point occurs at $\rho = 0.25$ fm^{-3}, with a binding energy per nucleon of -16.8 MeV.

2.1. *Equation of state for beta-stable matter*

The energy density for matter composed of interacting protons and neutrons, ultra-relativistic electrons and non-relativistic muons is given by [13]

$$\epsilon = \rho_p m_p c^2 + \rho_n m_n c^2 + B/A(\rho_p, \rho_n)\rho + \frac{(3\pi^2\rho_e)^{5/3}}{5\pi^2} + \rho_\mu m_\mu c^2 + \frac{\hbar^2}{2m_\mu}\frac{(3\pi^2\rho_\mu)^{5/3}}{5\pi^2}, \tag{5}$$

where $\rho = \rho_p + \rho_n$ is the nucleon density. In order to account for isospin asymmetry we introduce a parabolic approximation in terms of $\alpha = (\rho_n - \rho_p)/\rho$ [14]

$$B/A(\rho, \alpha) = (B/A)_{SNM}(\rho) + \alpha^2[(B/A)_{PNM}(\rho) - (B/A)_{SNM}(\rho)]. \tag{6}$$

Here SNM and PNM refer to symmetric nuclear matter and pure neutron matter respectively. We use the B/A_{PNM} and B/A_{SNM} obtained from the self-consistent calculations we have performed. The pressure is given by

$$P = P_{nucleons} + P_{leptons}, \tag{7}$$

$$P_j = \rho_j^2 \frac{\partial}{\partial \rho_j}\left(\frac{\epsilon_j}{\rho_j}\right). \tag{8}$$

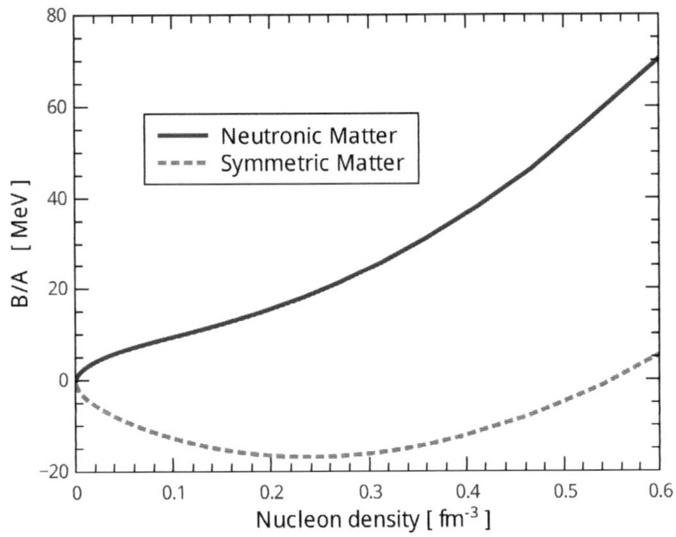

Figure 1. Energy per nucleon for symmetric (solid curve) and neutronic matter (dashed curve) based on the Argonne v_{18} potential.

Neutral beta-stable matter satisfies

$$\mu_n = \mu_p + \mu_e; \tag{9}$$
$$\rho_p = \rho_e + \rho_\mu; \tag{10}$$
$$\mu_e = \mu_\mu; \tag{11}$$

where $\mu_i = \partial\epsilon_i/\partial\rho_i$ represents the chemical potential of species i. Solving Eqs.(9)-(11) using Eq.(5) allows us to obtain the abundances of each specie ($\rho_i(\rho)$), hence the pressure as a function of the nucleon density $P(\rho)$ [15]. These considerations define the equation of state for beta-stable matter.

2.2. Neutron star structure

With a EoS for neutron star matter we are in condition to obtain the mass-radius behavior at equilibrium. To do so we solve the Tolman-Oppenheimer-Volkoff (TOV) equations for non-rotating stars,

$$\frac{dP(r)}{dr} = -\frac{Gm(r)\epsilon(r)}{r^2c^2}\left(1+\frac{P(r)}{\epsilon(r)}\right)\left(1+\frac{4\pi P(r)}{c^2m(r)}\right)\left(1-\frac{2Gm(r)}{rc^2}\right)^{-1}, \tag{12}$$
$$\frac{dm(r)}{dr} = 4\pi r^2\epsilon(r)/c^2, \tag{13}$$

where m is the enclosed mass up to a radius r and P is the local pressure. For the core of the star, where $\rho > \rho_0/5$, we have used the EoS for beta-stable matter obtained from the self-consistent calculations described above. For the crust and envelope we have used the EoS by Feynmam-Metropolis-Teller [16], Haensel-Zdunik-Dobaczewski [17] and Negele-Vautherin [18]. We have solved these equations considering different central densities.

In Fig. 2 we show the behavior of the mass of a neutron star and its radius where equilibrium is imposed. As observed, we obtain a maximum mass of roughly 1.8 $M_\odot$, not reaching the latest

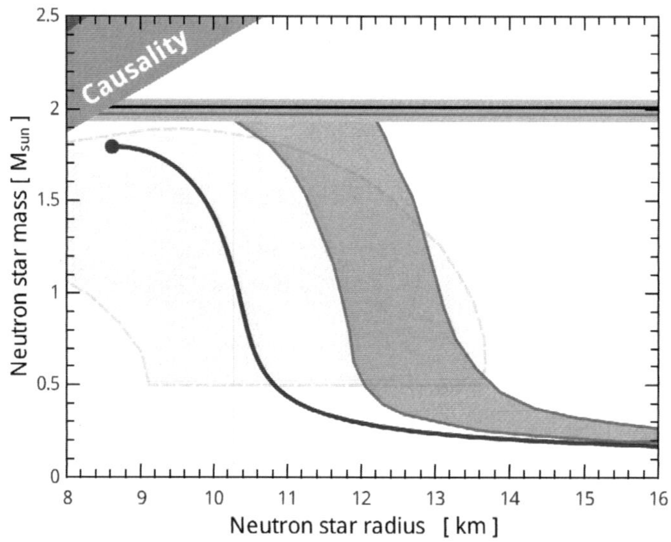

Figure 2. Neutron star mass expressed in solar mass units as a function of the star radius. The circle corresponds to the maximum mass reached within our approach. The brown region surrounded by solid lines is obtained from quiescent low-mass x-ray binary mass and radius observations [19]. The cyan region surrounded by dashed lines corresponds to the 90% confidence region for the mass-radius relation of NGC 6397 [20]. The two horizontal lines correspond to high mass neutron stars reported in Refs. [2, 3].

maximum masses observed of 2.1 $M_\odot$ [2, 3]. However, this shortcoming is not unexpected considering the simplicity of the model, where three-body forces and sub-hadronic degrees of freedom in the inner core are not included.

3. Neutron star cooling

Recent developments in astronomical observations have allowed increasing precision in measurements of neutron stars properties. These observations provide important information regarding the cooling mechanisms, particularly about its superfluid and superconducting properties.

The cooling evolution of neutron stars can be obtained solving the general relativistic equations of thermal evolution [9, 21]

$$\frac{e^{-\lambda-2\Phi}}{4\pi^2}\frac{\partial}{\partial r}\left(e^{2\Phi}L_r\right) = -Q - \frac{c_T}{e^{\Phi}}\frac{\partial T}{\partial t} \tag{14}$$

$$\frac{L_r}{4\pi\kappa r^2} = e^{-\lambda-\Phi}\frac{\partial}{\partial r}\left(Te^{\Phi}\right), \tag{15}$$

where Q es the neutrino emissivity, κ is the thermal conductivity, L_r is the local luminosity and T is the temperature. The function Φ is related to the local pressure and energy density by

$$\frac{d\Phi}{dr} = -\frac{1}{\epsilon(r)}\frac{dP}{dr}\left[1+\frac{P(r)}{\epsilon(r)}\right]^{-1}. \tag{16}$$

Additionally, $e^{-\lambda} = \sqrt{1-2Gm(r)/c^2r}$.

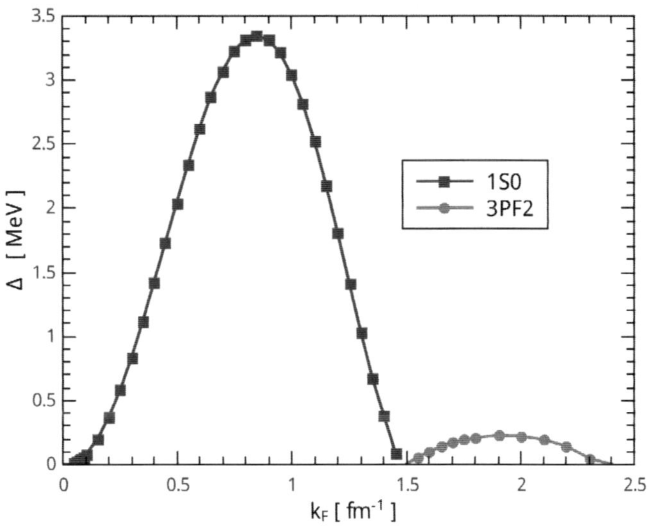

Figure 3. Energy gap as a function of the fermi momentum in neutronic matter for channels $^1S_0$ and $^3PF_2$.

We have solved Eqs. (14-15) using the code `NSCool` developed by D. Page [22]. Physical inputs such as neutrino emissivity, thermal conductivity and heat capacity are addressed in Refs. [21, 23]. We do not include internal heating, accretion, nor magnetic fields. The inputs are affected by superfluidity [24], a phenomenon which we are able to account for solving BCS gap equations in the $^1S_0$ and $^3PF_2$ channels in neutronic matter. To do so we have proceeded as in Ref. [12]. In this case the BCS gap equation for uncoupled channels is given by

$$\Delta(k) = -\sum_{k'} \langle k|U|k'\rangle \frac{\Delta(k')}{2E(k')}, \tag{17}$$

where $E(k)$ is the quasi-particle energy and

$$E(k) = \sqrt{\epsilon(k)^2 + |\Delta(k)|^2}; \qquad \epsilon(k) = e(k) - \mu; \tag{18}$$

and $e(k)$ is the sp energy. Here μ represents the chemical potential. Analog equations are for NN coupled channels. In Fig. 3 we plot the pairing gap energies as a functions of k_F for $^1S_0$ and $^3PF_2$ channels. It becomes evident that superfluidity in each channel occurs at non overlapping density domains. While $^1S_0$ superfluidity is restricted to $k_F \lesssim 1.5$ fm^{-1}, superfluidity in the $^3PF_2$ channel occurs at k_F between 1.5 and 2.5 fm^{-1}, with a peak of 0.2 MeV at $k_F \sim 1.9$ fm^{-1} ($\rho \sim 1.6\ \rho_0$).

We have studied the cooling evolution of a neutron star considering BHF solutions (accounting for dinucleons) and corresponding BCS gap solutions. In these calculations the critical temperature for superfluidity $T_c(k_F)$ is related to the gap energy at zero temperature by [25]

$$\frac{\Delta_{T=0}}{k_B T_c} \approx 1.764. \tag{19}$$

In Fig. 4 we plot the cooling curves for four neutron star masses. The rapid cooling of stars with 1.5 $M_\odot$ and 1.6 $M_\odot$ is triggered by the direct Urca process, which occurs only at high densities in the core of high-mass neutron stars. Our results suggest that the masses of the stars reported in Fig. 4 lie below 1.5 $M_\odot$, a region governed by the modified Urca process.

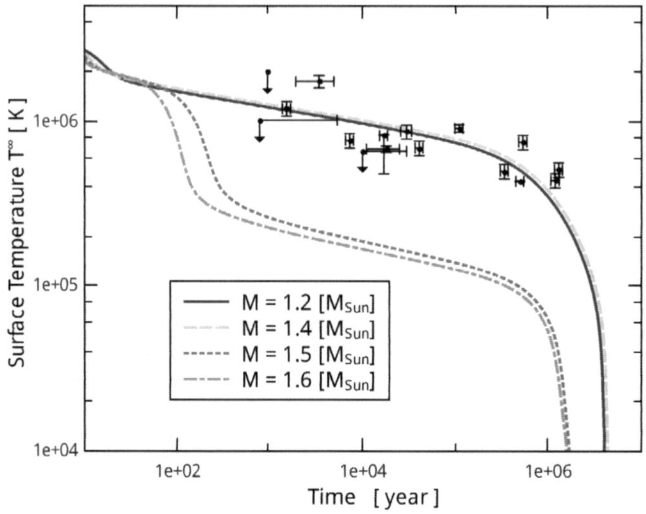

Figure 4. Surface temperature at infinity for three neutron star configurations with $^1S_0$ and $^3PF_2$ superfluidity. The observations points are reported in Ref. [26].

4. Summary and conclusions

We have investigated static and cooling properties of neutron stars. Emphasis has been placed on a microscopic description of nuclear matter, including superfluid states. An EoS has been calculated using the Argonne AV18 bare NN interaction for both symmetric and pure neutron matter. BCS pairing gap based in the same bare interaction have been calculated and incorporated in the cooling codes. We reach a maximum mass for neutron stars around 1.8 $M_\odot$. Cooling calculations show agreement between the observed temperature of neutron stars, as a function of time, if their masses lie below 1.5 $M_\odot$. The inclusion of three-body forces are claimed to be important to account for the correct saturation density for symmetric nuclear matter and to reach higher masses for neutron stars. Current codes we have developed are being upgraded to include the chiral N3LO potential with N2LO three-body forces [27]. These developments shall allow us to investigate the implications of three-body forces on structure and cooling properties of neutron stars.

References

[1] Haensel P, Potekhin A Y and Yakovlev D G 2007 *Neutron stars 1: Equation of state and structure* vol 326 (Springer Science & Business Media)
[2] Antoniadis J *et al.* 2013 *Sci.* **340** 448, 1233232
[3] Demorest P B, Pennucci T, Ransom S M, Roberts M S E and Hessels J W T 2010 *Nature* **467** 1081–1083
[4] Page D, Prakash M, Lattimer J M and Steiner A W 2011 *Phys. Rev. Lett.* **106** 081101
[5] Dong J M, Lombardo U and Zuo W 2013 *Phys. Rev. C* **87** 062801
[6] Li Z H, Lombardo U, Schulze H J J, Zuo W, Chen L W and Ma H R 2006 *Phys. Rev. C* **74** 047304
[7] Wiringa R B, Stoks V G J and Schiavilla R 1995 *Phys. Rev. C* **51** 38–51
[8] Baldo M, Burgio G F and Bombaci I 1997 *Astron. Astrophys.* **328** 274–282
[9] Page D, Geppert U and Weber F 2006 *Nucl. Phys. A* **777** 497–530
[10] Dean D J and Hjorth-Jensen M 2003 *Rev. Mod. Phys.* **75** 607–656
[11] Baldo M 1999 *Nuclear methods and the nuclear equation of state* vol 8 (World Scientific)
[12] Arellano H and Delaroche J P 2015 *Eur. Phys. J. A* **51** 1–17
[13] Blaschke D and Sedrakian D 2005 *Superdense QCD Matter and Compact Stars: Proceedings of the NATO Advanced Research Workshop on Superdense QCD Matter and Compact Stars* vol 197 (Springer Science & Business Media)

[14] Vidaña I, Providência C, Polls A and Rios A 2009 *Phys. Rev. C* **80** 045806
[15] Akmal A, Pandharipande V R and Ravenhall D G 1998 *Phys. Rev. C* **58** 1804–1828
[16] Feynman R P, Metropolis N and Teller E 1949 *Phys. Rev.* **75** 1561–1573
[17] Haensel P, Zdunik J L and Dobaczewski J 1989 *Astron. Astrophys.* **222** 353–357
[18] Negele J W and Vautherin D 1973 *Nucl. Phys. A* **207** 298–320
[19] Lattimer J M and Steiner A W 2014 *Astrophys. J.* **784** 123
[20] Heinke C *et al.* 2014 *Mon. Not. R. Astron. Soc.* **444** 443–456
[21] Gnedin O Y, Yakovlev D G and Potekhin A Y 2001 *Mon. Not. R. Astron. Soc.* **324** 725–734
[22] Page D Nscool: A neutron star cooling code URL `http://www.astroscu.unam.mx/neutrones/NSCool/`
[23] Yakovlev D G, Kaminker a D, Gnedin O Y and Haensel P 2001 *Phys. Rep.* **354** 1–155
[24] Yakovlev D G, Kaminker A D and Levenfish K P 1999 *Astron. Astrophys.* **343** 650–660
[25] Leggett A J 2006 *Quantum liquids: Bose condensation and Cooper pairing in condensed-matter systems* (Oxford University Press)
[26] Yakovlev D G, Gnedin O Y, Kaminker A D and Potekhin A Y 2008 *AIP Conf. Proc.* **983** 379–387
[27] Carbone A, Polls A and Rios A 2013 *Phys. Rev. C* **88** 044302

Acknowledgments

This work has been supported in part by FONDECYT under grant No 1120396. F. Isaule thanks CONICYT fellowship Beca Nacional, Contract No. 221320081.

Effects of angular correlations on particle-particle propagation in infinite nuclear matter

J Romero–Barrientos[1] and H F Arellano[2]

Departamento de Física – FCFM, Universidad de Chile, Av. Blanco Encalada 2008, Santiago, Chile

E-mail: [1]`jaromero@ing.uchile.cl`, [2]`arellano@dfi.uchile.cl`

Abstract. The effect of angular correlations on self-consistent solutions for single-particle (sp) potentials in infinite nuclear matter is investigated. To this end we treat explicitly the angular dependence of the particle–particle (pp) propagator in Brueckner–Hartree–Fock (BHF) equation for the g matrix. It is observed that the *exact* angular dependence of the pp propagator yields highly fluctuating structures, posing stringent difficulties in the actual search of self–consistent solutions for the sp energy. A perturbative approach is presented to evaluate the effect of the angular correlations in the self-consistent solutions. Solutions at Fermi momenta k_F in the range $1.20 - 1.75$ fm^{-1} are reported using Argonne v_{18} nucleon–nucleon potential. Although the sp potentials are sensitive to the treatment of the angular behaviour of the propagator, such sensitivity appears at momenta well above the Fermi surface. As a result, the saturation properties of symmetric nuclear matter differ marginally from those calculated using angle–averaged energy denominators in pp propagators.

1. Introduction

Infinite nuclear matter models have extensively been used for the study of saturation properties [1] as well as for the construction of density-dependent effective interactions to describe nuclear collisions [2]. To lowest order, the BHF approximation is able to account reasonably well for leading order short-range correlations of nuclear matter in normal state. In this model, interacting nucleons are conceived as single particles moving under a sp potential stemming from the interaction with all other nucleons. The determination of the physical properties of the system is reduced to the calculation of the sp potential. At this point, as discussed in Ref. [3], several approximations have been introduced aiming to circumbent difficulties arising from the non–spherical kernel of pp propagation, together with the multi-channel coupling of nucleon–nucleon (*NN*) states with different angular momentua.

In BHF, the propagation of intermediate states is driven by the pp propagator which takes the form $\Lambda \sim Q/e$, where Q is the Pauli blocking operator. The energy denominator e considers two nucleons moving in directions which are in general not identical. Both terms, Q and e, exhibit an explicit angular dependence, feature which remains even if multi–channel coupling is neglected. Among the most accurate approximations in use to account for Λ we mention the reduction of the propagator Q/e to the ratio of its angular averages, $\langle Q\rangle/\langle e\rangle$. More recently this approximation has been contrasted with the angular average of the ratio, $\langle Q/e\rangle$ [3], where the angular dependence of the energy denominator is retained up to second order in $\hat{K}\cdot\hat{q}$, with $\mathbf{K}$ and $\mathbf{q}$ the total and relative momenta of the interacting nucleons. In this work we examine more closely the approximation $\Lambda \approx \langle Q/e\rangle$, but treating exactly the intrinsic angular dependence of the energy denominator. As we shall see, the effect of this exact treatment manifest itself at high momenta with respect to the Fermi momentum k_F, with marginal incidence in the evaluation of the binding energy of the system.

Journal of Physics: Conference Series **720** (2016) 012022 doi:10.1088/1742-6596/720/1/012022

2. Theoretical Framework

In the BHF approximation in Brueckner–Bethe–Godstone theory for infinite nuclear matter at zero temperature, the g matrix depends on the density of the medium, characterized by k_F, and a starting energy ω [1]. In this framework the effective interaction g satisfies the integral equation

$$g(\omega) = v + v\frac{Q}{\omega - \hat{h}_1 - \hat{h}_2 + i\eta}g(\omega), \tag{1}$$

where v denotes the bare NN interaction, $\hat{h}_1$ and $\hat{h}_2$ the single particle energies and Q the Pauli blocking operator. The solution to this equation allows the evaluation of the mass operator

$$M(k;E) = \sum\langle\frac{1}{2}(\mathbf{k}-\mathbf{p})|g_{\mathbf{K}}(E+\epsilon_p)|\frac{1}{2}(\mathbf{k}-\mathbf{p})\rangle, \tag{2}$$

where $\mathbf{K} = \mathbf{k}+\mathbf{p}$ represents the total momentum of the interacting pair and $\epsilon_k = k^2/2m + U(k)$ is the sp energy of nucleons of mass m in terms of an auxiliary field U. Self–consistency requires that

$$U(k) = \mathcal{R}e\{M(k;\epsilon_k)\}. \tag{3}$$

This condition is achieved iteratively. In the continuous choice this condition is imposed to all momenta k. In infinite nuclear matter, the BHF equation takes the following form in momentum representation,

$$\langle\boldsymbol{\kappa}'|g_{\boldsymbol{K}}(\omega)|\boldsymbol{\kappa}\rangle = \langle\boldsymbol{\kappa}'|v|\boldsymbol{\kappa}\rangle + \int d\boldsymbol{q}\langle\boldsymbol{\kappa}'|v|\boldsymbol{q}\rangle\Lambda(\mathbf{K},\mathbf{q})\langle\boldsymbol{q}|g_{\boldsymbol{K}}(\omega)|\boldsymbol{\kappa}\rangle. \tag{4}$$

Here, the propagator takes the form

$$\Lambda(\mathbf{K},\mathbf{q}) = \frac{Q(\mathbf{K},\mathbf{q})}{\omega + i\eta - \frac{k_+^2}{2m} - \frac{k_-^2}{2m} - \Sigma} \tag{5}$$

with

$$\Sigma(K,q,x) \equiv U(k_+) + U(k_-),$$

$$k_\pm^2 = \frac{K^2}{4} + q^2 \pm qKx,$$

and $x = \hat{K}\cdot\hat{q}$. Furthermore, the Pauli blocking operator is represented by $Q(\mathbf{K},\mathbf{q}) = \Theta(k_+ - k_F)\Theta(k_- - k_F)$. This operator not only depends on the magnitudes of the momenta, but also on the angle between them.

A common practice to simplify the angular dependence of the propagator $\Lambda(\mathbf{K},\mathbf{q})$ is by treating separately the numerator Q from the denominator $e \equiv \omega - \frac{K^2}{4} - q^2 - \Sigma$. When the energy denominator is angle averaged the propagator takes the form $Q/\langle e\rangle$, and the Pauli blocking operator is treated in its full form [5, 6, 7, 8], including multi-channel coupling. The propagator can be approximated even further by angle averaging the Pauli operator [9], i.e. $Q \to \langle Q\rangle$, with

$$\langle Q\rangle = \frac{1}{4\pi}\int Q(\mathbf{Q},\mathbf{k})d\Omega_q = \Delta\,.$$

Here $\Delta = \min\{1, \max\{0, (K^2/4 - q^2 - k_F^2)/qK\}\}$. In this limit the propagator takes the form of ratio of averages, namely $Q/e \to \Delta/\langle e\rangle$. In this work, rather than using the ratio of averages we consider the average of the ratio Q/e. This line of thought is motivated after examining the angular behavior of the energy denominator [3]. In Fig. 1 we plot the difference $\delta\Sigma \equiv \Sigma - \langle\Sigma\rangle$, as function of x and q. Here $\langle\rangle$ denotes the angular average, $\langle\Sigma\rangle = \int_0^\Delta \Sigma(K,q,x)dx/\Delta$, with Σ calcultaed from the self–consistent U obtained within the ratio of the averages approach, using the continuous choice, for the Argonne v_{18} potential (AV18) [4]. Differences between Σ and its average $\langle\Sigma\rangle$ become evident. In this work we aim to assess the implications of these differences on the sp fields $U(k)$ when self–consistency is imposed.

Journal of Physics: Conference Series **720** (2016) 012022 doi:10.1088/1742-6596/720/1/012022

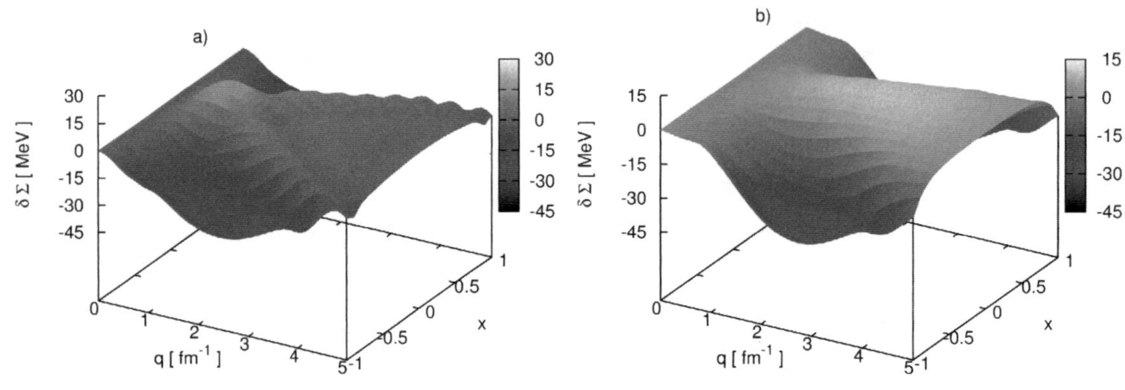

Figure 1. Surface plots of the difference $(\Sigma - \bar{\Sigma})$ as functions of q and x, with $K = 3k_F$. Surface (a) corresponds to $k_F = 1.50$ fm^{-1}, while (b) is for $k_F = 1.60$ fm^{-1}.

2.1. Treatment of the angular dependence of the propagator

In this study we work on the monopole approximation [3], when the propagator takes the form

$$\lambda(K,q) = \int d\hat{\mathbf{q}} Y_0^0(\hat{\mathbf{q}}) \frac{Q}{e + i\eta} Y_0^0(\hat{\mathbf{q}}) = \frac{1}{2} \int_{-1}^{1} \frac{Q}{e + i\eta} dx, \tag{6}$$

with x the cosine of the angle between $\mathbf{K}$ and $\mathbf{q}$. If we consider explicitly the form of the Pauli blocking operator Q, the angle–averaged propagator Q/e reduces to

$$\lambda(K,q) = \int_0^{\Delta} \frac{1}{e + i\eta} du. \tag{7}$$

We have found that the exact angular dependence in the energy denominator leads to highly fluctuating structures, posing severe difficulties in the search of the self–consistent solutions for the sp energy. To overcome these difficulties, we have adopted a perturbative strategy to evaluate the effect of angular correlations in the self–consistent solutions.

In general terms, the equation for the g matrix is given by

$$g = v + v\Lambda g, \tag{8}$$

where Λ denotes the exact propagator in the monopolar approximation. Let us denote Λ_0 the pp propagator approximated by the ratio of the averages, $\Lambda_0 = \Delta/\langle e \rangle$. Then

$$g = v + v(\Lambda - \Lambda_0)g + v\Lambda_0 g. \tag{9}$$

Let us define $\delta\Lambda \equiv (\Lambda - \Lambda_0)$, as a perturbative term and expand g as an infinite series in orders of $\delta\Lambda$

$$g = g_0 + g_1 + g_2 + \cdots . \tag{10}$$

Combining Eqs. (9) and (10) we obtain

$$g_0 + g_1 + \cdots = v + v\delta\Lambda(g_0 + g_1 + \cdots) + v\Lambda_0(g_0 + g_1 + \cdots) + \cdots . \tag{11}$$

Equating terms of the same order, we obtain

$$g_0 = v + v\Lambda_0 g_0 \tag{12a}$$
$$g_1 = v\,\delta\Lambda\, g_0 + v\Lambda_0 g_1. \tag{12b}$$

The first equation represents the BHF equation for g_0, when the ratio of averages is used. Combining Eqs. (12a) and (12b) we obtain

$$g_1 = g_0\,(\delta\Lambda)\,g_0. \tag{13}$$

Thus,

$$g \approx g_0 + g_1. \tag{14}$$

Therefore, to evaluate g_1 we need to calculate g_0 (following standard procedures) and compute the expectation value $\langle g_0\;\delta\Lambda\;g_0\rangle$. This procedure, though simple, exhibits some subtleties which we address in some detail. We denote by ϵ_0 the difference between the starting energy and the kinetic energy of the nucleons

$$\epsilon_0 = \omega - \frac{K^2}{4} - q^2, \tag{15}$$

and by $\langle\Sigma\rangle$ the sum of the angle–averaged sp potentials. With these definitions, we can write the exact propagator as

$$\Lambda = \frac{Q}{\epsilon_0 - \Sigma + i\eta}, \tag{16}$$

which is a function of q and x. For the propagator Λ_0 we have

$$\Lambda_0 = \frac{Q}{\epsilon_0 - \langle\Sigma\rangle + i\eta}, \tag{17}$$

which is a function of q but independent of x. The difference $\delta\Lambda$ then takes the form, when using the monopolar approximation,

$$\delta\Lambda = 2\pi\int_{-\Delta}^{\Delta} dx\left(\frac{1}{\epsilon_0 - \Sigma + i\eta} - \frac{1}{\epsilon_0 - \langle\Sigma\rangle + i\eta}\right). \tag{18}$$

By adding and substracting $\langle\Sigma\rangle$ in the denominator of the first integrand, and defining $e_0 \equiv \epsilon_0 - \langle\Sigma\rangle$, and $u_x \equiv \Sigma - \langle\Sigma\rangle$, Eq. (18) can be written as

$$\delta\Lambda = 2\pi\int_{-\Delta}^{\Delta} dx\left(\frac{1}{e_0 - u_x + i\eta} - \frac{1}{e_0 + i\eta}\right). \tag{19}$$

Note that e_0 does not depend on x. The real part of $\delta\Lambda$ reads

$$\delta\Lambda_R = 2\pi\frac{e_0}{e_0^2 + \eta^2}\int_{-\Delta}^{\Delta}\frac{u_x(e_0 - u_x)}{(e_0 - u_x)^2 + \eta^2}dx. \tag{20}$$

Considering

$$\Lambda_0 = \frac{1}{e_0 + i\eta} = \frac{e_0^2}{e_0^2 + \eta^2} - i\frac{\eta^2}{e_0^2 + \eta^2}, \tag{21}$$

then

$$\delta\Lambda_R = 2\pi\,\mathcal{R}e\{\Lambda_0\} \times \int_{-\Delta}^{\Delta}\frac{u_x(e_0 - u_x)}{(e_0 - u_x)^2 + \eta^2}dx. \tag{22}$$

Journal of Physics: Conference Series **720** (2016) 012022 doi:10.1088/1742-6596/720/1/012022

In the case of the imaginary part we have,

$$\delta\Lambda_I = 2\pi \int_{-\Delta}^{\Delta} \left[\frac{\eta}{(e_0 - u_x)^2 + \eta^2} - \frac{\eta}{e_0^2 + \eta^2} \right] dx. \tag{23}$$

Let us define

$$f(u_x) = \frac{\eta}{(e_0 - u_x)^2 + \eta^2} - \frac{\eta}{e_0^2 + \eta^2}, \tag{24}$$

which, after using the Plemelj formula in the limit $\eta \to 0$ leads to

$$f(u_x) = \delta(e_0 - u_x) - \delta(e_0). \tag{25}$$

Expanding around $u_x = 0$, we get

$$f(u_x) \approx u_x \delta'(e_0) + u_x^2 \frac{1}{2} \delta''(e_0) + O(u_x^3), \tag{26}$$

and replacing this result in Eq. (23) we obtain

$$\delta\Lambda_I = 2\pi \int_{-\Delta}^{\Delta} \left[u_x \delta'(e_0) + u_x^2 \frac{1}{2} \delta''(e_0) \right] dx. \tag{27}$$

The first term of Eq. (27) vanishes by definition. Therefore

$$\delta\Lambda_I = \pi \delta''(e_0) \int_{-\Delta}^{\Delta} u_x^2 dx. \tag{28}$$

Eqs. (22) and (28) can be used to evaluate g, in Eq. (14). With this, the g matrix is used to calculate the mass operator self–consistently. Convergence is imposed when fluctuations of $U(k)$ do not exceed 0.05 MeV on three consecutive iterations.

3. Results

We have calculated self–consistent solutions for $U(k)$ using Argonne v_{18} potential including all partial waves up to $J = 7$. Solutions where found considering the pp propagator in the form of the ratio of angular averages, and the monopole approximation as described above. In the case of SNM we obtained solutions for Fermi momenta between 1.20 and 1.75 fm^{-1}, in steps of 0.05 fm^{-1}. We have also investigated pure neutron matter, case in which the solutions where obtained in the range 1.20–1.75 fm^{-1}. In Fig. 2 we show the on–shell mass operator as a function of k for different values of k_F. The left–hand side panel shows the real part of the on-shell mass operator, while the right–hand panel shows the imaginary part. Blue (red) curves correspond to $k_F = 1.25(1.75)$ fm^{-1}.

In Fig. 3 we show a comparison between the self–consistent fields calculated using the angle–averaged propagator (red, continuous) and the ones calculated using the propagator which included the first order correction (blue, dashed). As we can observe, differences between the two are quite moderate and become only noticeable to the eye at momenta over 3.5 fm^{-1}. Calculation of the pp potentials allowed us to evaluate the energy per nucleon. The saturation points found for both the angle–average and monopolar approximations for the propagator are summarized in Table 1. Here we include the incompressibility of nuclear matter given by

$$K = k_F^2 \frac{\partial^2 (B/A)}{\partial k_F^2}, \tag{29}$$

where the second derivative is evaluated at the saturation momentum.

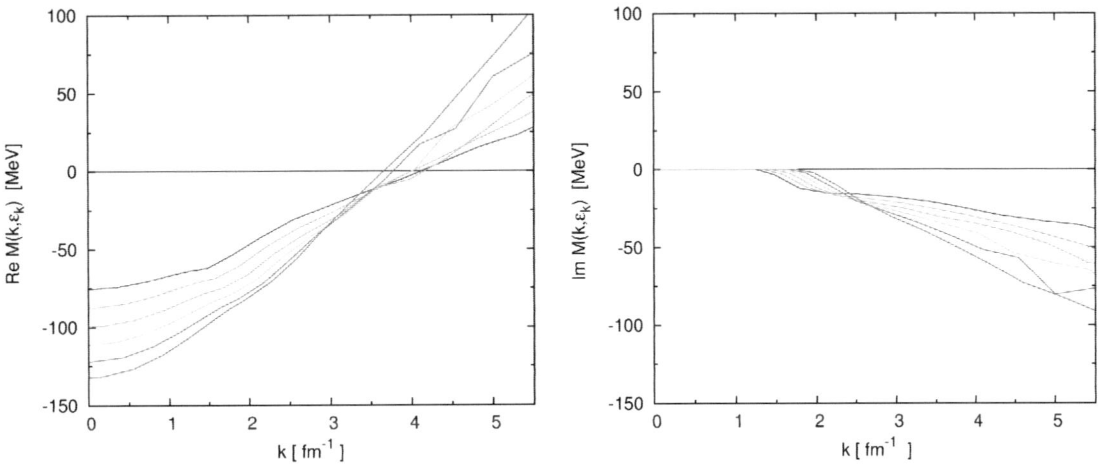

Figure 2. On–shell mass operator calculated using the monopole approximation for the pp propagator. Here $1.25 \leq k_F \leq 1.75$ fm^{-1}. Blue (red) curve denotes $k_F = 1.25(0.10)1.75$ fm^{-1}.

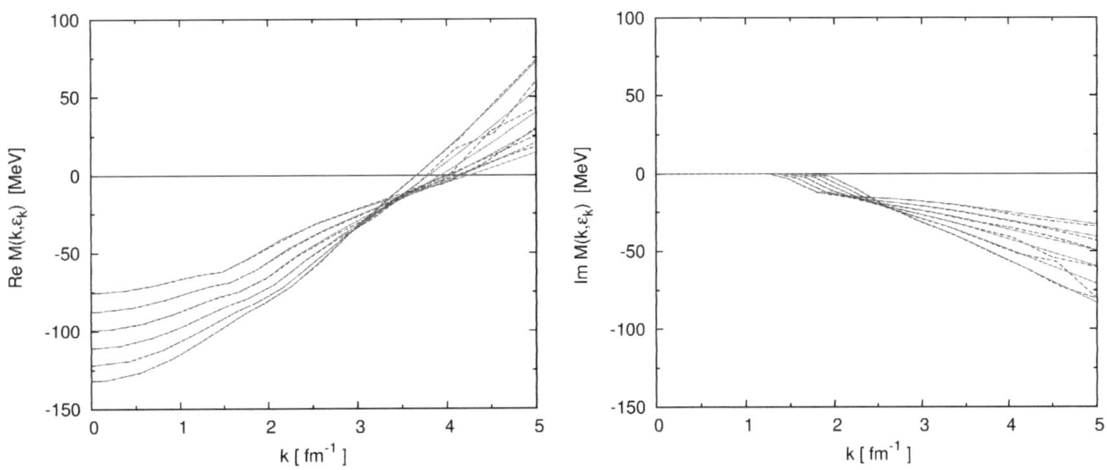

Figure 3. Comparison between the on–shell mass operator calculated using the angle–averaged propagator versus the ones calculated using $g \sim g_0 + g_1$.

Solutions	k_F [fm^{-1}]	B/A [MeV]	K [MeV]
Angle–averaged	1.520	−16.691	199.4
Monopole	1.521	−16.689	202.9

Table 1. Saturation properties for SNM found in this work. The third decimal place is shown only to illustrate the differences.

As we can see, the saturation energy and density found are essentially the same when using the angle–

average or the monopole approximation. The nuclear compressibility turned out to be more sensitive to the corrections in the propagator with a 3.5 MeV difference between the two.

In Fig. 4 we show the on–shell mass operator as a function of k for different values of k_F calculated for pure neutron matter. As for the SNM curves, blue (red) corresponds to $k_F = 1.20(1.75)$ fm^{-1}. In Fig. 5

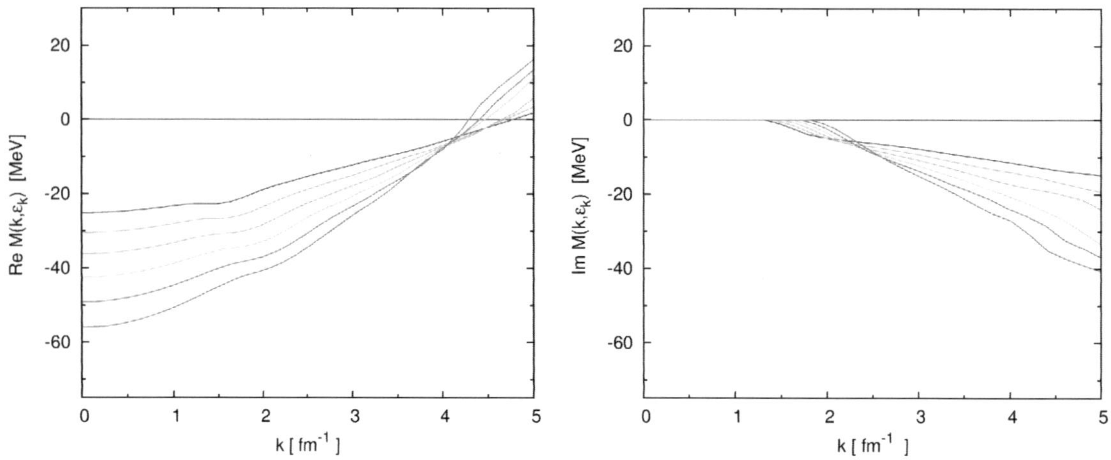

Figure 4. Same for Figure 2 but for pure neutron matter.

we show the comparison between the mass operator calculated using the angle–averaged pp propagator (red, continuous) and the mass operator calculated using the monopolar approximation (blue, dashed). Analogous to the SNM case, the differences between the two become noticeable to the eye at momenta over 3.5 fm^{-1}.

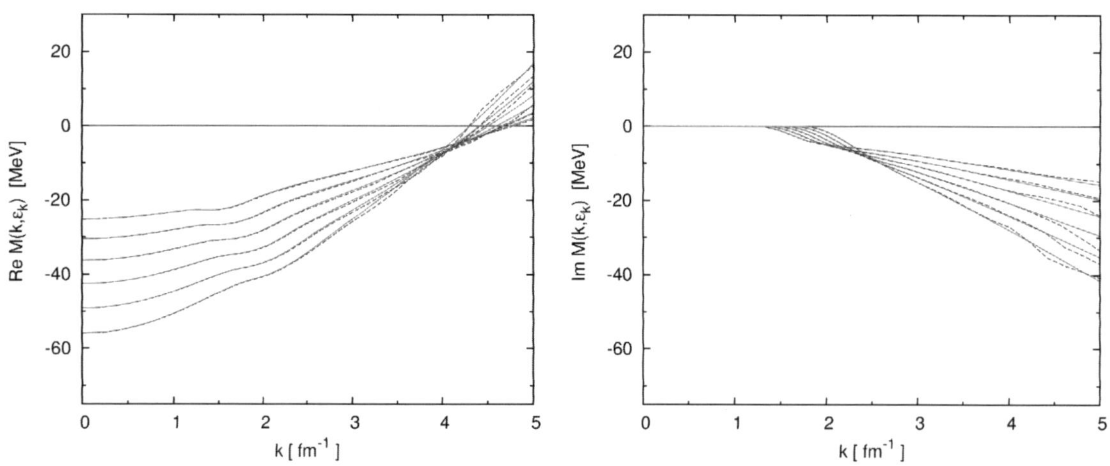

Figure 5. Comparison between the self–consistent fields calculated using the angle–averaged propagator versus the ones calculated using the first order correction for neutron matter

4. Summary and conclusions

We have investigated the effects of angular correlations resulting from considering the angular dependence of the pp propagator of the BHF approximation for symmetric and pure neutron matter. To accomplish this, we calculated the sp potentials for Fermi momenta between 1.20 and 1.75 fm^{-1}in the continuous choice. The bare *NN* potential used in this study was the Argonne v_{18} including all partial waves up to $J = 7$. The angular dependence of the pp propagator was treated like a perturbation, $\delta\Lambda = (\Lambda - \Lambda_0)$. The occurrence of possible $^1S_0$ and $^3S_1$–$^3D_1$ bound states was accounted for and its search and regularization was implemented in the calculations using the method outlined in Ref. [11]. Moderate effects are observed in the mass operator at momenta above $\sim$ 3.5 fm^{-1}, when the monopole approximation is used for the pp propagator. We find that both the ratio of angular averages and the average of the ratio yield saturation of nuclear matter at $k_F = 1.52$ $\mathrm{fm}^{-1}$and $B/A = 16.7$ MeV/nucleon. These results confirm that the ratio of averages is a good approximation for the calculation of the sp potentials. The incompressibility, on the other hand, differs by 3.5 MeV between the two solutions. We have also found that the effect of angular correlations in pp propagation in neutron matter is negligible.

Acknowledgments

This work has been supported in part by FONDECYT under grant No 1120396.

References

[1] Baldo M 1998 *Nuclear Methods and the Nuclear Equation of State* vol 8 (Singapore: World Scientific) chapter 1 pp 1–116
[2] Amos K et al 2002 *Adv. Nucl. Phys.* **25** 276–536
[3] Arellano H F and Delaroche J -P 2011 *Phys. Rev.* C **83.4** 044306
[4] Wiringa R B et al 1995 *Phys. Rev.* C **51** 38
[5] Sammarruca F, Meng X and Stephenson E J 2000 *Phys. Rev.* C **62** 014614
[6] Suzuki K, Okamoto R, Kohno M and Nagata S 2000 *Nucl. Phys.* A **665** 92
[7] Schiller E, Muther H and Czerski P 1999 *Phys. Rev.* C **59** 2934
[8] Cheon T and Redish E F 1989 *Phys. Rev.* C **39** 331
[9] Brueckner K A and Gammel J L 1958 *Phys. Rev.* **109** 1023
[10] Haftel M I and Tabakin F 1970 *Nucl. Phys.* A **158.1** 1–42
[11] Arellano H F and Delaroche J -P 2015 *Eur. Phys. J.* A **51** 7
[12] Carbone A et al 2013 *Phys. Rev.* C **88** 054326

Optical properties of Chern-Simons systems

Luis Huerta[1,2]

[1] Facultad de Ingeniería, Universidad de Talca, 3340000 Curicó, Chile.

[2] P4-Center for Research and Applications in Plasma Physics and Pulsed Power Technology, 7600713 Santiago, Chile.

E-mail: `lhuerta@utalca.cl`, `www.pppp.cl`

Abstract. Chern-Simons (CS) systems interacting with electromagnetic radiation are described by a term $\int F \wedge F$ added to the Maxwell action. In (3+1)D, this CS term is a boundary term affecting the system behaviour in its borders. We study the consequences of the above in the properties of electromagnetic radiation, in particular, by considering the interplay between magneto-electric properties and topology. Apart from a modified Kerr polarization rotation, compared to that found for the particular case of topological insulators, we also found two Brewster angles, for s and p polarization of reflected radiation, respectively. Energy distribution between reflected and transmitted radiation is also studied in terms of the magneto-electric properties and topological condition of the system.

Keywords: Chern-Simons theories, θ-vacuum, topological insulators, optical properties of surfaces.

1. Introduction

Chern-Simons (CS) theories arised in mathematics, in the study of topological invariants [1, 2]. Since then a variety of applications have been found from gauge field theory to cosmology [3]. On the other hand, a θ-term (a CS form in 3+1 dimensions), was introduced in QCD by Roberto Peccei and Helen Quinn in 1977 [4]. A decade after, the idea was generalized by Franck Wilczek to other contexts, particularly to condensed matter physics [5]. As he observed, the CS form in the action could have applications in describing phenomena where topology plays a role, such as those exhibiting Quantum Hall phenomena [6]. In recent years, the novel topological insulators have become a new example of the above [7, 8].

In (3+1)D, CS contribution is a boundary term, not affecting the behaviour of the system in the bulk. But this θ-term is important when the system is bounded [9], and nontrivial

Journal of Physics: Conference Series **720** (2016) 012023 doi:10.1088/1742-6596/720/1/012023

contributions arise from the boundary. An example of this is the modification in the Casimir energy inside a spherical θ-region [10]. Effects in gravity are currently under study [11].

In θ-systems, radiation experiences changes in its properties. Even the pure vacuum affects the radiation, if it is provided with nontrivial Chern-Simons properties -a θ-vacuum. In this latter case, an unexpected reflection makes the energy diminishes after crossing the interface, with the possibility of considering this for cosmological issues [12]. In the context of topological insulators, an equivalent phenomenon to Kerr-Faraday polarization rotation occurs in reflected and refracted waves at the surface of such materials [7, 13].

In this brief report, we study the polarization and energy distribution of the reflected and transmitted radiation at the surface of a θ-system. The existence of a standard s-Brewster angle (s-polarized reflected radiation) and a nonstandard Brewster angle (p-polarized reflected wave) is discussed. Finally, we consider the reflectance and transmittance of the θ-system surface.

2. The basic approach

For electromagnetic radiation in a θ-system, a CS term of the form

$$\frac{\theta}{2}\int F \wedge F \tag{2.1}$$

is added to the Maxwell action. In noncovariant notation, this amounts to a $\theta\mathbf{E}\cdot\mathbf{B}$ term. The field equations become [12]

$$\begin{aligned} \varepsilon\nabla\cdot\mathbf{E} &= \theta\delta(\Sigma)\mathbf{B}\cdot\mathbf{n} && (2.2)\\ \frac{1}{\mu}\nabla\times\mathbf{B}-\partial_t\mathbf{E} &= \theta\delta(\Sigma)\mathbf{E}\times\mathbf{n}, && (2.3) \end{aligned}$$

together with the two homogeneous Maxwell equations (we use units with the light speed $c = 1$; consequently, the electric permittivity, ε, and magnetic permeability, μ, become adimensional). The delta function stands for evaluation at the interface, $\mathbf{n}$ being the inwards unit normal to the interface surface Σ.

In the vicinity of the surface Σ, the field equations (2.2) and (2.3) imply that the normal component of $\mathbf{E}$ and the tangential component of $\mathbf{B}$ are discontinuous:

$$[\varepsilon\mathbf{E}_{\rm n}] = \theta\mathbf{B}_{\rm n}|_\Sigma\,, \qquad \left[\frac{1}{\mu}\mathbf{B}_\tau\right] = -\theta\mathbf{E}_\tau|_\Sigma\,. \tag{2.4}$$

Subindices n and τ stand for normal and tangent components of the fields, respectively, with respect to the interface Σ. The symbol [ ] must be interpreted as the difference between the fields evaluated immediately inside and immediately outside the θ-medium. In addition to those boundary discontinuities, continuity holds for the tangential component of $\mathbf{E}$ and the normal component of $\mathbf{B}$, thus ensuring consistency of (2.4).

3. The radiation fields at the interface

Consider a small region in the boundary of the θ-system where a linearly polarized planar electromagnetic wave incides from a non-θ surrounding region with an angle φ. The wave is reflected with the same angle and transmitted to the other medium with a different angle ψ, these angles measured as usual. The standard laws of reflection and refraction hold for θ-systems. By applying the boundary conditions for the fields at Σ, we derive the electric field amplitude for the refracted and reflected waves (the magnetic field is obtained from that by the standard relationship). As usual, we decompose the fields in parallel ($\|$, p) and perpendicular ($\perp$, s) components with respect to the incidence plane, and introduce the definitions $e_{i\|} \equiv E_{i\|}/E_i$, $e_{r\|} \equiv E_{r\|}/E_i$, $e_{t\|} \equiv E_{t\|}/E_i$, with $E_i = \left(E_{i\|}{}^2 + E_{i\perp}{}^2\right)^{1/2}$, with similar definitions for s-components of the fields. Then, we find for the transmitted wave,

$$\begin{pmatrix} e_{t\|} \\ e_{t\perp} \end{pmatrix} = \frac{2}{D} \begin{pmatrix} \eta s + 1 & -\theta \\ \theta s & \eta + s \end{pmatrix} \begin{pmatrix} e_{i\|} \\ e_{i\perp} \end{pmatrix} \tag{3.1}$$

and, for the reflected wave,

$$\begin{pmatrix} e_{r\|} \\ e_{r\perp} \end{pmatrix} = \frac{1}{D} \begin{pmatrix} (\eta s + 1)(\eta - s) + \theta^2 s & 2\theta s \\ 2\theta s & -(\eta s - 1)(\eta + s) - \theta^2 s \end{pmatrix} \begin{pmatrix} e_{i\|} \\ e_{i\perp} \end{pmatrix} \tag{3.2}$$

with $D \equiv (\eta s + 1)(\eta + s) + \theta^2 s$. We have introduced the magneto-electric parameter $\eta \equiv (n_2\mu_1)/(n_1\mu_2)$ (n is the index of refraction) and the so-called magnification factor $s \equiv \cos\psi/\cos\varphi$. To simplify notation, we have redefined $\theta/\eta_1 \rightarrow \theta$ ($\eta_1 = n_1/\mu_1$, and if the surrounding medium is vacuum space, θ coincides with the parameter in the action).

Figures 1 and 2 show the amplitude fields both for refracted and reflected waves compared to θ-vacuum. It is interesting to note, in Fig. 2, the Brewster angle for $\eta = 1.1$ and $\phi = 60^o$, for a value of $\theta = 0.504$. This issue will be discussed in more detail below.

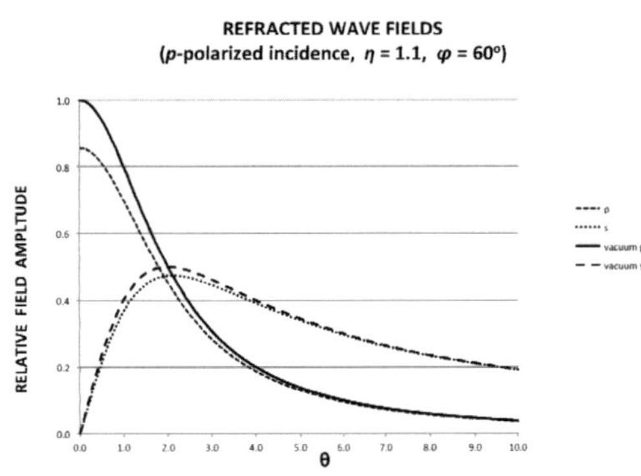

Figure 1: Field amplitude for the refracted wave. Comparison is made with θ-vacuum.

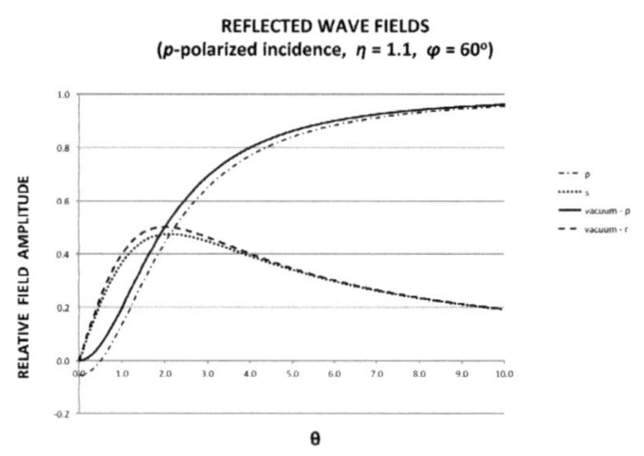

Figure 2: Field amplitude for the reflected wave. Note the Brewster angle, where p-component of the field vanishes, for $\theta = 0.5$.

Journal of Physics: Conference Series **720** (2016) 012023 doi:10.1088/1742-6596/720/1/012023

4. Polarization rotation

We see from (3.1) and (3.2), and unlike the non-θ systems, the p- and s-components of fields mix each other for both outcoming waves. Therefore, the reflected and refracted waves experience a polarization plane rotation, similar to the Kerr-Faraday rotation (Kerr case corresponds to the reflection of light by a magnetic surface) [14]. Figures 3 and 4 show those polarization angles (measured with respect to the incidence plane) for a p-polarized incident wave.

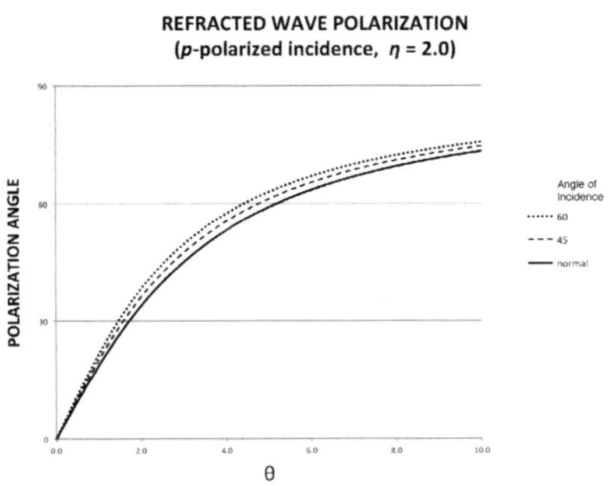

Figure 3: Refracted and reflected waves polarization for different incidence angles.

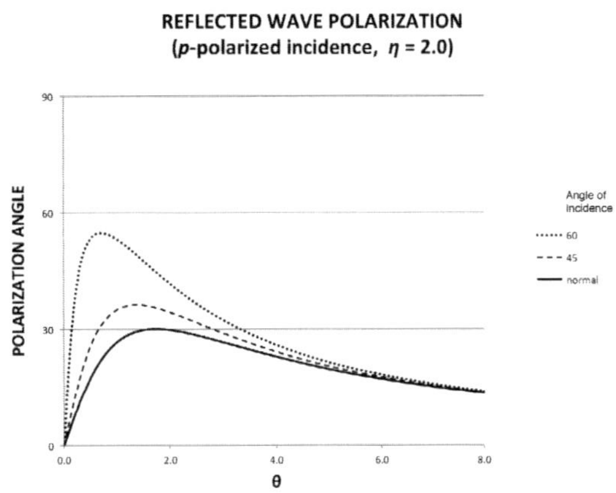

Figure 4: Refracted and reflected waves polarization for different incidence angles.

On the other hand, Figures 5 and 6 show the waves polarization rotation for different values of η (for $\eta = 1$ we have the θ-vacuum case).

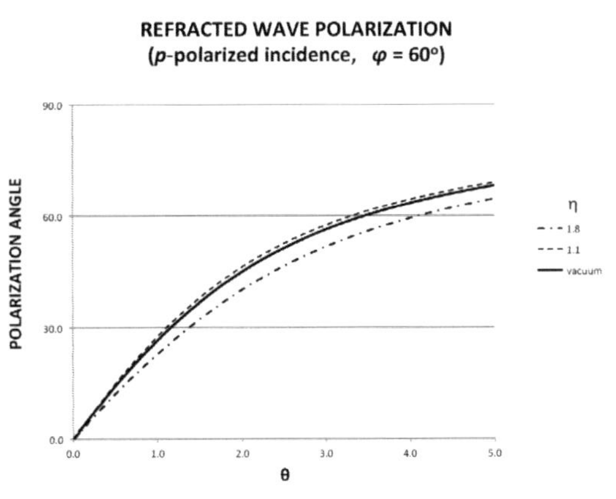

Figure 5: Refracted waves polarization for different system electric properties.

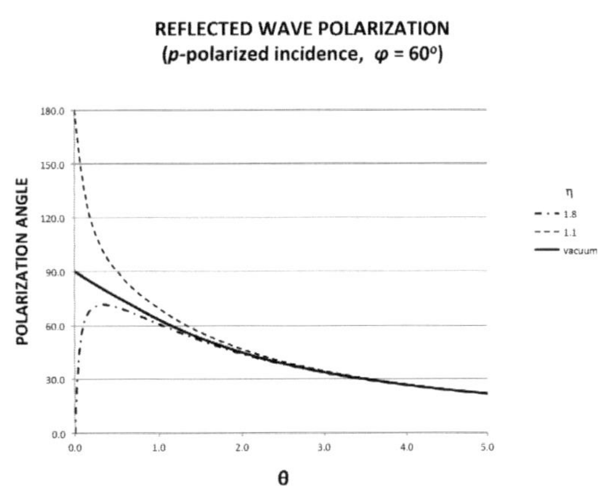

Figure 6: Reflected waves polarization for different system electric properties.

5. Brewster angle for θ-systems

The radiation reflected by a nonconducting surface could eventually become s-polarized, for the well-known Brewster angle of incidence, which we call the s-Brewster angle. The novelty here is that a nonstandard p-Brewster angle appears, where reflected wave becomes polarized in the plane of incidence.

The Brewster angles in terms of θ are shown in Figures 7 and 8. From Fig. 7, we see that even for rather small values of θ, the standard s-Brewster angle changes notoriously . For large values of θ compared to η, no s-Brewster angle is obtained (it goes to 90°) and, as an important difference with normal case, the standard s-Brewster angle depends on incident wave polarization, α_i.

On the other hand, in order to obtain the p-Brewster angle the incident wave must not be p-polarized and η has to be larger enough. Also, its existence is restricted to a definite domain of values of θ, different for each value of the magneto-electric parameter η, with a strong variation of the angle for the values of θ at the ends of the interval. Moreover, a tendency is observed for the angle to come closer to 90°. Actually, as η goes to 1 (vacuum case), p-Brewster angle will no longer exist.

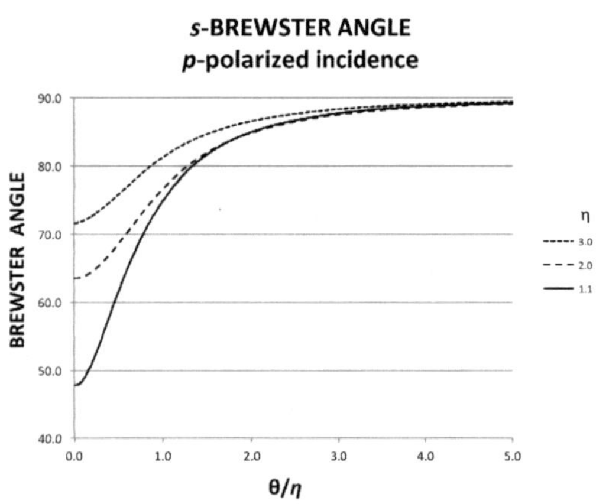

Figure 7: Standard s-Brewster angle for different values of η and when incident wave is p-polarized.

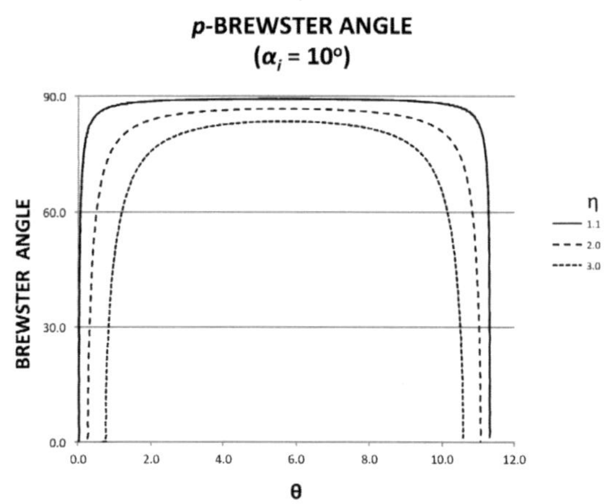

Figure 8: Nonstandard p-Brewster angle for different values of η.

6. θ-system reflectance and transmittance

Energy distribution between reflected and refracted waves is modified with respect to normal systems. Figures 9 and 10 show the situation in terms of θ.

For increasing values of θ, the material becomes progressively more opaque, this being increased by higher values of the magneto-electric parameter η.

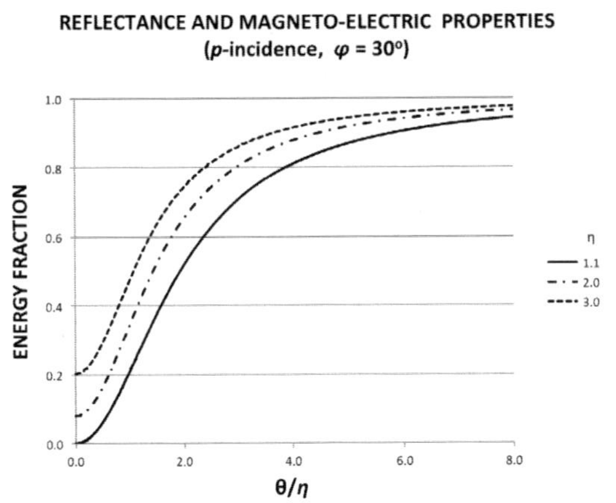

Figure 9: Reflectance as a function of θ/η for different values of magneto-electric parameter η. The incident wave is p-polarized.

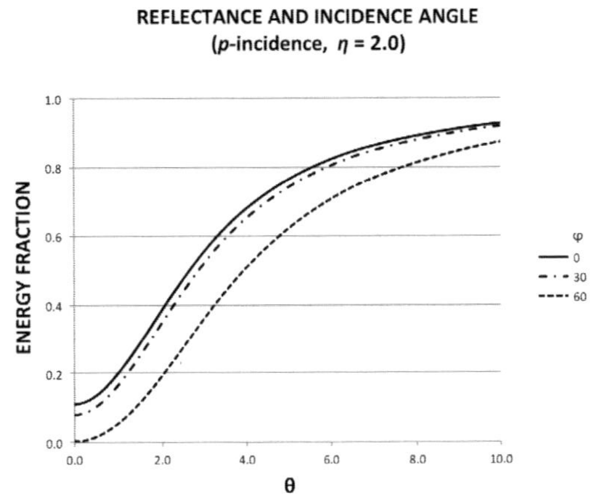

Figure 10: Reflectance in terms of θ for different values of the angle of incidence ϕ, including normal incidence. The incident wave is p-polarized.

7. Discussion

Results presented here show that electromagnetic radiation inciding in a bounded θ-system exhibits interesting behaviour in its polarization properties and the energy distribution of the reflected and refracted waves at the interface. In particular, compared to a θ-vacuum, where no matter is present ($\eta = 1$) but topology is not trivial, we find that magneto-electric properties modify significantly the system behaviour. In particular, for a small θ, the reflected wave polarization can be very different according to the value of η, and differing in both cases from the θ-vacuum case, where reflection polarization plane rotates in 90° with respect to the incident wave polarization. We also have two Brewster angles, both significantly depending on the interplay between the magneto-electric properties and θ parameter. A nonstandard p-Brewster angle arises, where reflected radiation becomes polarized in the plane of incidence. Finally, we find that, if θ is larger enough, the reflected wave could carry a larger part of the energy compared to normal non-θ systems. A more detailed examination and the analytical expressions for some of the results presented here are discussed in Ref. [15].

Acknowledgments

I am very grateful to J. Zanelli for discussions and suggestions and to Centro de Estudios Científicos (CECS) for warm hospitality. This work was partially supported by CONICYT grant PIA-1115. The P4-Center for Research and Applications in Plasma Physics and Pulsed Power Technology is partially supported by the Comisión Chilena de Energía Nuclear. This work is dedicated to my late daughter Mariana Huerta, twelve years after she participated as student in the XIII Simposio Chileno de Física at Concepción.

Journal of Physics: Conference Series **720** (2016) 012023 doi:10.1088/1742-6596/720/1/012023

References

[1] Chern S S and Simons J 1974 *Ann. Math.* **99** 48-69.

[2] Deser S, Jackiw R and Templeton S 1982 *Phys. Rev. Lett.* **48** 975; 1984 *Ann. Phys.* NY **140** 372.

[3] Zanelli J 2005 Lecture notes on Chern-Simons (super-)gravities *Preprint* arXiv:hep-th/0502193.

[4] Peccei R D and H. R. Quinn 1977 *Phys. Rev. Lett.* **38** 1440.

[5] Wilczek F 1987 *Phys. Rev. Lett.* **58** 1799.

[6] Girvin S M and MacDonald A H 1987 *Phys. Rev. Lett.* **58** 1252; Zhang S-C, Hansson T H and Kivelson S 1989 *Phys. Rev. Lett.* **62** 82; Zhang S-C 1992 *Int. J. Mod. Phys.* B **6** 25.

[7] Qi X-L, Hughes T L, Zhang S-C 2008 *Phys. Rev.* B **78** 195424, and references therein.

[8] See also Qi X-L and Zhang S-C 2011 *Rev. Mod. Phys.* **83** 1057, for other references.

[9] Huerta L and Zanelli J 2010 *Introduction to Chern-Simons Theories, Proc. of Sci.* ICFI2010 004.

[10] Canfora F, Rosa L and Zanelli J 2011 *Phys. Rev.* D **84** (in press) (*Preprint* arXiv:hep-th/1105.2490).

[11] Cambiaso M, Huerta L, Toloza A, Urrutia L F and Zanelli J 2014, work in progress.

[12] Huerta L and Zanelli J 2012 *Phys. Rev.* D **85** 085024.

[13] Maciejko J, Qi X-L, Drew H D, and Zhang S-C 2010 *Phys. Rev. Lett.* **105** 166803; Tse W-K and MacDonald A H 2010 *Phys. Rev.* B **82** 161104.

[14] Kerr J 1877 *Phil. Mag.* **3** 321; Kerr J 1878 *Phil. Mag.* **5** 161. See also Weinberger P 2008 *Phil. Mag. Lett.* **88** 897.

[15] Huerta L 2014 *Phys. Rev.* D **10** 105026.

Finite temperature effects and the validity of the Weinberg sum rules

Alejandro Ayala[1,3], C. A. Dominguez[3], M Loewe[2,3,4] and Y. Zhang[5]

[1] Instituto de Ciencias Nucleares, Universidad Nacional Autónoma de México, Apartado Postal 70-543, México Distrito Federal 04510, Mexico.
[2] Instituto de Física, Pontificia Universidad Católica de Chile, Casilla 306, Santiago 22, Chile.
[3]Centre for Theoretical and Mathematical Physics, and Department of Physics, University of Cape Town, Rondebosch 7700, South Africa.
[4] Centro Científico-Tecnológico de Valparaíso, Casilla 110-V, Valparaíso, Chile.
[5]CSIR National Laser Centre, PO Box 395, Pretoria 0001, South Africa.

E-mail: mloewe@fis.puc.cl

Abstract.
Using resent independent results from QCD sum rules for the thermal evolution of hadronic parameters in the vector and the axial-vector channels, we discuss the saturation of the two Weinberg sum rules. It turn out that both sum rules are quite well satisfied in a wide range from $T = 0$ up to $T/T_c \simeq 0.7 - 0.8$. At higher temperatures, coming closer to T_c, there is an asymmetry between both channels since in the vector case there is a leading order effect, proportional to T^2, due to a one loop pion contribution in the space-like region, which is absent in the axial-vector case. This leads then to a small deviation. More important, though, in this region the QCD sum rules for the hadronic parameters begin to have no solutions since the widths of the ρ and the a_1-mesons diverge signaling the occurrence of deconfinement. Close to and at T_c there are no pions left in the medium and chiral symmetry is restored so that the Weinberg sum rules are trivially satisfied.

1. Introduction

The two Weinberg sum rules (WSR) were formulated, in the $T = 0$ case, in the frame of the chiral $SU(2) \times SU(2)$ symmetry and current algebra [1]. They are given by

$$W_1 \equiv \int_0^\infty ds\, \frac{1}{\pi}\, [\mathrm{Im}\Pi_V(s) - \mathrm{Im}\Pi_A(s)] = 2\, f_\pi^2 \,, \tag{1}$$

$$W_2 \equiv \int_0^\infty ds\, s\, \frac{1}{\pi}\, [\mathrm{Im}\Pi_V(s) - \mathrm{Im}\Pi_A(s)] = 0 \,, \tag{2}$$

where $f_\pi = 92.21 \pm 0.14\,\mathrm{MeV}$ [2], and

$$\begin{aligned} \Pi^{VV}_{\mu\nu}(q^2) &= i \int d^4x\, e^{iqx} < 0|T(V_\mu(x)V^\dagger_\nu(0))|0 > \\ &= (-g_{\mu\nu}\, q^2 + q_\mu q_\nu)\, \Pi_V(q^2) \,, \end{aligned} \tag{3}$$

$$\begin{aligned}\Pi^{AA}_{\mu\nu}(q^2) &= i \int d^4 x \, e^{iqx} < 0|T(A_\mu(x) A^\dagger_\nu(0))|0 > \\ &= -g_{\mu\nu}\,\Pi_1(q^2) - q_\mu q_\nu\,\Pi_A(q^2)\,, \end{aligned} \tag{4}$$

with $V_\mu(x) =: \bar{d}(x)\gamma_\mu u(x)$: the conserved vector current in the chiral limit and $A_\mu(x) =: \bar{d}(x)\gamma_\mu\gamma_5\, u(x)$: the axial-vector current, being $q_\mu = (\omega, \vec{q})$ the four-momentum carried by the currents. The functions $\Pi_{V,A}(q^2)$ satisfy dispersion relations since they are free from kinematical singularities. In perturbative QCD (PQCD) they are normalized as

$$\mathrm{Im}\,\Pi_V(q^2) = \mathrm{Im}\,\Pi_A(q^2) = \frac{1}{4\pi}\left[1 + \mathcal{O}\left(\alpha_s(q^2)\right)\right]. \tag{5}$$

In the framework of QCD Eqs. (1)-(2) become effectively finite energy sum rules (in the chiral limit: $m_u = m_d = 0$)

$$W_{n+1}(s_0) \equiv \int_0^{s_0} ds\, s^n \frac{1}{\pi}\left[\mathrm{Im}\Pi_V(s) - \mathrm{Im}\Pi_A(s)\right] = 2\, f_\pi^2\, \delta_{n0}\,, \tag{6}$$

where $s_0 \simeq 1 - 3\,\mathrm{GeV}^2$ is the squared energy beyond which QCD is valid, and both sum rules have been combined. This result also follows from Cauchy's theorem in the complex squared energy plane, together with the assumption of quark-hadron duality. Given a Green function $\Pi(s)$, e.g. $\Pi_{V,A}(s)$, and an analytic function $f(s)$, Cauchy's theorem leads to

$$\int_0^{s_0} ds\, f(s)\, \frac{1}{\pi}\, \mathrm{Im}\Pi(s) = -\frac{1}{2\pi i}\oint_{|s|=s_0} f(s)\,\Pi(s)\, ds\,, \tag{7}$$

where the discontinuity across the real axis involves the hadronic sector, and the contour integral requires asymptotic information if the integration radius s_0 is large enough. According to quark-hadron (global) duality $\Pi(s)$ is assumed to be given by the QCD operator product expansion (OPE). A warning about this assumption, now referred to as potential duality violations associated with the QCD behaviour on or close to the positive real axis, was raised long ago in [3]. Different tests of the WSR were performed [4]-[5] using experimental data on τ-decays from the ARGUS collaboration at DESY [6], and later [7] from the ALEPH collaboration [8]. Both determinations led to a better saturation of the WSR if an integral kernel, e.g. a non-trivial $f(s)$ in Eq.(7), was introduced, e.g. such that $f(s_0) = 0$. This was interpreted in [7] as a possible signal of duality violations, on account of a considerably improved saturation of the WSR obtained with a *pinched* kernel [7], [9] $f(s) = (1 - s/s_0)$, i.e.

$$\begin{aligned} WP(s_0) &\equiv \int_0^{s_0} ds \left(1 - \frac{s}{s_0}\right)\frac{1}{\pi}\left[\mathrm{Im}\Pi_V(s) - \mathrm{Im}\Pi_A(s)\right] \\ &= 2\, f_\pi^2\,. \end{aligned} \tag{8}$$

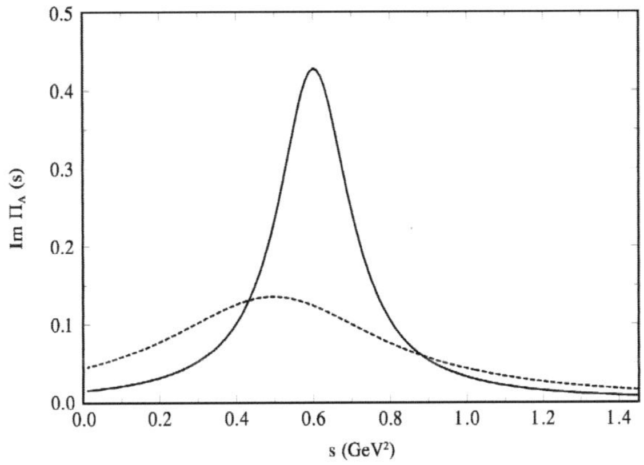

Figure 1. The vector spectral function at $T = 0$, Eq. (9) (solid curve), and at $T = 175$ MeV (dotted curve) with thermal parameters given in Eqs. (11)-(14). The FESR make use of only that part of the spectral function whose integral matches the QCD integral, according to quark-hadron duality

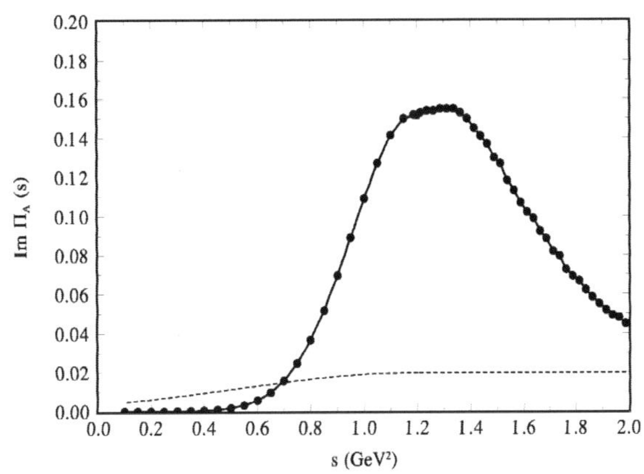

Figure 2. Solid curve is the axial-vector (a_1-resonance) spectral function at $T = 0$ fitted to the ALEPH data [8], shown with error bars the size of the data points. Dotted curve is the spectral function at $T = 175$ MeV with thermal parameters given in Eqs. (16)-(18). The FESR make use of only that part of the spectral function whose integral matches the QCD integral, according to quark-hadron duality

2. Thermal Hadronic Spectral Functions

First, let us concentrate on the vector channel. In a recent analysis [10] it was parameterized as a Breit-Wigner distribution

$$\frac{1}{\pi}\mathrm{Im}\Pi_V^{(+)}(s,T) = \frac{1}{\pi}\frac{1}{f_\rho^2}\frac{M_\rho^3\,\Gamma_\rho}{\left(s-M_\rho^2\right)^2 + M_\rho^2\,\Gamma_\rho^2}\,, \tag{9}$$

where the upper index (+) refers to the time-like region ($q^2 > 0$), and all parameters depend on the temperature. At $T = 0$ they are normalized as follows [2]: $f_\rho(0) = 5$, $M_\rho(0) = 0.776$ GeV, and $\Gamma_\rho(0) = 0.145$ GeV. This parametrization was then used in the first three FESR

$$(-)^{(N-1)}\quad C_{2N}\langle\hat{\mathcal{O}}_{2N}\rangle = 4\pi^2\int_0^{s_0} ds\, s^{N-1}\,\frac{1}{\pi}\mathrm{Im}\Pi(s)|_{\mathrm{HAD}}$$

$$-\quad \frac{s_0^N}{N}\,[1+\mathcal{O}(\alpha_s)]\quad (N=1,2,\cdots)\,, \tag{10}$$

and where the leading order vacuum condensates in the chiral limit are the dimension $d = 4$ gluon condensate and the dimension $d = 6$ four-quark condensate. From the above FESR we got for the continuum threshold, i.e. the onset of PQCD, $s_0 = 1.44\ \mathrm{GeV}^2$, and we checked that the resulting dimension $d = 4$ and $d = 6$ vacuum condensates were in agreement with independent determinations from data analysis. Alternative parametrizations lead to similar results. When extending this analysis to the finite temperature scenario we obtained the following results:

$$\frac{s_0(T)}{s_0(0)} = 1.0 - 0.5667\left(\frac{T}{T_c}\right)^a - 4.347\left(\frac{T}{T_c}\right)^b\,, \tag{11}$$

with $a = 11.38$, $b = 68.41$, $T_c = 197\,\mathrm{MeV}$ and

$$\frac{\Gamma_\rho(T)}{\Gamma_\rho(0)} = \frac{1}{\left[1-\left(\frac{T}{T_c}\right)^3\right]}\,, \tag{12}$$

$$\frac{M_\rho(T)}{M_\rho(0)} = 1 - \left(\frac{T}{T_M^*}\right)^{10}\,, \tag{13}$$

where $T_M^* = 222$ MeV, allowed to vary in the range $T_M^* = 210 - 240$ MeV, and finally

$$\frac{f_\rho(T)}{f_\rho(0)} = 1.0 - 0.3901\left(\frac{T}{T_c}\right)^c + 0.04155\left(\frac{T}{T_c}\right)^d\,, \tag{14}$$

with $c = 10.75$ and $d = 1.269$.

Let us consider now the axial-vector channel. A fit was made to the ALEPH data on τ-decay to be used as the $T = 0$ normalization. As in the vector channel, the first three FESR were employed to determine s_0 and to check consistency of the results for the dimension $d = 4, 6$ condensates. The parametrization leading to the best χ^2 is given by [10]

$$\frac{1}{\pi}\mathrm{Im}\Pi_A(s)|_{a_1} \;=\; C\,f_{a_1}\exp\left[-\left(\frac{s-M_{a_1}^2}{\Gamma_{a_1}^2}\right)^2\right]$$

$$(0\ \le s \le 1.2\,\mathrm{GeV}^2)\,, \tag{15}$$

Journal of Physics: Conference Series **720** (2016) 012024 doi:10.1088/1742-6596/720/1/012024

$$\frac{1}{\pi}\mathrm{Im}\Pi_A(s)|_{a_1} = C\, f_{a_1} \exp\left[-\left(\frac{1.2\ \mathrm{GeV}^2 - M_{a_1}^2}{\Gamma_{a_1}^2}\right)^2\right]$$

$$(1.2\,\mathrm{GeV}^2 \leq s \leq 1.45\,\mathrm{GeV}^2)\,, \tag{16}$$

$$\frac{1}{\pi}\mathrm{Im}\Pi_A(s)|_{a_1} = C\, f_{a_1} \exp\left[-\left(\frac{s - M_{a_1}^2}{\Gamma_{a_1}^2}\right)^2\right]$$

$$(1.45\,\mathrm{GeV}^2 \leq s \leq M_\tau^2)\,, \tag{17}$$

where $M_{a_1} = 1.0891$ GeV, $\Gamma_{a_1} = 568.78$ GeV, $C = 0.662$ and $f_{a_1} = 0.073$ (notice that the latter two parameters were split to facilitate comparison between f_{a_1} and f_ρ). Regarding our parametrization, this spectral function cannot be fitted with a normal Breit-Wigner expression, essentially due to the large width of the a_1-meson, and some background contamination. For instance, the early ARGUS data [6] was fitted with a Breit-Wigner modulated by a high degree polynomial [12]. Solving the FESR the onset of PQCD turns out to be identical to that in the vector channel, i.e. $s_0 = 1.44$ GeV^2, so that Eq.(17) is never needed. The finite temperature results for $s_0(T)$, $f_\pi(T)$, $f_{a_1}(T)$, and $\Gamma_{a_1}(T)$ can be written generically as

$$\frac{Y(T)}{Y(0)} = 1 + a_1 \left(\frac{T}{T_c}\right)^{b_1} + a_2 \left(\frac{T}{T_c}\right)^{b_1}\,, \tag{18}$$

where the corresponding coefficients are given in Table I.

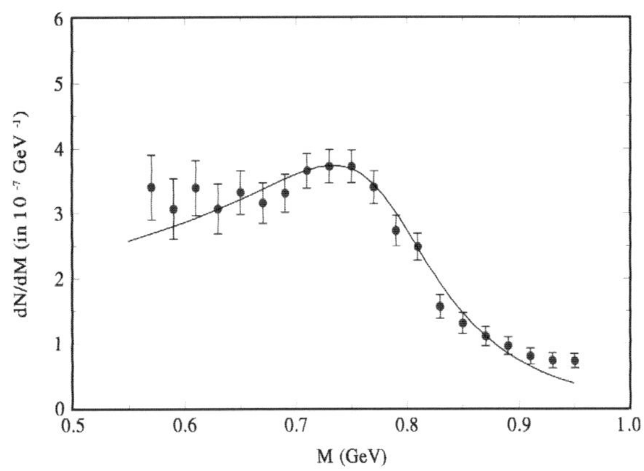

Figure 3. Experimental data on the dimuon production spectrum in In-In ion collisions around the ρ-peak from the NA60 Collaboration [14], compared with the prediction using the vector spectral function from [10] (solid line), reproduced here in Eqs.(9), (11)-(14).

Parameter	Coefficients in Eq. (18) a_1	a_2	b_1	b_2
$s_0(T)$	- 28.5	-0.6689	35.60	3.93
$f_\pi(T)$	- 0.2924	- 0.7557	73.43	11.08
$f_{a_1}(T)$	- 19.34	14.27	7.716	6.153
$\Gamma_{a_1}(T)$	2.323	1.207	20.24	7.869

Table 1. The values of the coefficients entering Eq. (18).

We kept the a_1 mass constant, since it hardly changes with temperature. This behavior can be ascribed to the very large width of the a_1 resonance. It should be noticed the (different) way the leptonic decay constants enter their respective spectral functions. However, the full T-dependence of these functions is the result of the aggregate behaviour of all their parameters. This T-dependence is consistent with a quark deconfinement scenario in both channels.
It should be mentioned that the FESR analyses leading to the above results [10]-[11] find that the hadronic integrals match the QCD integrals in the range $T \simeq 0 - 0.9\ T_c$. Beyond this temperature upper limit the integrals cease to have support. Since this upper limit is so close to T_c a smooth extrapolation should be acceptable.

The parametrization of the spectral function in the vector channel was later used to predict the dimuon production spectrum in heavy ion collisions at high energies [13]. Excellent agreement was obtained after comparing with data on In-In collision around the ρ-meson peak from the NA60 Collaboration [14]. Figure 3 shows the experimental data and the prediction using the vector spectral function from [10], as reproduced here in Eqs.(9), (11)-(14).

It should be clear from Figs. (1) and (2) that the vector spectral function evolves with increasing T independently of the axial-vector one. The latter involves, in addition to the a_1 resonance, the pion pole given by $f_\pi^2(T) \propto |\langle \bar{q}\, q \rangle|$. This contribution vanishes at $T = T_c$, signaling chiral-symmetry restoration, i.e. a transition from a Nambu-Goldstone to a Wigner-Weyl realization of chiral $SU(2) \times SU(2)$ symmetry. It should be recalled that the thermal quark condensate from lattice QCD has been used as input to the FESR in the axial-vector channel in [11]. As shown in [11] the T-dependence of the ratio$f_\pi(T)^2/f_\pi(0)^2$ follows closely that of $s_0(T)/s_0(0)$, vanishing at the same temperature within the accuracy of the method. This independent and clearly very different approach to $T = T_c$ exhibited by the thermal spectral functions does not suggest any potential chiral mixing as some authors suggested [17]
In the QCD sector the PQCD spectral functions for both $q^2 > 0$ (*annihilation*) and $q^2 < 0$ (*scattering*), as well as the dimension $d = 4$ gluon condensate are chiral symmetric, thus not contributing to Eq.(6). In the hadronic sector for $q^2 < 0$ (*scattering*) there is an asymmetric contribution from pion loops, which in the vector channel at leading order is [15]

$$\begin{aligned} \frac{1}{\pi}\mathrm{Im}\,\Pi_V^{(-)}|_{HAD}(s,T) &= \frac{2}{3\pi^2}\,\delta(s)\int_0^\infty y\, n_B\left(\frac{y}{T}\right) dy \\ &= \frac{T^2}{9}\,\delta(s)\,, \end{aligned} \tag{19}$$

where $n_B(z) = 1/(e^z - 1)$ is the Bose thermal function. The corresponding contribution in the axial-vector channel is a higher order two-loop effect, which is traditionally neglected.

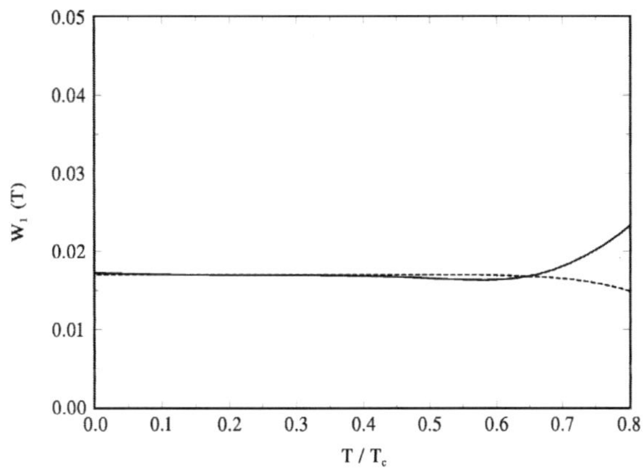

Figure 4. The first WSR, Eq.(6), at finite T. Solid (dash) line is the left (right) hand side of Eq.(6). The divergence at high T is caused by the asymmetric hadronic *scattering* contribution in the space-like region ($q^2 < 0$), which disappears at deconfinement ($T = T_c$).

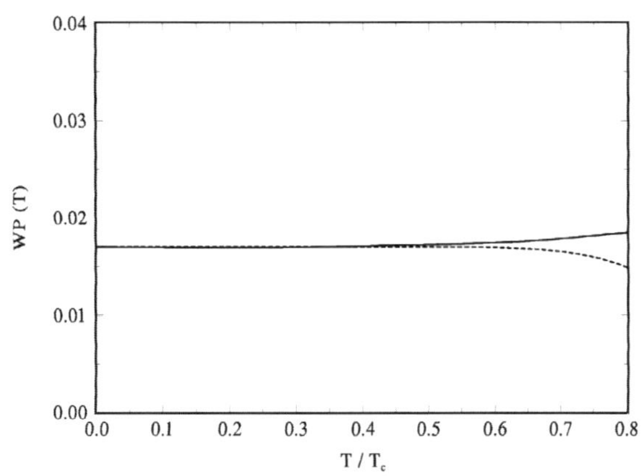

Figure 5. The pinched WSR, Eq. (8), at finite T. Solid (dash) line is the left (right) hand side of Eq.(6). The divergence at high T is caused by the asymmetric hadronic *scattering* contribution in the space-like region ($q^2 < 0$), which disappears at deconfinement ($T = T_c$).

3. Conclusions

Starting at $T = 0$ and using the results from the three FESR obtained in [10]-[11], i.e. $s_0(0)|_V = s_0(0)|_A = 1.44$ GeV^2 and the parametrizations Eq.(9) and Eqs.(15)-(17), the WSR, Eq.(6), and the pinched WSR, Eq.(8), are satisfied to better than 1%. The corresponding results at finite T are shown in Fig. 4 for the first WSR, $W_1(T)$, Eq.(6), and in Fig. 5 for the pinched WSR, $WP(T)$, Eq. (8). The solid (dash) lines correspond to the left (right) hand sides of the sum rules. Clearly, the upper limit of integration in the WSR, i.e. $s_0(T)$, is not strictly chiral-symmetric, as seen from Eqs.(11) and (18). The divergence between the left and the right hand sides of the WSR at high T is due to the asymmetric hadronic *scattering* term present in the vector spectral function, but loop suppressed in the axial-vector channel. The importance of this term is short lived, as it must disappear at $T = T_c$, since there would be no longer pions in the medium. In any case, since the source of this contribution is fully understood, it should not be interpreted as an invalidation of the thermal WSR.

As indicated in Figs. 4-5 the WSR are very well satisfied up to $T/T_c \simeq 0.75$. The source of the departure between the left hand side of the WSR and its right hand side, i.e. $f_\pi^2(T)$, or alternatively the thermal quark condensate $|\langle\bar{q}q\rangle(T)|$, is well understood. It is due to the two-pion loop contribution to the vector spectral function, growing like T^2, but disappearing at $T = T_c$. If this contribution would have been chiral symmetric, then it would have canceled out and the WSR would have been fully satisfied up to $T = T_c$.

The results for the behaviour of the WSR in the full temperature range can be interpreted as follows. First of all, at $T = 0$ the WSR make the highly non-trivial statement that the integrated difference between the vector and the axial-vector (including the pion) spectral functions vanishes. These functions could not be more different, i.e. the vector correlator has no pole, and the ground state resonance is narrow, while the axial-vector involves a pole and a very broad resonance almost twice as heavy. Next, there is no a-priori reason for the WSR to be satisfied at finite temperature. Their original derivation was done in the context of current algebra and chiral $SU(2) \times SU(2)$ symmetry. Today we reinterpret them in the framework of QCD, which does exhibit this chiral symmetry in the limit of vanishing light quark masses. In principle, an extension to finite T could lead to more than one scenario, including one in which the sum rules are not satisfied at all, except at $T = T_c$ where they must be trivially satisfied. The results obtained from thermal QCD FESR [10]-[11] clearly indicate that the vector spectral function evolves with T independently of the axial-vector spectral function. While both disappear at $T = T_c$, they do so following their own distinctive thermal evolution. In other words, there is no a-priori relation between these two channels, unless one were to demand it. Given the successful saturation of the WSR there seems to be no reason to justify such a demand. The behaviour of the spectral function parameters points to deconfinement, as the onset of PQCD, $s_0(T)$, and the resonance couplings decrease, while the widths increase substantially with increasing T. This results in the ρ and a_1 resonances becoming progressively broader, with decreasing couplings to their respective currents. At the same time, the pion decay constant/quark condensate decreases with increasing T, vanishing at T_c. The thermal behaviour of the hadron mass is irrelevant in the context of the approach to deconfinement. In fact, the mass only specifies the position of the real part of the resonance Green function in the complex squared energy s-plane. Its moving upwards or downwards with temperature is no indication of a deconfinement phase transition. It is the imaginary part of the Green function, i.e. the hadronic width, the relevant parameter in this context. Its considerable increase with increasing temperature thus serves as a phenomenological signal for deconfinement.

Alternative analyses of the thermal WSR have been performed recently [16], with somewhat different emphasis, e.g. in connection with potential chiral mixing. From our discussion, we would like to stress that we did not find any evidence for a possible mixing at finite temperature between the vector and axial-vector channel as some authors have claimed to be the case [17].

Acknowledgments
Support for this work has been received in part from DGAPA-UNAM under grant number PAPIIT-IN103811, CONACyT-México under grant number 128534, FONDECYT (Chile) under grants numbers 1130056, the National Research Foundation (South Africa) and the University of Cape Town Research Committee.

References

[1] S. Weinberg, Phys. Rev. Lett. 1967 **18**, 507.
[2] J. Beringer *et al.* (Particle Data Group) Phys. Rev. D 2012 **86**, 010001.
[3] R.Shankar, Phys. Rev. D 1977 **15**, 755.
[4] R. D. Peccei and J. Solà, Nucl. Phys. B 1987 **281**, 1 .
[5] C. A. Dominguez and J. Solà, Phys. Lett. B 1988 **208**, 131.
[6] H. Albrecht *et al.* (ARGUS Collaboration), Z. Phys. C 1986 **33**, 7.
[7] C. A. Dominguez and K. Schilcher, Phys. Lett. B 1999 **448**, 93.
[8] ALEPH Coll., R. Barate *et al.*, Eur. Phys. J. C 1988 **4**, 409.
[9] The use of *pinched* kernels in other QCD sum rules was independently proposed in K. Maltman, Phys. Lett. B 1998 **440**, 367 .
[10] A. Ayala, C. A. Dominguez, M. Loewe and Y. Zhang, Phys. Rev. D 2012 **86**, 114036. Notice that there is a misprint in Eq.(8) of this reference; the correct formula is given in Eqs.(15)-(17) of this paper.
[11] C. A. Dominguez, M. Loewe and Y. Zhang, Phys. Rev. D 2012 **86**, 034030 .
[12] C. A. Dominguez and J. Solà, Z. Phys. C 1988 **40**, 63 .
[13] A. Ayala, C. A. Dominguez, L. A. Hernandez, M. Loewe and A. J. Mizher, Phys. Rev. D 2013 **88**, 114028 .
[14] R. Arnaldi *et al.* (NA60 Collaboration), Phys. Rev. 2006 **96**, 16302 ; (NA60 Collaboration), Phys. Rev. Lett. 2008 **100**, 022302 ; S. Damjanovic for the NA60 Colaboration, J. Phys. G 2008 **35**, 104036 ; (NA60 Collaboration), Eur. Phys. J. C 2009 **59**, 607 ; (NA60 Collaboration), Eur. Phys. J. C 2009 **61**, 711 .
[15] A. I. Bochkarev and M. E. Shaposnikov, Nucl. Phys. B 1986 **268**, 220.
[16] N. P. M. Holt, P.M. Hohler and R. Rapp, Phys. Rev. D 2013 **87**, 076010 ; P. M. Hohler and R. Rapp, Phys. Lett. B 2014 **731**, 103 , and references therein.
[17] M. Dey, V. L. Eletsky, and B. L. Ioffe, Phys. Lett. B 1990 **252**, 620;

Aspects of the T-duality construction for the Supermembrane theory

M.P. Garcia del Moral[1], J.M. Pena[2], A. Restuccia[1]

[1]Departamento de Física, Universidad de Antofagasta, Aptdo 02800, Chile.
[2] Departamento de Física, Facultad de Ciencias,
Universidad Central de Venezuela, A.P. 47270, Caracas 1041-A, Venezuela

E-mail: maria.garciadelmoral@uantof.cl;
jpena@fisica.ciens.ucv.ve;alvaro.restuccia@uantof.cl

Abstract. In this note we explicitly show how the generalization of the T-duality symmetry of the supermembrane theory compactified in $M_9 \times T^2$ can be reduced to a parabolic subgroup of $SL(2, Z)$ that acts non-linearly on the moduli parameters and on the KK and winding charges of the supermembrane. This is a first step towards a deeper understanding of the dual relation between the parabolic type II gauged supergravity in nine dimensions.

1. Introduction

There is a renewed interest in finding a T-duality invariant target space formulation of String theory [1, 2, 3] and respectively a U-duality invariant action of M-theory [4, 5, 6, 7, 8]. Indeed, since String Theories and their corresponding effective Supergravity actions are limits of M-theory connected through dualities, M-theory should contain dualities as symmetries of the theory. This idea has been mainly realized in a bottom-up approach, by generalizing the geometry through the introduction of extra coordinates associated to the winding modes in supergravity actions in the context of Double Field Theory [9, 11, 10]. This approach was originally introduced by [12, 13, 14]. The reduction to normal coordinate space leads to gauged supergravity actions [15]. In this note we follow a different approach to realize it: we deep on the study of the U-duality invariant action of the Supermembrane worldvolume action [16, 17, 18]. This search included as an intermediate step, to obtain a generalization of T-duality for M-theory, in particular for supermembrane theory [17, 18]. This generalization acts locally on the theory and in the string theory limit, one recovers the standard T-duality transformation. Not less importantly, it also acts globally on the torus bundle with a monodromy description of the supermembrane theory compactified on $M_9 \times T^2$ target-space [19, 20, 16]. Previous search of the Supermembrane t-duality generalization were investigated in [21, 22, 23].

In this paper we are going to determine explicitly the construction of T-duality transformation, showing as a new result that it may be reduced to a nonlinear realization of a parabolic equivalence subgroup of $SL(2, Z)$ with a dependence on the Kaluza-Klein and winding modes of the theory. This fact will be of relevance when trying to obtain a deeper understanding of the role of T-dualities and gauged supergravities in nine dimensions. The paper is organized as follows: in section 2 we recall local T-duality transformation for the supermembrane and its String limit. Since in this note we are only interested in the T-duality construction, we will

Journal of Physics: Conference Series **720** (2016) 012025 doi:10.1088/1742-6596/720/1/012025

not discuss global aspects of T-duality. In section 3 the new result is introduced through the construction of T-duality transformation. Section 4 is devoted for conclusions.

2. T-duality

In this section we will consider the local and global action of a generalized T-duality at the M-theory level. In particular, at the level of the supermembrane theory it acts globally in a relevant way. In fact, when the supermembrane is compactified on $M_9 \times T^2$, and the T-duality acts on the moduli of the 2-torus and interchanges Kaluza-Klein charges of the mass operator to the winding ones W.

The T-duality transformation [24] is:

$$\text{for the moduli}: \quad \widetilde{\tau} = \frac{\delta\tau+\beta}{\gamma\tau+\delta}; \quad \mathcal{Z}\widetilde{\mathcal{Z}} = 1; \quad \text{for the charges}: \quad \widetilde{\mathcal{Q}} = \Lambda_0\mathcal{Q}, \quad \widetilde{W} = \Lambda_0^{-1}W, \tag{1}$$

where we have defined: $\mathcal{Z} := TA\widetilde{Y}$ and $\widetilde{\mathcal{Z}} := T\widetilde{A}Y$, with T denoting the supermembrane tension; τ is a complex moduli $\tau = \mathrm{Re}(\tau) + i\mathrm{Im}(\tau)$, $\mathrm{Im}(\tau) > 0$ (τ is the complex coordinate of the Teichmüller space for $g = 1$); $A = (2\pi R)^2\mathrm{Im}(\tau)$ the area of the target torus and $Y = \frac{R\,\mathrm{Im}(\tau)}{|q\tau-p|}$. We also have $\Lambda_0 = \begin{pmatrix} \delta & \beta \\ \gamma & \delta \end{pmatrix} \in SL(2,Z)$; $\mathcal{Q} = \begin{pmatrix} p \\ q \end{pmatrix}$ the Kaluza-Klein charges, and $W = \begin{pmatrix} l_1 & l_2 \\ m_1 & m_2 \end{pmatrix}$ the nontrivial windings of the supermembrane along the 2-torus of the targetspace. In all these expressions the tilde variables are the transformed quantities under T-duality. T-duality transformations act nonlinearly on the charges of the supermembrane, in distinction with the usual $SL(2,Z)$ transformation. This fact is very relevant for understanding the dual multiplet structure. The KK modes are mapped onto the winding modes and viceversa nonlinearly.

The dual radius of the torus moduli is defined as,

$$\widetilde{R} = \frac{|\gamma\tau+\delta||q\tau-p|^{2/3}}{T^{2/3}(\mathrm{Im}(\tau))^{4/3}(2\pi)^{4/3}R} \tag{2}$$

In the case $\mathcal{Z} = \widetilde{\mathcal{Z}} = 1$ the T-duality transformation becomes a symmetry and it imposes a relation between the tension, the moduli and the KK charges given by:

$$T_0 = \frac{|q\tau-p|}{R^3(\mathrm{Im}(\tau))^2}. \tag{3}$$

Then if we know the values of the moduli, it fixes the allowed tension T_0. Conversely, for a fixed tension T_0, the radius, the Teichmüller parameter of the 2-torus, and the KK charges satisfy (3). Also, for $\mathcal{Z} = 1$ the hamiltonian and the mass operator of the supermembrane with central charges is invariant under T-duality:

$$M^2 = T^2n^2A^2 + \frac{m^2}{Y^2} + T^{2/3}H = \frac{n^2}{\widetilde{Y}^2} + T^2m^2\widetilde{A}^2 + T^{2/3}\widetilde{H}, \tag{4}$$

with $H = \widetilde{H}$.

String T-duality transformation limit. It was possible to consider the string-like configurations from the physical configurations of the supermembrane with central charges [24, 16, 17]. In this condition the fields on the torus Σ, a compact Riemann surface, can be written as:

$$X^m = X^m(\tau, c_1\widehat{X}^1 + c_2\widehat{X}^2), \quad A^r = A^r(\tau, c_1\widehat{X}^1 + c_2\widehat{X}^2), \tag{5}$$

where c_1, c_2 are relative prime integral numbers because the global periodicity condition. X^m, A^r are scalar fields and they may always be expanded on a Fourier basis in term of a double periodic variable of that form. On these configurations we have

$$\{X^m, X^n\} = \{X^m, A^r\} = \{A^r, A^s\} = 0. \tag{6}$$

We then obtain the final expression for the mass contribution of the string states [24]:

$$M_{11}^2|_{SC} = (nT_{11}A)^2 + (\frac{m}{Y})^2 + 8\pi^2 R_{11}T_{11}|q\tau - p|(N_L + N_R) \tag{7}$$

where (p, q) are relatively prime and they may be interpreted as the wrapping of the membrane around the two cycles of the target torus.

The harmonic sector changes [24], that is:

$$dX_h = (qmd\widetilde{X}^1 + pd\widetilde{X}^2) + \widetilde{\tau}(-Qnd\widetilde{X}^1 + Pd\widetilde{X}^2), \tag{8}$$

where p, q and Q, P are the winding numbers of the supermembrane. This transformation leaves the hamiltonian invariant. Then, by restricting the worldvolume configurations of the M2 to those of the string [24], we exactly recover the mass operator of the IIB theory as was formerly conjectured by Schwarz [25], but it has a plus: in the supermembrane with central charges the pure membrane excitations are known. If now a T-duality is performed on the supermembrane M2 mass operator restricted to string-like configurations, then an SL(2,Z) non-perturbative multiplet of IIA is obtained [24].

Next we recover the standard T-duality transformations for the closed string operator as a limit of the T-duality for the Supermembrane. We define:

$$R_1 = \frac{R_{11}^{1/2}}{T_{11}^{1/6}}; \quad R_2 = R_{11}^{3/2}T_{11}^{1/6}\mathrm{Im}(\tau), \tag{9}$$

for the torus degenerating into a circle S^1. Since $R_{11} \to 0$, it implies $R_1 \to 0$ but R_2 is finite, so it corresponds to a closed curve that topologically is a circle. If we re-express the winding condition in terms of the new variables, we have:

$$\oint_{C_s} dX^1 = 2\pi R_1 l_s; \quad \oint_{C_s} dX^2 = 2\pi R_2 m_s. \tag{10}$$

Because $R_1 \to 0$, although l_s is taking finite, the first winding condition vanishes and the only residual winding condition is associated to the S^1 modulus is R_2. The former T-duality relations of the moduli in this limit become reduced to:

$$Z\widetilde{Z} = 1|_{string}, \quad \Rightarrow \quad T_{M2}^{4/3}R_2^3\widetilde{R}_2^3 = 1 \to \widetilde{R}_2 = \frac{\alpha'}{R_2}, \tag{11}$$

where $\widetilde{R}_2 = T_{11}^{1/2}\widetilde{A}^{1/2}\widetilde{Y}^{1/2}$. This defines for $T_{(p,q)} = T$ on the IIB string side precisely the duality relation of the strings. The transformation on the charges and windings are given by (1) and we finally obtain:

$$\{R; (l_1, m)\} \overset{T-duality}{\longrightarrow} \{\widetilde{R} = \frac{\alpha'}{R}; (m, l_1)\}, \tag{12}$$

where m is the common factor between the Kaluza-Klein charges (p, q).

3. Explicit construction of the Λ_0 transformation

In this section we are going to deep on the characterization of the T-duality the map Λ_0 of the supermembrane. In [16] (Appendix A), it was constructed the Λ_0 transformation as a nonlinear realization of $SL(2, Z)$ of the form

$$\Lambda_0 = \begin{pmatrix} \alpha & \beta \\ \gamma & \alpha \end{pmatrix}, \tag{13}$$

with the coefficients (α, β, γ) having a nontrivial dependence on the windings (l, m) and Kaluza Klein charges (p, q) for construction,

$$\begin{pmatrix} l \\ m \end{pmatrix} = \Lambda_0 \begin{pmatrix} p \\ q \end{pmatrix}. \tag{14}$$

Now we are going to show the existence of a parabolic class of solutions. Given the winding matrix W, its determinant $detW = n$ corresponds to the central charge of the supersymmetric algebra of the theory. We will assume in the following that n is an integer prime number. The equivalence class of the windings is defined by $\{WS\}$ where $S \in SL(2, Z)$. Given $\mathcal{Q} = \begin{pmatrix} p \\ q \end{pmatrix}$, the Kaluza Klein charges where (p, q) are relatively primes, then there always exists a parabolic transformation,

$$\Lambda_0 = \begin{pmatrix} 1 & k \\ 0 & 1 \end{pmatrix}, \tag{15}$$

such that $W = \Lambda_0 Q$, where $Q = \begin{pmatrix} p & P \\ q & Q \end{pmatrix}$ or $Q = \begin{pmatrix} q & U \\ q & V \end{pmatrix}$, where P, Q, U, V are integers such that

$$detW = n = detQ.$$

We are going to show that given (p, q) relatively primes, there always exists P, Q, U, V satisfying the above property.

3.1. Proof

Given W there always exists $S_1 \in SL(2, Z)$,[24] such that

$$WS_1 = \begin{pmatrix} \lambda_1 & \rho \\ 0 & \lambda_2 \end{pmatrix}, \tag{16}$$

with $\rho, \lambda_1, \lambda_2$ integers and satisfying that $\lambda_1\lambda_2 = n$. Let us consider WS_1S_2, where $S_2 = \begin{pmatrix} 1 & r \\ 0 & 1 \end{pmatrix} \in SL(2, Z)$, with r an integer to be determined. Then,

$$WS_1S_2 = \begin{pmatrix} \lambda_1 & \lambda_1 r + \rho \\ 0 & \lambda_2 \end{pmatrix}. \tag{17}$$

Finally consider

$$R = \begin{pmatrix} \lambda_1 & s \\ 0 & 1 \end{pmatrix} \in SL(2, Z), \tag{18}$$

then

$$RWS_1S_2 = \begin{pmatrix} \lambda_1 & \lambda_1 r + \lambda_2 s + \rho \\ 0 & \lambda_2 \end{pmatrix}. \tag{19}$$

Since n is prime, λ_1 and λ_2 are relatively primes. Moreover one of them is 1. There always exists r, s such that

$$\lambda_1 r + \lambda_2 s = -\rho, \tag{20}$$

hence

$$RWS_1S_2 = \begin{pmatrix} \lambda_1 & 0 \\ 0 & \lambda_2 \end{pmatrix}. \tag{21}$$

the same argument may be performed with the matrix Q:

$$\widetilde{R}Q\widetilde{S}_1\widetilde{S}_2 = \begin{pmatrix} \widetilde{\lambda}_1 & 0 \\ 0 & \widetilde{\lambda}_2 \end{pmatrix}. \tag{22}$$

Then either $\lambda_i = \widetilde{\lambda_i}$ or $\lambda_i = \widetilde{\lambda_j}$ with $i \neq j$ for $i, j = 1, 2$. In the first case $\lambda_i = \widetilde{\lambda_i}$

$$RWS_1S_2 = \widetilde{R}Q\widetilde{S}_1\widetilde{S}_2, \tag{23}$$

which imply

$$WS_1S_2\widetilde{S}_2^{-1}\widetilde{S}_1^{-1} = R^{-1}\widetilde{R}Q,$$

with $Q = \begin{pmatrix} p & P \\ q & Q \end{pmatrix}$ ans $S_1S_2\widetilde{S}_2^{-1}\widetilde{S}_1^{-1} \in Sl(2, Z)$, and $R^{-1}\widetilde{R} \in SL(2, Z)$.
Moreover,

$$\Lambda_0 = R^{-1}\widetilde{R} = \begin{pmatrix} 1 & \widetilde{s} - s \\ 0 & 1 \end{pmatrix}$$

with $k = \widetilde{s} - s$ as claimed.
If $\lambda_i = \widetilde{\lambda_j}, i \neq j$ the same argument follows by considering $Q = \begin{pmatrix} q & U \\ q & V \end{pmatrix}$.

Consequently it always exists a parabolic solution for the T-duality transformationΛ_0 with a nonlinear dependence on the windings and Kaluza-Klein charges of the theory.

4. Conclusions

In this work we deep on the T-duality characterization of the compactified Supermembrane theory on $M_9 \times T^2$. The T-duality transformation realizes non-linearly the SL(2,Z) with a nontrivial dependence on the windings and Kaluza-Klein charges of the theory. Moreover, the novelty of this work, shows that indeed the Λ_0 transformation always has a solution that belongs to the equivalence class of parabolic subgroup of $SL(2, Z)$. This is a nontrivial result, since the same statement does not necessarily hold for other inequivalent classes of T-duality transformations. This is an intermediate step towards a deeper understanding in terms of a fundamental invariants of M-theory, of the differences in the multiplet structure of the nine dimensional type IIA and type IIB gauged supergravities. The existence of this class of solutions is going to have very relevant implications toward the understanding of the structure of duals in type II gauged supergravities in nine dimensions as it will be discussed further in [26].

References

[1] Musaev E 2013 U-dualities in Type II string theories and M-theory *Preprint* hep-th/13113331
[2] Aldazabal G, Marques D and Nunez C 2013 Double Field Theory: A Pedagogical Review. *Class. Quant. Grav.* **30** 163001, (*Preprint* hep-th/13051907v2)
[3] Geissbuhler D, Marques D, Nunez C and Penas V 2013 Exploring Double Field Theory *J. High Energy Phys.* **JHEP1306(2013)101** (*Preprint* hep-th/1304.1472)
[4] Hull C M and Reid-Edwards R A 2006 Flux compactifications of M-theory on twisted tori *J. High Energy Phys.* **JHEP0610(2006)086** (*Preprint* hep-th/0603094)

[5] Berman D S and Thompson D C 2014 Duality symmetric string and M-theory *Phys. Rept.* **566** 1-60 (*Preprint* hep-th/13062643)

[6] Berman D S, Musaev E T and Thompson D C 2012 Duality invariant M-theory: gauged supergravities and Scherk-Schwarz reductions *J. High Energy Phys.* **JHEP1210(2012)174** (*Preprint* hep-th/12080020)

[7] Thompson D C 2011 Duality Invariance: From M-theory to Double Field Theory *J. High Energy Phys.* **JHEP1108(2011)125**. (*Preprint* hep-th/11064036)

[8] Blair C D, Malek E and Park J-H 2014 M-theory and Type IIB from a Duality Manifest Action *J. High Energy Phys.* **JHEP1401(2014)172** (*Preprint* hep-th/13115109)

[9] Hull C and Zwiebach B 2009 Double Field Theory *J. High Energy Phys.* bf JHEP0909(2009)099 (*Preprint* hep-th/09044664)

[10] Reid-Edwards R A 2009 Flux compactifications, twisted tori and doubled geometry *J. High Energy Phys.* **JHEP0906(2009)085** (*Preprint* hep-th/09040380)

[11] Shelton J, Taylor W and Wecht B 2005 Nongeometric flux compactifications *J. High Energy Phys.* **JHEP0510(2005)085** (*Preprint* hep-th/0508133)

[12] Duff M 1990 Duality Rotations in string theory *Nucl. Phys.* B **B335** 610

[13] Tseytlin A A 1990 Duality symmetric formulation of string world sheet dynamics *Phys.Lett.* B **242** 163-174

[14] Siegel W 1993 Superspace duality in low-energy superstrings *Phys.Rev.* D **48** 2826-37 (*Preprint* hep-th/9305073)

[15] Hull C M 2007 Global aspects of T-duality, gauged sigma models and T-folds *J. High Energy Phys.* **JHEP0710 (2007)057** (*Preprint* hep-th/0604178)

[16] Garcia del Moral M P, Pena J M and Restuccia A 2012 Supermembrane origin of type II gauged supergravities in 9D *J. High Energy Phys.* **JHEP1209(2012)063** (*Preprint* hep-th/12032767)

[17] Garcia del Moral M P, Pena J M and Restuccia A 2013 T-duality invariance of the supermembrane *Int. J. Geom. Meth. Mod. Phys.* **10** 1360010 (*Preprint* 12112434)

[18] Garcia del Moral M P 2012 Dualities as symmetries of the supermembrane theory *Preprint* hep-th/12116265

[19] Garcia del Moral M P 2014 On the Geometry of Sculpting-like Gauging Processes *Int. J. Geom. Meth. Mod. Phys.* 11 **9** 1460040 (*Preprint* hep-th/1311.4467v3)

[20] Garcia del Moral M P, Martin I, Pena J M and Restuccia A 2011 SL(2,Z) symmetries, supermembranes and symplectic torus bundles *J. High Energy Phys.* **JHEP1109(2011)068** (*Preprint* hep-th/11053181)

[21] Duff M J and Lu J X 1990 Duality Rotations in Membrane Theory *Nucl.Phys.* B **347** 394-419

[22] Russo J G 1997 T duality in M theory and supermembranes *Phys. Lett.* B **400** 37-42 (*Preprint* hep-th/9701188)

[23] Aldabe F 1996 T duality and gauge symmetry in supermembrane theory *Preprint* hep-th/9604110

[24] Garcia del Moral M P, Martin I and Restuccia A 2008 Nonperturbative SL(2,Z) (p,q)-strings manifestly realized on the quantum M2 *Preprint* hep-th/08020573

[25] Schwarz J H 1995 An SL(2,Z) multiplet of type IIB superstrings *Phys.Lett.* B **360** 13-18, Erratum-ibid. B **364** 252 (*Preprint* hep-th/9508143)

[26] Garcia del Moral M P, Pena J M and Restuccia A To appear.

Journal of Physics: Conference Series **720** (2016) 012026 doi:10.1088/1742-6596/720/1/012026

Inverse magnetic catalysis in the linear sigma model

A Ayala[1,3], M Loewe[2,3] and R Zamora[2]

[1] Instituto de Ciencias Nucleares, Universidad Nacional Autónoma de México, Apartado Postal 70-543, México Distrito Federal 04510, Mexico.
[2] Instituto de Física, Pontificia Universidad Católica de Chile, Casilla 306, Santiago 22, Chile.
[3]Centre for Theoretical and Mathematical Physics, and Department of Physics, University of Cape Town, Rondebosch 7700, South Africa.

E-mail: rrzamora@uc.cl

Abstract.
We compute the critical temperature for the chiral transition in the background of a magnetic field in the linear sigma model, including the quark contribution and the thermo-magnetic effects induced on the coupling constants at one loop level. For the analysis, we go beyond mean field aproximation, by taking one loop thermo-magnetic corrections to the couplings as well as plasma screening effects for the boson's masses, expressed through the ring diagrams. We found inverse magnetic catalysis, i.e. a decreasing of the critical chiral temperature as function of the intensity of the magnetic field, which seems to be in agreement with recent results from the lattice community.

1. Introduction

In recent years there has been an increasing interest in studying the QCD phase diagram in the presence of a magnetic field. Particularly, the effect of the magnetic field on the critical temperature for chiral phase transition, has been studied in different effective models [1, 2, 3, 4, 5, 6, 7] as well as in lattice QCD where initially it has been found that magnetic catalysis takes place, i.e. that the critical temperature for chiral phase transition becomes higher in the presence of a magnetic field [8, 9]. However, recent improved lattice calculations have found the opposite behavior [10, 11, 12].

In this work, we studied the sigma model with quarks, finding inverse magnetic catalysis in agreement with the latest lattice results. The main idea was to go beyond mean field approximation, by taking thermo-magnetic corrections to the couplings. In this way we found that the boson's coupling decreases with the magnetic field which is a substantial ingredient in order to find inverse magnetic catalysis. The idea of correcting the vertices has also been implemented in QCD [13], where the thermal and magnetic corrections to the quark-gluon vertex have been calculated in the Hard Thermal Loop(HTL) approximation. From the quark-gluon vertex we can obtain the dependence of the QCD coupling with the magnetic field. The result shows that the coupling decreases as function of the intensity of the magnetic field. The reader should go to the original reference [14] for a complete description of the technical details.

2. Efective Potential

The model is given by the Lagrangian

$$\begin{aligned} \mathcal{L} &= \frac{1}{2}(\partial_\mu \sigma)^2 + \frac{1}{2}(D_\mu \vec{\pi})^2 + \frac{\mu^2}{2}(\sigma^2 + \vec{\pi}^2) - \frac{\lambda}{4}(\sigma^2 + \vec{\pi}^2)^2 \\ &+ i\bar{\psi}\gamma^\mu D_\mu \psi - g\bar{\psi}(\sigma + i\gamma_5 \vec{\tau} \cdot \vec{\pi})\psi, \end{aligned} \tag{1}$$

where ψ is an SU(2) isospin doublet, $\vec{\pi} = (\pi_1, \pi_2, \pi_3)$ is an isospin triplet and σ is an isospin singlet, with

$$D_\mu = \partial_\mu + iqA_\mu, \tag{2}$$

is the covariant derivative. A^μ is the vector potential corresponding to an external magnetic field directed along the $\hat{z}$ axis,

$$A^\mu = \frac{B}{2}(0, -y, x, 0), \tag{3}$$

and q is the particle's electric charge. A^μ satisfies the gauge condition $\partial_\mu A^\mu = 0$. Since $A^3 = 0$, the gauge field only couples to the charged pion combinations, namely

$$\pi_\pm = \frac{1}{\sqrt{2}}(\pi_1 \mp i\pi_2). \tag{4}$$

The neutral pion is taken as the third component of the pion isovector, $\pi^0 = \pi_3$. The gauge field is taken as classical and thus we do not consider loops involving the propagator of the gauge field in internal lines. The squared mass parameter μ^2 and the self-coupling λ and g are taken to be positive.

To allow for an spontaneous breaking of symmetry, we let the σ field to develop a vacuum expectation value v

$$\sigma \to \sigma + v, \tag{5}$$

which can later be taken as the order parameter of the theory. After this shift, the Lagrangian can be rewritten as

$$\begin{aligned} \mathcal{L} &= -\frac{1}{2}[\sigma(\partial_\mu + iqA_\mu)^2\sigma] - \frac{1}{2}\left(3\lambda v^2 - \mu^2\right)\sigma^2 \\ &- \frac{1}{2}[\vec{\pi}(\partial_\mu + iqA_\mu)^2\vec{\pi}] - \frac{1}{2}\left(\lambda v^2 - \mu^2\right)\vec{\pi}^2 + \frac{\mu^2}{2}v^2 \\ &- \frac{\lambda}{4}v^4 + i\bar{\psi}\gamma^\mu D_\mu\psi - gv\bar{\psi}\psi + \mathcal{L}_I^b + \mathcal{L}_I^f, \end{aligned} \tag{6}$$

where $\mathcal{L}_I^b$ and $\mathcal{L}_I^f$ are given by

$$\begin{aligned} \mathcal{L}_I^b &= -\frac{\lambda}{4}\Big[(\sigma^2 + (\pi^0)^2)^2 \\ &+ 4\pi^+\pi^-(\sigma^2 + (\pi^0)^2 + \pi^+\pi^-)\Big], \\ \mathcal{L}_I^f &= -g\bar{\psi}(\sigma + i\gamma_5\vec{\tau}\cdot\vec{\pi})\psi, \end{aligned} \tag{7}$$

and represent the Lagrangian describing the interactions among the fields σ, $\vec{\pi}$ and ψ, after symmetry breaking. From Eq. (6) we see that the σ, the three pions and the quarks have masses given by

$$\begin{aligned} m_\sigma^2 &= 3\lambda v^2 - \mu^2, \\ m_\pi^2 &= \lambda v^2 - \mu^2, \\ m_f &= gv, \end{aligned} \tag{8}$$

respectively.

Using Schwinger's proper-time method, the expression for the one-loop effective potential for one boson field with squared mass m_b^2 and absolute value of its charge q_b at finite temperature T in the presence of a constant magnetic field can be written as

$$\begin{aligned} V_b^{(1)} &= \frac{T}{2}\sum_n \int dm_b^2 \int \frac{d^3k}{(2\pi)^3}\int_0^\infty \frac{ds}{\cosh(q_b Bs)} \\ &\times e^{-s(\omega_n^2+k_3^2+k_\perp^2\frac{\tanh(q_bBs)}{q_bBs}+m_b^2)}, \end{aligned} \tag{9}$$

where $\omega_n = 2n\pi T$ are boson Matsubara frequencies. Similarly, the expression for the one-loop effective potential for one fermion field with mass m_f and absolute value of its charge q_f at finite temperature T in the presence of a constant magnetic field can be written as

$$\begin{aligned} V_f^{(1)} &= -\sum_{r=\pm1} T\sum_n \int dm_f^2 \int \frac{d^3k}{(2\pi)^3}\int_0^\infty \frac{ds}{\cosh(q_f Bs)} \\ &\times e^{-s(\tilde{\omega}_n^2+k_3^2+k_\perp^2\frac{\tanh(q_fBs)}{q_fBs}+m_f^2+rq_fB)}, \end{aligned} \tag{10}$$

where $\tilde{\omega}_n = (2n+1)\pi T$ are fermion Matsubara frequencies. The sum over the index r corresponds to the two possible spin orientations along the magnetic field direction.

Including the v-independent terms, choosing the renormalization scale as $\tilde{\mu} = e^{-1/2}\mu$ and after mass and charge renormalization, it has been shown in Ref. [15] that the thermo-magnetic effective potential in the small to intermediate field regime, in a high temperature expansion can be written as

$$\begin{aligned} V^{(\text{eff})} &= -\frac{\mu^2}{2}v^2 + \frac{\lambda}{4}v^4 + \sum_{i=\sigma,\pi^0}\left\{\frac{m_i^4}{64\pi^2}\left[\ln\left(\frac{(4\pi T)^2}{2\mu^2}\right) - 2\gamma_E + 1\right]\right. \\ &- \left.\frac{\pi^2T^4}{90} + \frac{m_i^2T^2}{24} - \frac{T}{12\pi}(m_i^2+\Pi)^{3/2}\right\} + \sum_{i=\pi_+,\pi_-}\left\{\frac{m_i^4}{64\pi^2}\left[\ln\left(\frac{(4\pi T)^2}{2\mu^2}\right) - 2\gamma_E + 1\right]\right. \\ &- \frac{\pi^2T^4}{90} + \frac{m_i^2T^2}{24} + \frac{T(2qB)^{3/2}}{8\pi}\zeta\left(-\frac{1}{2},\frac{1}{2}+\frac{m_i^2+\Pi}{2qB}\right) \\ &- \left.\frac{(qB)^2}{192\pi^2}\left[\ln\left(\frac{(4\pi T)^2}{2\mu^2}\right) - 2\gamma_E + 1 + \zeta(3)\left(\frac{m_i}{2\pi T}\right)^2 - \frac{3}{4}\zeta(5)\left(\frac{m_i}{2\pi T}\right)^4\right]\right\} \\ &- \sum_{f=u,d}\left\{\frac{m_f^4}{16\pi^2}\left[\ln\left(\frac{(\pi T)^2}{2\mu^2}\right) - 2\gamma_E + 1\right] + \frac{7\pi^2T^4}{180} - \frac{m_f^2T^2}{12}\right. \\ &+ \left.\frac{(q_fB)^2}{24\pi^2}\left[\ln\left(\frac{(\pi T)^2}{2\mu^2}\right) - 2\gamma_E + 1\right]\right\}, \end{aligned} \tag{11}$$

where q is the absolute value of the charged pions' charge ($q = 1$), $q_u = 2/3$, $q_d = 1/3$ are the absolute values of the u and d quarks, respectively and γ_E is Euler's gamma. Though we take the quark masses as equal, the notation emphasizes that the effective potential is evaluated accounting for the different quark charges. We have introduced the leading temperature plasma screening effects for the boson's mass squared, encoded in the boson's self-energy Π. For the Hurwitz zeta function $\zeta(-1/2, z)$ in Eq. (11) to be real, we need that

$$-\mu^2 + \Pi > qB, \tag{12}$$

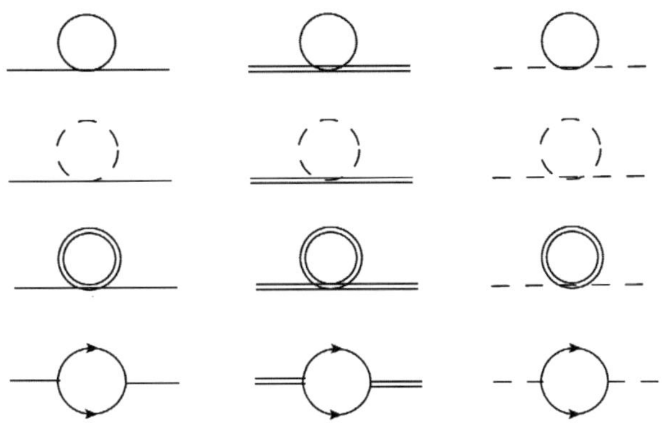

Figure 1. Feynman diagrams contributing to the one loop bosons' self-energies. The dashed line denotes the charged pion, the continuous line is the sigma, the double line represents the neutral pion and the continuous line with arrows represents the fermions.

condition that comes from requiring that the second argument of the Hurwitz zeta function satisfies $z > 0$, even for the lowest value of m_b^2 which is obtained for $v = 0$. Furthermore, for the large T expansion to be valid, we also require that

$$qB/T^2 < 1. \tag{13}$$

The diagrams representing the bosons' self-energies are depicted in Fig. 1. Each column corresponds to the diagrams contributing to the self-energy of a given boson. The total self-energy for any boson is identical to the other's and thus we concentrate on computing the diagrams in column (a). The contribution from the individual diagrams require of the expressions [hereby capital letters are used to denote four-momenta in Eucledian space, e.g. $K \equiv (\omega_n, \mathbf{k})$]

$$\begin{aligned}
\Pi_{a1}(m_\sigma^2) &= \lambda T \sum_n \int \frac{d^3k}{(2\pi)^3} D(K; m_\sigma^2), \\
\Pi_{a2}(m_{\pi^0}^2) &= \lambda T \sum_n \int \frac{d^3k}{(2\pi)^3} D(K; m_{\pi^0}^2) \\
\Pi_{a3}(m_{\pi^\pm}^2) &= \lambda T \sum_n \int \frac{d^3k}{(2\pi)^3} D_B(K; m_{\pi^\pm}^2), \\
\Pi_{a4}(P; m_f) &= -N_f g^2 T \sum_n \int \frac{d^3k}{(2\pi)^3} \\
&\times \ \mathrm{Tr}\ S_B(K; m_f) S_B(P-K; m_f),
\end{aligned} \tag{14}$$

where N_f is the number of fermions and the corresponding propagators are given by

$$\begin{aligned}
D(K; m_i^2) &= \frac{1}{K^2 + m_i^2}, \\
D_B(K; m_i^2) &= \int_0^\infty ds \frac{e^{-s(\omega_n^2 + k_3^2 + k_\perp^2 \frac{\tanh(qBs)}{qBs} + m_i^2)}}{\cosh(qBs)},
\end{aligned}$$

$$\begin{aligned} S_B(K;m_f) &= \int_0^\infty ds \frac{e^{-s(\tilde{\omega}_n^2+k_3^2+k_\perp^2\frac{\tanh(q_fBs)}{q_fBs}+m_f^2)}}{\cosh(q_fBs)} \\ &\times \left[(\cosh(q_fBs) - i\gamma_1\gamma_2\sinh(q_fBs))(m_f - \not{k}_\parallel) - \frac{\not{k}_\perp}{\cosh(q_fBs)}\right], \end{aligned} \quad (15)$$

and for the charged particle propagators we have used Schwinger's proper-time representation. For the computation of Π_{a3} we work in the *infrared limit*, namely, $P = (0, \mathbf{p} \to 0)$ and with the hierarchy of scales $qB, m_i^2 < T^2$. It has been shown [16] that this limit can be formally implemented by straightforward setting $P = 0$ in the third of Eqs. (14). The leading contribution at high temperature from each of these diagram is

$$\begin{aligned} \Pi_{a1}(m_\sigma^2) &= \lambda\frac{T^2}{12} \\ \Pi_{a2}(m_{\pi^0}^2) &= \lambda\frac{T^2}{12} \\ \Pi_{a3}(m_{\pi^\pm}^2) &= \lambda\frac{T^2}{12} \\ \Pi_{a4}(0;m_f) &= N_fg^2\frac{T^2}{6}, \end{aligned} \quad (16)$$

and therefore, considering the permutation factors, the total self-energy is given by

$$\begin{aligned} \Pi &= 3\Pi_{a1}(m_\sigma^2) + \Pi_{a2}(m_{\pi^0}^2) + 2\Pi_{a3}(m_{\pi^\pm}^2) + \Pi_{a4}(0;m_f) \\ &= \lambda\frac{T^2}{2} + N_fg^2\frac{T^2}{6}. \end{aligned} \quad (17)$$

3. One loop thermo-magnetic couplings and critical temperature

Let us now compute the one-loop correction to the coupling λ, including thermal and magnetic effects. Figure 2 shows the Feynman diagrams that contribute to this correction. Columns (a), (b), (c), (d), (e) and (f) contribute to the correction to the σ^4, $(\pi^0)^4$, $(\pi^+)^2(\pi^-)^2$, $\sigma^2\pi^+\pi^-$, $(\pi^0)^2\pi^+\pi^-$ and $\sigma^2(\pi^0)^2$ terms of the interaction Lagrangian in Eq. (7), respectively

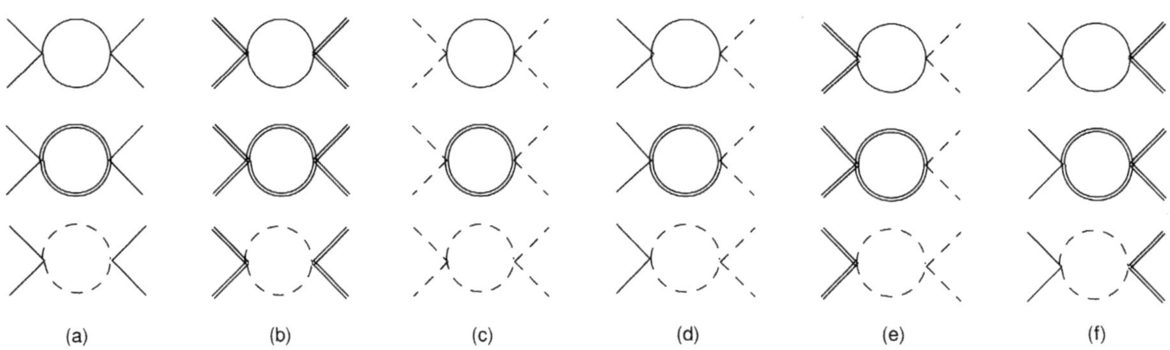

Figure 2. One-loop Feynman diagrams that contribute to the thermal and magnetic correction to the coupling λ. The dashed line denotes the charged pion, the continuous line is the sigma and the double line represents the neutral pion.

Once again we work in the *infrared limit*, namely, $P_i = (0, \mathbf{p} \to 0)$ and with the hierarchy of scales where $qB, m_i^2 < T^2$. Considering the permutation factors and the contribution from the

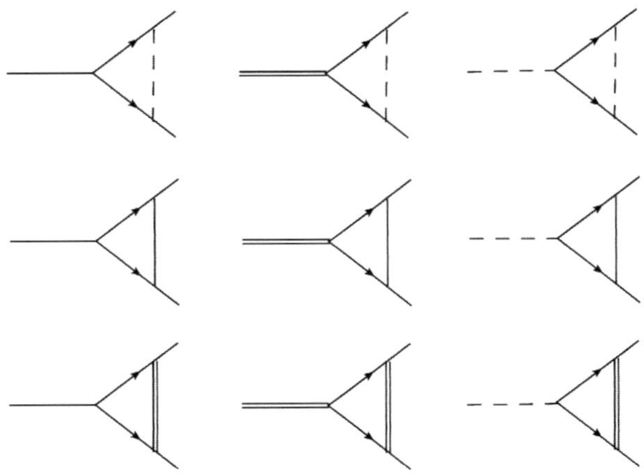

Figure 3. One-loop Feynman diagrams that contribute to the thermal and magnetic correction to the coupling g. The dashed line denotes the charged pion, the continuous line is the sigma, the double line represents the neutral pion and the continuous line with arrows represents the quarks.

s, t and u-channels, the correction to the self-coupling λ to one-loop order is given by

$$\lambda_{\rm eff} = \lambda\left[1 + 24\lambda\left(9I(0; m_\sigma^2) + I(0; m_\pi^2) + 4J(0; m_\pi^2)\right)\right], \tag{18}$$

where

$$I(0; m_i^2) = \frac{T}{8\pi}\frac{1}{(m_i^2 + \Pi)^{1/2}} - \frac{1}{16\pi^2}\left[\ln\left(\frac{(4\pi T)^2}{2\mu^2}\right) + 1 - 2\gamma_E + \zeta(3)\left(\frac{\sqrt{m_i^2 + \Pi}}{2\pi T}\right)^2\right], \tag{19}$$

$$J_{n\neq 0}(0; m_i^2) = -\frac{1}{16\pi^2}\left[\ln\left(\frac{(4\pi T)^2}{2\mu^2}\right) + 1 - 2\gamma_E + \zeta(3)\left(\frac{\sqrt{m_i^2 + \Pi}}{2\pi T}\right)^2\right] - \frac{(qB)^2}{1024\pi^6 T^4}\zeta(5). \tag{20}$$

Note that $\lambda_{\rm eff}$ depends on v through the dependence on the boson masses. Let us furthermore take the approximation where we evaluate $\lambda_{\rm eff}$ at $v = 0$. The rationale is that we are pursuing the effect on the critical temperature which is the temperature where the curvature of the effective potential at $v = 0$ vanishes.

We now turn to the calculation of the thermo-magnetic correction of the coupling g. Figure 3 shows the Feynman diagrams that contribute to this correction. We are interested in computing an effective value for this coupling, $g_{\rm eff}$, also for $v = 0$, in the same manner we did for $\lambda_{\rm eff}$. Columns (a), (b) and (c) contribute to the correction to the quark-σ, quark-π^0 and quark-$\pi^\pm$ terms of the interaction Lagrangian of Eq. (7), respectively. Therefore the correction to g at one-loop order is given by

$$g_{\rm eff} = g[1 + g^2 L(0,0)], \tag{21}$$

where

$$L(0,0) = \frac{g^3}{8\pi^2}\left[-2\gamma_E + 1 - \ln\left(\frac{T^2\pi^2}{2\mu^2}\right) - 7\left(\frac{m_\pi^2}{8\pi^2T^2}\right)\zeta(3) + 31\left(\frac{m_\pi^2}{8\pi^2T^2}\right)^2\zeta(5)\right.$$
$$\left. + \frac{3410}{9}\frac{(qB)^2}{(4\pi T)^4}\zeta(5)\right]. \quad (22)$$

In order to use a set of values for the couplings λ and g appropriate for the description of the phase transition note that the curvature of the effective potential vanishes for $v = 0$. Since the boson thermal masses are proportional to this curvature, they also vanish at $v = 0$. This observation provides a condition to obtain a relation between the model parameters at T_c that can be supplemented with information from the physical masses of the pion and sigma in vacuum [22].

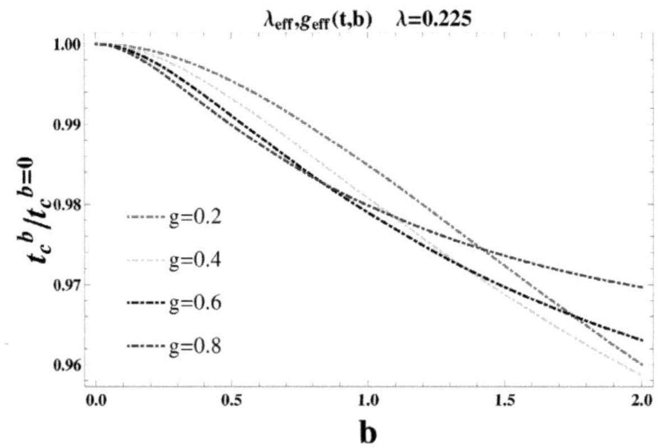

Figure 4. Color on-line. Effect of the full thermo-magnetic dependence of couplings on the critical temperature for a fixed value of the tree level $\lambda = 0.225$ and different values of the tree level g as a function of $b = qB/\mu^2$. In all cases the critical temperature is a decreasing function of b.

Let us now study the effect that the thermo-magnetic corrections to the couplings have on the critical temperature. We first look at the cases where we set the couplings to their tree level values and where only thermal effects are included. Figure 4 shows the critical temperature for the case where we consider the full thermo-magnetic dependence of the couplings, obtained from setting the second derivative of Eq. (11) equal to zero at $v = 0$, normalized to the critical temperature for vanishing magnetic field. Figure 4 shows the case when we set the tree-level coupling λ to a fixed value and vary the tree-level coupling g.

4. Conclusions

In summary, we have shown that when including the one-loop thermo-magnetic effects for the couplings in the linear sigma model with fermions interacting with an external magnetic field, the critical temperature for the chiral transition is a decreasing function of the field strength. This behavior is a direct consequence of the decrease of the boson self-coupling with the field strength. The effect of the fermions is marginal and the main contribution comes from the charged pions. We emphasize that the thermo-magnetic dependence of the couplings has been computed –as opposed to assumed– within the model itself.

Acknowledgments

Support for this work has been received in part from DGAPA-UNAM under grant number PAPIIT-IN103811, CONACyT-México under grant number 128534 and FONDECYT under grant numbers 1130056. R. Z. acknowledges support from CONICYT under Grant No. 21110295.

References

[1] M. Loewe, C. Villavicencio, R. Zamora 2014 *Phys. Rev.* D **89** 016004
[2] Weak magnetic field effects on chiral critical temperature in a nonlocal Nambu–Jona-Lasinio mordel. M. Loewe, F. Marquez, C. Villavicencio and R. Zamora, hep-ph 1410.7266.
[3] N. O. Agasian and S. M. Fedorov 2008 *Phys. Lett.* B **663** 445
[4] E. S. Fraga and A. J. Mizher 2008 *Phys. Rev.* D **78** 025016
[5] A. J. Mizher, M. N. Chernodub and E. S. Fraga 2010 *Phys. Rev.* D **82** 105016
[6] J. O. Andersen and R. Khan 2012 *Phys. Rev.* D **85** 065026
[7] A. Ayala, M. Loewe, A. J. Mizher, R. Zamora 2014 *Phys. Rev.* D **90** 036001
[8] V. V. Braguta, P.V. Buidovich, T. Kalaydzhyan, S. V. Kuznetsov and M. I. Polikarpov 2012 *Phys. Atom. Nucl.* **75** 488
[9] M. D'Elia and F. Negro 2011 *Phys. Rev.* D **83** 114028
[10] G. S. Bali, F. Bruckmann, G. Endrodi, Z. Fodor, S. D. Katz, S. Krieg, A. Schafer, and K. K. Szabo 2012 *J. High Energy Phys.* **02**, **044**
[11] G. S. Bali, F. Bruckmann, G. Endrodi, Z. Fodor, S. D. Katz and A. Schafer 2012 *Phys. Rev.* D **86** 071502
[12] G. S. Bali, F. Bruckmann, G. Endrodi, S. D. Katz and A. Schafer, arXiv:1406.0269 [hep-lat].
[13] A. Ayala, J.J Cobos-Martinez, M. Loewe, María Elena Tejeda-Yeomans and R.Zamora 2015 *Phys. Rev.* D **91** 016007
[14] A. Ayala, M. Loewe and R.Zamora 2015 *Phys. Rev.* D **91** 016002
[15] A. Ayala, L. A. Hernández, A. J. Mizher, J. C. Rojas, C. Villavicencio 2014 *Phys. Rev.* D **89** 116017
[16] A. Ayala, G. Piccinelli, A. Sánchez, M. E. Tejeda-Yeomans 2008 *Phys. Rev.* D **78** 096001
[17] A. Ayala, L. A. Hernández, J. López, A. J. Mizher, J. C. Rojas, C. Villavicencio 2013 *Phys. Rev.* D **88** 036010
[18] A. Ayala, A. Sánchez, G. Piccinelli, S. Sahu 2005 *Phys. Rev.* D **71** 023004
[19] D. J. Bedingham, hep-ph/0011012.
[20] T.-K. Chyi, C.-W. Hwang, W. F. Kao, G. L. Lin, K.-W. Ng, and J.-J. Tseng 2000 *Phys. Rev.* D **62** 105014
[21] L. Dolan and R. Jackiw 1974 *Phys. Rev.* D **9** 3320
[22] A. Ayala, A. Bashir, J.J. Cobos-Martinez, S. Hernandez-Ortiz, A. Raya, arXiv:1411.4953 [hep-ph].
[23] M. E. Carrington 1992 *Phys. Rev.* D **45** 2933

Journal of Physics: Conference Series **720** (2016) 012027 doi:10.1088/1742-6596/720/1/012027

The pressure of hot QCD

York Schröder

Grupo de Física de Altas Energías, Departamento de Ciencias Básicas,
Universidad del Bío-Bío, Casilla 447, Chillán, Chile

E-mail: yschroeder@ubiobio.cl

Abstract. When heated and/or compressed, strongly interacting matter exhibits a rich phase structure. In this talk, I will concentrate on its behavior under variations of the temperature, which is most relevant for phenomenological applications such as in cosmology, heavy-ion collisions, and astrophysics.

In particular, effective field theory methods can be used to combine lattice and continuum calculations, in order to obtain high-precision results for the relevant thermodynamic quantities such as the QCD pressure and equation of state. I will discuss the current status of this systematic approach to QCD thermodynamics, and point out the remaining (technical) problems.

1. Introduction

Quantum Chromodynamics (QCD) is the theory that describes the strong interactions in nature, which binds quarks into hadrons, such as protons or pions. Due to its nature of being a non-Abelian gauge theory, its equations are however extremely difficult to solve. While a large body of experimental information on details of the hadron spectrum exists [1], first-principle theoretical postdictions of these properties are rather sparse.

In the quest to check the theory of QCD versus its observed manifestations, we can identify three main types of approaches. First, one might just solve its equations. This is the realm of lattice QCD, where large-scale computer simulations are used to numerically approximate observables such as the hadron mass spectrum. As a result of worldwide efforts, multi-million-dollar investments, teraflop speeds, hardware development as well as algorithmic advances, the punchline of this line of investigation is that, starting from the (fractionally charged, partonic quark- and gluon-) QCD degrees of freedom, one indeed postdicts exactly the observed (integer-charged) low-lying hadron spectrum (see, e.g. [2]).

A second line of approaches centers around considering phenomenological models that are in some sense 'close' to QCD, but mathematically simpler to treat. Typical modifications include a change of space-time dimension, different symmetry groups, or modified particle content of the theory. These models are most successful when applied to specific narrow problems, but mostly lack generality.

Third, one might consider 'extreme' circumstances in which the QCD equations simplify, or at least allow for an application of systematically improvable approximation methods such as weak-coupling expansions. Such 'extreme' conditions can be realized e.g. when some parameters assume extreme values, such as the very high energies realized at particle colliders such as currently at the Large Hadron Collider at CERN. Indeed, the higher the collision energy, the cleaner the underlying interaction vertices can be seen, such that from the jet structure of the events one can map out precisely the underlying details of the theory. Of course, nowadays QCD contributions are not signals, but constitute (large and therefore important) background in searches for new particles.

Another such 'extreme' regime of QCD can be reached by heating the system up to very large temperatures T, where fundamental questions such as confinement and chiral symmetry breaking can be addressed. This is the realm of thermal field theory [3], in which problems that are phenomenologically relevant for e.g. cosmology (such as dark matter relic abundance), heavy-ion collisions (such as hydrodynamic expansion af a plasma) and for astrophysics (such as the properties of compact stars) are studied. Furthermore, due to asymptotic freedom, the high-temperature limit of QCD turns out to be tractable with analytic methods, namely weak-coupling expansions. In this limit, while staying strictly within QCD (as opposed to modelling some of its behavior), we hence gain the possibility of systematic improvements of theoretical approximations, whenever they are needed.

The thermodynamic properties QCD, such as certain features of its equilibrium phase diagram, can be studied in lattice Monte-Carlo simulations. One of the key results of thermal lattice QCD is the determination of the deconfinement temperature T_c at vanishing baryon density. These results basically confirm the expected transition from a hadron gas at low temperatures to a quark-gluon plasma above T_c, tracking the liberation of degrees of freedom via the QCD pressure, and confirming an approach to the asymptotic limit of an ideal gas, see Fig. 1.

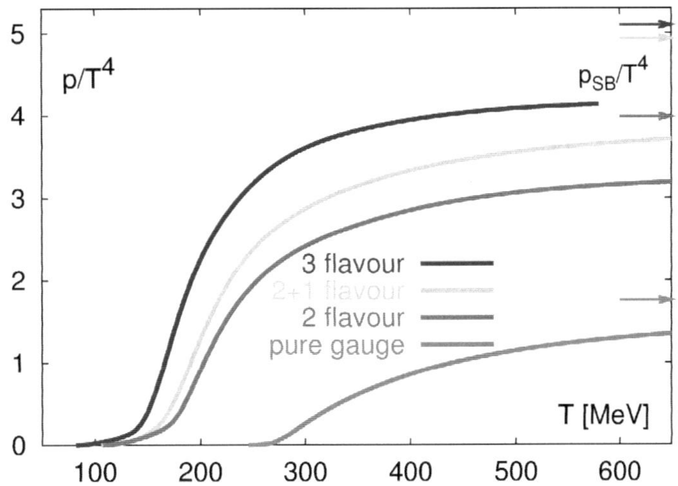

Figure 1. Lattice results for the scaled QCD pressure, for different flavor content of the theory [4]. The small arrows denote the respective values for an ideal (Stefan-Boltzmann) gas.

Further explorations of features of the equilibrium phase diagram at various values of temperatures as well as baryon-chemical potentials μ turn out to be more difficult, and typically involve some degree of modeling. Of particular interest are the equation of state (EoS), phase transition lines, their respective order, and locations of critical points, as well as medium properties such as spectral functions or correlation lengths. These questions are being actively pursued in a worldwide effort in order to obtain information relevant for early-universe cosmology (high T, low μ), heavy-ion physics (intermediate T and μ) as well as astrophysics (low T, intermediate μ).

Journal of Physics: Conference Series **720** (2016) 012027 doi:10.1088/1742-6596/720/1/012027

As has already been mentioned above, while hot QCD is a strongly coupled system near T_c where lattice simulations are needed, at very high temperature asymptotic freedom guarantees the gauge coupling g to become small, such that weak-coupling approaches can be used. In particular, effective field theory methods turn out to be most powerful in order to combine lattice and continuum calculations [5, 6], in order to obtain high-precision results for the relevant thermodynamic quantities such as the QCD pressure and equation of state [7]. These methods alleviate the notorious infrared problem of finite-temperature field theory [8], to an extent that higher orders in the weak-coupling expansion can now be systematically included. I will discuss some key features of this systematic approach to QCD thermodynamics using the best of both – lattice and continuum – approaches, and point out the remaining (technical) problems [9].

2. Effective field theory approach

As becomes apparent from the lattice data of Fig. 1, the task is to explain the deviation of the value of the QCD pressure of about 20% at $T \sim 4T_c$ from its ideal value, realized in the theoretical limit $T \to \infty$. In the imaginary-time path integral formalism at finite temperature, the pressure can be straightforwardly computed as the logarithm of the partition function in the thermodynamic limit,

$$p(T) = \lim_{V\to\infty} \frac{T}{V} \ln \int \mathcal{D}[A_\mu^a, \psi, \bar{\psi}] \exp\Big(-\frac{1}{\hbar}\int_0^{\hbar/T} \mathrm{d}\tau \int \mathrm{d}^{3-2\epsilon}x \, \mathcal{L}^{\mathrm{E}}_{\mathrm{QCD}} \Big) \, . \tag{1}$$

Due to the above-mentioned infrared problems, the loop expansion of $p(T)$ in the gauge coupling g turns out to be non-trivial, since it contains odd powers of the coupling and even logarithms thereof. After pioneering work in the seventies and eighties [10, 11, 12], the root cause of this non-analytic (in $\alpha_s = g^2/4\pi$) behavior has been well understood as coming from the dynamically generated scales in the thermal system: intercations make hot QCD a multi-scale system, where the relevant momentum scales $|k| \sim \pi T$, $|k| \sim gT$ and $|k| \sim g^2T$ are called hard, soft and ultrasoft, respectively. At large temperatures, the gauge coupling is small and hence these three scales are well separated, allowing for a clean effective theory setup that factorizes the partition function, such that the pressure can be re-written as the sum of contributions from these three scales [13],

$$p_{\mathrm{QCD}}(T) = p_{\mathrm{hard}}(T) + p_{\mathrm{soft}}(T) + p_{\mathrm{ultrasoft}}(T) \, . \tag{2}$$

Physics at those three different energy scales is governed by 4-dimensional (4d) thermal QCD, by a 3d adjoint Higgs model, and by 3d pure Yang-Mills theory, respectively. Soft and ultrasoft effects hence originate from 3-dimensional theories, a phenomenon known as dimensional reduction, as can be easily understood from the fact that long-range (small momentum) physics does not see the compact timelike dimension. As is common practice in effective field theory, couplings of the effective theory are related to the parameters of the original one (in this case T and g) through matching computations. It turns out that, after the matching has been done in the weak-coupling sense, the three parts of the pressure parametically start at different orders in the dimensionless 4d gauge coupling g,

$$p_{\mathrm{hard}}(T) \sim T^4(1 + g^2 + g^4 + g^6 + \ldots) \, , \tag{3}$$
$$p_{\mathrm{soft}}(T) \sim T^4(g^3 + g^4 + g^5 + g^6 + \ldots) \, , \tag{4}$$
$$p_{\mathrm{ultrasoft}}(T) \sim T^4(g^6 + \ldots) \, . \tag{5}$$

Now, as is well-known from effective theories, whenever a new physical scale enters a problem, one typically meets large logarithms (of the ratio of matching scales). Therefore, one might call the sum of the expansions shown above the 'physical leading order' of the pressure. Indeed, it is known that whenever one truncates the series before the order g^6, the series does not show signs of convergence at intermediate temperatures [?], a telltale sign that these large logarithms are missing. Only at parametric order g^6, for the first time contributions from all physical energy scales have entered, giving a true leading-order approximation.

3. Current status

Of all the coefficients shown in Eqs. (3-5), only the g^6 term of Eq. (3) remains unknown to date. The longest-distance input, $p_{\rm ultrasoft}$, is a non-perturbative contribution that has been determined with a clean lattice measurement [15] and matched from the lattice to the continuum ($\overline{\rm MS}$) scheme via numerical stochastic perturbation theory [16]. This alleviates the infrared problem mentioned above.

Contributions to $p_{\rm soft}$ and $p_{\rm hard}$ are purely perturbative, the g^6 coefficients entailing four-loop diagrams that are relatively straightforward to evaluate in the 3d theory [17], but much more involved in the 4d one [18]. Figure 2 shows typical diagrams at this level, all being of vacuum type. The open problem to be tackled is to develop a method that allows to systematically solve the large number of four-loop sum-integrals that appear in the short-distance piece $p_{\rm hard}$. Once this difficult but conceptionally clear perturbative contribution has been fixed and hence a result for the 'physical leading order' of the pressure is available, it would of course be valuable to check convergence of the series by evaluating the next-to-leading (NLO, g^7) order. Somewhat surprisingly, this step would not represent a huge technical complication, as it originates from the 3d theory only, allowing for an application of the full power of automated Feynman integral reduction algorithms (albeit at the five-loop level) that have been and are currently being developed for zero-temperature applications such as collider physics.

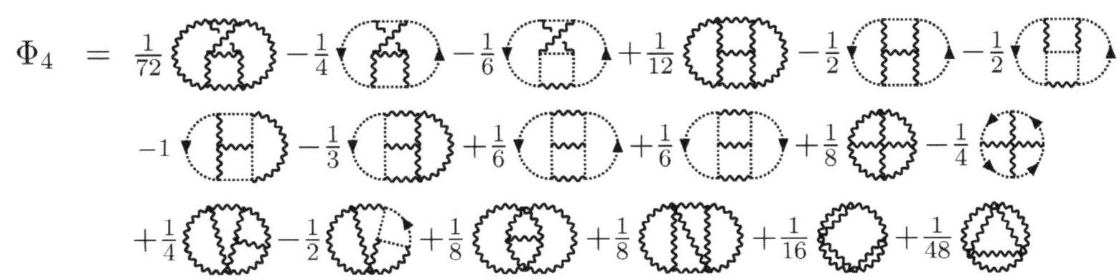

Figure 2. Some Feynman diagrams contributing to the four-loop pressure. Wiggly/dotted lines denote gluons/ghosts, respectively.

While working on the hard open problem of evaluating the required four-loop sum-integrals, one might be tempted to already show some results, in order to connect to phenomenology. To this end, as an attempt to strive for the currently best possible description of the pure-glue sector of hot QCD, the single unknown g^6 coefficient can be fixed by matching to available lattice data at intermediate temperatures $T \sim 3 - 5T_c$ [7]. The result is shown in Fig. 3, and can be argued to show a friendly functional behavior, smoothly connecting low-temperature data with the weak-coupling expansion at high temperatures.

Once the gluonic sector is under control, the effective theory framework allows for straightforward inclusion of fermions, since their effects are contained in the matching coefficients. To obtain an estimate of the equation of state (EoS) with physical quark masses, one can follow the strategy of unquenching: starting from the $N_f = 0$ results of Fig. 3, which correspond to $m_q = \infty$, one lowers N_f of them to their physical masses $m_{q,\rm phys}$ [7]. Schematically, the NLO result is $p(N_f = 0) \times \frac{[c_0+g^2\,c_2](N_f,m_i,\mu_i)}{[c_0+g^2\,c_2](N_f=0)}$ which, after going to physical units of energy by expressing $\Lambda_{\overline{\rm MS}}$ in units of MeV with the help of matching to a hadron resonance gas at low temperature, allows to plot the dimensionless ratios of Fig. 4. Let us mention that the deviation of the (dimensionless) EoS from a third, $(\frac{1}{3} - w(T))$, is often called the 'trace anomaly' or 'interaction measure'. One observes significant structure at tempreatures $T \sim 100 - 400$ MeV, a feature that does not come as a surprise since close to a second-order phase transition, one would expect e.g. the sound speed to scale with a critical exponent, $c(T) \sim (T - T_c)^{-\gamma}$.

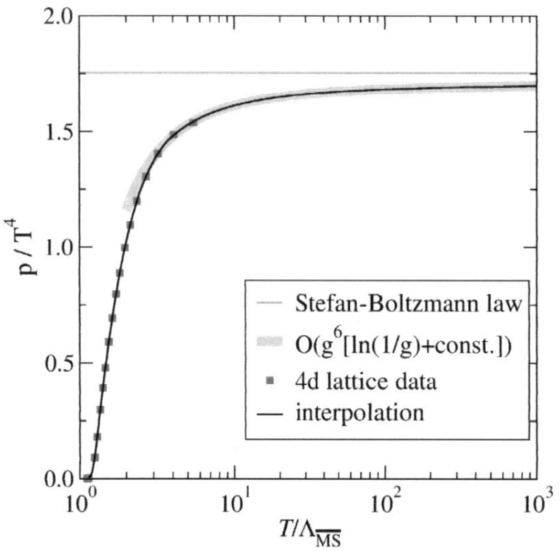

Figure 3. Interpolating curve (solid line) for the QCD pressure at $N_f = 0$ [7]. In the perturbative curve (grey band) the unknown g^6 constant has been adjusted to match lattice data [4] at $T \sim 4T_c$.

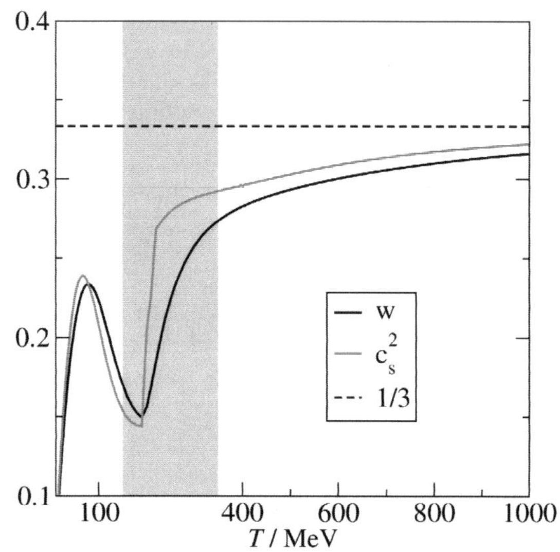

Figure 4. The equation-of-state parameter $w = \frac{p(T)}{e(T)}$ and the speed of sound squared $c_s^2 = \frac{p'(T)}{e'(T)}$ for QCD at $N_f = 4$ [7].

4. Conclusions

In summary, we have seen that phenomenologically relevant thermodynamic quantities of QCD, such as the pressure, can be determined to high accuracy with a mix of two main approaches: numerically at temperatures around 200 MeV, in the neighborhood of the critical one, and analytically at $T \gg T_c$. The latter approach involves multi-loop weak-coupling expansions and is efficiently organized in the framework of effective field theories, allowing for systematic improvements.

The relevant three-dimensional effective theories are interesting in their own right, opening up tremendous opportunities: fermions can be treated completely analytically, in contrast to the large problems they pose in the lattice approach. Furthermore, universality and superrenormalizability make these 3d theories an ideal playground for the development multi-loop methods.

Regarding the QCD pressure, the current status is that its 'physical leading order', in the sense defined above, remains unknown. A mild open problem being a matching of lattice- and continuum results for general N_c, the more difficult task to be completed is the evaluation of a large number of four-loop sum-integrals. However, the pressure shows a friendly functional behavior when the single yet-unknown coefficient is fitted.

Acknowledgments

This work was supported in parts by FONDECYT project number 1151281.

References

[1] K. A. Olive *et al.* [Particle Data Group Collaboration], *Review of Particle Physics*, Chin. Phys. C **38** (2014) 090001.
[2] S. Aoki *et al.* [PACS-CS Collaboration], *2+1 Flavor Lattice QCD toward the Physical Point*, Phys. Rev. D **79** (2009) 034503.
[3] N. P. Landsman and C. G. van Weert, *Real and Imaginary Time Field Theory at Finite Temperature and Density*, Phys. Rept. **145** (1987) 141.
[4] G. Boyd, J. Engels, F. Karsch, E. Laermann, C. Legeland, M. Lutgemeier and B. Petersson, *Thermodynamics of SU(3) lattice gauge theory*, Nucl. Phys. B **469** (1996) 419.
[5] F. Di Renzo, M. Laine, V. Miccio, Y. Schröder and C. Torrero, *The Leading non-perturbative coefficient in the weak-coupling expansion of hot QCD pressure*, JHEP **0607** (2006) 026.
[6] K. Kajantie, M. Laine, K. Rummukainen and Y. Schröder, *The Pressure of hot QCD up to g6 ln(1/g)*, Phys. Rev. D **67** (2003) 105008.
[7] M. Laine and Y. Schröder, *Quark mass thresholds in QCD thermodynamics*, Phys. Rev. D **73** (2006) 085009.
[8] A. D. Linde, *Infrared Problem in Thermodynamics of the Yang-Mills Gas*, Phys. Lett. B **96** (1980) 289.
[9] J. Möller and Y. Schröder, *Open problems in hot QCD*, Nucl. Phys. Proc. Suppl. **205-206** (2010) 218.
[10] E. V. Shuryak, *Theory of Hadronic Plasma*, Sov. Phys. JETP **47** (1978) 212 [Zh. Eksp. Teor. Fiz. **74** (1978) 408].
[11] J. I. Kapusta, *Quantum Chromodynamics at High Temperature*, Nucl. Phys. B **148** (1979) 461.
[12] T. Toimela, *The Next Term in the Thermodynamic Potential of QCD*, Phys. Lett. B **124** (1983) 407.
[13] E. Braaten and A. Nieto, *Free energy of QCD at high temperature*, Phys. Rev. D **53** (1996) 3421.
[14] J. P. Blaizot, E. Iancu and A. Rebhan, *On the apparent convergence of perturbative QCD at high temperature*, Phys. Rev. D **68** (2003) 025011.
[15] A. Hietanen, K. Kajantie, M. Laine, K. Rummukainen and Y. Schroder, *Plaquette expectation value and gluon condensate in three dimensions*, JHEP **0501** (2005) 013.
[16] F. Di Renzo, M. Laine, V. Miccio, Y. Schroder and C. Torrero, *The Leading non-perturbative coefficient in the weak-coupling expansion of hot QCD pressure*, JHEP **0607** (2006) 026.
[17] K. Kajantie, M. Laine, K. Rummukainen and Y. Schroder, *Four loop vacuum energy density of the SU(N(c)) + adjoint Higgs theory*, JHEP **0304** (2003) 036.
[18] K. Kajantie, M. Laine, K. Rummukainen and Y. Schroder, *The Pressure of hot QCD up to g6 ln(1/g)*, Phys. Rev. D **67** (2003) 105008.

Journal of Physics: Conference Series **720** (2016) 012028 doi:10.1088/1742-6596/720/1/012028

Stationary textures induced by spin-transfer torques: role of the angular dependence

Alejandro O. León

E-mail: aoleon@dfi.uchile.cl
Departamento de Física, Facultad de Ciencias Físicas y Matemáticas, Universidad de Chile, Casilla 487-3, Santiago, Chile.

Abstract. Magnetic systems forced with external fields or electric currents exhibit a rich spatiotemporal dynamics. A well known example is the one of spin-transfer torque driven textures, which includes switching, precessions, dissipative solitons, and periodic textures. Using different expressions that model the spin-transfer torque —angular dependence of the spin-transfer— we obtain analytic solutions for static spatially periodic states, study their stability, and elucidate the role that the angular dependence plays in the formation of textures. We demonstrate that the type of bifurcations changes from supercritical to subcritical, depending on the particular type of torque. Numerical simulations confirm this scenario. Thus, magnetoresistance measurements could permit to determine the form of the torque.

1. Introduction

Ferromagnetic materials subjected to dissipation and injection of energy or momentum self-organize in dissipative structures, such as solitons, skyrmions, periodic textures, and domain walls. The usual driving mechanisms are the couplings between the magnetization of the ferromagnetic material and external magnetic fields and electric currents. In 1996, Slonczewski[1] and Berger [2] predicted that a spin-polarized (SP) electric current exerts a torque in ferromagnetic nano-layers, effect known as spin-transfer torque [3, 4, 5]. Since then, the interest on excitations driven by SP currents has grown considerably, both from the scientific and the technological viewpoints [4]. A typical configuration to study spin-transfer torques is the multilayer nanopillar, or spin-valve, that consists on ferromagnetic conducting films separated by non-magnetic conductors as shown in Fig. 1(a). When an electric current is applied, it transfers spin angular momentum from one relatively thick layer with *fixed* magnetization to a thin film with *free* magnetization [4].

Spin-transfer torques are non-conservative effects, and therefore they can destroy, stabilize or destabilize states. Moreover, in the case in which a SP current injects enough energy to counterbalance the dissipation, the free magnetization switches or precesses with microwave frequencies, both effects have been studied experimentally [6, 7, 8, 9] and theoretically [10, 11, 12, 13]. Hence, electric currents permit to manipulate small scale magnets and use them as memory devices, or spin-wave emissors. The SP current can also stabilize configurations that are energy maxima. In this scenario, the free magnetization exhibits the phenomenology of the systems subjected to time-dependent driving forces [14]. This dynamic behaviors includes solitons, domain walls, and spatially periodic textures or *patterns*. Typical spin-transfer torque

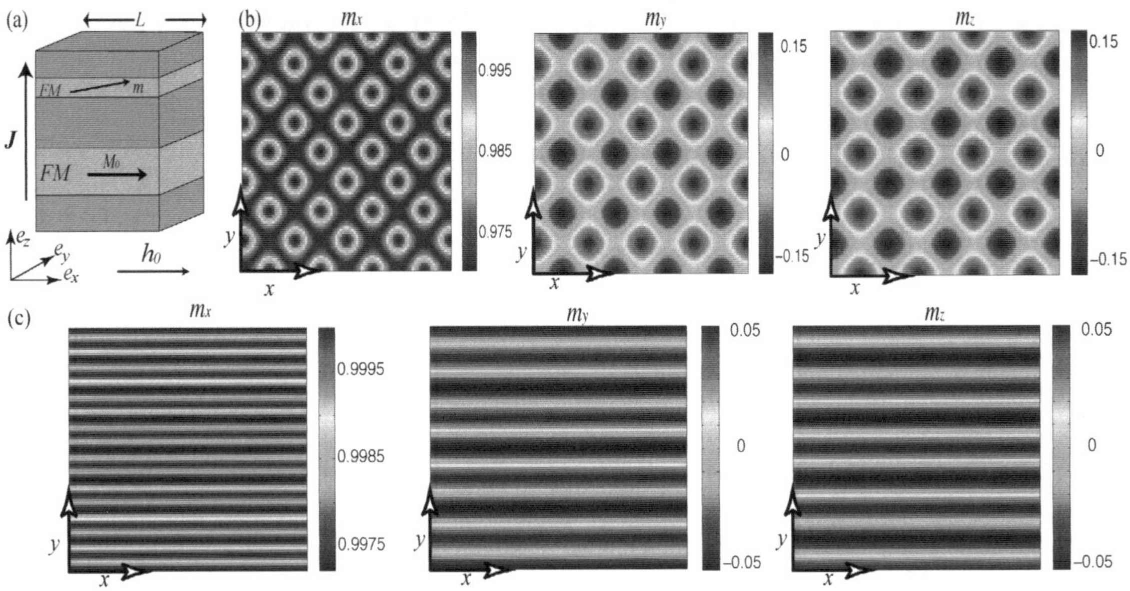

Figure 1. Nanopillar device and magnetic textures. a) Schematic spin-valve setup. The layers labeled as FM are the ferromagents, the rest of the structure is composed by nonmagnetic conductors; J is the SP electric current while h_0 is the external magnetic field, both effects are parallel to the easy-axes of the ferromagnetic layer under study. $\mathbf{M}_o$ stands for the magnetization of the fixed layer. b) Typical patterns induced by the spin-transfer torque effect. The parameters are $h_0 = -6$, $g = -0.4999$, $\beta_z = 1$, $\beta_x = 0.5$, $\alpha = 0.05$, and $dx = 0.126289$. The parameters of the torque are $\Delta = b = 0$, this is the sinus-approximation case. c) The same as b) with the Slonczewski torque, it is, $\Delta = 0$ and $b = 0.35$. Note the relation between the magnetic components $m_y \approx -m_z$. Since $m_x \approx 1-(m_y^2+m_z^2)/2$, the component m_x has spatial oscillations at twice the wavenumber of my and m_z.

induced patterns are shown in Figs. 1(b) and 1(c), they emerge due to an imbalance between the spin-polarized current and an external magnetic field [15]. The typical wavelength of these patterns is $20 - 30nm$.

In the classical approach, the free layer magnetization is described by the Landau-Lifshitz-Gilbert equation with an extra term that accounts for the interaction between the spin-polarized electric current and the ferromagnet [3, 4, 5]. Since the transport of the conducting electrons depends on the relative orientations of the free and fixed layers, the torque induced by the SP current can also depend on this orientation. This dependence is known as the *angular dependence of the spin-transfer torque*. Analytic expressions for this angular dependence have been proposed and studied for several geometries [1, 16, 19, 17, 18, 20]. In certain regions of the parameter space, the dynamics of the precessional states change quantitatively and qualitatively with the use of different torque terms, which permits to test the angular dependence for a given geometry [17, 18, 21, 22]. The textures predicted in Refs. [14, 15] are an alternative dynamical regime which permits to compare the predictions of the different spin-transfer models. In Ref. [15], a simplified version of the torque was used, and a more general study is required.

The aim of this article is to study the role of the angular dependence of the spin-transfer torque in pattern formation. We obtain analytic solutions for the spatially periodic patterns and elucidate their stability. It is found that the magnetic textures change strongly with the angular dependence. The manuscript is organized as follows: in Sec. 2, we describe the magnetization dynamics of the free layer. In Sec. 3, we obtain the amplitude equation for the patterns, and analyze the role of the angular dependence. The conclusions and remarks are left to the final section.

2. Free layer magnetization dynamics

Consider a square cross-section nanopillar device, with fixed layer magnetization $\mathbf{M}_0$ along the positive x-axis, this material is thicker than the free layer, and acts as a spin filter or spin polarizer for the electric current. Lateral sizes L in the transverse directions are around 100 nanometers for typical devices. Hereafter, we adimensionalyze the magnetization of the free layer $\mathbf{M} \to M_s\mathbf{m}$ and the external field $\mathbf{H_0} \to M_s\mathbf{h_0}$ by the saturation magnetization M_s; the time $t \to \gamma M_s t$ is written in terms of the gyromagnetic constant γ, and M_s; and the spatial coordinates $(x, y) \to l_{\text{ex}}(x, y)$ in terms of the *exchange length* $l_{\text{ex}} \equiv \sqrt{2A/(\mu_0 M_s^2)}$ where A is the exchange coupling in the ferromagnet. For example, in a $3nm$ thick Cobalt layer, $M_s \simeq 1.4\ 10^6 A/m$, and the characteristic time and space scales are $(\gamma M_s)^{-1} \simeq 3.2ps$ and $l_{\text{ex}} \approx 3.4nm$, respectively [5].

For thin films, the magnetic energy is given by [5]

$$E = \frac{\mu_0 M_s^2}{L^2} \int_0^L \int_0^L \left[-\mathbf{m} \cdot \mathbf{h_0} - \frac{1}{2}\beta_x m_x^2 + \frac{1}{2}\beta_z m_z^2 + \frac{1}{2}|\nabla \mathbf{m}|^2 \right] dxdy, \tag{1}$$

an external magnetic field $\mathbf{h_0} = h_0 \mathbf{e}_x$ points along the x-axis. The coefficients β_x and β_z are combinations of the anisotropy constants with respect to the appropriate axes, where β_x (β_z) favors (disfavors) the free magnetization in the x-axis (z-axis). Since the free layer is thin, the demagnetization field can be approximated by a shape anisotropy, and it is incorporated in the β_z coefficient.

The dynamic of the free layer magnetization is modeled by the dimensionless Landau-Lifshitz-Gilbert equation (LLG) with an extra term that accounts for the spin-transfer torque [5]

$$\frac{\partial \mathbf{m}}{\partial t} = -\mathbf{m} \times \mathbf{h}_{eff} + \alpha \mathbf{m} \times \frac{\partial \mathbf{m}}{\partial t} + \eta g_0\, \mathbf{m} \times (\mathbf{m} \times \mathbf{e_x}). \tag{2}$$

The first term of the right hand side of Eq. (2) accounts for the conservative precessions generated by the effective field,

$$\mathbf{h}_{eff} \equiv -\frac{L^2}{\mu_0 M_s^2} \frac{\delta E}{\delta \mathbf{m}} = (h_0 + \beta_x m_x)\, \mathbf{e}_x - \beta m_z \mathbf{e}_z + \nabla^2 \mathbf{m}. \tag{3}$$

The second and third terms of Eq. (2) are the phenomenological Gilbert damping and the spin-transfer torque, respectively. The dimensionless parameter g_0 is

$$g_0 \equiv \frac{\hbar J}{2d|e|\mu_0 M_s^2}, \tag{4}$$

where J is the electric current density, d the thickness of the layer and $e < 0$ the electric charge. The parameters J and g_0 are negative when the electrons flow from the fixed to the free layer. The dimensionless function $\eta = \eta(\mathbf{m} \cdot \mathbf{e_x})$ is known as the spin-torque efficiency, this function depends on the relative orientation of the ferromagentic films, and the device geometry and the physical properties of the layers, it has the general form of [4, 17, 18]

$$\eta(\mathbf{m} \cdot \mathbf{e_x}) \equiv \frac{(1 - b^2)}{1 - b + \Delta(1 + b)} \left(\frac{1}{1 + bm_x} + \frac{\Delta}{1 - bm_x} \right). \tag{5}$$

The coefficient Δ measures the anisotropy between ferromagnetic layers of the spin-valve. In the case of an symmetric structure (identical ferromagnetic layers and leads) Δ vanishes. Additionally, for short leads, Δ can be neglected. In these cases, the efficiency is just $\eta = (1+b)/(1+bm_x)$, which is term originally proposed by Slonczewski [1, 16]. The constant b accounts for the strength of the spin scattering at the free layer interface. Notice that the η function is $\eta(1) = 1$ in the parallel configuration. The approximation $\eta(m_x) \approx \eta(1) = 1$ for $m_x \approx 1$, which is equivalent to impose $b = 0$ in the formula (5), is known as the sinus spin-transfer torque, because the torque $||\eta g_0\, \mathbf{m} \times (\mathbf{m} \times \mathbf{e_x})|| \sim |\sin(\Psi)|$, where Ψ is the angle between the free and fixed magnetizations. This approximation is widely used in literature. We will compare the SP-driven dynamics in the case of a general torque ($b, \Delta > 0$), a symmetric torque ($\Delta = 0$), and a sinus torque ($b = 0$).

Notice that the LLG Eq. conserves the magnetization norm $\|\mathbf{m}\| = 1$, since $\mathbf{m}$ and $\partial_t \mathbf{m}$ are perpendicular. This permits us to use a spherical representation of the free magnetization

$$\mathbf{m} = \sin(\theta)\left[\cos(\phi)\mathbf{e}_x + \sin(\phi)\mathbf{e}_y\right] + \cos(\theta)\mathbf{e}_z. \tag{6}$$

Introducing this representation in Eq. (2), one obtains the following set of equations

$$\begin{aligned}
\partial_\tau \theta &= -(h_0 + \alpha\eta g_0)\sin(\phi) + (\alpha h_0 - \eta g_0)\cos(\phi)\cos(\theta) + \frac{\alpha}{2}\sin(2\theta)\left[\beta_z + \beta_x \cos^2(\phi)\right] \\
&\quad -\frac{\beta_x}{2}\sin(\theta)\sin(2\phi) + \sin(\theta)\nabla^2\phi + 2\cos(\theta)\nabla\phi\cdot\nabla\theta + \alpha\nabla^2\theta - \frac{\alpha}{2}\sin(2\theta)\left(\nabla\phi\right)^2, \\
\sin(\theta)\partial_\tau\phi &= (\eta g_0 - \alpha h_0)\sin(\phi) - (\alpha\eta g_0 + h_0)\cos(\phi)\cos(\theta) - \frac{1}{2}\sin(2\theta)\left[\beta_z + \beta_x\cos^2(\phi)\right] \\
&\quad -\alpha\frac{\beta_x}{2}\sin(\theta)\sin(2\phi) + \alpha\sin(\theta)\nabla^2\phi + 2\alpha\cos(\theta)\nabla\phi\cdot\nabla\theta - \nabla^2\theta + \frac{1}{2}\sin(2\theta)\left(\nabla\phi\right)^2,
\end{aligned} \tag{7}$$

where $\tau = t/\left(1+\alpha^2\right)$, and $\eta = \eta(\sin(\theta)\cos(\phi))$. The simplest equilibria of Eq. (2) are $\mathbf{m} = \pm\mathbf{e}_x$, which represent a free magnetization parallel ($+$) or anti-parallel ($-$) to the fixed magnetization $\mathbf{M}_0$. Both states correspond to extrema of the free energy E. In spherical coordinates, the parallel and anti-parallel states are $(\theta,\phi) = (\pi/2, 0)$, and $(\theta,\phi) = (\pi/2,\pi)$ respectively. From the technological viewpoint, when both equilibria are stable, the spin-valve can be seen as a two-states system capable of saving information. Resistance measurements permit to read such data from the valve, and the spin-transfer torque induced switching allows one to write on the device. This is the basis of the spin-transfer torque based RAM memories [4].

For the rest of the manuscript we will concentrate on the parallel equilibrium $\mathbf{m} = \mathbf{e}_x$.

3. Periodic patterns

Note that the linear stability analysis around the $m_x = 1$ state is independent of the parameters b and Δ because the angular dependence is constant around the parallel state, $\eta(m_x) = \eta(1) + \mathcal{O}(m_y^2 + m_z^2) \approx 1$. The instabilities of this state have been subject of several studies the last decades [3, 4, 5, 15]. There are three main types of bifurcations: the Andronov-Hopf instability that generates self-oscillations, the stationary instability in which the system goes to other stationary equilibra, and the spatial instability (see Ref. [15] for more details); this last bifurcation is the motivation of the of this work. The spatial instability is characterized because the perturbations that destabilize the parallel state are non-uniform and generally give rise to periodic patterns. In the case of a spin-valve, when $g_0, h_0 < 0$, the SP current favors the parallel state while the external field disfavors it. The competition of these opposing effects is the physical origin of the spatial instability. Considering a critical perturbation of the form

$$\begin{pmatrix} \theta \\ \phi \end{pmatrix} \approx \begin{pmatrix} \pi/2 \\ 0 \end{pmatrix} + e^{\lambda t}\left[A_0 e^{i\mathbf{q}\cdot\mathbf{r}} + \bar{A}_0 e^{-i\mathbf{q}\cdot\mathbf{r}}\right]\begin{pmatrix} 1 \\ 1 \end{pmatrix}, \tag{8}$$

where the groth rate is $\lambda = \beta_z/2 + g_0$. The norm q of the wave-vector $\mathbf{q}$ is controlled by the external magnetic field through $q = -h_0 - \beta_x - \beta_z/2$, and it exists for $h_0 < -\beta_x - \beta_z/2 < 0$. The complex amplitude A_0 is a constant fixed by the initial conditions, and $\bar{A}_0$ is the complex conjugate of A_0. The spatially periodic perturbations are amplified on time for $g_0 > -\beta_z/2$, or decay to zero for $g_0 < -\beta_z/2$. Notice that since the nanopillar has a square cross-section, there can be several admitted modes with wave-numbers q, which will grow and interact for $g_0 > -\beta_z/2$.

In the critical situation $g_{0,c} \equiv -\beta_z/2$, there is no exponential grow or decay of the perturbations, and the modes of formula (8) exhibit a slow dynamics given by the nonlinear corrections. At the onset of the spatial bifurcation, it is possible to obtain the equations for the critical modes by introducing the following ansatz

$$\begin{pmatrix} \theta \\ \phi \end{pmatrix} = \begin{pmatrix} \pi/2 \\ 0 \end{pmatrix} + \left[A(t)e^{iqx} + B(t)e^{iqy} + \bar{A}(t)e^{-iqx} + \bar{B}(t)e^{-iqy}\right] \begin{pmatrix} 1 \\ 1 \end{pmatrix} + \mathbf{W}(A, B, \bar{A}, \bar{B}, \mathbf{r}), \tag{9}$$

where the amplitudes $A(t)$ and $B(t)$ are slowly varying functions of time, and $\mathbf{W}$ is a small correction that appears due to the nonlinear nature of the problem, and it depends nonlinearly on the amplitudes A and B. We have considered just two modes by the sake of simplicity, and the amplitudes $A(t)$ and $B(t)$ are the envelops of the spatial oscillations along the $x-$axis and $y-$axis, respectively. Replacing the above ansatz in Eq. (7), linearizing in $\mathbf{W}$, and after imposing a solvability condition [23] , we obtain

$$\begin{aligned} \frac{dA}{dt} &= (g_0 - g_{0,c})A - \Gamma A\left(|A|^2 + 2|B|^2\right), \\ \frac{dB}{dt} &= (g_0 - g_{0,c})B - \Gamma B\left(|B|^2 + 2|A|^2\right), \end{aligned} \tag{10}$$

where the coefficient of the nonlinearity is

$$\Gamma = \frac{3}{2}\frac{\beta_z b}{1-b^2}\frac{(1-b)^2 - \Delta(1+b)^2}{1-b+\Delta(1+b)}. \tag{11}$$

The above set of equations describes the growth of the pattern for $g_0 > g_{0,c}$, with a nonlinear interaction between modes, and a nonlinear saturation (or nonlinear gain) for $\Gamma > 0$ ($\Gamma < 0$). Let us start with the case of a nonlinear saturation $\Gamma > 0$. Notice that the set of Eqs. (10) is phase invariant, it is, they remain unchanged under transformations of the form $(A, B) \rightarrow (Ae^{i\psi_1}, Be^{i\psi_2})$ for arbitrary real constants ψ_1 and ψ_2. This motivates the use of reduced real valued variables $a_1(t) = \sqrt{\Gamma}|A(t)|$, and $a_2(t) = \sqrt{\Gamma}|B(t)|$ for the case of nonlinear saturation,

$$\begin{aligned} \frac{da_1}{dt} &= \epsilon a_1 - a_1\left(a_1^2 + 2a_2^2\right), \\ \frac{da_2}{dt} &= \epsilon a_2 - a_2\left(a_2^2 + 2a_1^2\right). \end{aligned} \tag{12}$$

where $\epsilon \equiv g_0 - g_{0,c}$ For negative ϵ, the spatially periodic perturbations decay to the homogeneous solution $(a_1, a_2) = (0, 0)$, which is linearly stable. Hence, the parallel state is stabilized by the current for negative enough values of g_0.

Figure 2 shows the phase portrait of Eq. (12) for positive values of ϵ. The 4 steady states of the system are the parallel state $(a_1, a_2) = (0, 0)$, the roll patterns $(a_1, a_2) = (\sqrt{\epsilon}, 0)$ and $(a_1, a_2) = (0, \sqrt{\epsilon})$, and finally the square texture $(a_1, a_2) = (\sqrt{\epsilon/3}, \sqrt{\epsilon/3})$. Given the amplitude

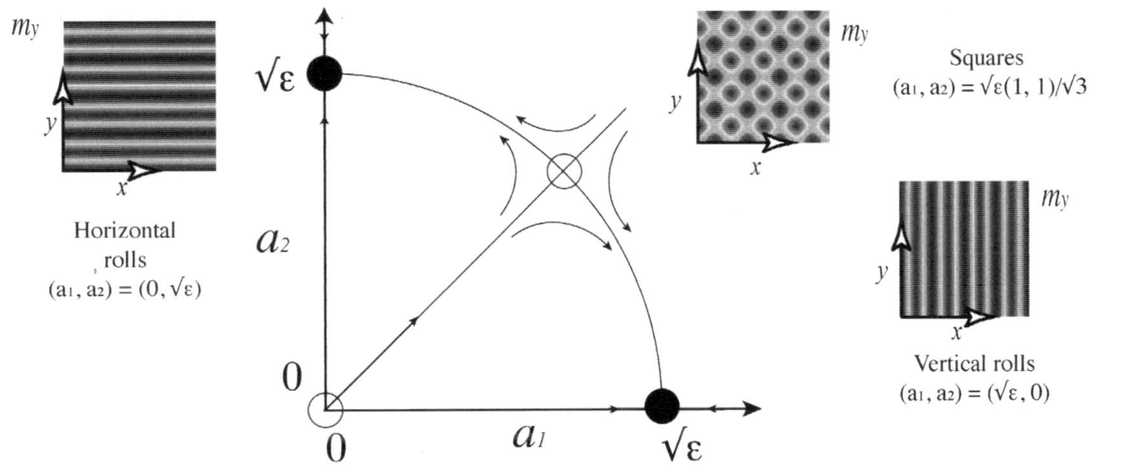

Figure 2. Phase portrait of amplitude Eq. (12), which describes the pattern formation driven by a Slonczewski spin-transfer torque. Insets show the magnetic component m_y of the equilibrium states. For $g_0 > g_{0,c}$ the vertical and horizontal roll equilibria (black circles) are stable for the Slonczewski form of the torque. The Square pattern and the uniform solution (empty circles) are unstable.

Eq. (12), the linear stability analysis of the patterns becomes a trivial task. This analysis reveals that the roll states are stable (black circles in Fig. 2) and the square is unstable (empty circle). Thus, after transients, the magnetization converges to the horizontal or vertical rolls of the form

$$m_x \approx 1 - 4\left(g_0 - g_{0,c}\right)\left[\frac{1-b^2}{\beta_z b}\frac{1-b+\Delta(1+b)}{(1-b)^2-\Delta(1+b)^2}\right]\left(\frac{1-\cos\left(\mathbf{q}\cdot\mathbf{r}\right)}{3}\right),$$
$$m_y \approx m_z \approx 2\sqrt{g_0-g_{0,c}}\sqrt{\frac{2}{3}\frac{1-b^2}{\beta_z b}\frac{1-b+\Delta(1+b)}{(1-b)^2-\Delta(1+b)^2}}\cos\left(\mathbf{q}\cdot\mathbf{r}\right), \quad (13)$$

where $\mathbf{q}\cdot\mathbf{r} = qx$ or $\mathbf{q}\cdot\mathbf{r} = qy$. Fig. 1(c) shows this state. This roll-like functions are the typical magnetic configurations for the most general spin-torque.

In the next subsections, we discuss the different types of instabilities that arise with each spin-transfer torque term. These transitions correspond to supercritical ($\Gamma > 0$), quintic supercritical ($\Gamma = 0$), and subcritical ($\Gamma < 0$).

3.1. Slonczewski limit ($\Gamma > 0$)

If the spin-valve has equal ferromagnetic films and equal leads, then $\Delta = 0$ and the $\eta(m_x)$ function is the same as proposed originally by Slonczewski in Refs. [1, 16]. In this case, the saturation coefficient takes the form

$$\Gamma = \frac{3}{2}\frac{\beta_z b}{1+b}. \quad (14)$$

Figure 1(c) illustrates the rolls for $b = 0.35$. This type of bifurcation is known as a supercritical transition. It is worth noting that the saturation is independent of the external field h_0 and easy-axis anisotropy constant β_x. Figure 3(a) compares the numerical solution of the Landau-Lifshitz-Gilbert equation with the predicted amplitude $2|A|(\epsilon) \sim 2\left(\epsilon/\Gamma\right)^{1/2}$ of rolls $m_y \approx -m_z \approx 2|A|\cos(\mathbf{q}\cdot\mathbf{r})$. In order to integrate the LLG equation, space was discretized using a finite differences centered scheme of order six, and the temporal evolution was obtained with a fifth order variable step size Runge-Kutta.

The magnetoresistance δr permits to compare predicted textures with experimental data. Hence, it is an adequate physical quantity for characterizing magnetic configurations in spin-valves. For uniform magnetized free layers, the magnetoresistance depends only on the relative orientation between the free and fixed layers, $\delta r = \delta r(m_x)$. For nonuniform magnetic states, the average indicator $\delta r \equiv \delta r(m_x^{av})$ is commonly used [24], where m_x^{av} is the average magnetization along the x-axis. Consequently, for symmetric spin-valves the average resistance will be independent of external field and anisotropy coefficient β_x.

3.2. Sinus-aproximation ($\Gamma = 0$)

In the sinus approximation ($b = 0$), the function η is constant and the cubic coefficients of the set of Eqs. (10) vanishes. In order to quantify the saturation, it is necessary to include higher order corrections into the amplitude equation. This case was studied in detail in Ref. [15]. Let us summarize the main results of Ref. [15], and compare them with the predictions of other type of spin-transfer torques. In a similar way of the previous section, we consider a two modes ansatz of the form

$$\begin{pmatrix} \theta \\ \phi \end{pmatrix} \approx \begin{pmatrix} \pi/2 \\ 0 \end{pmatrix} + R\left[A(t)e^{i\mathbf{q_1}\cdot\mathbf{r}} + B(t)e^{i\mathbf{q_2}\cdot\mathbf{r}} + \bar{A}(t)e^{-i\mathbf{q_1}\cdot\mathbf{r}} + \bar{B}(t)e^{-\mathbf{q_2}\cdot\mathbf{r}}\right]\begin{pmatrix} 1 \\ 1 \end{pmatrix} + \mathbf{W}'(A, B, \bar{A}, \bar{B}, \mathbf{r}), \tag{15}$$

where the wave-vectors $\mathbf{q_1}$ and $\mathbf{q_2}$ have norm q, and their orientation is fixed by the border of the sample. The prefactor R is the characteristic scale for patterns induced by a sinus torque, which is defined as

$$R \equiv \sqrt[4]{\frac{4\beta_z}{(6\beta_x + 3\beta_z - 2q^2)^2}}. \tag{16}$$

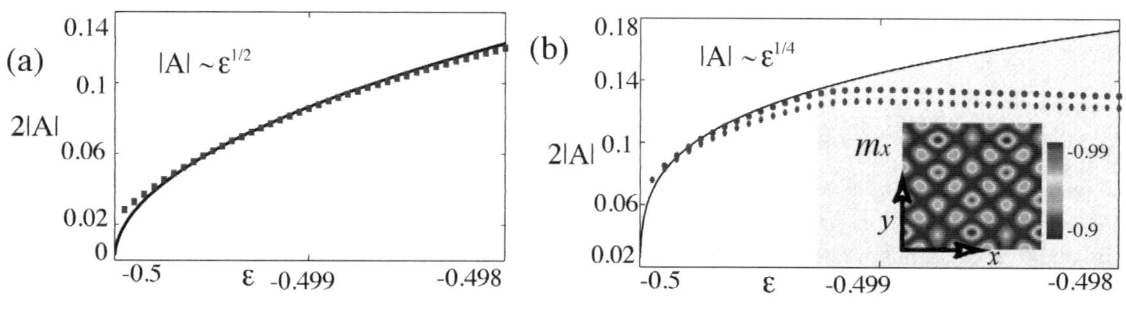

Figure 3. Comparison between analytic calculations (solid lines) and micromagnetic simulations (crosses and circles). (a) Pattern amplitudes for the Slonczewski spin-transfer torque with $b = 0.55$, and $\Delta = 0$. The solid line is the solution roll solutions of Eq. (10), while crosses $+$ and circles $\bullet$ stand by the maximum of the Fourier transform of m_z and m_y, respectively. It is worth noting that at dominant order $|m_y| \approx |m_z| \approx |\phi| \approx |\theta - \pi/2|$. The fields (m_y, m_z) were obtained with the direct integration of the LLG Eq. (2) with $h_0 = -6$, $\beta_x = 0.5$, $\beta_z = 1$, $\alpha = 0.05$, and spatial step size $dx = 0.126289$. (b) The same as in (a) for a sinus spin-transfer torque ($b = 0$). At the onset of the spatial instability, the amplitudes of follow the law $|A| \sim \epsilon^{1/4}$ predicted by the amplitude equation (17), nevertheless far from the bifurcation the square pattern becomes unstable giving rise to the stationary pattern of the inset.

Replacing the ansatz (15) in the LLG equation and imposing solvability condition, we get

$$\begin{aligned}\frac{dA}{dt} &= \epsilon A - \frac{8}{3}\frac{2-3D}{1-D}A|A|^2|B|^2 - A|A|^4 - \frac{8}{3}\frac{1}{1-D}A|B|^4,\\ \frac{dB}{dt} &= \epsilon B - \frac{8}{3}\frac{2-3D}{1-D}B|B|^2|A|^2 - B|B|^4 - \frac{8}{3}\frac{1}{1-D}B|A|^4,\end{aligned} \tag{17}$$

with $\epsilon \equiv g_0 - g_{0,c}$. Note that the modes interaction is mediated by quintic nonlinearities. This type of transition is known as quintic supercritical bifurcation. The parameters D accounts for the competition between the external magnetic field and the anisotropies

$$D \equiv \frac{2}{3}\frac{q^2}{2\beta_x + \beta_z} = \frac{2|h_0| - 2\beta_x - \beta_z}{6\beta_x + 3\beta_z}. \tag{18}$$

The above set of equations admits roll solutions of the form $(|A|,|B|) = (\epsilon^{1/4}, 0)$ and $(|A|,|B|) = (0, \epsilon^{1/4})$. The phase portraits of the set of Eqs. (17) are shown in Fig. 4. These solutions are linearly stable when the field $|h_0|$ is small compared with the anisotropy constants (or equivalently $D < 1$). For intense fields ($D > 1$) the rolls become unstable. In the last case, the square-like equilibrium $A = B = (\epsilon/9)^{1/4}$ is linearly stable. Figure 1(c) shows a square pattern obtained for $h_0 = -6$, where $D > 1$. The phase portrait of this case is in Fig. 4(b). For the critical value of the external field $h_{0,c} = -2(2\beta_x + \beta_z)$ the quintic saturation vanishes, and the analysis of pattern formation requires higher order corrections.

Since the saturation is given by the quintic order nonlinearities, then the equilibrium states grow as $m_y \sim m_z \sim (g_0 - g_{0,c})^{1/4}$ as illustrated in Fig. 3(b). Moreover, the m_x magnetization component follows the law $m_x \sim (g_0 - g_{0,c})^{1/2}$. Note that the square pattern of Fig. 1(c) becomes unstable when g_0 is surpasses a critical value; in this situation additional modes grow and interact. In the context of dynamical systems this bifurcation is known as a secondary instability of the underlying pattern. A different pattern appears from this instability as shown in the inset of Fig. 3(b).

In brief, the case of the sinus approximation: the equilibrium magnetization depends on the external field and the anisotropies; there is a transition for a critical value of the external field, in this transition the roll solutions interchange stability with the square-like pattern. Finally, the margnetoresistence depends on the square-root of the current $\delta r \sim (g_0 - g_{0,c})^{1/2}$.

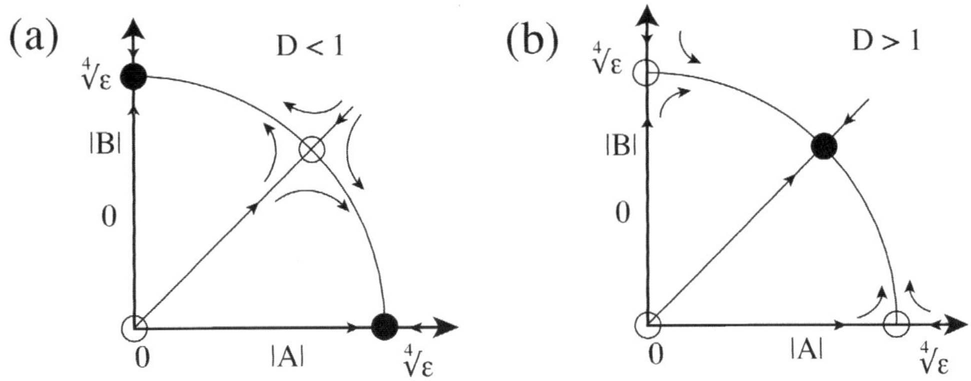

Figure 4. Phase portrait of amplitude Eq. (17), where a sinus-approximation for the spin-transfer torque has been considered. Black circles stand for stable pattern equilibria, while empty circles are for unstable states. a) Small external field ($D < 1$) favors the formation of rolls. b) High external magnetic field ($D > 1$) favors the square equilibrium.

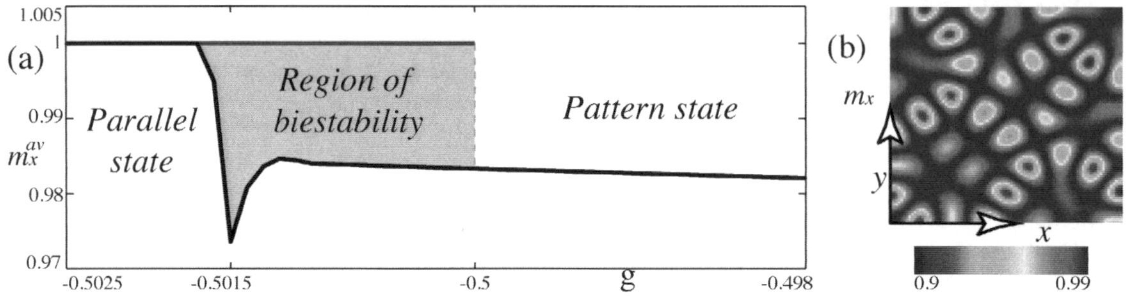

Figure 5. Hysteresis loop for asymmetric spin-valves. (a) The spatial average m_x^{av} of m_x, at equilibrium, is obtained integrating the LLG Eq. (2) with $b = 0.55$, $\Delta = 0.2$, $h_0 = -6$, $\beta_x = 0.5$, $\beta_z = 1$, $\alpha = 0.05$, and spatial step size $dx = 0.126289$. (b) m_x component of the equilibrium magnetization for $g_0 = g_{0,c}$.

3.3. Hysteresis ($\Gamma < 0$)

A completely different scenario occurs for highly asymmetric spin-valves. If the coefficient Δ surpasses the critical value $\Delta_c = (1-b)^2/(1+b)^2$, the cubic nonlinearities amplify the pattern and higher order corrections must be included in Eq. (10). In such case the spatial instability is subcritical, and patterns exist even for $g_0 < g_{0,c}$. Physical systems with subcritical spatial instabilities are characterized by an hysteresis region close to the critical point (in this case $g_0 = g_{0,c}$), it is, both the parallel state and patterns are stable. Figure 5 illustrate the hysteresis for the spatial average of the magnetization component m_x at equilibrium, m_x^{av}. Hence, in the case of asymmetric enough spin-valves, the magnetoresistence has a discontinuity.

From the viewpoint of dynamical systems, subcritical spatial instabilities generate a wide variety of states, such as domain walls, localized patterns, and interacting solitary structures [23], among others. The self-organization of highly asymmetric spin-valves forced with spin-polarized currents is an open problem, which could be interesting for memory technologies.

4. Conclusions and remarks

We have studied the formation of periodic textures in spin-valves, with a general type of spin-transfer torque. Using the weakly nonlinear analysis, we have found that the type of spatial instability changes from supercritical, to quintic supercritical and subcritical bifurcations depending on the form for the spin-transfer. The sinus approximation gives a pattern growing law of $(g_0 - g_{0,c})^{1/4}$, the magnetoresistance depends on the anisotropies and the external field. In opposition, for the Slonczewski spin-transfer torque and the general spin-transfer torque, the textures grow as $(g_0 - g_{0,c})^{1/2}$, and the magnetoresistance is independent of the external field and the easy-axis anisotropy β_x. Moreover, for highly asymmetric spin-valves, the spatial instability that creates patterns becomes subcritical, which generate hysteresis loops. Numerical simulations of the Landau-Lifshitz-Gilbert equation confirm these predictions. We expect that the formation of textures will permit to have an additional dynamical test to the spin-transfer torque terms, as well as possible application of pattern states in technology.

Acknowledgments

I thank Marcel G. Clerc for fruitful discussions. I gratefully acknowledge finantial support from Becas Conicyt 2012, Contract No. 21120878.

References

[1] Slonczewski J C, J. 1996 *Mag. Mat. Mag.* **159** L1.

[2] Berger L 1996 *Phys. Rev. B* **54** 9353.
[3] Ralph D C, and Stiles M D 2008 *J. Magn. Mag. Mater.* **320** 1190.
[4] Stiles M D, and Miltat J, in *Spin Dynamics in Confined Magnetic Structures*, edited by B. Hillebrands, A. Thiaville (Springer, Berlin, 2006) Vol. 3, Chap. 7
[5] Mayergoyz I D, Bertotti G, and Serpico C, *Nonlinear Magnetization Dynamics in Nanosystems* (Elsevier, Oxford, 2009).
[6] Katine J A, Albert F J, Buhrman R A, Myers E B, and Ralph D C 2000 *Phys. Rev. Lett.* **84** 3149.
[7] Kiselev S I, Sankey J C, Krivorotov I N, Emley N C, Schoelkopf R J, Buhrman R A, and Ralph D C 2003 *Nature (London)* **425** 380.
[8] Kiselev S I, Sankey J C, Krivorotov I N, Emley N C, Rinkoski M, Perez C, Buhrman R A, and Ralph D C 2004 *Phys. Rev. Lett.* **93** 036601.
[9] Kiselev S I, Sankey J C, Krivorotov I N, Emley N C, Garcia A G F, Buhrman R A, and Ralph D C 2005 *Phys. Rev. B.* **72** 064430.
[10] Lee K J, Deac A, Redon O, Nozieres J P, and Dieny B 2004 *Nature Materials* **3** 877.
[11] Bertotti G, Serpico C, Mayergoyz I D, Magni A, d'Aquino M, and Bonin R 2005 *Phys. Rev. Lett.* **94** 127206.
[12] Berkov D V, and Gorn N L 2008 *J. Phys. D* **41** 164013.
[13] Slavin A, and Tiberkevich V 2009 *IEEE Trans. Magn.* **45** 1875.
[14] León A O, and Clerc M G 2015 *Phys. Rev. B* **91** 014411.
[15] León A O, Clerc M G, and Coulibaly S 2014 *Phys. Rev. E* **89** 022908.
[16] Slonczewski J C 2002 *J. Mag. Mat. Mag.* **247** 324.
[17] Xiao J, Zangwill A, and Stiles M D 2004 *Phys. Rev. B* **70** 172405.
[18] Xiao J, Zangwill A, and Stiles M D 2005 *Phys. Rev. B* **72** 014446.
[19] Xiao J, Zangwill A, and Stiles M D 2007 *Eur. Phys. J. B* **59** 415.
[20] Barnas J, Fert A, Gmitra M, Weymann I, and Dugaev V K 2005 *Phys. Rev. B* **72** 024426.
[21] Kim W, Lee S-W, and Lee K-J 2011 *J. Phys. D* **44** 384001.
[22] Lee S-W, and Lee K-J 2010 *IEEE Trans. Magn.* **46** 2349.
[23] Pismen L M *Patterns and Interfaces in Dissipative Dynamics*, (Springer Series in Synergetics, Berlin Heidelberg 2006).
[24] Lee K J, Deac A, Redon O, Nozieres J P, and Dieny B 2004 *Nature Materials* **3** 877.

Journal of Physics: Conference Series **720** (2016) 012029 doi:10.1088/1742-6596/720/1/012029

Synthesis and characterization of chemically deposited CdS thin films without toxic precursors.

A. Fernández-Pérez[1] and M. G. Sandoval-Paz [2],

[1] Departamento de Fśica, Facultad de Ciencias, Universidad del Bío-Bío, Collao 1202, Casilla 5-C, Concepción, Chile.

[2] Departamento de Física, Facultad de Ciencias Físicas y Matemáticas, Universidad de Concepción, Casilla 160-C, Concepción, Chile

E-mail: arturofe@ubiobio.cl

Abstract. Al doped and undoped CdS thin films (CdS:Al) were deposited on glass, copper and bronze substrates by chemical bath deposition technique in an ammonia-free cadmium-sodium citrate system. The structural and optical properties of the CdS films were determined by X-ray diffraction (XRD), scanning electron microscope (SEM), and simultaneous transmission-reflection spectroscopy. It was found that the properties of the films depend on the amount of Al in the growth solutions and deposition time. The increase in Al content in the reaction solution led to a smaller crystallite size and higher energy band gap that varies in the range 2.42 eV - 2.59 eV depending on the Al content.

1. Introduction

Over the years, the study of optical and electrical properties of Cadmium Sulphide (CdS) has been the subject of intense research due to the wide variety of technological applications, such as opto-electronic devices [1] and thin film solar cells [2]. For thin film solar cells based on CdTe and Cu(InGa)Se2 (CIGS) absorber layers, CdS is the most commonly used optical window material because of its high optical transparency , wide band gap (2.42 eV), n-type conductivity and thickness of about 100 nm [3].

Now, one of the most inexpensive and scalable methods to prepare CdS thin films is chemical bath deposition (CBD), which uses a controlled chemical reaction to grow thin films by precipitation. To achieve this, substrates are vertically immersed in an alkaline solution containing the chalcogenide source, the metal ion and a complexing agent. Many studies on the deposition of CdS using this technique employ a highly toxic material which is ammonia in the chemical bath as complexing agent and hydroxide source [4, 5].

In terms of introducing impurities in the films, doping agents on CdS thin films synthesized by CBD alter their optical, electrical and structural properties, as it has been reported in [6, 7, 8, 9]. These authors reported a decrease in the band gap of doped CdS films for some specific ratios between the concentrations of doping agents and Cd. It is important to note that in all these studies, ammonia is one of the main components of the chemical bath. In this sense, the aim of this research is to study optical, structural and mechanical properties of CdS:Al thin films that were grown by CBD in an ammonia-free system and to provide information on the surface morphology of CdS films grown by CBD on metal substrates.

2. Experimental procedure

Several sets of CdS films were grown by chemical bath deposition (CBD) on glass, copper and bronze substrates (25 mm × 75 mm × 1 mm). Before the process of deposition, the glass substrates were previously cleaned with deionized water and then dried at room temperature. On the other hand, copper and bronze substrates were previously cleaned with ciclohexane, then rinsed with deionized water and dried at room temperature. The formula to prepare CdS films is described in [10], referred as AF films, and it is based on a mixture of the following precursors dissolved in an aqueous solution: Cadmium Chloride, 0.05 M; Sodium Citrate, 0.5 M; Potassium Hydroxide 0.5 M; Thiourea 0.5 M. Also, pH 10 Borate buffer and deionized water is added to the beaker. The films were grown by vertically immersing the substrates in the aqueous solution at a temperature of 70C for 120 minutes.

In the case of glass substrates, the doping of the CdS films was performed by adding Aluminium chloride ($AlCl_3$) to the mixture with a different molar ratio in the solution, $R = [Al]/[Cd]$ (0.00, 0.05, 0.10), where the initial concentration of Cd at 0.5 M remains constant.

In order to determine the surface morphology, the films were analyzed by a scanning electron microscope (SEM) using a Jeol microscope model 6610LV. The nanomechanical properties were determined through nanoindentation technique using a HysitronTriboindenter model TI-900 with a Berkovich indenter tip. Additionally, the structural properties of the CdS:Al films were determined through X-ray diffraction measurements (XRD) using a Bruker Endeavor D4 unit with Cu-Kα radiation, $\lambda = 0.15406$ nm (Voltage 40 kV with 20 mA flux and measuring angle (2) varying from 20° to 70°).

Optical properties of the films were determined using a reflectometer Filmetrics model F10-RT, which measured the reflectance and transmittance simultaneously in the wavelength range from 380 - 1050 nm. The film thickness was calculated using a software *Filmeasure*, which uses a mathematical model [11] to adjust the spectral data in order to determine the films thickness.

3. Results and Discussion

In the glass substrates, the CdS:Al films were yellowish, homogeneous, and their color slightly changed for different Al contents becoming more orange-colored as Al concentrations increased. Besides, it is observed that the more Al content, the better the films adhered to the substrate.

Conversely, the films deposited on copper and bronze substrates had a low adhesion and they were inhomogeneous. CBD applied to grow CdS films on metals produces oxides during the deposition that prevent a pure CdS film from growing unless we perform a thin film deposition on metal and characterize the surface morphology using SEM and Nanoindentation.

In Fig. 1 it is shown the SEM micrographs of CdS films deposited on copper and bronze substrates. We observed a vertical orientation of CdS grown on metal sheets due to the metal lamination process. The films were inhomogeneous and they did not reflect specularly, but they had yellowish regions as well as in the glass substrates, implying that a CdS film is grown by CBD process. This is different from the CdS:Al films deposited on glass substrates which were homogeneous, hard, and they adhered better to the substrate. In Fig. 2, we show in the SEM images of CdS:Al films. A dense layer of small grains that had less than 60 nm in diameter was observed and large particles of about 200 300 nm were embedded in the surface, which were absorbed by the it because of colloids formed in the solutions. In this case, CdS can appear heterogeneously in the solution, via an ion-by-ion process forming the film first layer and then through the cluster-by-cluster process, as it is reported in [10].

In the Figs. 3 and 4 it is shown 2D images obtained from the data of the tip scanning performed in the Nanoindenter. Nanoindentation consists in the penetration of a Berkovich tip that applies many cycles of load and unload to the CdS film measuring the hardness and the elastic modulus of each sample. The data is resumed in the Table 1.

The films bad adhesion to the metal substrates is deduced from the data because the measured

Figure 1. SEM images with different magnifications of CdS films deposited on metal substrates. (a) CdS on Bronze (100X); (b) CdS on Bronze (10000X); (c) CdS on Copper (100X); (d) CdS on Copper (10000X).

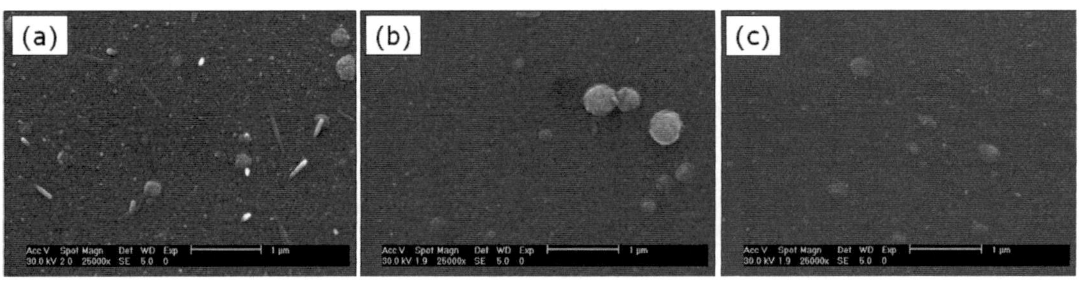

Figure 2. SEM images of CdS:Al films deposited on glass substrates. (a) R=0.00; (b) R=0.05; (c) R=0.10.

hardness and elastic modulus are close to the reported values of bronze and copper. Then, the Berkovich tip penetrated the substrates and the lamination process of the metal sheet fabrication used in the deposition resulted in many regions with no deposition shown as the vertical lines in the SEM images. On the other hand, CdS:Al films grown on glass are homogeneous and had a good adhesion to the substrate. We observed that the elastic modulus and hardness measured using the nanoindentation technique had similar values. Besides, the penetration depth of the indentation can be controlled taking into account that the films have a thickness of about 100 nm. The tips penetration depth is 10 to 20 nm, therefore the indentation process was carried out in the films and not in the glass.

Structural properties of CdS:Al films were studied by X-ray diffraction and the diffraction

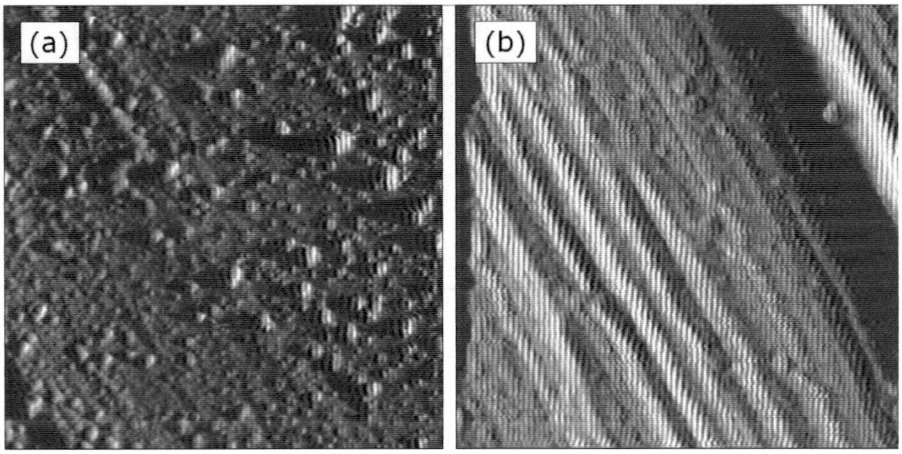

Figure 3. 2D images of CdS films deposited on metal substrates obtained from the tip scanning in the Nanoindenter in an area of 15 μm x 15 μm (a) CdS on Bronze; (b) CdS on Copper

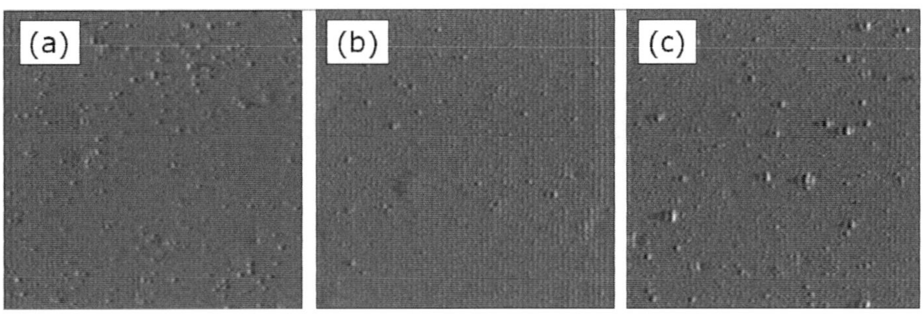

Figure 4. 2D images of CdS:Al films deposited on glass substrates obtained from the tip scanning in the Nanoindenter in an area of 15 μm x 15 μm (a) R=0.00; (b) R=0.05; (c) R=0.10.

Table 1. Hardness and elastic modulus of CdS thin films grown on different substrates obtained from nanoindentation.

Film Substrate	Hardness (GPa)	Elastic modulus (GPa)
Bronze	123.00 ± 15.01	2.112622 ± 0.341433
Cooper	180.19 ± 29.12	1.910959 ± 0.284137
Glass (R=0.00)	81.76 ± 6.82	3.292905 ± 0.343220
Glass (R=0.05)	87.13 ± 7.07	3.150583 ± 0.079346
Glass (R=0.10)	81.32 ± 8.26	3.091134 ± 0.103674

Table 2. Average interplanar distance $d_{(002)}$ and crystallite size D of CdS:Al thin films obtained from XRD.

Molar ratio [Al]/[Cd] in solution	Average interplanar distance (nm)	Crystallite size (nm)
R=0.00	3.34365	32.58835
R=0.05	3.32884	21.45239
R=0.10	3.34973	16.09833

patterns are shown in Figs. 5, 6 and 7. In all the samples the observed peak at $2\theta = 26.63°$ corresponds to the hexagonal structure of the CdS (wurtzite) and belongs to the plane (002), which has the lowest surface energy in CdS. Besides, an increase in the full width at half maximum (FWHM) is observed as the Al doping increases. This result can be due to Al atoms inducing dislocations and/or defects inside the crystalline structure of CdS and this causes a modification in the XRD pattern. According to the Bragg's law $n\lambda = 2d\sin\theta$ the average interplanar distance $d_{(002)}$ was calculated, where λ is the X-ray wave length and θ is the peak angle. Besides, we obtained the crystallite size through the Scherrer equation $D = \frac{0{,}9\lambda}{\beta\cos\theta}$, where β is the FWHM in radians. These results are resumed in Table 2.

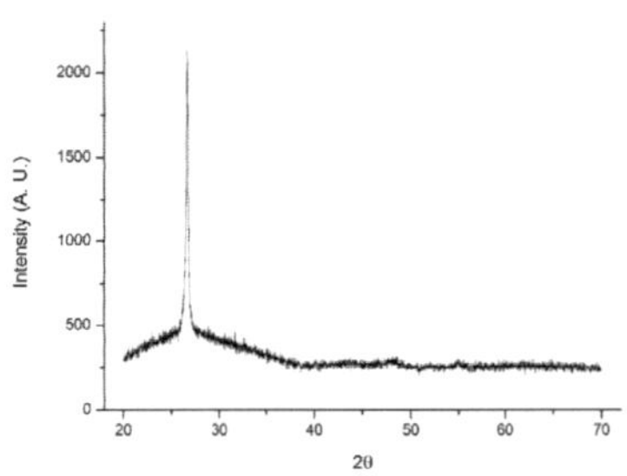

Figure 5. X-ray diffraction pattern of undoped CdS thin film (R=0.00).

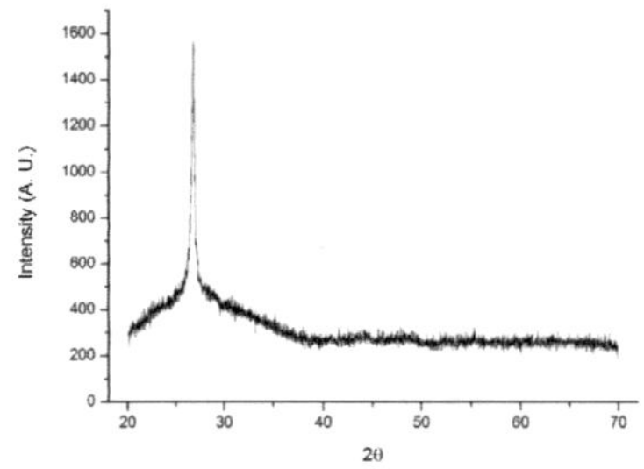

Figure 6. X-ray diffraction pattern of CdS:Al thin film (R=0.05).

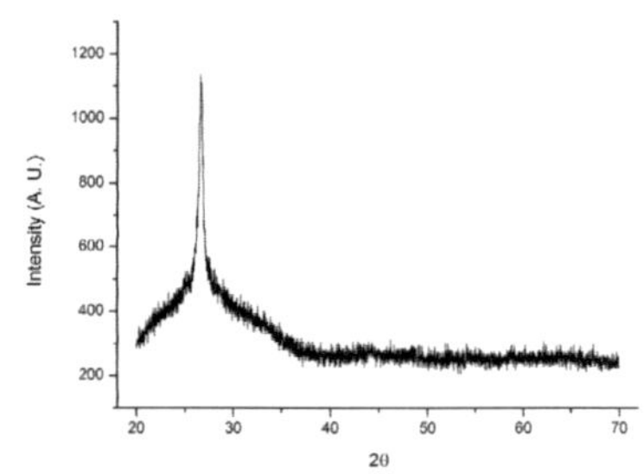

Figure 7. X-ray diffraction pattern of CdS:Al thin film (R=0.10).

The ionic radius of Al^{3+} is 0.53 Åand it is smaller than the ionic radius of Cd^{2+} (0.95 Å) [12]. Therefore, the difference in the interplanar distance may be due to Al^{3+} ions were replaced substitutionally in the lattice with the Cd^{2+} ions producing a decrease in the interplanar distance , which is observable when R=0.05. Except when the ratio [Al]/[Cd] is 0.10, Al^{3+} ions are also incorporated in the interstitial states of the lattice, which implies that the interplanar distance begins to increase. We also observed a decrease in the crystallite size when the Al content in the bath solution increased, as it was also reported in [8] with CdS:Al films synthesized by CBD using ammonia as one of the precursors.

Optical properties of the CdS:Al thin films were measured with a light incidence of 0°, obtaining absorbance, transmittance and reflectance spectrum simultaneously. We also calculated the thickness using the software *Filmeasure* that numerically fitted the spectrum, taking into consideration the film structure deposited on the substrate and a mathematical model for the dielectric constants, in this case the Lorentz oscillator was used. The absorbance, reflectance and tranmittance of the CdS:Al films are shown in Figs. 8, 9 and 10. Film thicknesses are resumed in the Table 3. We did not present the spectra of thin films grown on metal

Table 3. CdS:Al film thickness calculated with software *Filmeasure* using the measured spectra.

Molar ratio [Al]/[Cd] in solution	Film thickness (nm)
R=0.00	73.13
R=0.05	99.91
R=0.10	162.70

substrates because the non uniformity of the films and the deformation of copper and bronze do not let us perform accurate measurements with this apparatus.

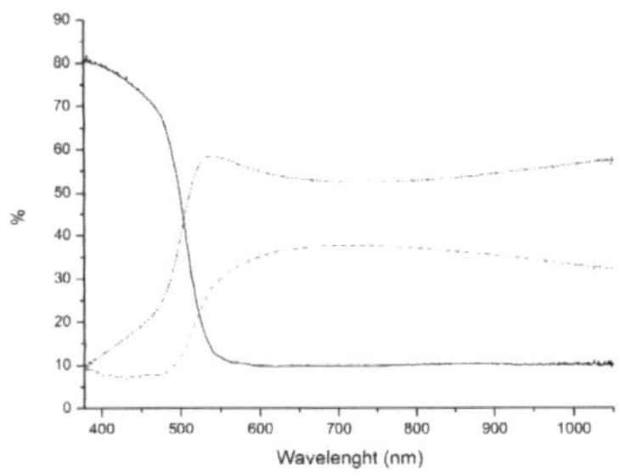

Figure 8. Absorbance (Black ——), reflectance (Red - - - -) and transmittance (Blue — · · —) spectrum of undoped CdS thin film (R=0.00).

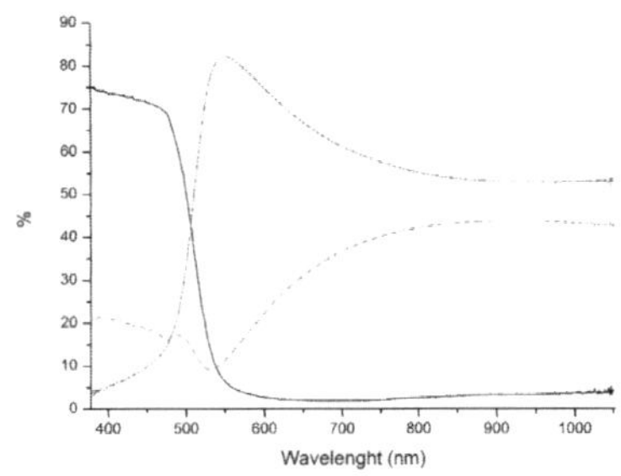

Figure 9. Absorbance (Black ——), reflectance (Red - - - -) and transmittance (Blue — · · —) spectrum of CdS:Al thin film (R=0.05).

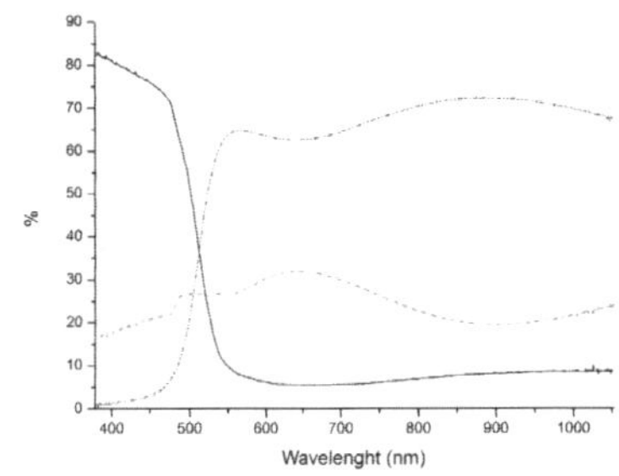

Figure 10. Absorbance (Black ——), reflectance (Red - - - -) and transmittance (Blue — · · —) spectrum of CdS:Al thin film (R=0.10).

In order to calculate the band gap for CdS:Al films (the energy gap between the valence band and the conduction band), we used the measured absorbance α. This value depends on the incident photon energy $h\nu$ in a semiconductor as:

$$\alpha(h\nu) = K\,(h\nu - E_g)^{n/2} \tag{1}$$

where K is a constant, ν is the photon frequency, E_g is the optical band gap, and n is equal to 1 for direct band gap materials such as CdS [18]. Then, the band gap was determined for each film by plotting α^2 vs $h\nu$ and then extrapolating the straight-line portion of the curve to intercept the energy axis as it is shown in Fig. 11. The band gap value of R=0.00, R=0.05 and R=0.10 thin films were found to be 2.374, 2.369, and 2.357 eV, respectively.

We observed a decrease in the band gap of the films as the Al content in the solution increased. These results are similar to those reported in [13]. We believe that this is related to previous reports of Lokhande and Pawar in [9], who suggested that Al^{3+} ions substitute Cd^{2+} in the lattice. Also the sulfur deficiency gave rise to donor levels that became degenerate and merged with the conduction band extending the conduction band to the band gap and therefore reducing the latter.

4. Conclusions

CdS films were deposited on bronze and copper substrates by CBD process. The films obtained were inhomogeneous due to metals highly malleability. The deposition took place in regions of

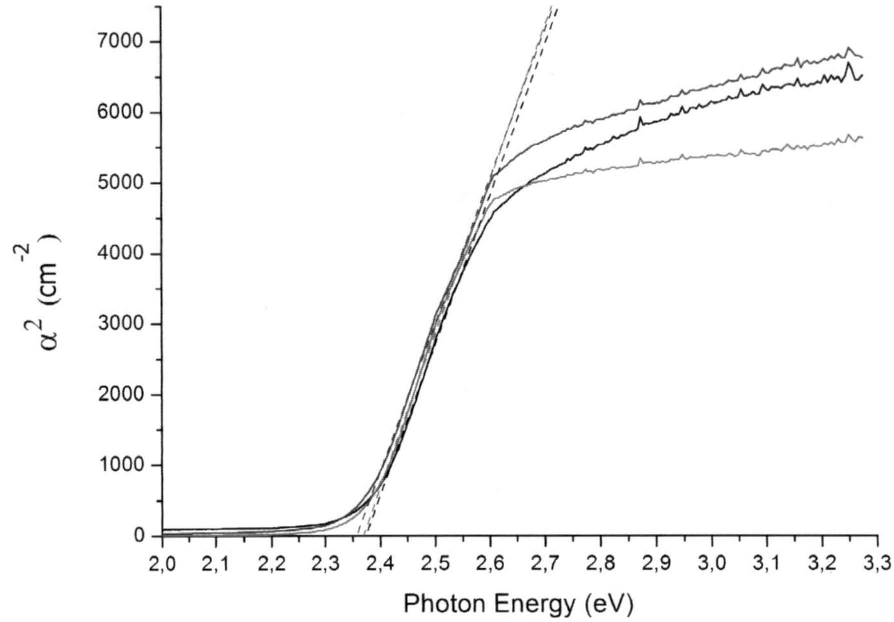

Figure 11. Plot of α^2 vs vs $h\nu$ for R=0.00 (Black), R=0.05 (Red) and R=0.10 (Blue). Dashed lines are the linear fit of the straight-line portion of the curves.)

the substrates shaped by a lamination process leaving many regions with no CdS deposition. We believe that CBD is not a good technique to synthesize semiconductors thin films on metal, and is highly recommendable to employ another technique to carry out this process, such as physical vapor deposition. SEM images and 2-D nanoindentation images show a very inhomogeneous film and the hardness or elastic modulus cannot be determined because the Berkovich tip penetrates the substrates and removes the CdS.

For CdS:Al films deposited on glass, we obtained a film with (002) hexagonal phase and there was no modification in the crystalline structure by Al doping. We only observed a decrease in the crystallite size and a change in the interplanar distance because Al^{3+} ions were replaced substitutionally or interstitially depending on the Al content in the solution. No considerable differences are observed in the hardness and elastic modulus of films, which suggest that a low concentration of Al in the films composition does not modify the mechanical properties of the thin films.

The optical properties of the films deposited on glass showed a decrease in the band gap energy when the Al content increased due to Al^{3+} ions, which induced donor levels when merging with the conduction band and consequently the band gap was reduced. The reflectance spectrum was also modified, probably because of the high reflectivity of Al which alters the color of the films and the light reflectance in the UV region.

References

[1] Zhao J, Bardecker J A, Munro A M, Liu M S, Niu Y, Ding I, Luo J, Chen B, Jen A K Y and Ginger D S 2006 *Nanoletters* **6**(3) 463-67.
[2] Shirakata S, Ohkubo K, Ishii Y and Nakada T 2009 *Sol. Energy Mater. Sol. Cells* **93** 988-92.
[3] Repins I, Contreras B, Egaas C, Dehart J, Scharf C, Perkins B and Noufi B 2008 *Prog. Photovoltaics: Res. Appl.* **16** 235.
[4] Metin H and Esen R 2003 *J. Cryst. Growth* **258**(1-2) 14148.
[5] Bhattacharya R N 2008 *ECS Transactions* **13**(17) 173-76.

[6] Lee J H, Yi J S, Yang K J, Park J H and Oh R D 2003 *Thin Solid Films* **431-432** 344-48.
[7] Khallaf H, Chai G, Lupan O, Chow L, Park S and Schulte A 2009 *Appl. Surf. Sci.* **255** 4129.
[8] Khallaf H, Chai G, Lupan O, Chow L, Park S and Schulte A 2008 *J. Phys. D: Appl. Phys.* **41** 185304.
[9] Lokhande C D, Pawar S H 1982 *Solid State Commun.* **44** 1137.
[10] Sandoval-Paz M G, Sotelo-Lerma M, Mendoza-Galvan A and Ramrez-Bon R 2007 *Thin Solid Films* **515** 3356-62.
[11] Zawaideh E 1999 U.S. Patent No. 5889592.
[12] Shannon R D 1976 *Acta Cryst.* **A32** 751-67.

Natural dyes as sensitizers to increase the efficiency in sensitized solar cells

Bayron Cerda, R. Sivakumar, M. Paulraj

Departament of physics, Faculty of physical and mathematical science, University of Concepcion, Concepcion, Chile

E-mail: bayron.nath@gmail.com, sivakrk2005@gmail.com, paulraj@udec.cl

Abstract. A dye-sensitized solar cell (DSSC) is a sandwich type solar cell consisting of a photoelectrode, a counter electrode and a liquid electrolyte. The photo electrode comprises of a titanium dioxide semiconducting thin film grown over a glass substrate which in-turn has a transparent thin conducting layer of tin oxide film doped with fluorine (FTO) coated over it. The aim of this work is to develop photoelectrodes with different dyes to increase the efficiency of this type of solar cells. Dyes obtained from fresh sources of maqui, black myrtle, spinach and a dye mixture of spinach and spinach-maqui-myrtle were used. The technique used for the extraction of the dyes was maceration for one day, in methanol. Colourants and photoelectrodes were studied using, UV-vis spectrophotometer for their spectral properties. Their photovoltaic properties such as efficiency, fill factor, open circuit voltage and short circuit current were studied using a solar simulator and source meter unit.

1. Introduction

Since the development of Gratzel's Dye Sensitized Solar Cell (DSSC) in 1991[1],This has had considerable attention because it is environmentally friendly and has low production costs. A DSSC is composed of a nanocrystalline semiconductor porous electrode with absorbed dye, a counter electrode, and an electrolyte containing ions of iodide and tri-iodide. In DSSC, sensitizing dyes plays a key role in absorption of light and the transformation of solar energy into electrical energy. Many metal complexes and organic dyes have been synthesized and have been used as sensitizers. Although DSSC have provided high relative efficiency, use of precious metals like ruthenium (Ru) complexes is its disadvantages: noble metals are considered as limited resources, so its production is costly. Moreover, the organic dyes are not only cheaper but have been reported to show efficiency of 9.8 %. However often natural dyes have presented problems because of its complicated synthesis and low yield. However, natural dyes found in flowers, leaves and fruits can be extracted by simple procedure. Due to its cost efficiency, non-toxicity and retrogradations, natural dyes have been a popular research topic. Until now various natural dyes have been used as sensitizers in DSSC, such as carotene, chlorophyll, etc.

2. Experiment

2.1. Sensitizers preparing natural dyes

For all dyes methanol was used as the extraction solvent. The extracted dyes were obtained by the following steps: fresh fruit and vegetables were washed with distilled water. They were then cut into small pieces, immersed in methanol at room temperature and left to macerate for a day. This was then filtered using a vacuum pump and the filtrate was used as a sensitizer.

2.2. Preparation of dye sensitized solar cells
FTO conductive glass sheets were first ultrasonically cleaned using a mixture containing 10 ml of ethanol, 10 ml of propanol and 10 ml of distilled water for 30 minutes.

Titanium dioxide (TiO_2) paste was prepared using Degussa P-25, whose average TiO2 particle size was 26 nm. P-25 (0.14 g) was set in a shaker, and HNO_3 (0.6 ml) was added into the shaker. Later, polyethylene glycol (PEG) was added into it. The obtained TiO_2 paste was coated using Doctor Blade method onto the FTO clean glass substrate. The obtained substrate was set in an electric oven for annealing at 450C for 30 min in order to eliminate water vapour and the polymer binder. This formed the pre-prepared TiO_2 photoelectrode which is to be sensitized. Sensitization was done by immersing this electrode in a methanol solution containing the extracted dye for 1 day in dark.
platinum (Pt) counter electrodes were fabricated by spreading a drop of 5 mM chloroplatinic acid hexahydrate ($H_2PtCl_6.6H_2O$) in isopropyl alcohol onto the FTO surface followed by heating it at 450C for 30 min. Pt electrode was used because it is most resistant to corrosive attacks by the redox couple (I^-/I^{-3}) and facilitates a reversible redox reaction to occur. Since only very small quantity of Pt is been used in making the counter electrode, this still reduces its cost at the same time maintaining its transparency for photon entry. The dye sensitized TiO_2 photoelectrode and the Pt counter electrode was assembled to form a sandwich solar cell. Solution composing; 0.6 M butylmethylimidazolium iodide (BMII), 0.05 M I_2, 0.1 M LiI, and 0.5 M tert-butylpyridine in 1:1 acetonitrile/valeronitrile was used as the redox electrolyte.

2.3. Measurements
UV-Vis spectrophotometer was used to determine the absorbance of the natural dyes extracted and the sensitized photoelectrodes. The photovoltaic parameters were obtained from the current-voltage (I-V) curves that were measured using a solar simulator and a Keithley source meter.

3. Result and discussion
We tried to use 3 types of natural dyes as sensitizers for DSSCs. Table 1 absorbance data shows absorption of dyes extracted using methanol from the maqui fruit, black myrtle and spinach.

Natural dye	$\lambda_{max}(nm)$	$V_{oc}(V)$	$I_{sc}(A)$	FF	$\eta\%$
maqui	546	0.06	0.00002	4.73	0.006
black myrtle	542	0.40	0.00002	4.89	0.040
spinach	437	0.10	0.00001	3.03	0.005
spinach-black myrtle	437	0.39	0.00001	0.59	0.009

Table 1. Photoelectrochemical parameters

The main components of spinach is chlorophyll, the for maqui it was anthocyanins.

3.1. Photovoltaic properties
Photovoltaic properties of these assembled DSSCs using these natural colorontes as sensitizer were studied by measuring their current-voltage curves under white light irradiation from a solar simulator. The performance of these DSSCs were assessed by their open circuit voltage V_{oc}, short circuit current I_{sc}, fill factor (FF) and conversion efficiency η; obtained parameters are listed in the table 1.

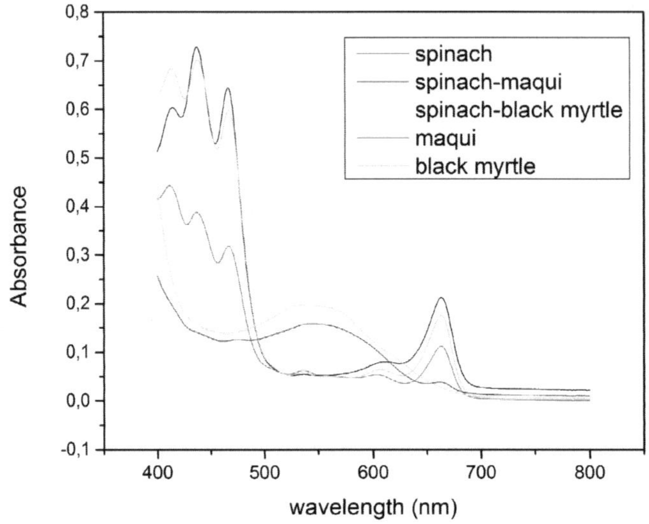

Figure 1. Absorbance spectrum of the natural dyes

Figure 1 shows the absorbance of the natural dyes extracted. We can see that spinach has two main peaks one to the 660 nm and other to 450 nm, these peaks corresponds to chlorophyll peaks a and b. Black myrtle, show a peak at 550 nm. Maqui showw a peak at 550 nm. Spinach-maqui and spinach-black myrtle have the same peaks as spinach.

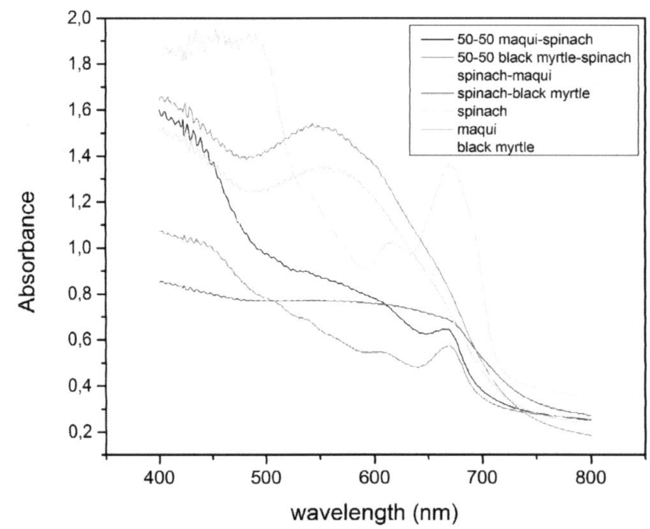

Figure 2. Absorbance spectrum of the sensitized photoelectrodes

Figure 2 shows the absorbance of the sensitized photoelectrode. Spinach shows absorption peaks at 400, 500 and at 650 nm. The maqui has good absorption between 500 and 650 nm. Black myrtle has a similar to that of maqui with good absorption between 500 and 650 nm. Maqui, spinach-maqui and black myrtle have the same absorption in 500-650 nm. 50-50 maqui-spinach and 50-50 black myrtle-spinach have the same absorption in 670 nm and 400-450 nm. Spinach- black murtle has a good absorption in 400-700 nm.

4. Conclusion

DSSCs were assembled using the dyes extracted from maqui, black myrtle, spinach and even a mixture of some of them. For maqui three absorption peaks at 476, 546 and 661 nm were found. For black myrtle one absorption peak at 542 nm was found. For spinach six absorption peaks at 414, 437, 466, 535, 612 and 662 nm were found. For spinach-maqui mixture 6-absorption peaks 414, 436, 466, 535, 602 and 662 nm were found and for spinach-black myrtle mixture 6-absorption peaks 413, 438, 467, 535, 607 and 663 nm were found. The efficiencies of the DSSC were also studied. Extract used from black myrtle showed the highest efficiency among others.

5. Acknowledgment

The authours thank the following funding agency, FONDECYT Regular (No. 1130802) and FONDECYT postdoctoral (No. 3130370), Government of Chile for their financial support to carry out this work.

References

[1] B. ORegan, M. Grtzel, Nature 335 (1991) 737.

Theoretical study of intrinsic defects in CdTe

E. Menéndez-Proupin[1], W. Orellana[2]

[1] Departamento de Física, Facultad de Ciencias, Universidad de Chile, Las Palmeras 3425, 780-0003 Ñuñoa, Santiago, Chile

[2] Departamento de Ciencias Físicas, Universidad Andres Bello, República 220, 037-0134 Santiago, Chile

E-mail: emenendez@uchile.cl

Abstract. The quantum states and thermodynamical properties of the Cd and Te vacancies in CdTe are studied by first principles calculations. It is shown that the band structure of a cubic 64-atoms supercell with a Te vacancy is dramatically different from the band structure of the perfect crystal, suggesting that it cannot be used as model to calculate isolated defects. This flaw is solved modeling the Te vacancy within a cubic 216-atoms supercell. However, even with this large supercell, the 2− charge state relaxes to an incorrect distorted structure. This distortion is driven by partial filling of the conduction band induced by the k-point sampling. The correct structures and formation energies are obtained by relaxation with restriction of system symmetry, followed by band-filling correction to the energy, or by using a larger supercell that allows sampling the Brillouin zone with a single k-point.

1. Introduction

Cadmium telluride (CdTe) is a semiconductor crystal employed as photovoltaic material which is growing rapidly in acceptance, becoming a cheap alternative to the silicon solar cells. CdTe has advantages for photovoltaic applications, such as a direct band gap, allowing to absorb solar energy in thin films. Its band gap of 1.5 eV at room temperature is in the range for maximum conversion efficiency of a single absorber solar cell [1]. CdTe is also widely used for radiation detectors [2]. However, the nature of the defects in CdTe thin films has been subject of controversy for long time, and its identification frequently relies on thermodynamics arguments and ab initio calculation of the defects and impurities formation energies. Density functional theory (DFT) is the most popular framework for ab initio calculations.

To explain the computational method, let us define the formation energy of V_{Te}. It can be obtained computing the difference between the total energies of a crystal supercell containing n CdTe units, $E(\mathrm{Cd}_n\mathrm{Te}_n)$, and the energy of the same supercell with a missing Te, $E(\mathrm{Cd}_n\mathrm{Te}_{n-1})$, added with the Te chemical potential $E(\mathrm{Te}) + \Delta\mu_{\mathrm{Te}}$

$$\Delta H_f(\mathrm{V}^q_{\mathrm{Cd}}) = E(\mathrm{Cd}_{n-1}\mathrm{Te}_n) + E(\mathrm{Cd}) - E(\mathrm{Cd}_n\mathrm{Te}_n) + \Delta\mu_{\mathrm{Cd}} + q[E_V + E_F] + \Delta E_{size} + \Delta E_{q.p.}. \quad (1)$$

To obtain the formation energy of the tellurium vacancy, V_{Te}, it suffices to exchange Cd with Te in Eq. (1). Here, $E(\mathrm{Cd})$ is the energy of Cd in a reference state, in this case it is the energy per atom of bulk Cd. $\Delta\mu_{\mathrm{Cd}}$ is the chemical potential relative to the reference state, and it depends on the thermodynamic equilibrium conditions. In Cd-rich conditions $\Delta\mu_{\mathrm{Cd}} = 0$, while in Cd-poor conditions $\Delta\mu_{\mathrm{Cd}}$ equals the CdTe formation energy [3]. For a charged vacancy, q

is the number of electrons donated to the environment, and $E_V + E_F$ is the electron chemical potential expressed as the sum of the valence band maximum (VBM) E_V and the Fermi level E_F relative to the VBM. ΔE_{size} and $\Delta E_{q.p.}$ are size and gap corrections that are explained in sections 2.2 and 2.3, respectively.

Most calculations of CdTe point defects have been investigated using the 64-atom supercell (SC64) of the perfect crystal [2–8]. This supercell is $2 \times 2 \times 2$ times the 8-atoms conventional cubic unit cell. The SC64 supercell is the standard framework nowadays to model defects in diamond- or zincblende-type semiconductors. It is a popular choice because of its cubic shape, that maximizes the distance between the defect and its periodic images, and allows easy charge corrections. In addition, it is affordable for high level methods beyond DFT. This setup is sufficient to model some defects like V_{Cd}. However, for V_{Te} we will show that the SC64 supercell is too small to obtain a reliable band structure, while thermodynamic calculations are hampered by uncontrollable size effects that are incompatible with the notion of diluted defects. We observe that changes in the conduction band are so dramatic that destroy the reference to apply gap corrections to the formation energy of negatively charged states. We show that this problem is solved using a 216-atom supercell of the perfect crystal (SC216), which is $3 \times 3 \times 3$ times the 8-atom cubic unit cell. In addition, the band filling induced by the use of a k-point grid causes a spurious distortion in the double negatively charged state 2−. We show that this size effect can be discarded by analyzing the band diagram and applying the band-filling correction. According to our results, a large supercell with only one k-point is enough to avoid this problem. For CdTe, this large supercell can be $4 \times 4 \times 4$ times the cubic unit cell, that contains 512 atoms for the perfect crystal (SC512).

2. Methodology

2.1. Electronic structure calculations

We have performed DFT calculations by using a plane-wave projector augmented wave [9, 10] scheme, as implemented in the Vienna Ab Initio Simulation Package (VASP) [11]. We have used mostly the generalized gradient approximation (GGA) for the exchange-correlation functional as proposed by Perdew, Burke, and Ernzerhof (PBE) [12]. In certain cases, we have also used the hybrid functional proposed by Heyd, Scuseria and Ernzerhof (HSE06) [13]. This functional generally allows one to obtain better band gap energies and structural properties than those obtained with PBE, but at the cost of a great increment in computer time.

We have used a plane-wave cutoff energy of 460 eV for the wavefunctions. To sample the reciprocal space, we have used Γ-centered $3 \times 3 \times 3$ and $2 \times 2 \times 2$ k-point grids for the SC64 and SC216 supercells, respectively. This setup allows convergences of 0.5 meV/atom and 0.4 kbar in total energy and pressure, respectively, as tested for variable cell relaxations of the CdTe zincblende unit cell with $8 \times 8 \times 8$ k-point grid. In fact, a reduced cutoff of 285 eV allows a convergence within 10 meV/atom in total energy and 2 kbar in pressure, but the properties derived from relative energies, e.g., lattice constants and bulk moduli are well converged. The CdTe lattice constant obtained with variable cell relaxation with a cutoff of 285 eV differs by 0.4% from the converged value, but when the energy is fitted with the Birch-Murnaghan equation of state, the difference fells to 0.005 %. The energy difference between the zincblende and wurtzite structures is 3.5 meV/atom considering both cutoffs. The $n \times n \times n$ supercell size is n times the converged theoretical lattice constant of the perfect crystal (6.6212 Å). We have used a cutoff of 285 eV to relax the point defect structures with fixed supercell size, and a cutoff of 460 eV to compute the total energies. The highest precision is needed because the energy error is amplified by the number of atoms in the supercell, e.g., for the neutral Cd vacancy in SC64 supercell, the formation energy computed with both cutoffs differs by 0.03 eV. We have verified that for higher cutoff energy, the forces on the atoms in the relaxed geometry keep below 0.01 eV/Å.

Journal of Physics: Conference Series **720** (2016) 012031 doi:10.1088/1742-6596/720/1/012031

The calculated CdTe heat of formation is found to be of 0.93 eV, in good agreement with the available experimental value of 0.96 eV [2]. The energy of bulk Cd has been obtained by relaxing the 2-atom unit cell of the experimental structure ICSD #52793 available from the MedeA software environment [14]. The Brillouin zone was sampled by a $22 \times 22 \times 10$ k-point grid. The energy of bulk Te has been obtained relaxing the crystallographic 3-atoms unit cell ICSD #23060. The Brillouin zone was sampled by a $8 \times 8 \times 5$ k-point grid and tested with a $15 \times 15 \times 10$ grid.

2.2. Size corrections

An isolated defect either populates the conduction band minimum (CBM) or create holes at the VBM. However, in a finite supercell, when a k-point mesh is used, the electron or hole created by the point defect fills half a band, considering the spin degeneracy. The single-particle contribution to the total energy is affected by the band dispersion. The correction is to replace the energies of the defect band, either of electrons or holes, by the corresponding band extremum. For donors, the correction can be expressed as [15]

$$\Delta E_{b.f.} = -\sum_{n,k} \Theta(e_{n,k} - e_{min}) w_k \eta_{n,k} (e_{n,k} - e_{min}), \tag{2}$$

where indexes n and k denote the bands and k-points, w_k are the k-point weights, $e_{n,k}$ are the single particle energies with minimum e_{min}, $\eta_{n,k}$ are the occupation numbers, and Θ is the Heaviside function. The correction for acceptors is

$$\Delta E_{b.f.} = \sum_{n,k} \Theta(e_{max} - e_{n,k}) w_k (S - \eta_{n,k})(e_{max} - e_{n,k}), \tag{3}$$

where $S = 1$ ($S = 2$) stands for spin polarized (unpolarized) calculations. e_{max} is the maximum of the acceptor band.

In VASP calculations, the average of the electrostatic potential is set to zero. Hence, the potential energy reference is undetermined by an additive constant, which is different in supercells containing different number of atoms or different charge. The change has two sources: a) the presence of excess charge, b) the missing or appearance of extra atoms. For charged systems, the additive constant enters in the Hartree term of the total energy per cell:

$$E_H = \int e[n_0(\mathbf{r}) - N(\mathbf{r}) + \Delta n(\mathbf{r})](v(\mathbf{r}) + \Delta v) d^3\mathbf{r},$$

where e is the electron charge, $n_0(\mathbf{r})$ is the electron density for a neutral system, $\Delta n(\mathbf{r})$ is the excess electron density. $N(\mathbf{r})$ is the nuclear charge, and Δv is the additive potential implicitly added to make zero the average potential, with respect to the potential of the system without defect and neutral. The contribution of the additive potential to the Hartree energy is equal to

$$\Delta E_H = \int e\Delta n(\mathbf{r}) \Delta v d^3\mathbf{r} = \left[\int \Delta n(\mathbf{r}) d^3\mathbf{r}\right] (e\Delta v) = (-q)\Delta V. \tag{4}$$

Here, q is the number of electrons donated $\int \Delta n(\mathbf{r}d^3\mathbf{r})/(-e)$, and $\Delta V = e\Delta v$ is the change of electron potential energy. ΔV can be estimated by the core level shifts of atoms far from the defect with respect to the core levels in the calculation with the perfect crystal. The correction to the total energy due to potential alignment (p.a.) is

$$\Delta E_{p.a.} = q\Delta V. \tag{5}$$

Periodic boundary conditions induce an spurious interactions of the charged defect with the periodic replicas. The leading terms can be corrected by the Makov-Payne formula

$$\Delta E_{M-P} = \frac{\alpha_M q^2}{2\epsilon V^{1/3}} + \frac{2\pi q Q}{3\epsilon V} + O(V^{-5/3}),$$

where α_M is a Madelung constant, ϵ is the dielectric constant, V is the supercell volume, and Q is a quadrupole moment. We have used a simplified version due to Lany and Zunger [15]

$$\Delta E_{L-Z} = \frac{2}{3}\frac{\alpha_M q^2}{2\epsilon V^{1/3}}. \tag{6}$$

We have used the static dielectric constant $\epsilon = 12.272$, obtained from density functional perturbation theory calculations and including local field effects [16, 17].

In summary, the supercell size corrections are given by the sum of band-filling, potential alignment, and Lany-Zunger corrections

$$\Delta E_{size} = \Delta E_{b.f.} + \Delta E_{p.a.} + \Delta E_{L-Z}. \tag{7}$$

2.3. Gap corrections

Semilocal exchange-correlation functionals (LDA, GGA) cause a severe underestimation of the valence to conduction band gap, and this affects the formation energies. Let us consider the gap error

$$\Delta E_g = E_g^{exp} - E_{gap}^{DFT} = \Delta E_C - \Delta E_V.$$

The above equation established a partition of the gap error in shifts applied to the valence and conduction band.

For shallow acceptors, where active holes occupy perturbed host states, the hole levels undergo the same shift as the valence band edge. This has been verified in quasiparticle calculations. For shallow donors, the active electron levels are perturbed states of the conduction band, undergoing the shift. Including fractional occupations, the shifts of single particle levels and total energies can be expressed as

$$\Delta E_g^{(1)}(q) = \begin{cases} z_e(X,q)\Delta E_C & \text{for donors,} \\ z_h(X,q)(-\Delta E_V) & \text{for acceptors,} \end{cases} \tag{8}$$

where z_e (z_h) is the number of electrons introduced (removed) by the defect near the CBM (VBM).

A second gap correction modifies the reference of the Fermi level

$$\Delta E_g^{(2)}(q) = q\Delta E_V. \tag{9}$$

The total gap correction is

$$\Delta E_{q.p.} = \Delta E_g^{(1)}(q) + \Delta E_g^{(2)}(q). \tag{10}$$

With our computational setup that includes the PBE functional, the VBM and CBM energies are 1.789 and 2.388 eV, respectively. For gap corrections we have used the generalized DFT with the hybrid functional HSE06 [13] including the spin-orbit coupling. The corrected VBM and CBM energies are 1.773 and 2.977 eV, respectively. With this approximation the band gap is 1.204eV, which is still lower than the experimental value 1.6 eV [18].

Journal of Physics: Conference Series **720** (2016) 012031 doi:10.1088/1742-6596/720/1/012031

3. Results and discussions

3.1. Cadmium vacancy

Figure 1 shows the calculated formation energies of five charge states of V_{Cd}. As can be seen, only the 1− and 2− charge states have the lower formation energies, being stable for the Fermi level within the CdTe band gap. The transition level is $\varepsilon(-/2-) = 0.36$ eV. The transition level $\varepsilon(0/-) = -0.01$ eV is virtually at the VBM within the error if the calculation. These transition levels are close to the values obtained with the GGA+U^{SIC} approach [8] and with hybrid functional PBE0 [2] and HSE [7]. In the later cases, the transition level $\varepsilon(0/-) \sim 0.23$ eV, but this result is affected by the neglect of the spin-orbit interaction, that raises the VBM by $\sim$ 0.3 eV. The neglect of the spin-orbit contribution to the VBM also affects the GGA calculations of Ref. [4].

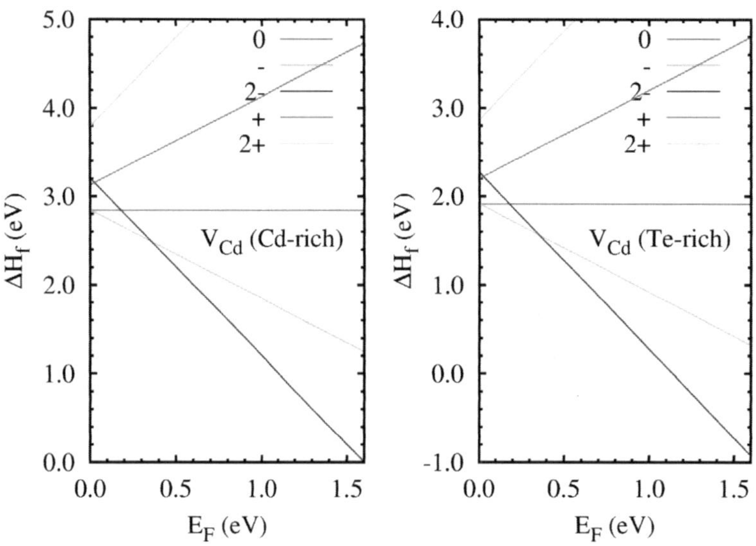

Figure 1. (Color online) Formation energies as a function of the Fermi energy for the Cd vacancy computed with a 64-atom supercell (SC64).

The absolute values of the GGA formation energies (present work) are about 0.7 − 1 eV smaller than the values obtained with GGA+U^{SIC} and with PBE0. These differences are related to the different precision of the functionals in the semiconductor CdTe, the metal Cd and the semimetal Te. The formation energy contains the term $E(Cd)$, and let us note that $E(Cd) + E(Te) + \Delta H(CdTe) = E(CdTe)$. The formation energy $\Delta H(CdTe)$ is -1.13 eV and -1.78 eV within PBE0 and GGA+U^{SIC}, respectively. These numbers imply larger values for the sum $E(Cd) + E(Te)$.

From the electronic point of view, V_{Cd} in the neutral state has two holes delocalized in the valence band. Upon charging with two electrons, i.e., the state V_{Cd}^{2-}, the valence band gets fully occupied. In all the charge states considered here the relaxation of the atoms are quite small, consistently with the delocalized character of the holes. These features allow one to calculate the formation energies with the rather small SC64 supercell and to include the size effects as corrections.

3.2. Tellurium vacancy

In this section, we present calculations of V_{Te} with the supercells SC64 and SC216. First, we analyze the band structure of the defective supercells, and show the problems to apply band-filling and gap corrections with the SC64 model.

Journal of Physics: Conference Series **720** (2016) 012031 doi:10.1088/1742-6596/720/1/012031

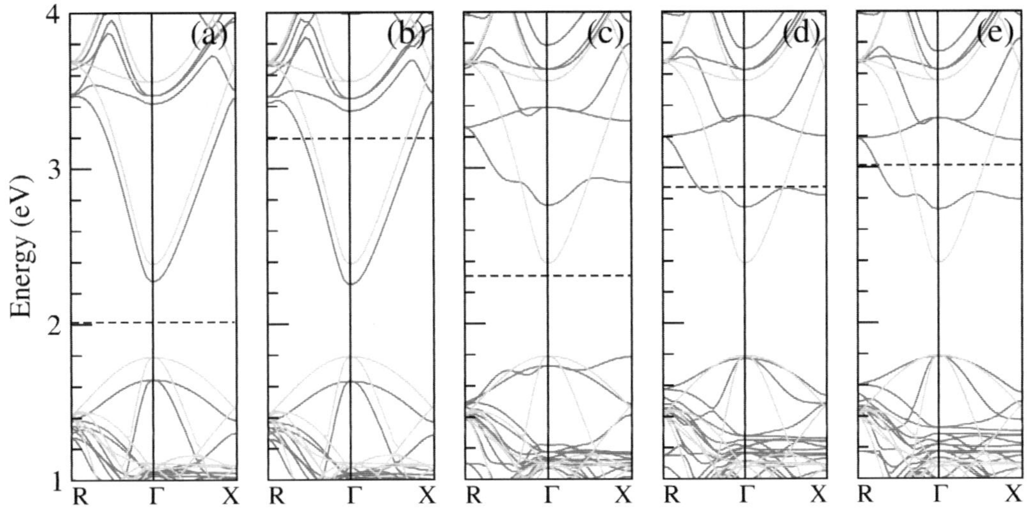

Figure 2. Band structure calculations for the Te vacancy in CdTe performed with a 64-atom supercell. The red lines are the bands of the defective supercell in charge states 2+ (a), 1+ (b), 0 (c), 1- (d), and 2- (d). The green lines are the bands for the perfect crystal. The dashed lines indicate the Fermi levels.

Figure 2 shows the band structure calculations of the perfect CdTe and V_{Te} in the SC64 supercell. The atomic positions were relaxed keeping the T_d symmetry, while distorted configurations will be considered separately using larger supercells. The vacancy in the SC64 supercell is strictly a periodic model of V_{Te} in a concentration of one every 64 atoms or 10^{20} cm^{-3}. Even when the band diagram is not valid in aperiodic situations, it allows a certain order in the visualization of the eigenstates. In Figure 2 we see that for the charge states $0, 1-$, and $2-$, the conduction band (CB) is dramatically modified with a significative increment of the band gap. This global change produced by the vacancy is incompatible with the notion that an impurity should introduce small perturbations on the crystal band structure. This example shows that band diagrams, which are in principle not adequate to study aperiodic systems, are still useful to uncover flaws in defect models. From the practical side of thermodynamic calculations, the drastic alteration of the conduction band makes impossible to apply gap corrections, resulting in a false stability of the negatively charged states in n-type doping conditions. The only solution to avoid the exaggerated influence of V_{Te} is to explore configurations using a larger supercell.

Figure 3 shows the band diagram of the perfect crystal and the V_{Te} defect in different charge states computed with the SC216 model. The T_d symmetry of the atomic arrangement around the vacancy has been preserved during the relaxation as for SC64. Distorted configurations with lower energy will be studied separately below. For all the charge states considered, the effects of the vacancy are small band shifts and level splits. In addition, one can recognize an empty-defect band over 3.2 eV for $q = 2-, 1-, 0$. As we will see later, the defect band becomes active, driving the so called Jahn-Teller distortions. The origin of the defect band can be understood as follows: Each Te-Cd covalent bond is populated by 0.5 Cd electron and 1.5 Te electron. When V_{Te} is formed, four sp^3 bonding orbitals and six electrons are removed. This process leaves two electrons from the four surrounding Cd 5s orbitals. These orbitals combine to form four defect bands: one mixed with the host valence band (VB) that gets occupied by the electrons left by the Cd broken bonds, and three empty isolated bands ($\sim 3.1 - 3.5$ eV). Of these three bands, the lowest two are degenerate. In states $1-$ and $2-$, the defect levels keep empty and the extra

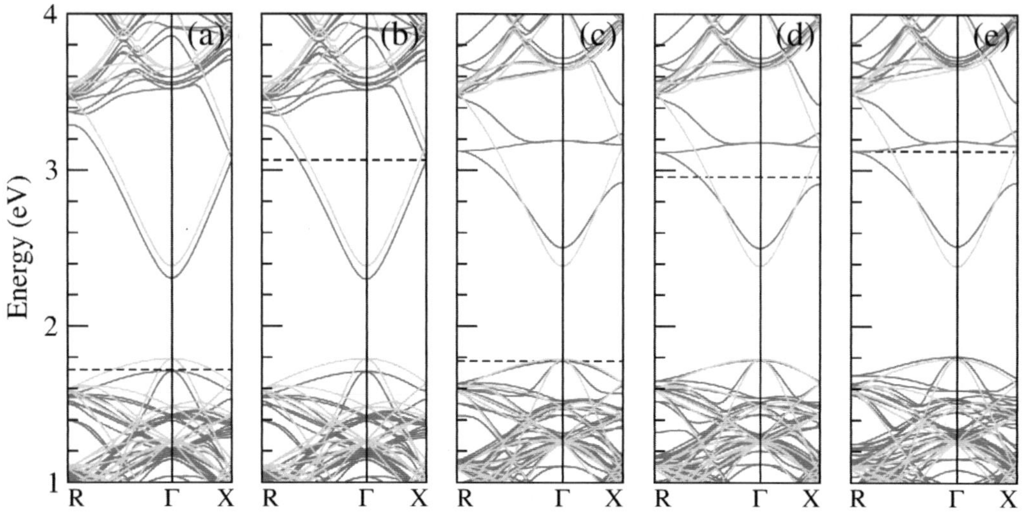

Figure 3. Band structure calculations for the Te vacancy in CdTe performed with a 216-atom supercell. The red lines are the bands of the defective supercell in charge states 2+ (a), 1+ (b), 0 (c), 1- (d), and 2- (d). The green lines are the bands for the perfect crystal. The dashed line indicates the highest occupied level.

electrons occupy perturbed host states [19], that in the supercell calculation appear similar to the CB.

The neutral state shows a similar band structure if the atomic positions keep the T_d symmetry. However, the total energy and the band diagram are modified due to a symmetry breaking (C_{2v} distortion) that lowers the total energy by 0.3 eV. The distortion leads to formation of a Cd_2 dimer with a bond length of 2.86 Å. Figure 4 shows the band diagrams of both symmetric and distorted configurations, and the geometry of the later. The origin of the distortion is not due to the Jahn-Teller mechanism, because the symmetric state does not present degeneracy. As discussed above for the 2− state, the missing Te atom leaves four broken bonds with 0.5 electron each one. The energy is reduced when two Cd atoms loss 0.5 electrons and move to adopt a quasi-planar bonding structure with their three Te neighbors. The rules of covalent bonding are satisfied if these two Cd atoms distribute their 1.5 electrons in three sp^2 orbitals that make σ bonds with the Te sp^3 orbitals. The other two Cd atoms, which also have $0.5 \times 2 = 1$ electron in dangling bonds, trap the electron released by the first Cd pair, getting closer to form a doubly occupied covalent bond. This bond corresponds to the VB top in Fig. 4(b), where the corresponding charge density (square of the wave function) is shown in Fig. 4.

For the 2− charge state, the total energy is apparently minimized by a D_{2d} distortion leading to the formation of two perpendicular Cd dimers. The band energy is minimized by a downshift of the defect band, that becomes occupied, while the dispersive host-like conduction band gets empty. This mechanism lowers the total energy by 0.6 eV. However, the self-consistent energy must be corrected by subtracting the band filling energy, which is -0.87 eV for the undistorted configuration, but it is negligible for the distorted one. Therefore, the symmetric configuration has lower energy. In order to test this assertion, we have inserted the distorted configuration into a larger supercell composed of $4 \times 4 \times 4$ conventional unit cells, i.e. the SC512 model. Using only the Γ point, the band filling error is null. As expected, the distorted configuration in the SC512 relaxes to the non-distorted structure. Therefore, the distortion around the vacancy for the 2− state is a supercell-size effect driven by the conduction band filling in the SC64 and

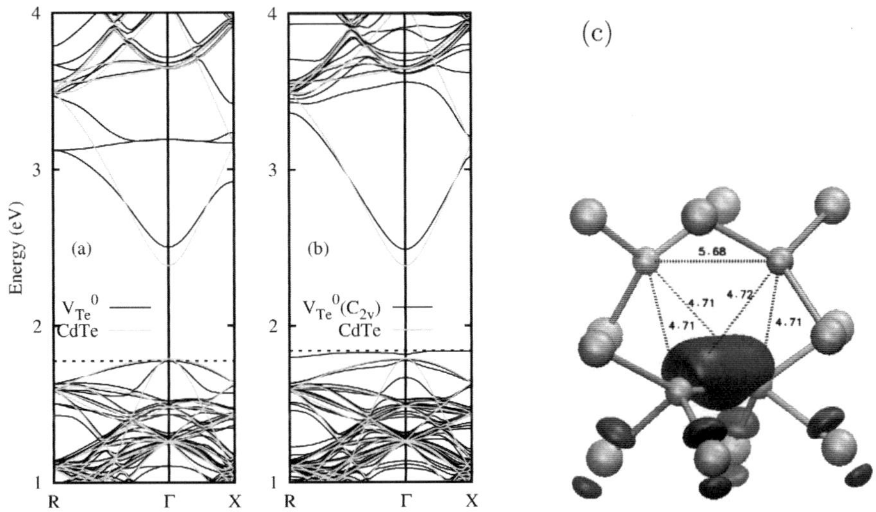

Figure 4. (Color online) Stable geometries and band diagrams for the neutral Te vacancy without (a) and with (b) a C_{2v} distortion, using the SC216 supercell. The solid black (green) lines are the bands of the defective (perfect) supercell, and the dashed lines represent the highest occupied level. (c) Charge density isosurfaces for the valence-band maximum of the distorted Te vacancy in neutral charge state. Pink and cyan balls represent Te and Cd, respectively.

SC216 models.

For the neutral state, there is no need of band filling correction because there is neither hole in the VB nor electron in the CB. Therefore the distorted configuration is more stable. The calculation with the SC512 model confirms this result.

It seems contradictory that for the 1+ state, with one missing electron, the highest occupied level jumps into the CB, and for the 2+ state the highest occupied level returns to the VB. This behavior is due to strong relaxation of the four neighboring Cd atoms into planar configurations $CdTe_3$, favoring a sp^2-like hybridization. Within a sp^2 shell structure, each Cd atom shares its 2 valence electrons with 3×1.5 electrons coming from its three neighboring Te, resulting in 0.5 electron in excess of the sp^2 shell. Adding the four Cd atoms, this sums 2 electrons. The net effect is that two electronic levels are removed (holes). Therefore, the closed-shell structure is recovered in the charge state 2+, while for the state 1+ there is one electron in the CB. For the state 1+, the formation energy must be corrected for the conduction band filling, as well as for the errors in the VB and CB edges, considering $z_e = 1$ and $z_h = 2$ in Eq. 8, respectively.

Figure 5 shows the formation energy plots as function of the Fermi level. For the gap corrections we have approximated the VBM and CBM energies by the values obtained using the HSE06 hybrid functional including the spin-orbit interaction. As this method gives a band gap of 1.2 eV [8], this is the maximum value for the Fermi level in Fig. 5. It can be seen that the stable charge states are the 2+ and the neutral state, with the thermodynamic transition level $\varepsilon(2+/0) = 0.69$ eV.

Recent calculations using the HSE06 method [5,6] with the supercell SC64 suggest that V_{Te} is stable only in the state 2+. These authors attribute the high stability of the charged state to the large Cd displacements around the vacancy. We think they may have missed the energy downshift of the neutral state due to C_{2v} distortion. In any case, we consider unreliable the results obtained with the SC64 model.

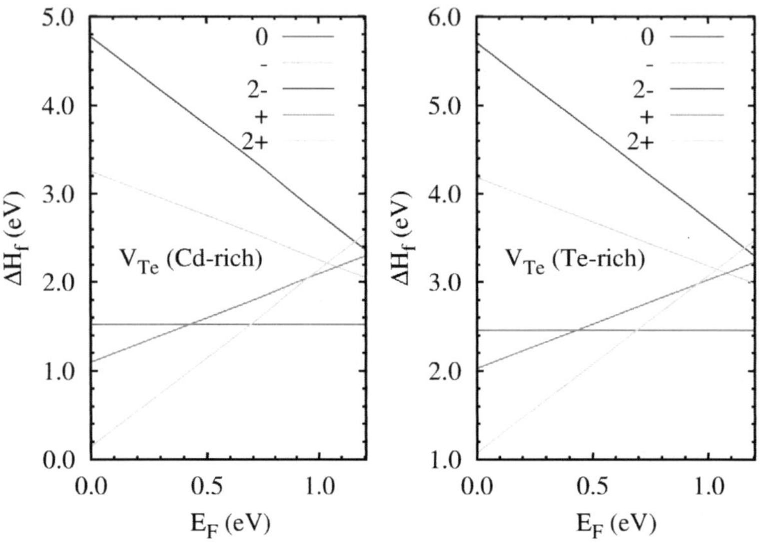

Figure 5. (Color online) Formation energies as a function of the Fermi energy for the Te vacancy computed with a 216-atom supercell (SC216).

4. Conclusions

Our results for V_{Cd}, are consistent with the literature [2, 4, 7, 8]. The main difference in our results is the missing of a stable neutral state. We attribute this difference to the location of the VBM, which is upshifted by the inclusion of the spin-orbit interaction.

For V_{Te}, our results challenge the previous studies [2–6], which have been obtained using the SC64 model. On the one hand, we have shown that the band structures of the neutral and negative charge states present unphysical features, such as enlargement of band gap and strong modification of the conduction band dispersion. The study of the band diagram of the supercell calculation is a useful tool to uncover these unphysical states. On the other hand, we have shown that for the charged vacancy V_{Te}^{2-}, the band-filling error associated to the k-point sampling can drive the system to a distorted configuration, as in Ref. [4]. After band-filling corrections are applied, the undistorted configuration has lower energy. Our calculations with the $\Gamma-$point sampling using the SC512 model, are free of the band-filling error, confirming that the symmetric configuration has lower energy. For a general case, if calculations with the SC512 model were not affordable, we suggest a careful analysis of the band-filling errors in the distorted and undistorted configurations, provided that optimized undistorted configurations can be obtained. High level calculations like hybrid functionals, Green functions GW, or Quantum Monte Carlo are probably not affordable for such large supercells, and may need to be done for the SC64 model. In these cases, GGA reoptimizations of the distorted structures inserted in a large supercell with a single k-point, are recommended to evaluate the veracity of the distortions.

Acknowledgments

This work was supported by the FONDECYT Grant No. 1130437. The authors thankfully acknowledge the computer resources, technical expertise and assistance provided by the Madrid Supercomputing and Visualization Center (CeSViMa) and Red Española de Supercomputación, where part of these calculations were done, and Prof. P. Wahnón for her hospitality and useful discussions at Universidad Politécnica de Madrid. WO acknowledges support from CONICYT-PIA under the Grant Anillo ACT-1107.

References

[1] Shockley W and Queisser H J 1961 *J. Appl. Phys.* **32** 510
[2] Biswas K and Du M H 2012 *New J. Phys.* **14** 063020
[3] Wei S H and Zhang S B 2002 *Phys. Rev. B* **66**(15) 155211
[4] Du M H, Takenaka H and Singh D J 2008 *Phys. Rev. B* **77** 094122
[5] Ma J, Kuciauskas D, Albin D, Bhattacharya R, Reese M, Barnes T, Li J V, Gessert T and Wei S H 2013 *Phys. Rev. Lett.* **111**(6) 067402
[6] Yang J H, Park J S, Kang J, Metzger W, Barnes T and Wei S H 2014 *Phys. Rev. B* **90**(24) 245202
[7] Run X, Xu H T, Tang M Y and Wang L J 2014 *Chin. Phys. B* **23** 077103
[8] Menéndez-Proupin E, Amézaga A and Hernández N C 2014 *Physica B* **452** 119 – 123
[9] Blöchl P E 1994 *Phys. Rev. B* **50** 17953
[10] Kresse G and Joubert D 1999 *Phys. Rev. B* **59** 1758
[11] Kresse G and Furthmüller J 1996 *Phys. Rev. B* **54** 11169–11186
[12] Perdew J P, Burke K and Ernzerhof M 1996 *Phys. Rev. Lett.* **77** 3865–3868
[13] Heyd J, Scuseria G E and Ernzerhof M 2006 *J. Chem. Phys.* **124** 219906
[14] MedeA® Materials Design, Inc., Santa Fe, New Mexico, USA, 2013.
[15] Lany S and Zunger A 2008 *Phys. Rev. B* **78** 235104
[16] Wu X, Vanderbilt D and Hamann D R 2005 *Phys. Rev. B* **72**(3) 035105
[17] Baroni S, de Gironcoli S, Dal Corso A and Giannozzi P 2001 *Rev. Mod. Phys.* **73** 515–562
[18] Chelikowsky J R and Cohen M L 1976 *Phys. Rev. B* **14** 556–582
[19] Lany S and Zunger A 2005 *Phys. Rev. B* **72** 035215

Journal of Physics: Conference Series **720** (2016) 012032 doi:10.1088/1742-6596/720/1/012032

Z method calculations to determine the melting curve of silica at high pressures

F. González-Cataldo, S. Davis, G. Gutiérrez

Grupo de NanoMateriales, Departamento de Física, Facultad de Ciencias, Universidad de Chile, Casilla 653, Santiago, Chile.

E-mail: `fgonzalez@lpmd.cl`

Abstract. The Z method is a novel technique that allows to calculate the melting temperature of materials at different pressures from the microcanonical ensemble. In this work, we apply this method to study the melting behavior of silica at high pressures, determining melting temperatures and dynamical properties.

1. Introduction

Silica is one of the most studied materials. It makes up most of the composition of the Earth's lower mantle and is one of the basic components of rocky bodies in space. Knowing the properties of silica at different pressures is not only important for physics, but for earth sciences and astrophysics, where the wide variety of extrasolar planets discovered has put challenges in our understanding of how planets are formed and how their interiors evolve. Models for planetary interiors (see figure 1) are constrained by the bulk composition and distribution of chemical species, and they need to include equations of state, such as phase transitions and melting curves at high pressure, for perovskites, magnesiowustites and silicates, which are among the most common rock-forming elements.

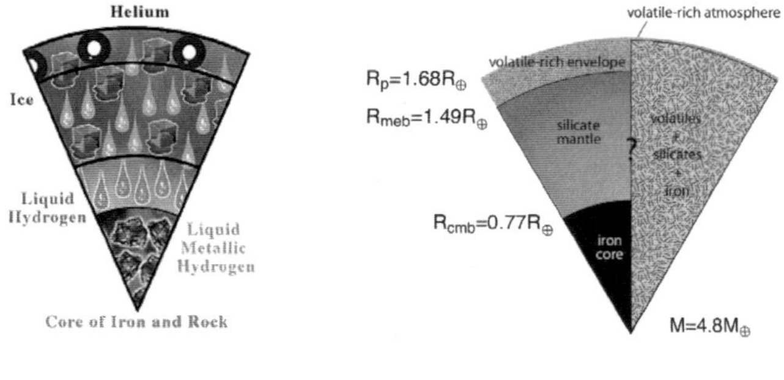

Figure 1. Left: interior of a gas-giant-like planet (Jupiter, Saturn). Right: interior of a super-Earth-like planet (Earth, CoRoT-7b, Kepler 20b) [1, 2].

Since the discovery of stishovite in 1962 by Stishov and Popova, various experimental and theoretical studies have characterized the so-called post-stishovite phases, including $CaCl_2$, α-PbO_2, pyrite, and the recently predicted phases, cotunnite and Fe_2P phases [3, 4]. However, the

Journal of Physics: Conference Series **720** (2016) 012032 doi:10.1088/1742-6596/720/1/012032

melting curve of silica for the pressures corresponding to these phases is poorly constrained, or simply not known. The latter point is of considerable interest for understanding the physical evolution of planetary interiors, such as the process of differentiation of terrestrial planets into a dense metallic core surrounded by a rocky mantle, where the dynamics is driven by the gravitational separation of a liquid phase.

Here we present a study of the melting behavior of silica at high pressures, obtained from first principles molecular dynamics simulations using the Z method. We calculate the melting point of three high pressure phases of silica (pyrite-, cotunnite-, and Fe_2P-type SiO_2) at different pressures.

2. The Z method

The theoretical rationale of the Z method is that, in the NVE (microcanonical) ensemble, with N the number of particles, V the volume, and E the energy of the system, there is a maximum energy E_{LS} that can be given to the crystalline system before it melts. Increasing the energy beyond this point, at which the temperature is the limit of the superheating temperature T_{LS}, the solid spontaneously melts, but due to the increase in potential energy, namely the latent heat of fusion, the temperature decreases. The interesting fact is that the final temperature after melting seems to coincide with the melting temperature T_m obtained from other methods.

Althought we still lack of a thermodynamical explanation for the Z method, it has been empirically observed [5–23] that the method works, and the procedure is as follows: at a fixed volume, the (E, T) points from different simulations draw a 'Z' shape, as shown in figure 2. In these Z–shaped curves the sharp inflection at the higher temperature corresponds to T_{LS} and the one at the lower temperature to T_m. Thus, knowledge of the lower inflection point for different densities allows the determination of the melting curve for a particular range of pressures.

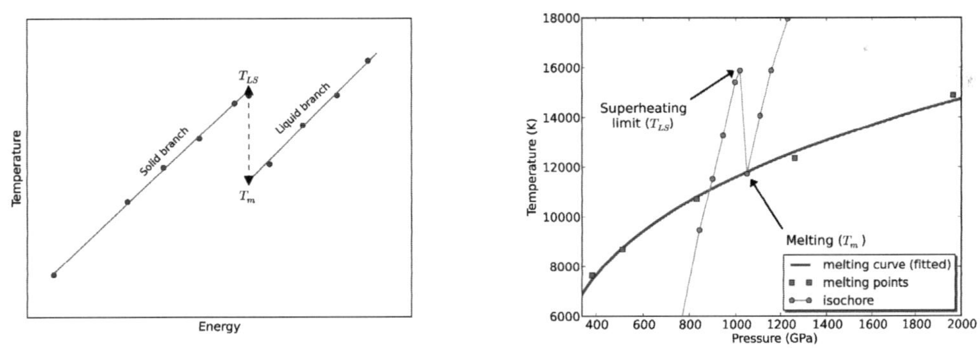

Figure 2. Schematic representation of the isochoric lines in Z-method: Mean temperature versus total energy (left) and temperature versus pressure (right).

3. Ab initio molecular dynamics simulations

The first principles molecular dynamics (FPMD) simulations were preformed in the framework of the Kohn–Sham density functional theory. We worked under the NVE-ensemble, using the Born-Oppenheimer molecular dynamics (BOMD) method, as implemented in the VASP code [24], employing PAW pseudopotential [25], and the Perdew-Burke-Ernzerhof (PBE) exchange-correlation functional [26]. For the plane wave expasion of the wavefunctions, we used a cutoff energy of 900 eV, and k-points convergence tests were performed in order to determine the amount of k-points necessary to perform accurate calculations. The results of these tests yielded that the Γ point was enough for the sampling in all cases. Due to the high temperatures explored, the fast dynamics of the ions needs a small time step, therefore a time step of $\Delta t = 0.2$ fs was used in all calculations. A total of 5000 steps (1 ps) was used in the simulations, which we discuss in section 4.2.

4. Results

4.1. Size dependance

Since the primitive cell of the Fe_2P-type silica has 9 atoms [3], the most simple supercell ($2\times2\times2$) contains 72 atoms, a size which might be still insuficient to lead to reliable results. Therefore, we have carried out calculations comparing the predictions of the Z method using 72 and 243 atom cells. In figure 3 we show the temperature as a function of the pressure for both systems. We observe that only a slight difference in the predictions is obtained in the Fe_2P-type silica when the size of the system is increased. At 1920 GPa, the mean temperature is slightly greater in

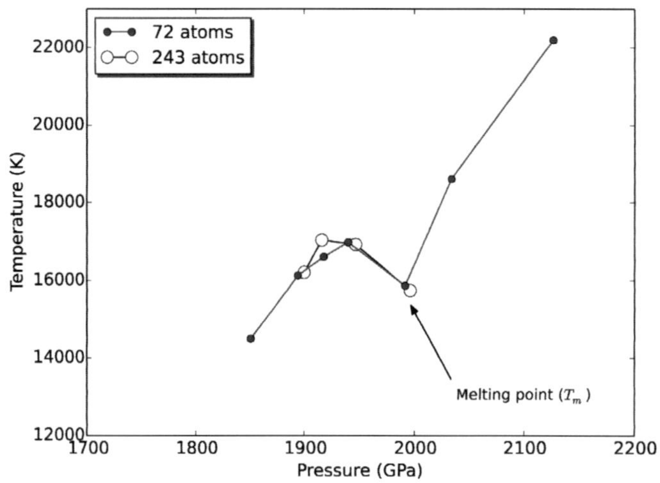

Figure 3. Size dependance of the calculations: isochore for silica in the Fe_2P phase at $V = 2.93$ $Å^3$/atom ($\rho = 11.33$ g/cm^3).

the system with 243 atoms, but this does not affect the prediction of the melting temperature, which corresponds to $T_m = 15800$ K in both cases. Therefore, we conclude that using a 72 atom cell is enough for our purposes.

4.2. Simulation time dependance

As noticed by Alfè *et al.* [5], the waiting times before the transition to liquid takes place are variable, and sometimes considerably long. Starting from the perfect crystal, and as we know from energy equipartition, the initial temperature T_i drops rapidly about the half of its initial value, and it then fluctuates about a value T_{sol}, and if the energy is greater than E_{LS}, it drops again and fluctuates about a lower steady value T_{liq} when the system melts. However, if T_i is high enough, the system will not be able to remain in solid state and will melt spontaneously, reaching T_{liq} in the first steps of the simulation. The same happens with low initial temperatures, for which T_{sol} will never drop to T_{liq}. Therefore, long simulation times are not needed to get a rough shape of the Z curve and, therefore, a good estimation for T_m.

The evolution of temperature as a function of time in a 96-atom cell of silica in the pyrite phase is shown in figure 4. The initial velocities are randomly assigned, so that they correspond to the a given temperature, leading to different values of instantaneous temperatures in the simulation but the same mean temperature. Homogeneous melting occures for the sample initiated at 26000 K during the first 200 steps (40 fs), while the structure remains as a superheated solid when initiated at 23000 K, where the crystal lattice sites are still identifiable (figure 4, left). The mean temperatures of the superheated solid and the final liquid are $T_{\text{sol}} = 9821$ K and $T_{\text{liq}} = 8815$ K, while the mean pressures are $P_{\text{sol}} = 375$ GPa and $P_{\text{liq}} = 391$ GPa.

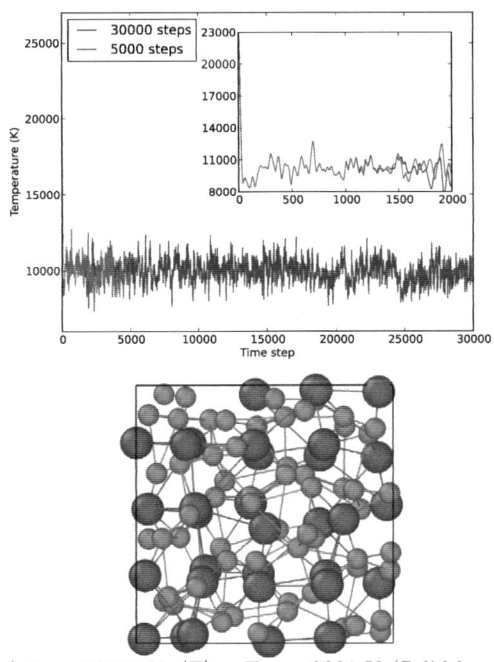

(a) $T_0 = 23000$ K. $\langle T \rangle = T_{\text{sol}} = 9821$ K (Solid branch)

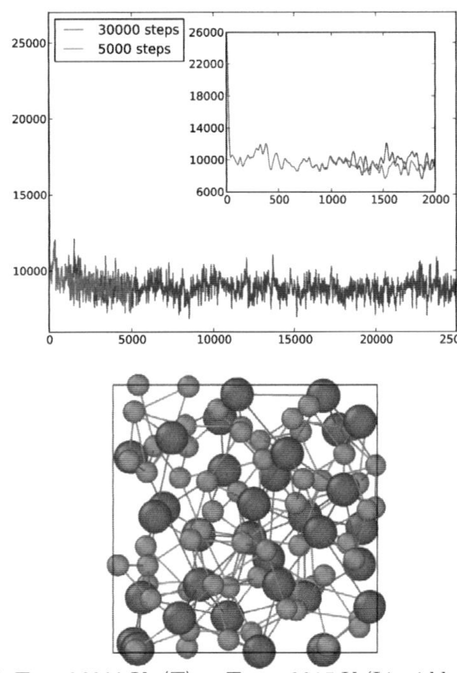

(b) $T_0 = 26000$ K. $\langle T \rangle = T_{\text{liq}} = 8815$ K (Liquid branch)

Figure 4. Evolution of temperature as a function of time in four independent simulation runs for a volume $V = 5.06$ Å^3/atom ($\rho = 6.58$ g/cm^3) of the pyrite structure: two initiated at 23000 K (left) and two initiated at 26000 K (right).

4.3. Mean Squared Displacement

In figure 5 we show an isochore of pyrite-like silica, where we observe a temperature drop around $P \sim 380$ GPa. The highest temperature in the solid branch, 9821 K, corresponds to a simulation initiated at 23000 K (the same shown in figure 4, left), which drops to 8815 K when initiated at 26000 K, which we associate to melting. We analyzed the mean squared displacement (MSD) for all simulations in the isochore, which are shown in figure 6. We observe that there is a clear difference in behavior of atoms when the sample is initiated at $T_0 = 26000$ K, respect to lower initial temperatures. Both Si and O show a diffusive behavior, which is steeper for highest temperature, an evidence that the sample is already molten. However, when the sample is initiated at $T_0 = 23000$ K, only the oxigen atoms diffuse, with a slope in the MSD curve that is very small. The MSD shows that superheated solids present this anomalous behavior:

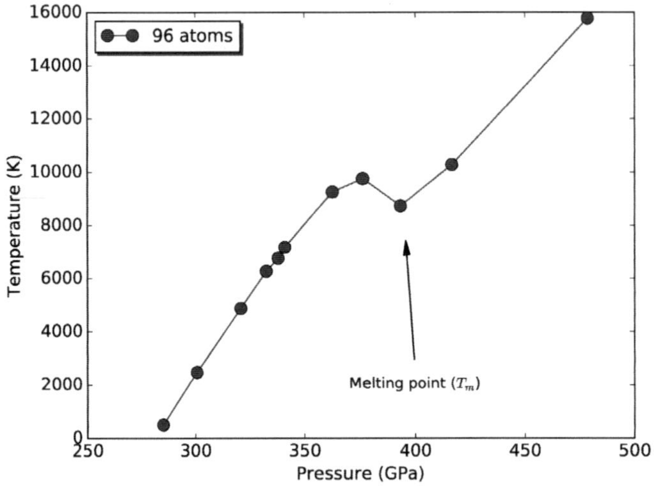

Figure 5. Isochoric of the pyrite-type silica. $V = 5.06$ Å^3/atom ($\rho = 6.58$ g/cm^3). The melting point corresponds to a simulation intiated at $T_0 = 26000$ K.

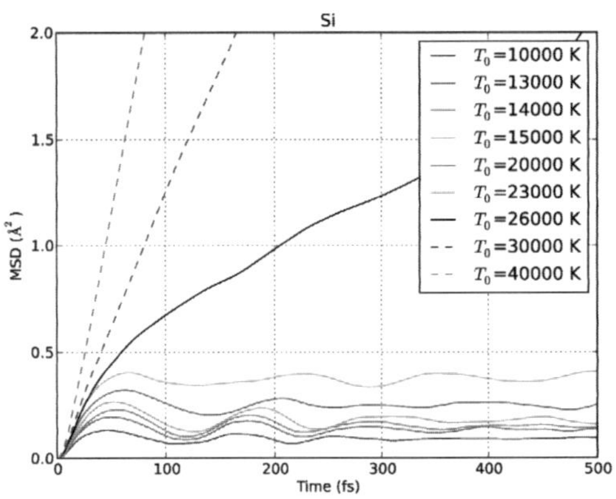

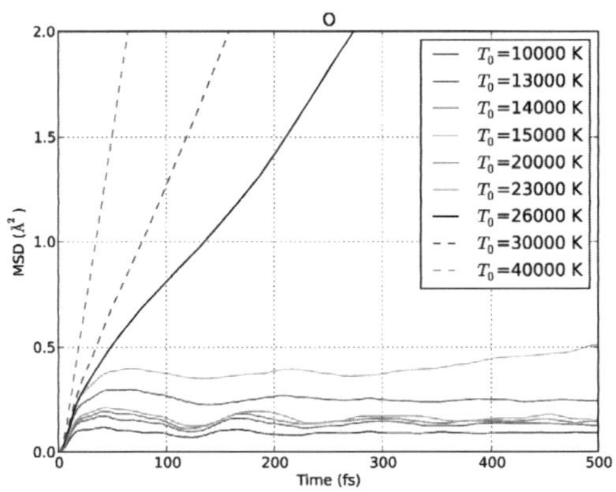

Figure 6. Mean squared displacement of Si and O atoms in the pyrite-type silica crystal as the initial temperature T_0 is increased.

a non-flat slope MSD curve or an almost-flat zero-slope MSD that adquires a non-zero slope, which sometimes makes it difficult to differentiate between a solid and a liquid.

4.4. Melting points

Using the Z method and the MSD analysis, we have obtained some high pressure melting temperatures for silica, which are depicted in table 1.

5. Conclusions

By using the Z method, we have determined some melting temperatures of silica for pressures higher than 300 GPa. These findings have considerable importance for the field of planetary science, since there still are no constraints for the solid–liquid phase diagram of silica at pressures concerning the interior of massive rocky planets (greater than 1000 GPa), which enable us to

Table 1. Melting temperatures (T_m) of silica in different crystalline phases at high pressures.

Structure	T_m (K)	P (GPa)	V/atom (Å^3)
Pyrite	8815	391	5.06
Cotunnite	12886	1273	3.43
Fe_2P	16922	1946	2.93

build more precise models for their interiors. However, further research must be done in order to constrain the melting points in the Z method, since the precision of the temperature obtained depends on the amount of simulations performed close to the T_{LS}–T_m drop, and the waiting times considered to determine if the system has reached the final equilibrium temperature. Further studies should consider the different structures of silica and determine whether how the melting temperature predicted by the Z method depends on the initial structure used.

Acknowledgments

F.G.-C. acknowledges CONICYT PhD fellowship No. 201090712 and a short-term fellowship from Universidad de Chile. G.G. thanks partial support from CONICYT-PIA grant ACT-1115, Chile.

References

[1] Valencia D, Ikoma M, Guillot T and Nettelmann N 2010 *Astronomy and Astrophysics* **516** A20
[2] Wagner F W, Tosi N, Sohl F, Rauer H and Spohn T 2012 *Astronomy & Astrophysics* **541** A103
[3] Tsuchiya T and Tsuchiya J 2011 *Proceedings of the National Academy of Sciences of the United States of America* **108** 1252–5
[4] Wu S, Umemoto K, Ji M, Wang C Z, Ho K M and Wentzcovitch R M 2011 *Physical Review B* **83** 184102
[5] Alfè D, Cazorla C and Gillan M J 2011 *Journal of Chemical Physics* **135** 19
[6] Belonoshko A, Skorodumova N, Rosengren A and Johansson B 2006 *Physical Review B* **73** 012201
[7] Belonoshko A, Davis S, Skorodumova N, Lundow P, Rosengren A and Johansson B 2007 *Physical Review B* **76** 064121
[8] Belonoshko A, Burakovsky L, Chen S, Johansson B, Mikhaylushkin A, Preston D, Simak S and Swift D C 2008 *Physical Review Letters* **100** 135701
[9] Belonoshko A, Arapan S, Martonak R and Rosengren A 2010 *Physical Review B* **81** 054110
[10] Belonoshko A and Rosengren a 2012 *Physical Review B* **85** 174104
[11] Benazzouz B K, Zaoui A and Belonoshko A B 2013 *American Mineralogist* **98** 1881–1885
[12] Bouchet J, Bottin F, Jomard G and Zérah G 2009 *Physical Review B* **80** 094102
[13] Burakovsky L, Chen S P, Preston D L, Belonoshko A, Rosengren a, Mikhaylushkin a S, Simak S I and Moriarty J a 2010 *Physical Review Letters* **104** 255702
[14] Burakovsky L, Chen S P, Preston D L and Sheppard D G 2014 *Journal of Physics: Conference Series* **500** 162001
[15] Davis S, Belonoshko A, Johansson B, Skorodumova N V and van Duin A C T 2008 *The Journal of chemical physics* **129** 194508
[16] Davis S, Belonoshko A B, Rosengren A, van Duin A C and Johansson B 2010 *Central European Journal of Physics* **8** 789–797
[17] Davis S, Belonoshko A, Johansson B and Rosengren A 2011 *Physical Review B* **84** 064102
[18] Davis S and Gutiérrez G 2013 *Modelling and Simulation in Materials Science and Engineering* **21** 075001
[19] Finney A R and Rodger P M 2011 *Phys. Chem. Chem. Phys.* **13**(44) 19979–19987
[20] Li D, Zhang P, Yan J and Liu H Y 2011 *EPL (Europhysics Letters)* **95** 56004
[21] Li D, Zhang P and Yan J 2014 *Scientific reports* **4** 4707
[22] Moriarty J A, Hood R Q and Yang L H 2012 *Physical Review Letters* **108** 036401
[23] Sun J, Martinez-Canales M, Klug D D, Pickard C J and Needs R J 2013 *Physical Review Letters* **111** 175502
[24] Kresse G and Furthmüller J 1996 *Phys. Rev. B* **54** 11169
[25] Blochl P E 1994 *Phys. Rev. B* **50** 17953
[26] Perdew J P, Burke K and Ernzerhof M 1996 *Phys. Rev. Lett.* **77** 3865

Sn doped CdTe as candidate for intermediate-band solar cells: A first principles DFT+GW study

Mauricio A. Flores[1], **Eduardo Menéndez-Proupin**[2]

[1,2]Departamento de Física, Facultad de Ciencias, Universidad de Chile, Las Palmeras 3425, 780-0003 Nuñoa, Santiago, Chile.

[1]Departamento de Matemáticas, Facultad de Ciencias, Universidad de Chile, Casilla 653, Santiago, Chile.

E-mail: mauricio.flores@ug.uchile.cl

Abstract. In this work, we investigate the electronic properties and defect formation energies of Sn doped CdTe combining first principles density-functional theory and many body GW calculations. Due to the Sn dopant, an isolated impurity band is formed in the middle of the forbidden band gap of CdTe allowing the absorption of sub-bandgap photons via an intermediate-band. Our results suggest CdTe:Sn as a promising candidate for the development of third-generation intermediate-band solar cells with theoretical efficiencies up to 63.2%.

1. Introduction

In traditional semiconductor photovoltaic cells, electrons are directly excited from the valence band (VB) to the conduction band (CB). Inserting a partially filled intermediate-band (IB) into the forbidden band gap enables two additional transitions to the unoccupied bands allowing the absorption of sub-bandgap photons. Just as plants break the water molecule in the chlorophyll function, two photons are used to produce an electron-hole pair of high free energy using a semi-filled isolated intermediate-band (IB) as an intermediate step [1, 2, 3]. This process not only increases the photocurrent without degrading the open-circuit voltage, but also reduces energy losses due to thermal relaxation of optically excited carriers [4]. This kind of solar cell presents efficiencies higher than those established by the Shockley-Queisser thermodynamic limit of 31% for single-junction solar cells [5]. The theoretical limiting efficiency of intermediate-band solar cells (IBSCs) is 63.2%, comparable to the efficiency for optimized triple-junction solar cells [6, 7].

The quest for finding optimal IBSCs been focused on alloys that naturally exhibit an IB [8, 9, 10] and arrays of quantum dots (QDs) that form an IB [11, 12, 13]. Walukiewicz et. al. [14, 15] were the first to produce an IB material based on highly mismatched alloys (HMAs). Wang and co-workers [4] demonstrated an overall 50% increase in power conversion efficiency in an IB solar cell based on ZnTe:O, where the intermediate-band position is tunable with alloy [4]. Ahsan et. al. [16] experimentally confirmed a two-photon excitation in an intermediate-band solar cell based on GaNAs alloy. Recently, a Sn-doped thin film of $CuGaS_2$ chalcopyrite were synthesized based on theoretical band structure calculations [17].

In the present work, we investigate the defect formation energies, charge transition levels, and theoretical band-structures of Sn doped CdTe using the highly accurate DFT+GW approach [18, 19, 20]. Our calculations show that an isolated intermediate-band in the middle of the

Journal of Physics: Conference Series **720** (2016) 012033 doi:10.1088/1742-6596/720/1/012033

forbidden gap is formed in Sn doped CdTe, which could be a promising candidate for high-efficiency photovoltaic cells.

2. Methods

2.1. Computational methods

First-principles total energy and band structure calculations were performed using the generalized gradient approximation density functional of Perdew, Burke, and Ernzerhof (PBE) [21] as implemented in the plane-wave based Quantum-ESPRESSO package [22]. Electron-ion interactions were described by GBRV ultrasoft pseudopotentials [23]. The valence wave functions were expanded in a plane-wave basis with cutoff energy of 390 eV. To avoid finite-size effects as much as possible, the defect calculations are performed within a 512-atom cubic supercell where all the atoms are fully relaxed until the forces on every atom were less than 0.025 eV/Å. The Brillouin zone was sampled with the Γ point only.

Many body GW calculations were carried out within the Vienna Ab Initio Simulation Package (VASP) [24, 25]. We use the one-shot G_0W_0@HSE approach, because some improvements can be achieved via the use of a better starting point [26]. The number of bands where chosen to be 1600 for bulk CdTe. Due to the slow convergence of the quasiparticle energies, an extrapolation to an infinite number of states is required. A scheme based on the $1/N$ convergence behavior of the quasiparticle energy, where N is the size of the basis set, was carried out for this purpose. Spin-orbit coupling was included as a correction.

2.2. Formation energies

The formation energy of a defect X in charge state q is defined as [27]

$$\Delta H_f[X^q] = E(q, \mathbf{R}_q) - E[bulk] - \sum_i n_i\mu_i + q(E_{\mathrm{VBM}} + E_F) + E_{corr}, \tag{1}$$

where $E(q, \mathbf{R}_{q'})$ is the total energy of a system in charge state q and atomic positions $\mathbf{R}_{q'}$, and E$[bulk]$ is the total energy for the perfect crystal using an equivalent supercell [28, 27]. The integer n_i indicates the number of atoms of type i that have been added or removed from the supercell, and μ_i are the corresponding chemical potentials of these species. E_F is Fermi energy referenced to the valence band maximum E_{VBM} in the perfect supercell, corrected by aligning the core potential of atoms far from the defect with that in the nondefective bulk supercell, and E$_{corr}$ is a correction term that accounts for the electrostatic interactions between supercells. The scheme proposed by Lany and Zunger [29] was applied to calculate charge corrections.

The chemical potential μ_i for ith atom species can be written as

$$\mu_i = \mu_i^{bulk} + \Delta\mu_i, \tag{2}$$

where μ_i^{bulk} is the chemical potential of species i in solids, and $\Delta\mu_i$ is a relative chemical potential referenced to μ_i^{bulk}. In the case of CdTe, if the system is under thermal equilibrium and there is no Te or Cd precipitation, we have

$$\Delta\mu_{\mathrm{Cd}} + \Delta\mu_{\mathrm{Te}} = \Delta H_f[\mathrm{CdTe}], \tag{3}$$

$$\Delta H_f[\mathrm{CdTe}] < \Delta\mu_{\mathrm{Te}} < 0, \tag{4}$$

$$\Delta H_f[\mathrm{CdTe}] < \Delta\mu_{\mathrm{Cd}} < 0, \tag{5}$$

where ΔH_f[CdTe] is the formation energy of CdTe. ΔH_f[CdTe] is calculated to be -0.90 eV, in good agreement with the experimental value of -0.96 eV.

Most of CdTe crystals are grown by techniques which suppresses the loss of Cd during growth, resulting in a Te-rich material with high concentration of Cd vacancies (V_{Cd}) and Te anti-sites (Te_{Cd}). Experimental measurements show that residual oxygen has a concentration typically orders of magnitudes higher than other residual impurities and comparable to intrinsic defects [30]. In order to put reasonable limits on the chemical potentials of Sn defects, we consider restrictions that prevent the formation of phases with the host atoms and the precipitation of the bulk phases of Tin. The relative chemical potential $\Delta\mu_{\mathrm{Sn}}$ is restricted by

$$\Delta\mu_{\mathrm{Sn}} + 2\Delta\mu_{\mathrm{O}} \leq \Delta H_f[\mathrm{SnO_2}] \quad (6)$$

The formation energy of SnO_2 is calculated to be -6.01 eV with PBE functional, in excellent agreement with the experimental value of -5.99 eV. In order to obtain the relative chemical potential $\Delta\mu_{\mathrm{O}}$, we consider CdTe to be in thermodynamic equilibrium with $CdTeO_3$ in the Te-rich limit [31, 32]. We find $\Delta\mu_{\mathrm{O}} = -1.88$ eV. From eq. (6) we get $\Delta\mu_{\mathrm{Sn}} \leq -2.25$ eV.

2.3. DFT + GW formalism

One shortcoming of the density functional theory is its failure to produce accurate excited states. Among other things, it gives rise to the well-known underestimation of the quasiparticle gap [33, 34] which render difficult the accurate description of midgap charged defect states [29, 19]. PBE severely underestimates the ionization potentials (IPs) for most semiconductors, while the commonly used hydrid functional of Heyd, Scuseria, and Ernzerhof (HSE) [35] good produces band gaps of small-gap materials, but tends to underestimate both IPs and electron affinities (EAs) [36], and fails in producing accurate binding energies of d-states in II-VI semiconductors [37].

The thermodynamic transition level $\epsilon(q/q+1)$ from charge state q to charge state $q+1$ is the IP of the system in the state q, i.e,

$$\epsilon(q/q+1) = E(q, \mathbf{R}_q) - E(q+1, \mathbf{R}_{q+1}) - E_{\mathrm{VBM}} \quad (7)$$

By adding and subtracting $E(q+1, \mathbf{R}_q)$, equation (1) can be rewritten as

$$\epsilon(q/q+1) = [E(q+1, \mathbf{R}_q) - E(q+1, \mathbf{R}_{q+1})] + [E(q, \mathbf{R}_q) - E(q+1, \mathbf{R}_q)] - E_{\mathrm{VBM}} \quad (8)$$

The first term corresponds to the relaxation energy $E(q+1, \mathbf{R}_q) - E(q+1, \mathbf{R}_{q+1})$ and can be calculated at the DFT level (the problem of the discontinuity in the chemical potential does not arise as the electron number remains constant). The second term, $E(q, \mathbf{R}_q) - E(q+1, \mathbf{R}_q)$, which defines the vertical excitation, is not well described within density functional theory, but a correction can be calculated using the ab initio GW method.

Our calculations are made within the DFT+GW formalism developed in Refs. [18, 19]. However, due to the high computational cost of performing GW calculations on charged supercells large enough to avoid spurious interactions between neighboring cells, we approximated the valence-band and conduction-band offsets by those of the bulk CdTe.

3. Results and discussion

3.1. Formation energies of Cd vacancy and Sn impurities in CdTe

The dominant intrinsic defect is believed to be Cd vacancy. It behaves as a dominant doubly charged acceptor and is responsible for the p-type conductivity and low resistivity of undoped CdTe [38, 39]. Hence, we also include (V_{Cd}) in the present study.

Sn substituted on both the Cd and Te sublattices, and the possibility of occupying an interstitial site were considered. Figure 1 shows the defect formation energies of Sn impurities and Cd vacancy as a function of E_F under Te-rich growth conditions. Our results show that Sn atoms should reside on Cd sites, in agreement with previous theoretical studies [40, 41].

For samples of undoped CdTe, the Fermi energy is pinned near the middle of the band gap and (V_{Cd}) will be in the double negatively charge state. In order to preserve the crystal electroneutrality, (Sn_{Cd}) should tend to bind with (V_{Cd}) forming ($Sn_{Cd}-V_{Cd}$) complexes.

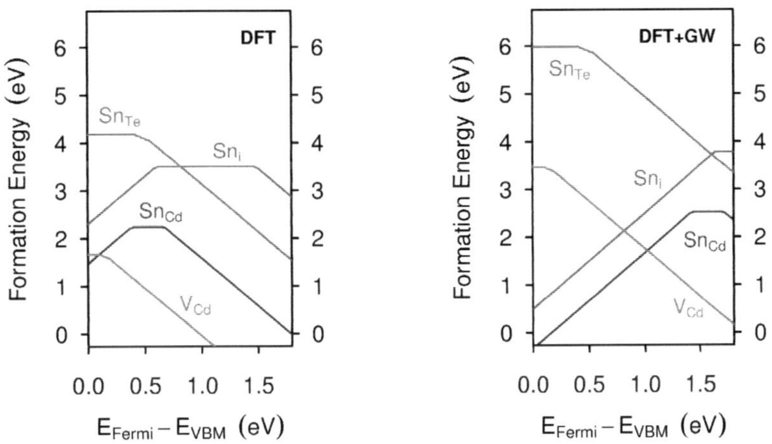

Figure 1. Formation energies of Cadmium vacancy and Sn impurities in various charge states under Te-rich conditions as a function of the Fermi level inside the band gap in CdTe. The slope of an energy line indicates the charge state of the defect. (Left) Formation energies calculated at DFT-PBE level. (Right) Formation energies calculated within the DFT+GW formalism.

3.2. Formation of an isolated intermediate-band in CdTe:Sn

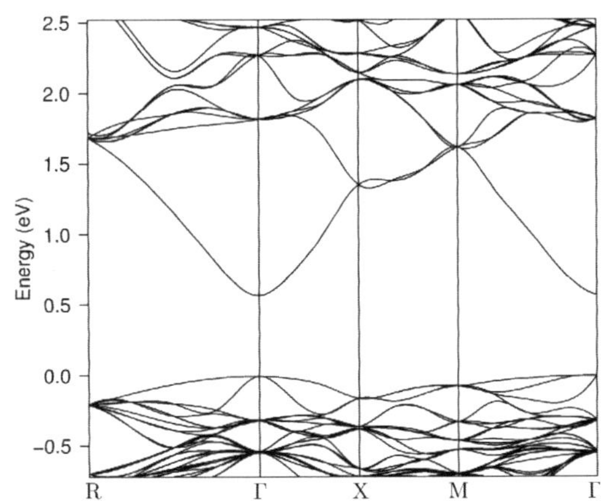

Figure 2. Kohn-Sham band structure of CdTe calculated using a 216 atom supercell. The VBM is set to be zero.

Figure 2 shows the theoretical band structure of perfect CdTe calculated using a 216-atom supercell. As shown in Figure 3, in both CdTe:$(Sn_{Cd})^{+2}$ and CdTe:($Sn_{Cd}-V_{Cd}$) an isolated

Journal of Physics: Conference Series **720** (2016) 012033 doi:10.1088/1742-6596/720/1/012033

intermediate-band is formed in the middle of the forbidden band gap. The impurity band, derived mainly from Sn 5*s* states, is unoccupied. The IB scheme requires that the IB must be partially filled in order to allow three direct bandgap transitions (VBM→IB, IB→CBM, VBM→CBM). However, the filling of the IB vary with light intensity, cell voltage, density of IB-states, concentration of native defects, and its position in the forbidden band gap. Strandberg and Reenaas [42] have shown that partial filling can be achieved by optical pumping or photofilling when the cell is exposed to concentrated light. A photofilled IBSC can give efficiencies comparable to those of a prefilled IBSC.

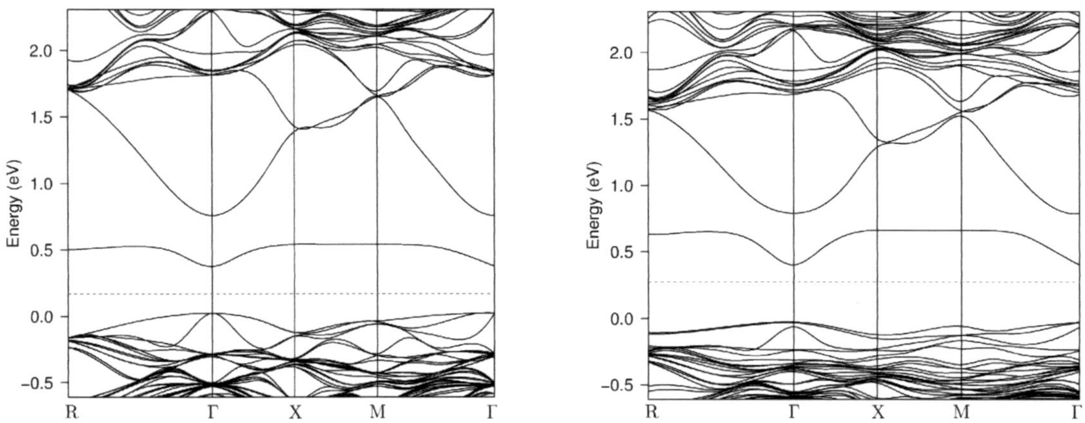

Figure 3. (Left) Konh-Sham band-structure of CdTe:$(Sn_{Cd})^{+2}$. (Right) Band-structure of CdTe:$(Sn_{Cd}-V_{Cd})$ in the neutral charge state. The dotted red line indicates the position of the average electrochemical potential. The VBM is set to be zero.

According to Shockley-Read-Hall theory [43, 44], impurity states closer to the midgap are harmful as non-radiative recombination centers. The transition of a delocalized CB electron to a localized impurity state causes the impurity level to cross the CB (taking into account lattice vibrations). The energy of the electron is therefore transformed into potential energy of the impurity in a single step [2, 45]. Fortunately, the non-radiative recombination can be prevented through charge redistribution [46] or if the density of impurities are highly enough to exceed the Mott transition [47].

4. Conclusions

In summary, we have studied the defect formation energies, charge transition levels, and band-structures of Sn doped CdTe combining density-functional theory and many body GW calculations. An isolated impurity band derived mainly from Sn 5s states is formed in the middle forbidden band gap of bulk CdTe, allowing three direct bandgap transitions (VBM→IB, IB→CBM, VBM→CBM). Our results suggest that CdTe:Sn could be a promising candidate for the development of third-generation high-efficiency intermediate-band solar cells.

Acknowledgments

This work was supported by FONDECYT Grant No. 1130437.

References

[1] P Wannón and C Tablero 2002 *Phys. Rev. B.* **65** 165115
[2] A Luque, A Martí, and C Stanley 2012 *Nature Photon.* **6** 146
[3] V Mlinar 2013 *Nanotechnology* **24** 042001
[4] W Wang, A Lin, and J Phillips 2009 *Appl. Phys. Lett.* **95** 011103
[5] W Shockley, H J Queisser 1961 *J. Appl. Phys.* **32** 510
[6] A Martí and G Araujo 1996 *Sol. Energy Mater. Sol. Cells.* **43** 204
[7] A Luque and A Martí 1997 *Phys. Rev. Lett.* **78** 5014
[8] P Palacios, P Wannón, S Pizzinato, and J Conesa 2006 *J. Chem. Phys.* **124** 14711
[9] P Palacios, I Aguilera, K Sanchez, J Conesa, and P Wannón 2008 *Phys. Rev. Lett.* **101** 046403
[10] K Sanchez, I Aguilera, P Palacios, and P Wannón 2008 *Phys. Rev. B.* **79** 165203
[11] A Martí, L Cuadra, and A Luque IEEE, 2000 *Proc. 28th IEEE Photovoltaics Specialists Conf.* 904–943
[12] A Luque, A Martí, N López, E Antolín, and E Cánovas 2005 *Appl. Phys. Lett.* **87** 083505
[13] Martí A *et al.* 2008 *Thin Solid Films.* **516** 6716
[14] K Yu, W Walukiewicz, J Wu, W Shan, J Beeman, M Scarpulla, O Dubon, and P Becla 2003 *Phys. Rev. Lett.* **91** 246403
[15] K Yu, W Walukiewicz, J Ager III, D Bour, R Farshchi, O Dubon, S Li, I Sharp, and E Haller 2006 *Appl. Phys. Lett.* **88** 092110
[16] Ahsan N *et al.* 2012 *Appl. Phys. Lett.* **100** 172111
[17] C Yang, M Qin, Y Wang, D Wan, F Huang, and J Lin 2013 *Scientific reports* **3** 1286
[18] M Hedstrom, A Schindlmayr, G Schwarz, and M Scheffler 2006 *Phys. Rev. Lett.* **97** 226101
[19] P Rinke, A Janotti, M Scheffler, and C G Van de Walle 2009 *Phys. Rev. Lett.* **102** 026402
[20] A Malashevich, M Jain, and S G Louie 2014 *Phys. Rev. B.* **89** 075205
[21] J P Perdew, K Burke, and M Ernzerhof 1996 *Phys. Rev. Lett.* **77** 3865
[22] P Giannozzi S Baroni N B *et al.* 2009 *J. Phys.: Condens. Matter* **21** 395502
[23] K F Garrity, J W Bennett, K M Rabe, and D Vanderbilt 2014 *Comput. Mater. Sci.* **81** 446
[24] G Kresse and J Furthmüller 1996 *Comput. Mater. Sci.* **6** 15
[25] G Kresse and J Furthmüller 1996 *Phys. Rev. B.* **54** 11169
[26] F Fuchs, J Furthmüller, F Bechstedt, M Shishkin and G Kresse 2007 *Phys. Rev. B.* **76** 115109
[27] C Freysoldt, B Grabowski, T Hickel, J Neugebauer, G Kresse, A Janotti, and C G Van de Walle 2014 *Rev. Mod. Phys.* **86** 253
[28] C G Van de Walle, D Laks, G Neumark, and S Pantelides 1993 *Phys. Rev. B.* **47** 9425
[29] S Lany and A Zunger 2008 *Phys. Rev. B.* **78** 235104
[30] S Awadalla, A Hunt, K Lynn, H Glass, C Szeles, and S-H Wei 2004 *Phys. Rev. B* **69** 075210
[31] R Diehl, and B Nolaeng 1984 *J. Cryst. Growth.* **66** 91
[32] M Flores and E Menéndez-Proupin unpublished "First principles DFT+GW study of oxygen doped CdTe"
[33] J Perdew and M Levy 1983 *Phys. Rev. Lett.* **51** 1884
[34] L J Sham and M Schluter 1983 *Phys. Rev. Lett.* **51** 1888
[35] J Heyd, G E Scuseria, and M Ernzerhof 2006 *J. Chem. Phys.* **124** 219906(E)
[36] Y Hinuma, A Grüneis, G Kresse, and F Oba 2014 *Phys. Rev. B.* **90** 155405
[37] Y Zhanga, X Yuanb, X Suna, and W Zhangb 2014 *Solid State Commun.* **152** 588
[38] M Fiederle, C Eiche, M Salk, R Schwarz, K Benz, W Stadler, D Hofmann, and B Meyer 1998 *J. Appl. Phys.* **84** 6689
[39] M Berding 1999 *Phys. Rev. B* **60** 8943
[40] J Jaffe 2006 *J. Appl. Phys.* **99** 033704
[41] V Babentsov, J Franc, H Elhadidy, A Fauler, M Fiederle, and R James 2007 *J. Mater. Res.* **11** 3249
[42] R Strandberg and T Reenaas 2009 *J. Appl. Phys.* **105** 124512
[43] W Shockley and W Read 1952 *Phys. Rev.* **87** 835
[44] R Hall 1952 *Phys. Rev.* **87** 387
[45] D Lang and C Henry 1975 *Phys. Rev. Lett.* **35** 1525
[46] C Tablero 2009 *Physica B.* **404** 4023
[47] N Mott 1968 *Rev. Mod. Phys.* **40** 677

Spin-orbit coupling effects in gold clusters: The case of Au_{13}

Mauricio A. Flores[1] **and Eduardo Menéndez-Proupin**[2]
[1,2]Departamento de Física, Facultad de Ciencias, Universidad de Chile,
Las Palmeras 3425, 780-0003 Nuñoa, Santiago, Chile.
[1]Departamento de Matemáticas, Facultad de Ciencias, Universidad de Chile,
Casilla 653, Santiago, Chile.

E-mail: mauricio.flores@ug.uchile.cl

Abstract. Non relativistic description of the chemical bonding for heavy atom systems is unrealistic. However, only few relativistic DFT calculations have been reported for gold clusters. In this work, we investigate the effect of the spin-orbit coupling in the structure, relative stability and HOMO-LUMO gap of the lowest-lying isomers of neutral Au_{13} clusters. Our results show that spin-orbit coupling increases the energy gap between 2D and 3D isomers and a relativistic bond contraction appears. Moreover, the relative energy of some isomers change when the spin-orbit coupling is included.

1. Introduction

One of the largest challenges in the study of metal clusters is the determination of their ground-state structures. Neutral gold clusters up to 10 atoms adopt planar configurations, but the precise size for which they change from 2D to 3D structures still remains as an open issue mainly due to the scarcity of experimental study. It has been pointed out that the preferred planarity of small gold clusters is correlated with strong hybridization of the atomic $5d$ and $6s$ orbitals due to relativistic effects [1].

Relativistic effects can strongly influence the chemical and physical properties of heavy elements. For the sixth row elements (around W to Bi) relativistic effects are comparable to the usual shell-structure effects [2]. Non relativistic description of the chemical bonding for heavy atom systems is unrealistic. In a many-electron atom relativistic effects tend to contract and stabilize the s and p shells, and expand and destabilize the d and f shells [2]. In the case of gold the $6s$ orbital is strongly contracted (by about 17%), whereas the $5d$ orbitals slightly expand [3]. Spin-orbit (SO) interactions split the d-band into $d_{3/2}$ and $d_{5/2}$ subbands, increasing the overlap of s and d bands leading to appreciable s-d hybridation in the bonding molecular orbitals [4, 5].

A maximum of relativistic effects has been identified at gold [2, 6]. Inertness and many of its unusual properties such that yellow color, nobility, and trivalency derive from relativistic effects [7]. Spin-orbit coupling has a significant effect on the cohesive energy and the relative stability of the isomers of gold clusters [8]. Its electronegativity is a result of the relativistic contraction of the valence $6s$ and $6p$ orbitals. Using first-principles calculations, Takeuchi et al. [9] have shown that, due to the strong relativistic effects, it is energetically favorable for the top layer of the (100) surface of Au to transform from the ideal square lattice of a fcc metal to a slightly distorted and contracted hexagonal-close-packed arrangement.

In the present work, we report the results of first principles density functional theory calculations of Au_{13} clusters with and without spin-orbit coupling. We investigate the stability, HOMO-LUMO gap, interatomic distances, and the role of the semicore orbitals when are considered as valence electrons.

2. Methodology

First-principles total energy calculations were performed using the generalized gradient approximation density functional of Perdew, Burke, and Ernzerhof (PBE) [10] as implemented in the plane-wave based Quantum-ESPRESSO package [11]. Electron-ion interactions were described using PAW pseudopotentials [12] from the PS Library 0.3.1 [13].

In order to elucidate the role of semicore states, two pseudopotentials are considered, namely PP1 and PP2. The first one contains 11 valence electrons ($5d^{10}6s^1$) and the second 19 valence electrons ($5s^25p^65d^{10}6s^1$). We used a plane wave energy cutoff of 40 Ry and 60 Ry, respectively to represent Kohn-Sham wavefunctions, and 500 Ry to represent the charge density. To have negligible interactions between periodic images of clusters, DFT calculations have been performed using a large cubic supercell with an edge of 24 Å. The Brillouin zone was sampled with the Γ point only. Structural optimizations of the lowest-energy isomers were carried out using the Broyden-Fletcher-Goldfarb-Shanno (BFGS) algorithm until the Hellman-Feynman forces are less than 0.001 Ry/bohr.

3. Results and discussion

An exhaustive search of the potential energy surface was performed at DFT level using a novel global optimization algorithm [14]. Equilibrium geometries for the six lowest lying structures of neutral Au_{13} clusters are presented in the Figure 1.

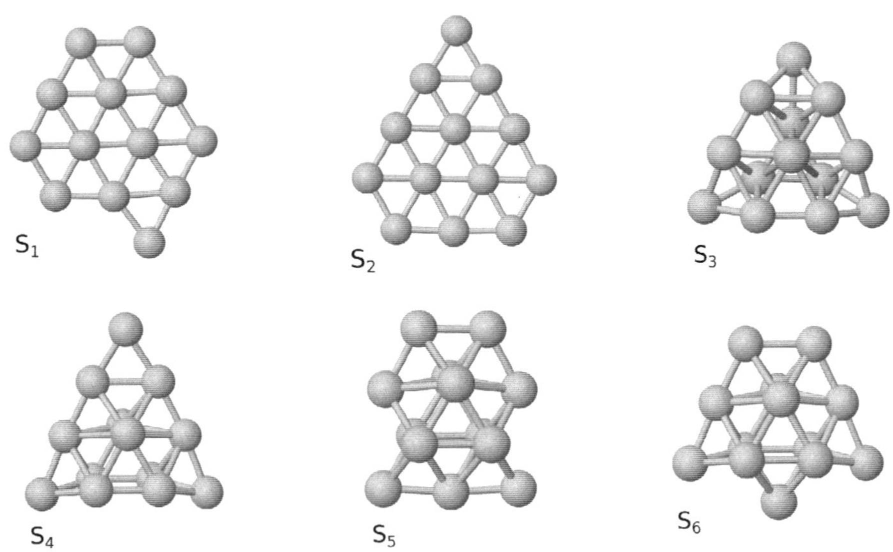

Figure 1. Structures the six lowest lying isomers of Au_{13} after geometry optimization.

Relative energies are depicted in Figure 2 (Left). All methods considered in this work agree that the lowest structure of the Au_{13} is the planar isomer S_1 different from that predicted by previous DFT [15] and revTPSS [16] calculations. When spin-orbit coupling is taken into account the energy gap between planar and lowest-lying 3D isomers is increased. This effect is stronger when the semi core states are treated as valence electrons (PP2 + SO). We note that spin-orbit effects change the relative stability between isomers S_5 and S_6 predicted from the DFT calculations without spin-orbit coupling.

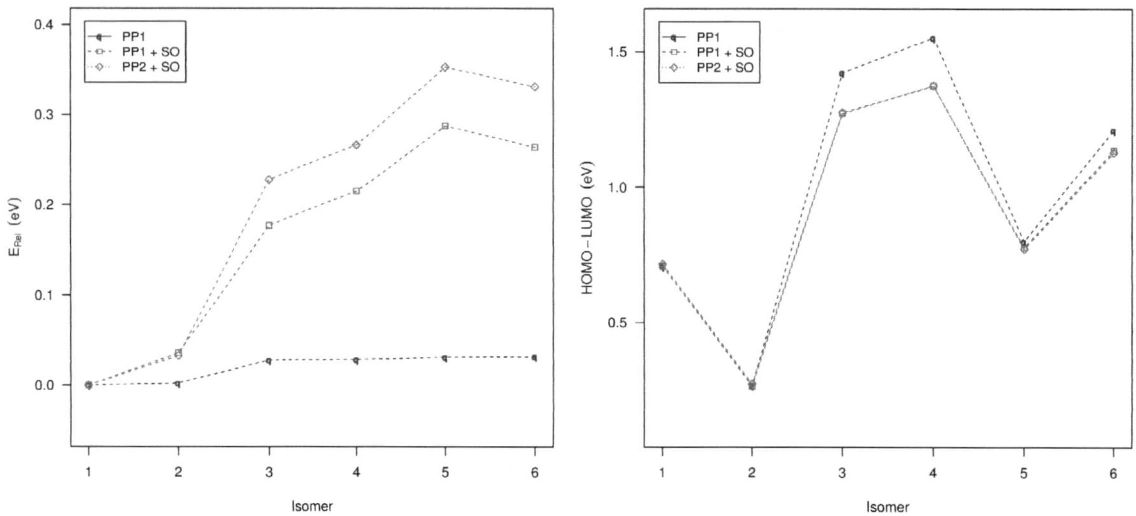

Figure 2. (Left) Relative energies compared to isomer S_1 obtained at DFT level with and without including the spin-orbit coupling. (Right) The HOMO-LUMO energy gaps of the lowest lying isomers of Au_{13} obtained with and without including the spin-orbit coupling.

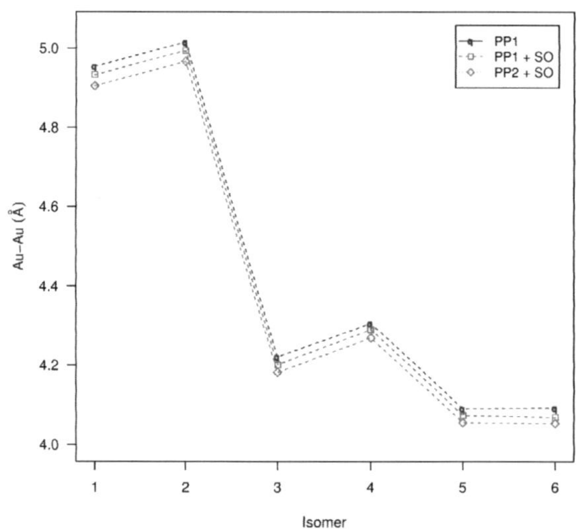

Figure 3. Average interatomic distance obtained for all Au_{13} isomers after geometry optimization with and without including the spin-orbit coupling.

Figure 2 (Right) shows the effect of the spin-orbit coupling on the HOMO-LUMO energy gap. It is reported that a large energy gap between the highest occupied molecular orbital (HOMO) and the lowest unoccupied molecular orbital (LUMO) implies high stability and small energy gap implies low stability. The energy gap of all 3D isomers decreases when SO is included while the HOMO-LUMO gap of planar clusters remains unchanged. Surprisingly, the most stable isomers with and without spin-orbit are S_4 and S_5. The HOMO-LUMO gap is unaffected when the semi core states are considered as valence electrons.

Figure 3 shows the average interatomic distance obtained for all isomers after geometry optimization. The first observation is that a relativistic bond contraction appears. It is dependent of the cluster shape and is stronger for planar isomers. The second observation is that the effects of SO coupling on the average atomic distances are monotonic, contrary to previous results reported for Pt clusters [17, 18] where relativistic effects are also very strong. We note that bond contractions are increased when the semi core states are considered as valence electrons.

4. Conclusions

We performed a exhaustive search of the potential energy surface of gold at DFT level reporting the six lowest-energy isomers of Au_{13}. Our relativistic DFT calculations show that spin-orbit coupling change the relative stability between some isomers predicted from DFT calculations without spin-orbit coupling. The energy gap of all 3D isomers decreases when SO is included while the HOMO-LUMO gap of planar clusters remains unchanged. A relativistic bond contraction dependent of the cluster shape appears. The inclusion of semi core 5s and 5p states in the valence shell increases the the relative energy stability of planar isomers and the relativistic bond contractions.

References

[1] H Häkkinen, M Moseler, and U Landman 2002 *Phys. Rev. Lett* **89** 033401
[2] P Pyykkö 1988 *Chem. Rev.* **88** 563
[3] Schwerdtfeger P *et al.* 1989 *Journal of chemical physics* **91** 1762
[4] F Ramos, R Arratia-Perez, and G Malli 1987 *Physical Review B* **35** 3790
[5] Qteish A *et al.* 1987 *Physical Review B* **72** 155317
[6] N Kaltsoyannis 1997 *J. Chem. Soc., Dalton Trans.* 1
[7] P Pyykkö 2012 *Annual review of physical chemistry* **63** 45
[8] Y Shi, Z Li, and K Fan 2010 *J. Phys. Chem. A* **114** 10297
[9] N Takeuchi, C Chan, and K Ho 1991 *Phys. Rev. B* **43** 14363
[10] J P Perdew, K Burke, and M Ernzerhof 1996 *Phys. Rev. Lett.* **77** 3865
[11] P Giannozzi S Baroni N B *et al.* 2009 *J. Phys.: Condens. Matter* **21** 395502
[12] P Blöchl 1994 *Phys. Rev. B* **50** 17953
[13] A Dal Corso 2014 *Comp. Material Science* **95** 337
[14] M Flores (in preparation) "Ground-state structures of 13-atom transition-metal clusters: Y - Cd"
[15] B Assadollahzadeh and P Schwerdtfeger 2009 *J. Chem. Phys.* **131** 064306
[16] M Johansson, I Warnke, A Le, and F Furche 2014 *J. Phys. Chem. C* **118** 29370
[17] M Huda, M Niranjan, B Sahu, and L Kleinman 1991 *Phys. Rev. A* **73** 053201
[18] A Sebetci 2009 *Phys. Chem. Chem. Phys* **11** 921

Synthesis of solar cells sensitized using natural photosynthetic pigments & study for the cell performance under different synthesis parameters

Simon Roa *, Sivakumar Radhakrishnan, Paulraj Manidurai*

Department of Physics, Faculty of Physical and Mathematical Sciences, University of Concepción, P.O. Box 160-C, Concepción, Chile.

E-mail: `sroadaz8@gmail.com`, `paulrajm@gmail.com`

Abstract. In this study we used photosynthetic pigments extracted from spinach and purple cabbage for their potential application in dye sensitized solar cells (DSSC). Pigments were extracted by dissolving small amounts of each one of these plant products in methanol and distilled water. The extraction was also done at two different temperatures (70°C and 80°C respectively). This was to assess for the solvent that promotes better extraction of the pigments. A parallel study was also carried out using a mixture of both these dyes in 1:1 ratio. Good absorption, about 60 % to 80 % was obtained for spinach pigments diluted in methanol in the visible range between 400-480nm, and between 9 % to 15 % for purple cabbage pigments in the wavelength range between 480-630 nm when extracted using distilled water at 80°C. In contrast, the diluted mixture in methanol shows good absorption of 20 % and 32 % for wavelengths in the range 400-480nm. Solar cells sensitized using these natural dyes were studied for their photovoltaic properties by measuring current-voltage behavior. Efficiencies ranging from 0.011% to 0.0719% were observed. Mixture of spinach & purple cabbage pigments extracted using methanol was found to have the highest efficiency of 0.0719%.

1. Introduction

The solar cells produced today are generally divided into several types such as monocrystalline silicon (Si) solar cells, polycrystalline Si solar cells, amorphous Si solar cells and dye-sensitized solar cells (DSSC). The so-called "solar cells" generally refer to cells made of crystalline silicon and its popularity is due to its higher photoelectric conversion efficiency compared to DSSC. However, the production cost of solar cells based on glassy silicon are higher than the DSSC. Simply because of its advantage of lower manufacturing costs, the DSSC are highly interesting [1]. With its simple and easy process of manufacture, DSSC are easy to mount and the cost is only about one-third of what it costs to manufacture solar cells based on silicon.
In the future, the cost can be further reduced for DSSC having the same efficiency as Si based solar cells. The predicted total cost will be only one tenth. Efficiency of DSSC is mainly determined by the sensitizer (photosynthetic pigment in this case) used. The marketed dyes are mainly synthetic chemicals, such as N719 and N3 dyes, both of which have a satisfactory conversion efficiency. However, these dyes contains some heavy metals, which are expensive and produce environmental contamination. Moreover, the synthesis process is very complex and expensive. With the purpose of replacing these dyes, many types of natural dyes have been

actively studied and tested. However, natural dyes usually gave poor photovoltaic response [2] due to its weak binding energy with TiO_2 thin films and low charge transfer due to optical absorption in the entire region of the visible spectrum, but these dyes are very cheap and can be easily prepared in comparison to the synthetic dyes.
The energy conversion efficiency is less than 1%, which is typical for most natural extracts. Several studies published in recent years have made remarkable progress in the use of organic dyes for DSSCs. Natural dyes have high value at low cost. In this work, extracts from spinach leaves and red cabbage were used as light sensitizers for preparing DSSC. Effectiveness of dyes extracted from these two plants under different conditions, namely, variation in temperature and solvent medium used for extraction, is been discussed.

2. Experimental Details

Synthesis of solar cells sensitized using natural dyes were divided into four stages: Extraction of the photosynthetic pigments, deposition of TiO_2 thin films, preparation of electrolyte and finally the DSSC assembly.

2.1. Extraction of photosynthetic pigments

Mainly two different extraction methods were adopted:

- Spinach and purple cabbage leaves were diluted in methanol and the extraction was done at room temperature.
- Spinach and purple cabbage leaves were diluted in deionized water and the extraction was done at two different temperatures; [70°C and 80°C].

In both cases, 10 g of each plant product was taken and was then dissolved in 40 mL of the respective solvents [i.e., methanol or deionized water].

Table 1. Methodology used for extracting the photosynthetic pigments.

Sample	Spinach [g]	Purple Cabbage [g]	solvent, Volume [mL]	Temperature [°C]
1	10	—	Methanol, 40	Room
2	—	10	Methanol, 40	Room
3	10	10	Methanol, 80	Room
4	10	—	Deionized water, 40	70
5	—	10	Deionized water, 40	70
6	10	—	Deionized water, 40	80
7	—	10	Deionized water, 40	80
8	10	10	Deionized water, 80	70

Once the extraction was done the samples were filtered out and were stored in dark.

2.2. Preparation of TiO_2 thin films

Titanium Dioxide Powder (EMSURE®, purity 99,5%, 1.5 gm), 100 mL of acetylacetone, 1.8 mL of distilled water and 50 mL of TRITON X-100, were taken and mixed in a mortar to form a paste. This paste was deposited over a glass substrate coated with FTO (fluorine doped tinoxide), on the FTO side. The films were deposited over an area of 2.5 cm x 2.5 cm, using "*Doctorblade*" method [3]. After allowing it to dry for 1 hour, they were sintered at 450°C for 30 minutes.

Journal of Physics: Conference Series **720** (2016) 012035 doi:10.1088/1742-6596/720/1/012035

2.3. Preparation of electrolyte

The electrolyte or the redox couple was prepared using 0.6 M butylmethylimidazolium iodide (BMII), 0.05 M I_2, 0.1 M LiI, and 0.5 M tert-butylpyridine in 1:1 acetonitrile/valeronitrile.

2.4. DSSC assembly

To initiate the DSSC assembly, TiO_2 layer deposited on FTO should be sensitized first, using the natural dye extracts. For this, the TiO_2 films were immersed into 20 mL of the dye solution in dark, over night. Then the following day, they were rinsed with ethanol and left to dry. This formed the photoanode. A thin layer of platinium coated over another FTO coated glass substrate, formed the counter electrode. It was synthesized by centrifugation process by using chloroplatinic acid solution (H_2PtCl_6 .$6H_2O$) (0.005 M in isopropanol). Then this Pt-FTO electrode was sintered at 400°C for 20 min.

These two electrodes were assembled face to face. Subsequently, the electrolyte that was already prepared was injected into this sandwich layer. A thin layer of surlyn (25 μ m) was used to seal this liquid electrolyte and also to hold the electrodes intact, avoiding short circuiting, fig. 1.

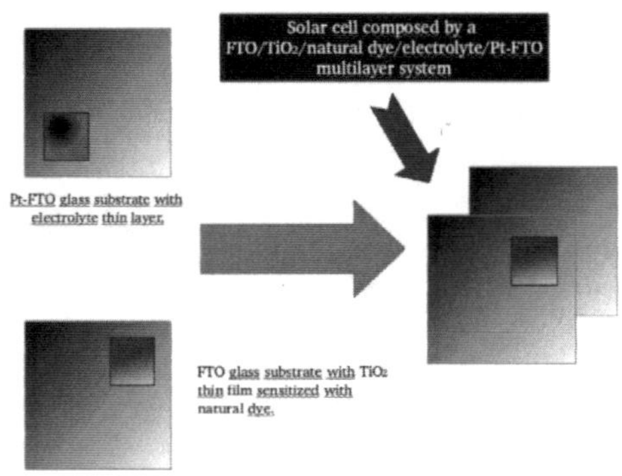

Figure 1. Scheme depicting a basic DSSC assembly.

Figure 1. depicts a basic DSSC aasembly. The configuration shown in fig. 2, basically corresponds to the cell structure. To measure cell efficiency their I-V characteristics were studied using a solar simulator unit coupled with an Keithley source meter.

cells were irradiated using 1sun, AM 1.5, light source from the solar simulator. Thus a multilayer system (Figure 2): FTO/TiO_2/natural-dye/electrolyte/Pt-FTO, was obtained as dye sensitized solar cell.

3. Results and Discussions

3.1. Optical properties of natural dyes

Absorbance spectra of the dye coated photoanode samples were measured in visible range between 400 to 700nm, fig. 3 and 4. For spinach, significant absorption was obtained for the case of sample 1 where the solvent used was methanol, fig. 3a. Four principal absorption peaks 413.2 nm, 466.2 nm, 437.2 nm and 662.4nm were identified using Lorentzian curve fitting. Absoption peaks located at 413.2 nm and 662.4 nm correspond to charateristic absorption peaks of chlorophyll-a, and absorption peaks located at 466.2 nm 437.2 nm correspond to characteristic absorption peaks of carotenoid. On the other hand, extractions done using aqueous medium (sample 4 and 6), showed no significant absorption peaks.

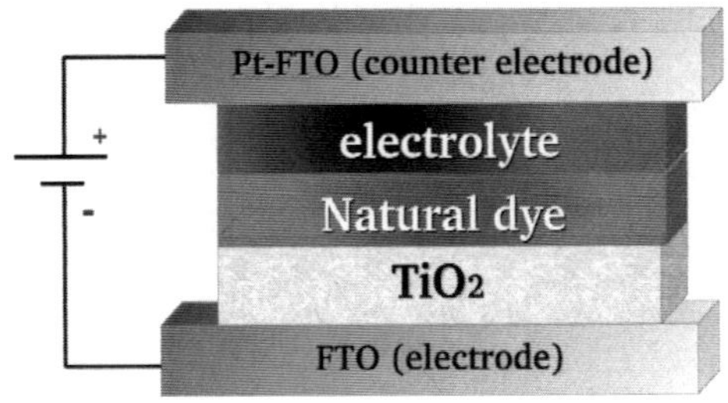

Figure 2. Dye sensitized solar cell composed by a FTO/TiO_2/natural-dye/electrolyte/Pt-FTO multilayer system.

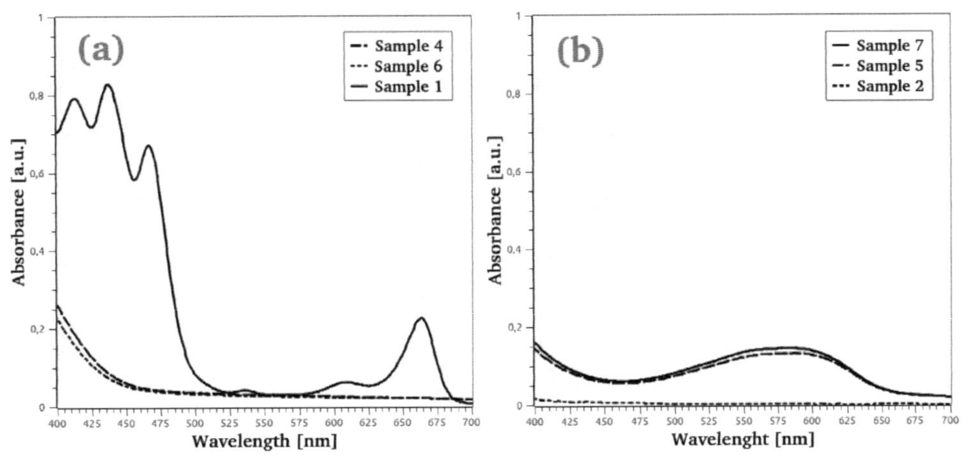

Figure 3. Absorption spectra obtained for natural dyes extracted from spinach (a) and purple cabbage (b).

For pigments extracted from purple cabbage, absorption peaks for samples 5 and 7 (fig. 3b) showed a broad peak in the range between 475nm to 640nm. Whereas for sample 2 no significant absorption peak was observed eventhough the solvent used for extraction was methanol. As the absorption spectrum was pronounced for sample 7, we took this for the study, in this case. The absorption peak seemed like a superposition of two or more peaks, thus an Lorentzian fitting was used to deconvolute the mixed peaks, one at 540.8 nm and another at 608.1nm.

According to literature [6], these peaks corresponds to characteristic absorption peaks of phycoerythrin and ficoeritrocianina pigments (blue colour) and phycocyanin (red colour) pigments respectively. That explains the superposition or mixture of two pigments and the origin of purple colour for purple cabbage.

For the mixture samples, absorption spectra for sample 3 is similiar to sample 1 and 8 similar to sample 7, except that the absorption peaks are less intense, but conclusions are identical to the cases of the samples 1 and 7, so in case of mixture diluted in methanol a better extraction of chlorophyll-a and carotenoids were obtained and a better extraction of phycoerythrin and phycocyanin were obtained in case of mixture diluted in deionized water, that can be seen in Figures 3 and 4.

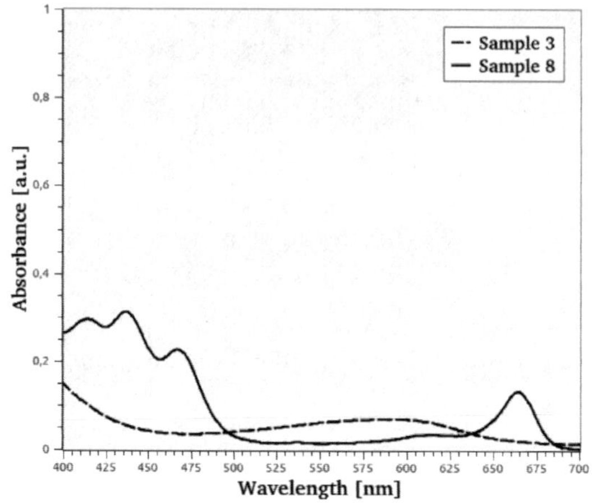

Figure 4. Absorption spectra obtained for natural dye composed of spinach and purple cabbage mixture.

3.2. Optical properties of sensitized TiO_2 layers

All the sensitized TiO_2 samples showed increased absorption in the spectrum range between 400nm to 620nm compared to the un-sensitized TiO_2 samples, fig. 5.

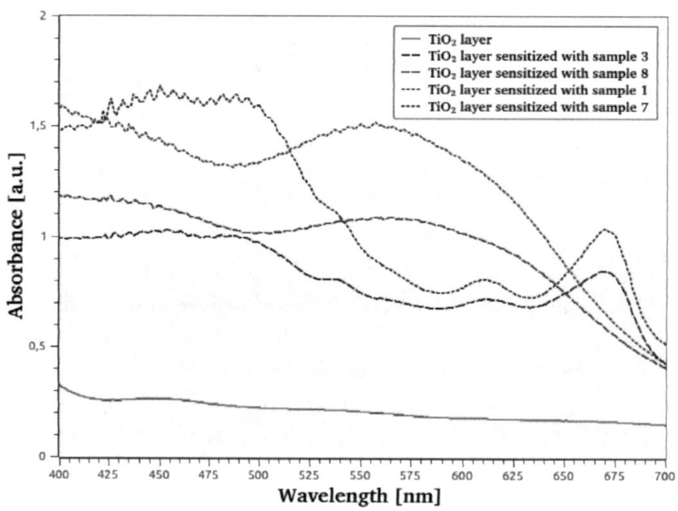

Figure 5. Absorption espectra of TiO_2 layer sensitized with different natural dyes.

However TiO_2 layers sensitized with samples 1 and 3 correspond to pigments diluted in methanol, which presents better absorption in visible range compared to TiO_2 layer sensitized with samples 7 and 8 that corresponds to pigments diluted in deionized water. In the later case, the absorption begins to decay to after 620nm.

TiO_2 layers sensitized with samples 1 and 3 showed characteristic absorption peaks in the wavelengths range 620 nm - 700 nm, whereas higher wavelength peaks over 620nm were not observed for sample 7 and 8.

3.3. I-V characteristics and solar cell efficiency

Solar cell efficiency (η) is the percentage of absorbed sunlight converted to electrical energy when a solar cell is connected to an electrical circuit. This term is calculated using the following relation, under standard conditions (STC);

$$\eta = \frac{P_{max}}{I \times A_c} \tag{1}$$

Where, P_{max}, is the maximum power; I, Light irradiance on the cell (in W/m^2); Ac, Surface area of the solar cell that is been irradiated by the light

STC specifies a temperature of 25°C and irradiance of 1000 W/m^2 with a spectral air mass 1.5 (AM 1.5).

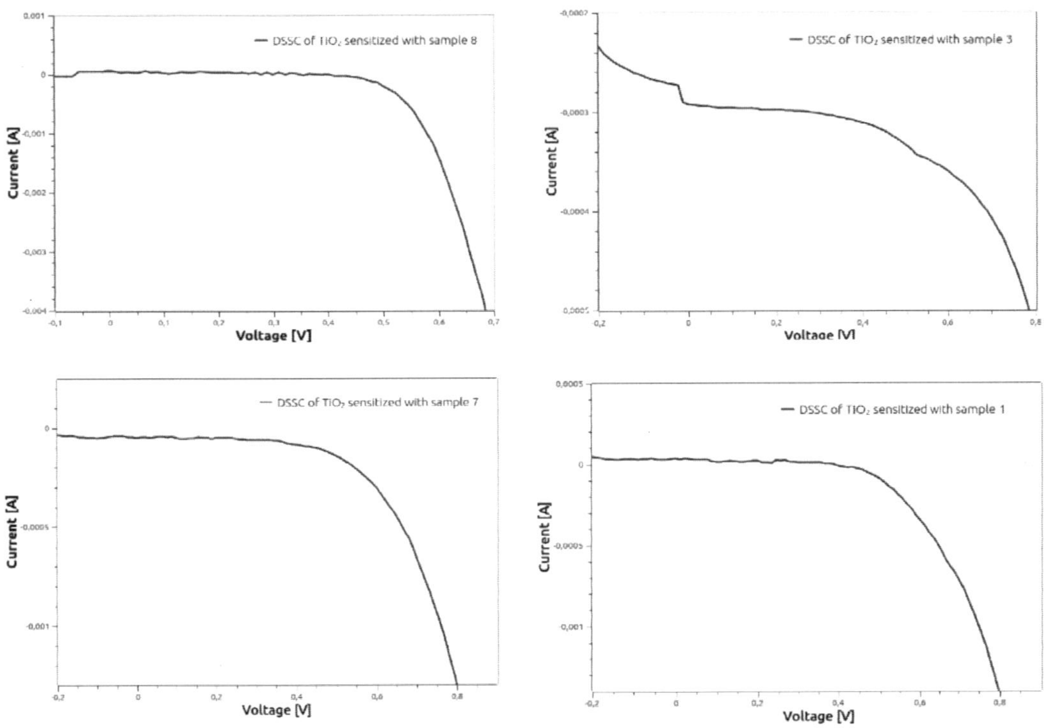

Figure 6. I-V curves of dye sensitized solar cells.

Table 2. I-V characteristics of solar cells measured by PET.

TiO_2 sensitized layer	V_{max} [V]	I_{max} [mA]	P_{max} [mW]	Efficiency [%]
Sample 3	0.27	0.270	0.072	0.072
Sample 8	0.11	0.619	0.069	0.017
Sample 1	0.26	0.048	0.012	0.012
Sample 7	0.23	0.048	0.011	0.011

The current-voltage characteristics and efficiency of each cell shown in fig. 6, were measured using a solar cell simulator PET (Photo Emission Tech, Inc., USA, Model CT80AAA) incorporated with a 1000 W/cm^2 Xenon arc lamp. The sweep voltage was from -0.3 to 0.8V over an effective area of ($1cm^2$) of the cell. Measurements show efficiencies ranging from 0.011 to 0.072% where lowest efficiency obtained 0.011 %, was for the cell corresponding to the sample composed of TiO_2 layer sensitized with sample 7 and highest efficiency 0.072 %, for the cell composed of TiO_2 layer sensitized with sample 3. Table 2 shows some main electrical parameters obtained for each cell in terms of current-voltage behavior.

4. Conclusion

Pigments were extracted from spinach and purple Cabbage using methanol and deionized water. When using methanol, the extraction was done at room temperature, whereas when using deionized water the extraction was done at 70°C and 80°C. Extracts done using methanol as the solvents showed better absorption of visible photon. The extracted dyes were used to sensitize the TiO_2 layer, which was used as the photoanode to assemble the DSSC. These TiO_2 layers were prepared using doctor blade method over an FTO coated glass substrate. The counter electrodes were prepared by platinizing the FTO coated glass substrate. After injecting an electrolyte inbetween the sandwiched electrodes, it formed a DSSC. The cells were further studied for their efficiencies under standard conditions. A composite extract of spinach and purple cabbage in methanol showed the maximum efficiency of 0.72 %.

5. Acknowledgement
The authors would like to thank FONDECYT Postdoctoral program N 3130370; FONDECYT regular N 1130802 CONICYT/FONDAP/15110019, Government of Chile for financial support.

6. References

[1] H. Chang, H.M. Wu, T.L. Chen, K.D. Huang, C.S. Jwo, Y.J. Lo, Dye-sensitized solar cell using natural dyes extracted from spinach and ipomoea, *Journal of Alloys and Compounds*, Volume 495, Issue 2, 16 April 2010, Pages 606-610.

[2] Daoyong Yu, Guoliang Zhu, Shuang Liu, Baosheng Ge, Fang Huang, Photocurrent activity of light-harvesting complex II isolated from spinach and its pigments in dye-sensitized TiO2 solar cell, *International Journal of Hydrogen Energy*, Volume 38, Issue 36, 13 December 2013, Pages 16740-16748.

[3] Mann, George (1952), Print: A Manual for Librarians and Students Describing in Detail the History, Methods, and Applications of Printing and Paper Making, London: *Grafton*, p. 62, OCLC 3331032.

[4] C. Klein, Md. K. Nazeeruddin, D. Di Censo, P. Liska, and Michael Grtzel, Amphiphilic Ruthenium Sensitizers and Their Applications in Dye-Sensitized Solar Cells, Laboratory for Photonics and Interfaces, Institute of Chemical Sciences and Engineering, Swiss Federal Institute of Technology, CH-1015 Lausanne, Switzerland.

[5] http://www.acuariodearrecife.com/parametros/iluminacion-2.htm

[6] Bsaca Loya Guadalupe Adriana, Tesis en proceso, Caracterización Físico-Química de Ficoeritrina en Interfases, Tesis de Doctorado, Doctorado en Polímeros y Materiales, Universidad de Sonora.

Effect of TiO_2 blocking layer on TiO_2 nanorod arrays based dye sensitized solar cells

R Sivakumar and M Paulraj

Department of Physics, Faculty of Physical Sciences and Mathematics, University of Concepcion, P.O. Box. 160-C, Concepcion, Chile

Email: paulrajm@gmail.com, sivakrk05@yahoo.com

Abstract. Highly ordered rutile titanium dioxide nanorod (TNR) arrays (1.2 to 6.2 μm thickness) were grown on TiO_2 blocking layer chemically deposited on fluorine doped tin oxide (FTO) substrate and were used as photo-electrodes to fabricate dye sensitized solar cells (DSSC's). Homogeneous layer of TiO_2 on FTO was achieved by using aqueous peroxo-titanium complex (PTC) solutions via chemical bath deposition. Structural and morphological properties of the prepared samples were investigated using X-ray diffraction (XRD), scanning electron microscopy (SEM) measurements. TNR arrays (6.2 μm) with TiO_2 blocking layer showed higher energy conversion efficiency (1.46%) than that without TiO_2 blocking layer. The reason can be ascertained to the suppression of electron-hole recombination at the semiconductor/electrolyte interface by the effect of TiO_2 blocking layer.

1. Introduction

In DSSC's, the TiO_2 photo-electrode essentially collects and transports the photo-generated electrons [1, 2]. However, the electron transport in nanoparticle based device is suffered by trap limited diffusion process, a slow recombination step that results in low efficiency. Also the large grain boundaries of TiO_2 nanoparticles lead to fast recombination of the photo-generated electrons with I_3^- in the electrolyte prior to their collection at the electrode. Hence, to provide uninterrupted transport of photo-generated electrons in TiO_2, one dimensional (1D) nanostructure such as nanorods [3], nanowires [4], and nanotubes [5] etc. have been designed and demonstrated for their potential application in DSSC's. It has been proved that the electron transport in 1D nanostructure is expected to be several orders of magnitude faster than that of nanoparticle network [6, 7]. Several researchers fabricated thick 1D nanostructure arrays and achieved higher energy conversion efficiency [8, 9]. Another approach to improve the charge transport properties and collection efficiency is by introducing TiO_2 seed layer on the FTO layer prior to the growth of 1D nanostructure [10-15]. Recently, Qian et al. [16] and Yang et al. [17] derived TiO_2 blocking layers from peroxo-titanium solution and studied their effect on dye sensitized solar cells. Increase in photocurrent by the introduction of PTS overlayer was mainly attributed to the retardation of the back-transport reaction in the FTO/TiO_2 interfaces. In a systematic approach, hereby we made an endeavour to introduce TiO_2 blocking layer through per-oxo titanium method on TNR arrays based DSSCs.

Journal of Physics: Conference Series **720** (2016) 012036 doi:10.1088/1742-6596/720/1/012036

2. Materials and methods

2.1. Materials

Fluorine doped tin oxide (FTO) glass plates (sheet resistance 7Ω/□), Chloroplatinic acid hexahydrate (H_2PtCl_6. $6H_2O$; ≥ 37.50% Pt basis), Lithium Iodide (99.9% trace metals basis), Iodine (≥ 99.99% trace metals basis), 4-tert-Butyl pyridine (96%) were purchased from Sigma Aldrich. Di-tetrabutylammonium cis-bis(isothiocyanato)bis(2,2′-bipyridyl-4,4′-dicarboxylato) ruthenium (II) (Ruthenizer 535-bisTBA), 1-methyl-3-propylimidazolium iodide and Surlyn polymer frame (Meltonix 1170-25) were purchased from Solaronix. Titanium (IV) chloride ($TiCl_4$) was purchased from Merck. All other chemicals used were of analytical grade, purchased from Merck and used without further purification.

2.2. Electrode preparation

2.2.1. Preparation of TiO_2 blocking layer on the FTO substrate. Titanium (IV) chloride was chosen as titanium resource for the preparation of stable transparent peroxo-titanium complex (PTC) solution. Briefly, 0.5 mL of $TiCl_4$ was dissolved in 50 mL of deionized water under the conditions of 0-3 °C and then 2 mL of H_2O_2 was added. The mixture turned to orange red with the addition of H_2O_2 and then to clear red after stirring for one hour at room temperature. Accordingly, $TiCl_4$ reacts with H_2O_2 to form stable PTC solution. Deposition of TiO_2 layer was carried out by vertical immersion of the FTO glass substrate in the PTC solution for 12 hour with constant stirring at room temperature (25 °C). After the deposition time, the substrate was withdrawn from PTC solution and rinsed with distilled water, dried and annealed at 450 °C for 30 minutes. On calcination, the deposited per-oxo titanium complex forms uniform layer of TiO_2 on FTO substrate and it is designated as FTO_{PTC}.

2.2.2. Fabrication of TNR arrays. TiO_2 nanorod arrays were grown on FTO_{PTC} substrate using Teflon-lined stainless steel autoclave (45 mL volume, Parr Instrument Co.) following the hydrothermal method [3]. Briefly, 0.30 mL of $TiCl_4$ was added to a mixture solution containing 5 mL of deionized water and 5 mL of hydrochloric acid (37 wt%) in a teflon container and was stirred for 5 minutes for homogenization. FTO_{PTC} glass plates were placed at an angle against the wall of the Teflon-liner with the conducting side facing downwards and the teflon liner was placed in a stainless steel autoclave to initiate the hydrothermal reaction. The reaction was carried out at 150 °C in an electric oven for 4, 8, 12, 20, 30 and 40 hours. For comparison TNR arrays were grown on bare FTO glass plates for 20 hours at 150 °C. After the set reaction period, the samples were taken out and rinsed extensively with deionized water and left to dry in ambient air. Finally, all the samples were annealed at 450 °C for 30 minutes.

2.3. Analytical measurements

X-ray diffraction patterns were recorded using a Bruker D4 Endeavor advance diffractometer with Cu Kα radiation at a scanning rate of 2°/min. Morphology and microstructure analysis of the samples were performed using a JEOL JSM-6380LV scanning electron microscope (SEM). The samples under study were first sputtered with gold for 3 min. to obtain a layer of 150 A° thick using an Edwards S 150 Sputter Coater.

2.4. Fabrication of dye sensitized solar cells

TNR array electrodes were dipped in dye solution (3×10^{-4} M) in ethanol for 24 h. After the dye adsorption was complete, the electrode were withdrawn from the solution and dried under a stream of argon. The platinum counter electrode and the dye-coated TiO_2 electrode was then put together with a thin transparent film of Surlyn polymer frame (25 μm). The sandwiched electrodes were tightly held and then heated to 130 °C, so that the Surlyn frame would seal the two electrodes intact. A thin layer of electrolyte, consisting 0.6 M propyl methylimidazolium iodide (PMII), 0.05 M I_2, 0.1 M LiI, and 0.5 M tert-butylpyridine in 1:1 acetonitrile/valeronitrile, was introduced into the inter-electrode space from the counter electrode side through predrilled holes. The drilled holes were sealed with a microscope cover slide to avoid leakage of the electrolyte solution.

2.5. Photovoltaic Measurements

Solar simulator device (Model: CT80AAA, PET Photo Emission Tech., Inc., USA) incorporated with a 300 W xenon arc light source was used for measuring the photovoltaic properties. A 1 sun, AM 1.5G, light irradiance of 100 mW/cm^2 was achieved over the surface of the test cell. During each measurement, temperature of the platen which holds the cell was controlled at 25°C using a control system attached to the solar simulator. The power of simulated light was calibrated using a reference silicon photodiode (20mm x 20mm) provided with the instrument. The test cell was placed at the same position as the calibration cell. Photocurrent-voltage plots for all cells were recorded using a Keithley 2400 digital source meter, from +950 to −300 mV, in steps of 10 mV.

3. Results and discussion

Figure 1 a and b shows the scanning electron microscopic images of bare FTO and FTO_{PTC}. A homogeneous layer of TiO_2 was observed for the TiO_2 blocking layer deposited over bare FTO (Fig. 1b). Small grains with dense structure and well surface coverage are observed for chemically deposited TiO_2 blocking layer. Accordingly, this blocking layer consisting of anatase TiO_2 nanocrystal formed upon high temperature treatment of per-oxo titanium complex [18]. Apparently, TNR arrays grown on FTO_{PTC} (Fig. 2a) and bare FTO (Fig. 2c) reveal uniform surface coverage and the shape of the nanorod is tetragonal with square top facets as expected for rutile TiO_2 structure. From cross sectional SEM images, the thickness of TNR arrays grown on FTO_{PTC} were found to be 1.2 μm, 2.5 μm, 4.2 μm and 6.2 μm for the hydrothermal duration set to 4, 8, 12 and 20 h respectively. For comparison, TNR arrays grown on bare FTO results in similar length, 6.1 μm (Fig. 2d). On prolonging the hydrothermal reaction to 30 and 40 h, no increase in nanorod length was observed in FTO_{PTC} and bare FTO.

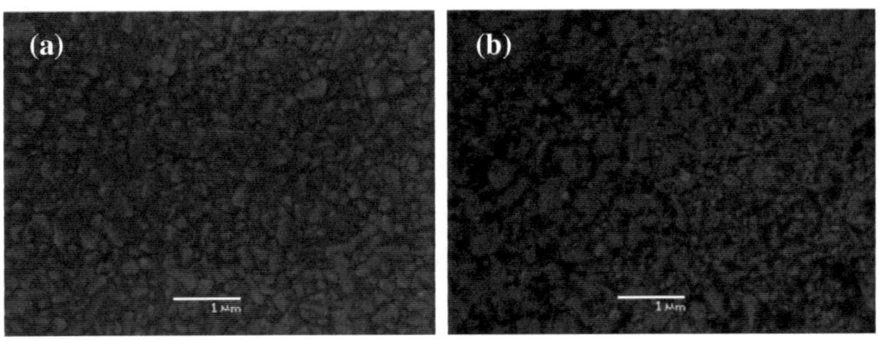

Figure 1. Scanning electron microscopic image of (a) Bare FTO and (b) FTO_{PTC}.

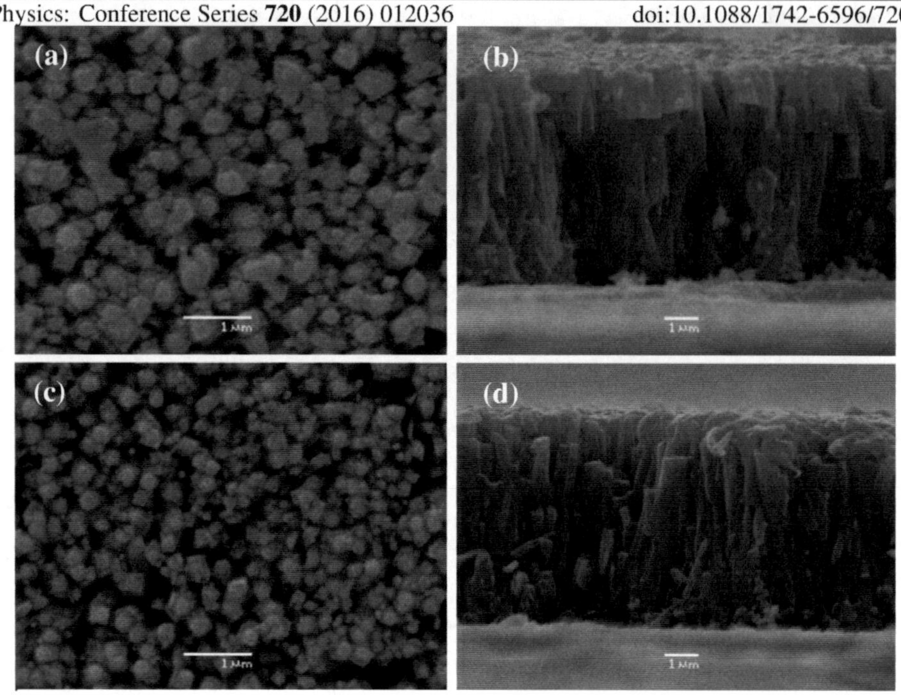

Figure 2. SEM images of TiO_2 nanorods grown on FTO_{PTC} and bare FTO. (a) and (c) are the top view of TNR arrays grown on FTO_{PTC} and FTO, respectively at 150 °C for 20 hr; (b) and (d) are the corresponding cross sectional view.

Figure 3 shows the X-ray diffraction patterns of bare FTO and FTO_{PTC} and TNR arrays grown on FTO_{PTC} and bare FTO. Diffraction peaks of bare FTO (Fig. 3a) correspond to tetragonal rutile phase of SnO_2. The diffraction peaks of TiO_2 crystallites were not observed for TiO_2 blocking layer deposited FTO (Fig. 3b) due to the strong diffraction peaks of FTO. However, after the hydrothermal growth of TNR arrays on FTO_{PTC} and bare FTO, peaks corresponding to tetragonal TiO_2 with rutile phase were observed (Fig. 5c-g). The intensity of the peak (002) increases as the hydrothermal duration increases. Significantly, the enhancement of diffraction peak (002) is a clear indication of directed growth of TNR arrays perpendicular to the FTO substrate. Absence of diffraction peaks at (110), (111), and (211) is an evidence for uni-directional growth of TNR arrays with respect to the substrate surface.

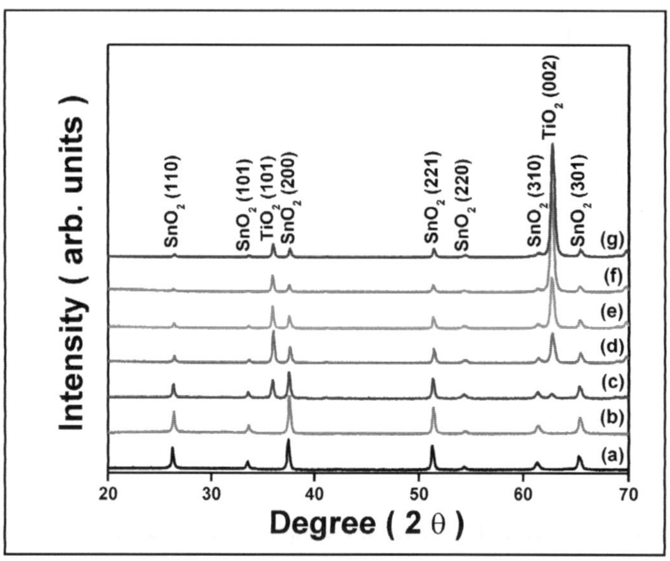

Figure 3. X-ray diffraction patterns of (a) bare FTO and (b) FTO_{PTC}; (c), (d), (e) and (f) are diffraction patterns of TNR arrays grown on FTO_{PTC} for 4 h, 8 h, 12 h and 20 h at 150 °C respectively; (g) TNR arrays grown on bare FTO for 20 hr at 150 °C.

Figure 4 shows the photocurrent-voltage characteristics of DSSC fabricated using TNR array electrodes grown over FTO_{PTC} and bare FTO. The cell parameters such as, short circuit photocurrent density (J_{SC}), open-circuit voltage (V_{OC}), fill factor (FF), and the power conversion efficiency (η) are summarized in Table 1. It was observed that the efficiency increases as the thickness of the TNR arrays increases. DSSC assembled with 6.2 μm thick TNR arrays grown on FTO_{PTC} exhibits a photocurrent of 3.56 mA/cm^2, open circuit voltage of 0.79 V and overall energy conversion efficiency of 1.46%. Whereas, TNR arrays (6.1 μm) grown on bare FTO showed marginally low efficiency (1.19%). It is noteworthy that the DSSCs assembled from TNR arrays with TiO_2 blocking layer show higher V_{OC} (0.79 V) than that without TiO_2 blocking layer (0.74 V). This indicates that TiO_2 blocking layer plays a vital part on the performance of rutile TiO_2 nanorod based DSSC's. This is because, TiO_2 blocking layer covering the vacant spaces unoccupied by TNR arrays on FTO are effective in blocking the direct contact of the redox electrolyte and FTO and diminishes the dark current that arises from reduction of I_3^- at the FTO/electrolyte interface. Also the increase in J_{SC} proves the lower recombination rate of the whole photo anode in the presence of TiO_2 blocking layer. We note that such a per-oxo titanium route presented here offers a significant advantage for the DSSCs.

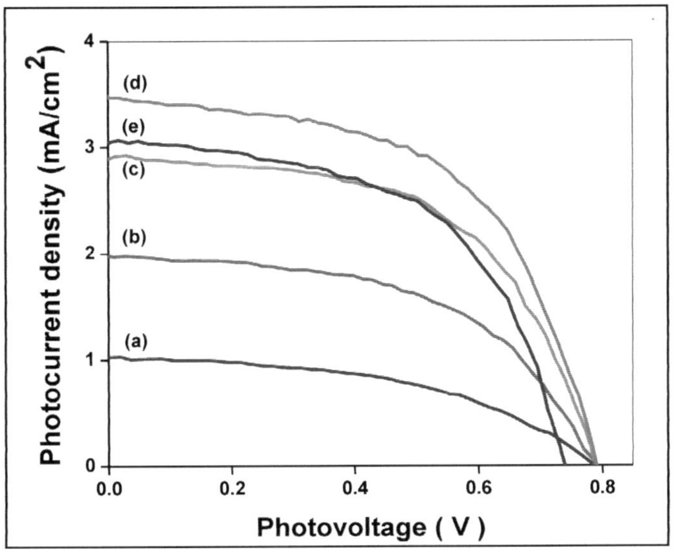

Figure 4. Photocurrent density (J) - voltage (V) characteristics of DSSC's assembled with TNR arrays (a) 1.2 μm, (b) 2.5 μm, (c) 4.2 μm and (d) 6.2 μm grown on FTO_{PTC} respectively, and (e) 6.1 μm on bare FTO.

Table 1. Photovoltaic performance parameter of DSSC's assembled with TNR arrays grown on FTO_{PTC} and bare FTO.

Photoelectrode	J_{SC} (mA)	V_{OC} (V)	FF	η (%)
TNR(1.2 μm)/FTO_{PTC}	1.02	0.786	0.47	0.38
TNR(2.5 μm)/FTO_{PTC}	1.98	0.787	0.49	0.76
TNR(4.2 μm)/FTO_{PTC}	2.92	0.789	0.52	1.20
TNR(6.2 μm)/FTO_{PTC}	3.56	0.791	0.52	1.46
TNR(6.1 μm)/FTO	3.04	0.739	0.53	1.19

4. Conclusions

TiO_2 nanorod arrays were grown on TiO_2 blocking layer chemically deposited on FTO substrate from per-oxo titanium complex solution. The peroxo-titanium route provides oxygen rich environment to enhance the Ti-O-Ti linkage which avoids the atomic incorporation of chlorine. DSSC assembled using TNR arrays (6 μm) in the presence of TiO_2 blocking layer produces 1.46% efficiency which is higher than that assembled in the absence of TiO_2 blocking layer (1.19 %). This enhancement was attributed to the effect of TiO_2 blocking layer on suppressing the dark current that arises from the direct contact of electrolyte with the FTO. These findings provide a new route to preparing highly pure TiO_2 nanocrystals on FTO and further extend their application into solid state DSSCs.

Acknowledgement
RS would like to thank FONDECYT Postdoctoral program N° 3130370; MP for FONDECYT regular N° 1130802 & CONICYT/FONDAP/15110019, Government of Chile for financial support.

References

[1] Bai Y, Sdqή I M, Angelis F D, Bisquert J and Wang P 2014 *Chem. Rev.* **114** 10095
[2] Lee K, Mazare A and Schmuki P 2014 *Chem. Rev.* **114** 9385
[3] Liu B and Aydil E S 2009 *J. Am. Chem. Soc.***131** 3985
[4] Feng X, Shankar K, Varghese O K, Paulose M, Latempa T J and Grimes C A 2008 *Nano Lett.* **8** 3781
[5] Xu C, Shin P H, Cao L, Wu J and Gao D 2010 *Chem. Mater.* **22** 143
[6] Hendry E, Koeberg M, O' Regan B and Bonn M 2006 *Nano Lett.* **6** 755
[7] Law M, Greene L E, Johnson J C, Saykally R J and Yang P D 2005 *Nat. Mater.* **4** 455
[8] Lv M, Zheng D, Ye M, Xiao J, Guo W, Lai Y, Sun L, Lin C and Zuo J 2013 *Energy Environ. Sci.* **6** 1615
[9] Roh D K, Chi W S, Jeon H, Kim S J and Kim J H 2014 *Adv. Funct. Mater.* **24** 379
[10] Wang H E, Chen Z, Leung Y H, Luan C, Liu C, Tang Y, Yan C, Zhang W, Zapien J A, Bello I and Lee S T 2010 *Appl. Phys. Lett.* **96** 263104
[11] Wei Z, Yao Y, Huang T and Yu A 2011 *Int. J. Electrochem. Sci.* **6** 1871
[12] Sun P, Zhang X, Liu X, Wang L, Wang C, Yang J and Liu Y 2012 *J. Mater. Chem.* **22** 6389
[13] Wang J, Zhang T, Wang D, Pan R, Wang Q and Xia H 2013 *J. Alloy. Compd.* **551** 82
[14] Liao M. Y, Fang L, Xu C L, Wu F, Huang Q L and Saleem M 2014 *Mat. Sci. Semicon. Proc.* **24** 1
[15] Jyoti D and Mohan D 2014 *Prog. Nanotech. Nanomat.* **3** 57
[16] Qian D, Li Y, Zhang Q, Shi G and Wang H 2011 *J. Alloy. Compd.* **509** 10121

[17] Yang S, Hou Y, Zhang B, Yang X H, Fang W Q, Zhao H J and Yang H G 2013 *J. Mater. Chem. A* **1** 1374
[18] Sankapal B R, Sartale S D, Steiner M C L and Ennaoui A 2009 *C. R. Chimie* **9** 702

Dynamic Characteristics of X-pinch Experiments Conducted in a Small Capacitive Generator: Refractive Optical Observations.

Adolfo Sepúlveda[2,3], Cristian Pavez[1,2,*], José Pedreros[2], Gonzalo Avaria[1,2], Patricio San Martín[1,2], and Leopoldo Soto[1,2]

[1]Comisión Chilena de Energía Nuclear – Chile

[2]Center for Research and Applications in Plasma Physics and Pulsed Power – P4 Chile

[3]Facultad de Ingeniería, Universidad Mayor, Santiago, Chile

**Email*: cpavez@cchen.cl

Abstract. Among the dense plasmas configurations of interest for applications as a portable intense source of X-rays, the X-pinches are the most attractive by their brightness, source size, short duration and space localization, being particularly reproducible when they are conducted with fast pulsed power generators. In recent time, several characteristics of the dynamics and emission have been reproduced in compact generators (typically capacitive generators) of low current rise-rate (less than 0.5 kA/ns). In this work, a preliminary characterization of the dynamic of X-pinch plasma conducted in a small capacitive generator is reported. In order to obtain the plasma dynamics and quantitative information of the plasma density, the dark field Schlieren technique and interferometry were implemented. The experiments were carried out on the multipurpose generator (1.2 μF, 345 J, 47.5 nH, T/4=375 ns and Z=0.2 Ω in short circuit) capable to produce currents up to 122 kA with 500 ns quarter period, when a charging voltage of 24 kV and metallic X-pinches are used as load. The electrical behavior of the discharge and the X-ray emission are monitored with a Rogowski coil and filtered PIN diodes respectively. For the refractive optical diagnostics a 532 nm frequency-doubled Nd-YAG laser was used. As from a single Schlieren record per shot, a sequence with the time evolution of the plasma is constructed. From the images, a similar dynamic of X-pinches conducted in fast generators of high current is observed, where structures such as coronal plasma, plasma flares and plasma jets are identified. The plasma dynamics observed from a VUV gated pinhole image system is compared with registered dynamic with refractive optical techniques.

Journal of Physics: Conference Series **720** (2016) 012037 doi:10.1088/1742-6596/720/1/012037

1. Introduction

Several dense transient plasmas configurations have been explored to produce small size (μm) and short durations (ns) X-ray sources: plasma focus [1], vacuum spark [2, 3], X-pinches [4] , among others [5]. In particular in the X-pinch configuration, two (or more) thin metallic wires are arranged between two faced electrodes forming a 'X' shaped configuration with only one contact point at the crossing of the wires. This kind of configuration has captured the attention of different research groups because of its brightness, source size, short duration and space localization, being particularly reproducible when fast pulsed power generators are used [5]. The application of a current pulse across the wires results in the ablation of the wire surface by Joule effect creating a hot and low-density coronal plasma surrounding a cold and dense wire core. The current bulk flows by the coronal plasma where the plasma experiences a local self-induced magnetic field, which serves to locally confine and pinch the plasma toward the axes of the individual wires. Moreover, as a result of the current flow along leg of the X-pinch, a distribution of global magnetic field in all space is produced, which presents a greater intensity in the region close to the cross point of the wires. Thus the X-pinch dynamics is governed by the characteristics of the precursor plasma and its interaction with the global magnetic field. The strong global magnetic field at the cross point of the wires compresses the plasma on axis, forming a hot and dense micro-pinch. Under these conditions, the plasma yields an intense and small soft X-ray source of short duration ideal for studies and applications, as for instance, for applications to high-resolution radiography of biological specimens [6]. The X-pinch hot spot has been widely studied, in contrary to the low-density coronal plasma research which has not been in the priorities of the scientific community, particularly in the low current generators. Characterization of the coronal plasma has been performed with complementary techniques such as interferometry and time-resolved self-emission imaging or dark-field Schlieren [7, 8].
In this direction and considering the encouraging results reported in previous work using our compact capacitive generator [9], we report preliminary characterization of the plasma dynamics using refractive optical diagnostics, which are contrasted with those obtained with time-gated self emission pinhole images in the VUV range.

2. Experimental set-up

The X-pinch experiments shown here were conducted with a multipurpose generator, whose electrical characteristics are reported in reference [9, 10]. The generator is capable to deliver currents about 122 kA, with quarter period of 500 ns when a 24 kV charging voltage and metallic X-pinches load is used. Two different materials were diagnosed, aluminum wire of 25 μm for the self emission pinhole images and tungsten wire of 13 μm for the refractive optical diagnostics.
Three types of X-pinch assemblies were tested, which were determined by the electrode configurations available, as experience is gained in the assembling technique. In Figure 1(a) and Figure 1(b) the X-array is formed with two wires hanging in parallel connecting the electrodes by the edge of the apertures. Then, a conical metallic stopper is introduced in the lower aperture (anode) fixing and pressuring the wires against the electrode. At the first scheme, the upper free ends of the wires are twisted in 360° while the second one in 180°, with only one touch point. In Figure 1(c) the X-array is formed introducing both ends of a single wire through holes available in the upper and lower electrodes. After that, the upper electrode is rotated a little bit more than 180° forming an X. In the three assemblies, the inter-electrodes spacing is 10 mm.
In order to map the plasma structures linked to dynamics of the coronal plasma, namely; instabilities, precursor plasma, plasma flares and subsequent plasma jet formation, the dark-field Schlieren technique and interferometry were implemented. For the Schlieren technique a circular stopper of 0.7

mm was placed at the focal point of a lens of 300 mm of focal length, allowing a minimum detectable angular deflection of 0.001 radians. To the characterization of the plasma density, a Mach-Zehnder interferometer was implemented, in which only a section of the X-pinch (almost complete) was imaged on the CMOS detector in order to obtain a better spatial resolution. For this purpose a lens of focal length f=250 mm was used, which led to magnification of m=2.1.

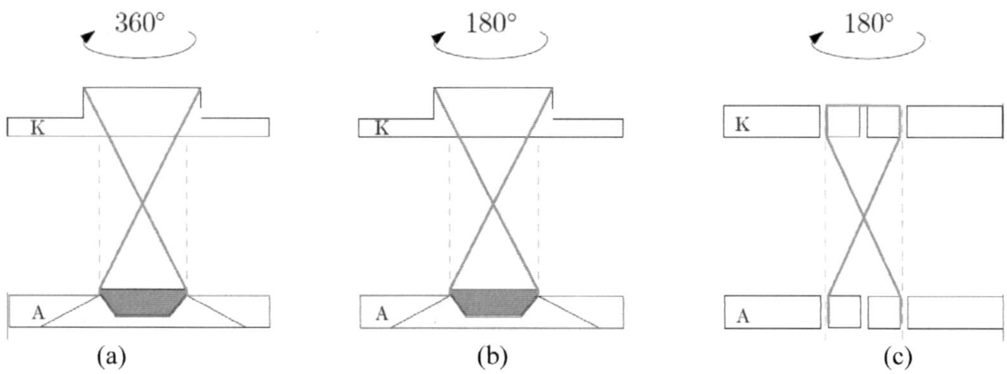

Figure 1: Schematic representation of X-array assembly: (a) Two parallel wires rotated 360°, (b) Two parallel wires rotated 180°, and (c) Only one wire rotated 180°.

In a first stage, only dark-field Schlieren technique for twisted X-pinch (rotated at 360°) was implemented. In a second stage, where one of the electrodes is rotated 180° to form the X-array, both diagnostics, Schlieren and interferometry were available simultaneously. In all refractive optical diagnostics a 532 nm frequency-doubled Nd-YAG laser was used. In Figure 2, a schematic of the experimental set-up is shown, including the additional diagnostics such as filtered PIN diode and filtered PCD detector. For the Schlieren diagnostic only (without interferometry), the experimental setup is similar, but in this case the reference beam is blocked.
Additionally, a gated imaging system based on a four-stripe MCP camera coupled to a vertical pinhole array of 50 μm/each was implemented. The pinhole array was assembled 26 cm away from the plasma and 14.8 cm in front of the MCP detector, resulting in a plasma image of magnification 0.58 over each stripe.

3. Experimental results

Figure 3 shows a sequence of dark-field Schlieren images which was built with single records per shot, for Tungsten wires of 13 μm and charge voltage of 24 kV. From Fig. 3(a) to Fig. 3(d) the X-pinches were formed with the rotation of the two parallel wires in 360°, as shown in Figure 1(a). From the image sequence is easily recognizable a coronal plasma around each wire, which is present from an early stage of the discharge along the entire wire. Furthermore, we observed plasma flares that migrate from the coronal plasma close to the cross point region which feed a central plasma column. It should be noted that this central plasma column is longer than the distance between the cross-point of the wires and the intersection point of last observable plasma jet (from the Schlieren technique) with the symmetry axis.

This could lead one to believe that only a portion of coronal plasma is ejected toward the axis and feeds the plasma column. Nevertheless, from the pinhole image sequence in the VUV range (Figure 4), it is possible observe how the coronal plasma, according to how it is coming to the symmetry axis, is shaping the upper end of jet like plasma structure. On the other hand, the plasma characteristics that migrates from the furthest regions of the X-pinch, where the global magnetic field is lower, presents density gradients that are not solved for the Schlieren optical system.
In Fig. 3(e), a plasma column gap of approximately 1.2 mm in the micro-pinch region is observed. This gap is responsible for the appearance of the diode effect, which produces a strong electric field that accelerates the electrons against the virtual anode, producing an intense hard x-ray pulse.

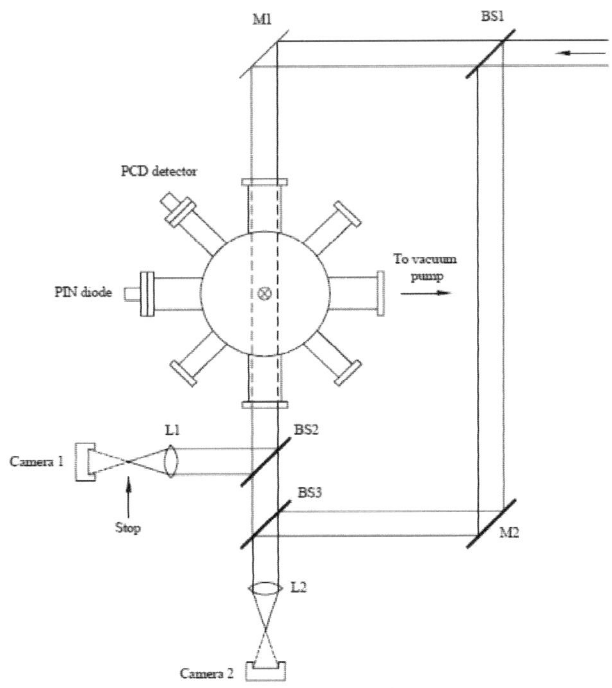

Figure 2: Experimental set-up for dark-field Schlieren, interferometry and complementary diagnostics.

Journal of Physics: Conference Series **720** (2016) 012037 doi:10.1088/1742-6596/720/1/012037

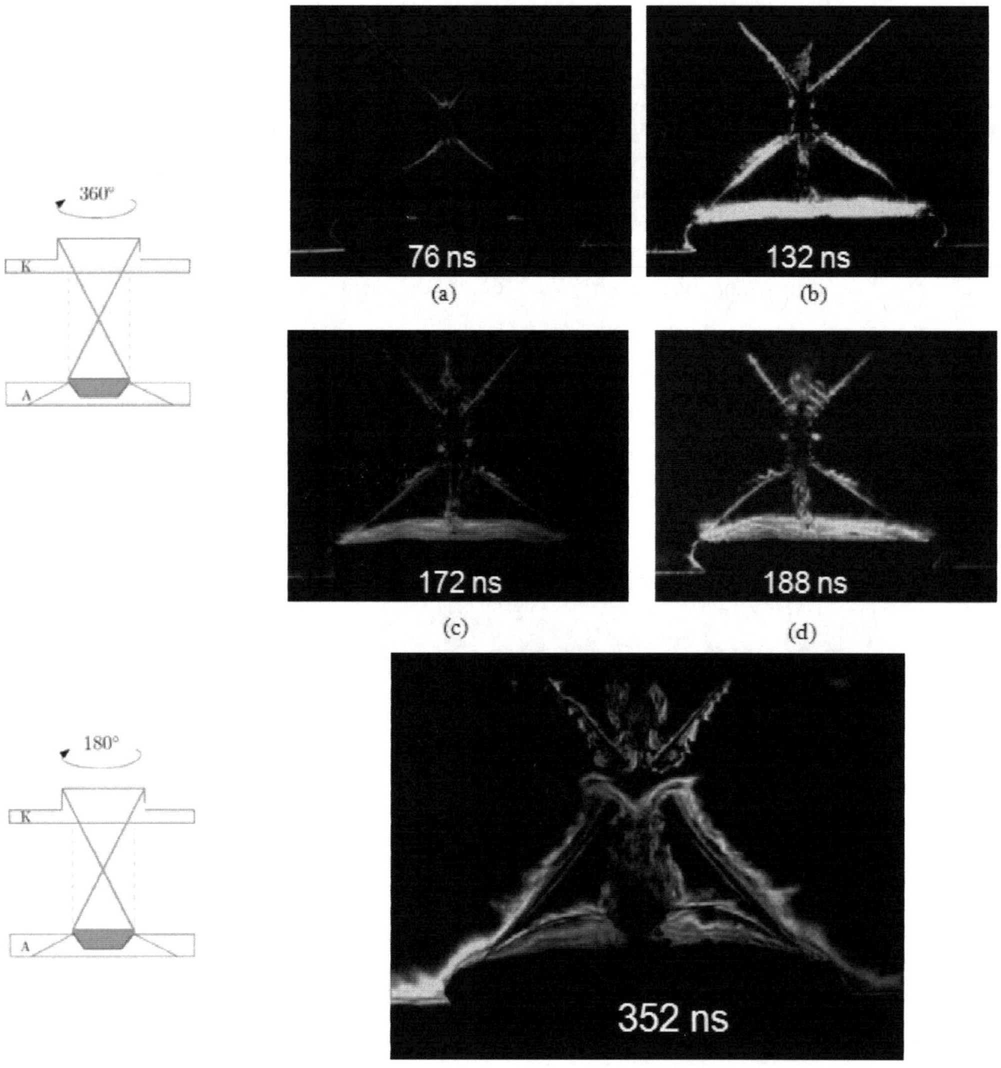

Figure 3: (a-d) Schlieren sequence of 13 μm Tungsten X-pinches for the wires assembly of fig 1(a); (e) Schlieren recording for 13 μm Tungsten X-pinches using the wires assembly of fig 1(b).

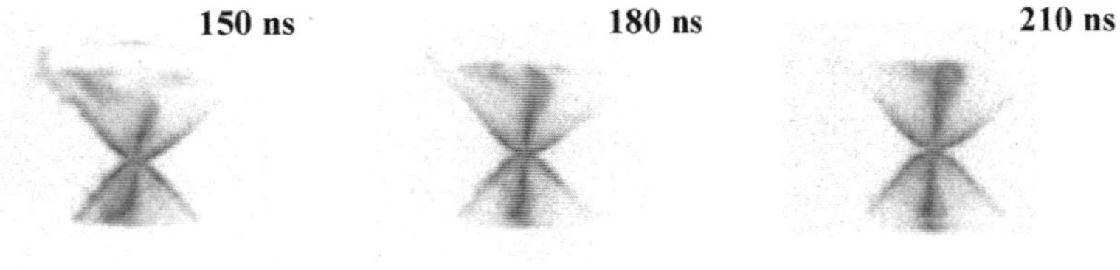

Figure 4: Pinhole image sequence (for the same shot) in the VUV spectral range, for a typical shot of Al X-pinch of 25 μm.

The interferometry diagnostic presented at this work was implemented on a 13 μm Tungsten X-pinch. The X-pinch is formed with the assembly of wires shown in Figure 1(c). Figure 5 shows an example of a simultaneous record of a Schlieren and plasma interferogram in an early stage of the discharge. In order to characterize and improve the spatial resolution at the region around of the cross point of the wires, an interferometric optical system with higher magnification was implemented (m=2.1). The time is taken with respect to the discharge onset time.
It is important to point out that in this experimental stage, the laser stability was affected by the electromagnetic noise, which resulted in the record of low-contrast interferograms due to the intensity variations of the laser. For this reason and in order to use the obtained information, we decided to apply techniques of digital image processing on the original interferogram. Frequency filtering techniques to isolate the carrier frequency were used, with which a synthetic interferogram is constructed. Subsequently this interferogram is decoded to obtain both the interferometric phase and the areal density. The entire process is shown in the picture sequence of Figure 6.

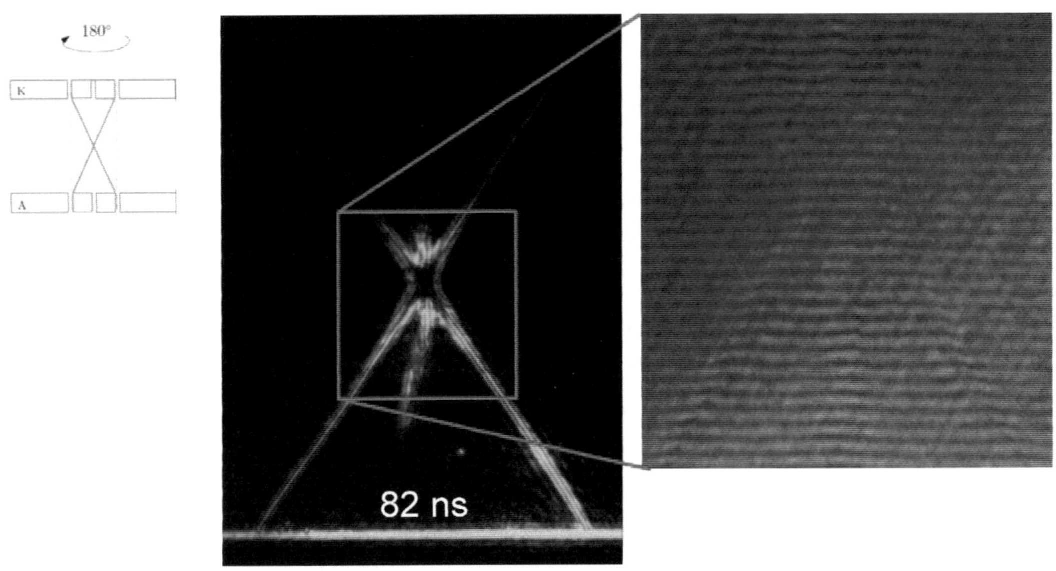

Figure 5: Dark-field Schlieren and interferometry simultaneously of 13 μm Tungsten X-pinch.

Journal of Physics: Conference Series **720** (2016) 012037 doi:10.1088/1742-6596/720/1/012037

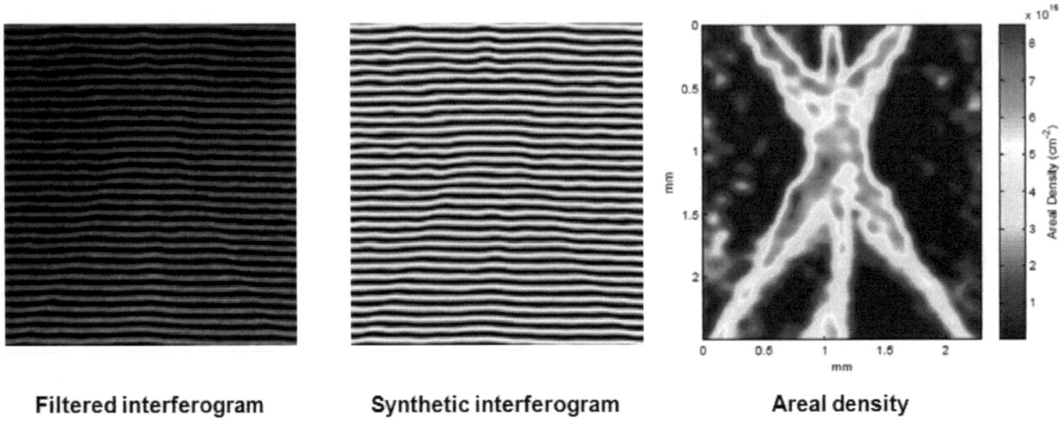

Figure 6: Digital processing for interferogram of 13 μm Tungsten X-pinch.

From the decoded information it is possible recognize consistencies between the information recorded both the Schlieren and at the interferogram, in particular those that account for the existence of a low-density coronal plasma around of wires and a jet-like plasma at middle plane. However, the interferogram does not show an opaque structure corresponding to the dense core at the centre of the each wire as other authors have reported in studies of this kind [5, 7, 8]. This effect could be due to the beam diffraction at the thin and dense plasma structure, as it is observed between the X-pinch arms. It should be pointed out that in dark-field Schlieren records of the X array (without plasma) before the generator triggering, the diffraction effects were not observed. For a deep characterization of the plasma density in X-pinch experiments conducted with the "multipurpose generator", an extensive interferometric study is necessary.

4. Remarks and conclusions

A preliminary set of refractive optical observations, on different X-pinch configurations have been conducted. The plasma dynamics observed from Schlieren diagnostics are compared with the dynamics registered by means of a gated VUV pinhole image system. From the diagnostics set employed and experimental results obtained, an approach to the plasma dynamics on X-pinch experiments carried out in a small capacitive generator of low current can be summarized as following: (i) Experiments with metallic wire conducted on the multipurpose generator showed similar characteristics to those observed in higher currents and high current rate generators ($\geq 10^{12}$ A/s), namely: density modulation along the wire and migration of precursor plasma because of the global magnetic field (similar dynamics). (ii) From the four-frame MCP diagnostic, a plasma jet and a tiny compressed structure with high emission of soft X-ray were observed at the crossing point of the wires. (iii) From the Schlieren sequence, flare-like plasma structures coming from the coronal plasma are observed, particularly from the regions close to the cross point of the wires, where the global magnetic field is higher. At later times, a clear break of the plasma structure at the cross point of the wires is observed, leaving a gap that form virtual electrodes, where an intense electric field is formed.

Further experiments with complementary diagnostics are being designed in order to characterize the dynamics of the cross-point region of the wires, with a special interest in the characteristics of the micro-pinch in the successive bursts of the hard X-ray during the rise current, as is reported in [9].

The authors would like to acknowledge financial support by PIA CONICYT grant ACT 1115 and FONDECYT grant 1151471. G. Avaria acknowledges the financial support from FONDECYT Iniciacion 11121587.

References

[1] C. Pavez, J. Pedreros, M Zambra, F Veloso, J Moreno, A Tarifeño-Saldivia and L. Soto (2012), "Potentiality of a small and fast dense plasma focus as hard x-ray source for radiographic applications", *Plasma Phys. and Control. Fusion* ***54*** 105018.
[2] E. Wyndham, H. Chuaqui, M.Favre, L. Soto and P.Choi (1992), "Beam and Hot-spot Formation in a Low Impedance Line Driven Vacuum Spark Discharge", *Journal of Applied Physics* ***71***, 4164.
[3] H. Chuaqui, M. Favre, R. Saavedra, E. Wyndham, L. Soto, P. Choi, and C. Dumitrescu-Zoita, (1998), "Observations of Vacuum Spark Dynamics from its soft X-Ray Emission", *IEEE. Transactions on Plasma Science* ***26***, 1162.
[4] D. H. Kalantar and D. A. Hammer (1995), "The x-pinch as a point source of x rays for backlighting", *Rev. Sci. Instrum.* ***66***, 779
[5] M. Haines (2011), "A review of the dense Z-pinch", Plasma Phys. Control. Fusion **53**.
[6] B. Moo Song, T. A. Shelkovenko, S. A. Pikuz, M. A. Mitchell, K. M. Chandler, and D. A. Hammer (2004), *IEEE Trans Nuclear Sci* ***51*** 2514-2519.
[7] I. H. Mitchell , R. Aliaga-Rossel, R. Saavedra, H. Chuaqui, M. Favre, and E. S. Wyndham (2000), "Investigation of the plasma jet formation in X-pinch plasmas using laser interferometry", *Phys. Plasmas*, Vol. 7, no. 12, pp. 5140-5147.
[8] S. C. Bott, D. M. Haas, U. Ueda, Y. Eshaq, R. Madden, G. Collins, and F. N. Beg (2008), "High-Resolution laser Schlieren imaging of coronal plasma evolution in 80-kA x-pinches", *IEEE Trans Plasma Scie* **36** 1274-1275.
[9] N. Cabrini, C. Pavez, G. Avaria, P. San Martín, F. Veloso, B. Zuñiga, A. Sepúlveda, and L. Soto, "Wire array experiments in a low impedance and low current generator", *Journal of Physics Conference Series*, 15th Latin American Workshop on Plasma Physics 2014.
[10] A. Tarifeño, C. Pavez, M. Cardenas, L. Soto (2008), *Phys. Scr.* **T131** 014028 (5pp).

Development of plasma needle to be used for biomedical applications

B Bora[1,2*], J Jain[1,3], M J Inestrosa-Izurieta[1,2], G Avaria[1,2], J Moreno[1,2], C Pavez[1,2], K. Marcelain[4,5], R. Armisen[4,5], L Soto[1,2]

[1] Comisión Chilena de Energía Nuclear, Santiago, Chile.
[2] Center for Research and Applications in Plasma Physics and Pulsed Power, P4, Chile
[3] Universidad de Talca, Chile.
[4] Instituto de Ciencias Biomédicas, Facultad de Medicina, Universidad de Chile, Independencia 1027, Santiago, Chile
[5] Centro de Investigación y Tratamiento del Cáncer, Facultad de Medicina, Universidad de Chile, Independencia 1027, Santiago, Chile

E-mail: bbora@cchen.cl

Abstract. Plasma needle is a novel design of a plasma source at atmospheric pressure to achieve a non-thermal plasma jet. The advantage of the plasma needle is that it can be operated in open air, outside a vessel. The plasma that is generated with the plasma needle is small (about one millimetre) and non-thermal, the temperature of the neutral particles and ions is in about room temperature and suitably can interact with living biological cell without damaging the cell. In this work, we report the development of a plasma needle, which is operated by a dc power source and produced a stable plasma jet on water surface. Argon gas is used to operate the plasma needle. The preliminary electrical diagnostics of the plasma needle shows that the discharge is filamentary in nature. For diagnostic of the plasma jet produced by the developed plasma needle, the produced plasma jet is directed to water surface and characterization are carried out by means of electrical discharge characteristics and optical emission spectroscopy. In this work, preliminary results of the diagnostic will be presented.

1. Introduction

The plasma needle is a type of non-thermal atmospheric glow discharge; it has a single-electrode configuration and it is a novel design of a plasma source at atmospheric pressure to achieve a non-thermal plasma jet[1]. The advantage of the plasma needle is that it can be operated in open air, outside a vessel. The plasma that is generated with the plasma needle is small (about one millimetre) and non-thermal, the temperature of the neutral particles and ions is in about room temperature and suitably can interact with living biological cell without damaging the cell. Important properties of this type of plasma are that it operates near room temperature, allows treatment of irregular surfaces and has a small penetration depth. These characteristics give the needle great potential for use in the biomedical field. Several experiments have shown that the plasma needle is capable of bacterial decontamination [2] and of localized cell removal without causing necrosis to the treated or neighbouring cells [3]. Areas of detached cells were made with a diameter of 0.1mm, which indicates that the precision of the treatment can be very high. It is believed that plasma particles, such as radicals and ions, and also emitted UV light, interact with the cell membranes and cell adhesion molecules and therefore cause detachment of the cells [3,4]. In this work, we report the development of a plasma needle, which is operated by a dc power source and produced a stable plasma jet which can interact with living cell. Argon gas is used to operate the plasma needle. The preliminary electrical diagnostics of the plasma needle shows that the discharge is filamentary in nature. For diagnostic of the plasma jet produced by the developed plasma needle, the produced plasma jet is directed to water surface and characterization are carrying out by the means of electrical discharge characteristics and optical emission spectroscopy. In this work, preliminary results of the diagnostic are presented.

2. Experimental setup

The plasma needle consist of a metallic needle covered by several well machined Teflon pieces for flow of argon and electrical connection. A photograph of the plasma needle along with the components are shown in figure 1. The current have been measured by a set of Tektronix current probes (CT1 and TCP 312A along with TCPA 300). The current have been also estimated by measuring the voltage drop across a 45 Ω resistance connected in series with the discharge. The discharge voltages at the electrode is measured by a Tektronix voltage probe (P6015A). The voltage and current data are recorded with the help of Tektronix oscilloscope (MDO 3034).

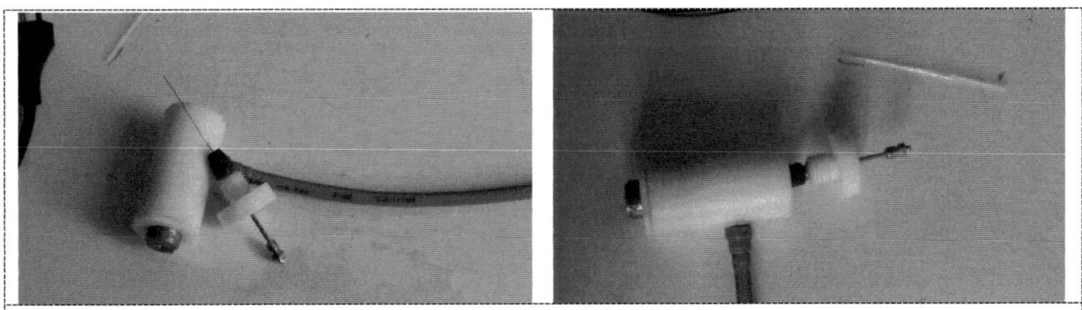

Figure 1.Photograph of the plasma needle along with the other components.

3. Result and discussion

A typical image of the plasma jet interacted with human hand is shown in figure 2. However, for details characterization of the plasma jet, the plasma jet is operated on the water surface to mimic the interaction of the plasma with living cell and stability of the plasma jet. The plasma jet operated on the water surface are shown in figure 3.

Figure 2. Photograph of the plasma jet interacted with human hand.

Figure 3. Photograph of the plasma jet operated on water surface.

Figure 4 shows the electrical discharge characteristics of the plasma needle operated over water surface at two different power supply voltage of 3 kv and 5 kV. From the preliminary electrical characteristics of the plasma needle, it have been observed that the discharge is filamentary in nature. It have been observed that the increased in the voltage of the power supply increases the frequency of the discharges without considerably affecting the discharge voltage and current. A typical optical emission spectra (OES) are shown in figure 5. Several prominent Ar lines are observed in the spectra.

Journal of Physics: Conference Series **720** (2016) 012038 doi:10.1088/1742-6596/720/1/012038

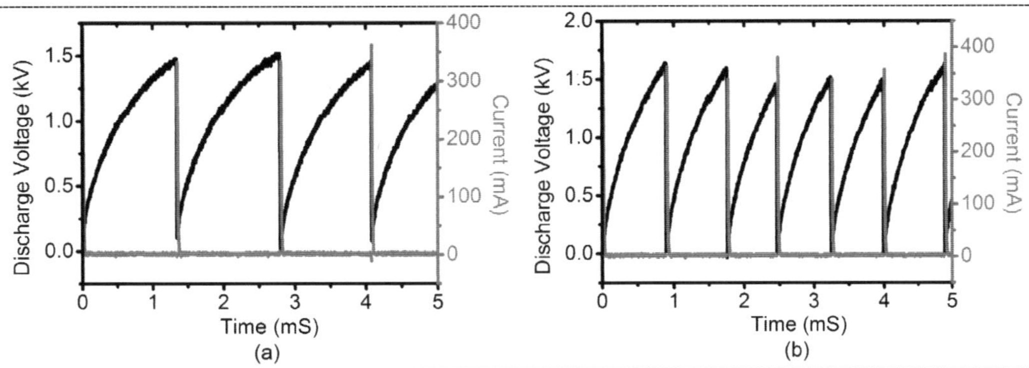

Figure 4. Electrical discharge characteristics of the plasma needle for power supply voltage of (a) 3 kv and (b) 5 kv.

Based on the intensity of the emission spectral lines it is possible to estimate the excitation temperature of the plasma. The calculation of temperature can be made by either the intensity ratio of two lines or Boltzmann plot method. The error related to measurements can be reduced by considering several lines by Boltzmann plot method [5, 6].

Considering I_1 and I_2 as the intensity peaks at two different emissions at wavelength λ_1 and λ_2, the ratio is given by [5],

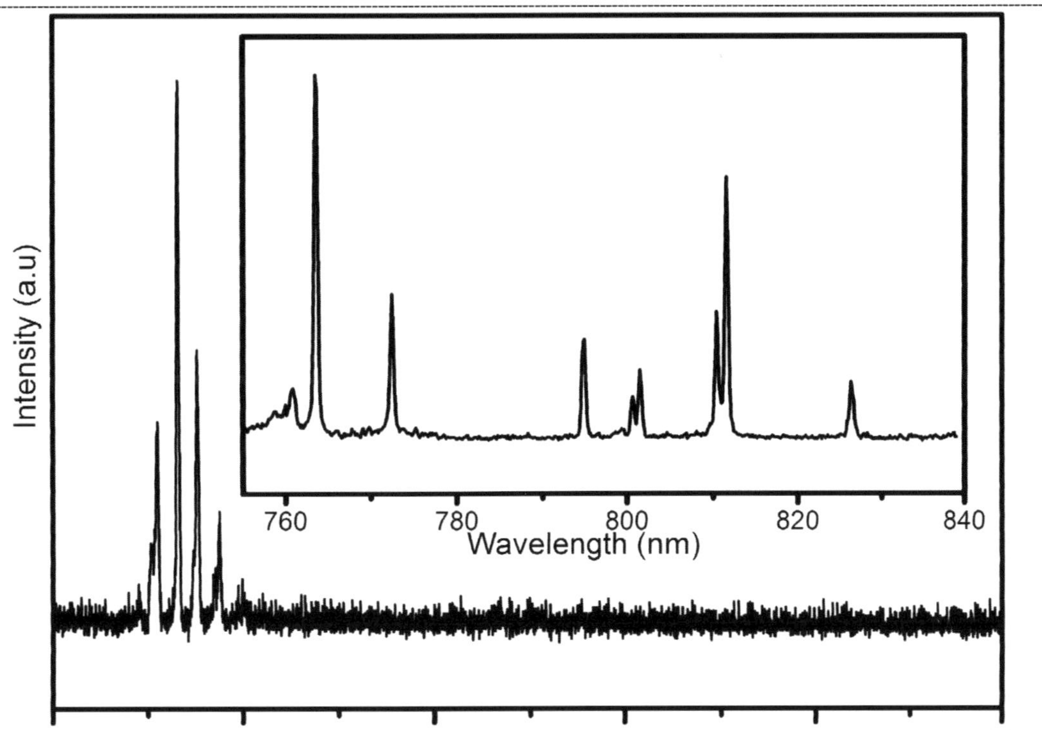

Figure 5. Photograph of the plasma jet operated on water surface.

$$\frac{I_1}{I_2} = \frac{A_1 g_1 \lambda_2}{A_2 g_2 \lambda_1} \exp\left[\frac{E_2 - E_1}{kT}\right]$$

Where, A is the Einstein coefficient of spontaneous emission (transition probability), g is the statistical weight of the upper level of transition, E is the energy of the upper level of the transition, k is Boltzmann's constant and T is the temperature. The plasma temperature can be calculated by above equation on the basis of line intensity ratio by measuring the spectral intensity of two lines. However, the Boltzmann plot method is more accurate and reliable than the line intensity ratio method since several spectral lines are involved in Boltzmann plot method.
In general, the intensity I_k for a particular spectral line can be express as [5],

$$\ln\left[\frac{I_k \lambda_k}{g_k A_k}\right] = \ln\left(l \frac{hc}{4\pi} \rho_0\right) - \frac{E_k}{kT}$$

where l is the path length, h is Planck's constant, c is the light velocity and ρ_0 is the density of atoms. By plotting the term $\ln(I_k\lambda_k/g_kA_k)$ as a function of E_k, the slope $1/kT$ of which gives temperature, since the first term in the right hand side are constant and does not depend on the particular spectral line. A typical Boltzmann plot of the emission spectra of the plasma jet produced by the plasma needle is shown in Fig. 6. The excitation temperature calculated from the OES using Boltzmann's plot method are found to be approximately 0.6 ev. It have been observed that the variation of the excitation temperature with changing the discharge voltages is almost negligible. However, details studies is underway.

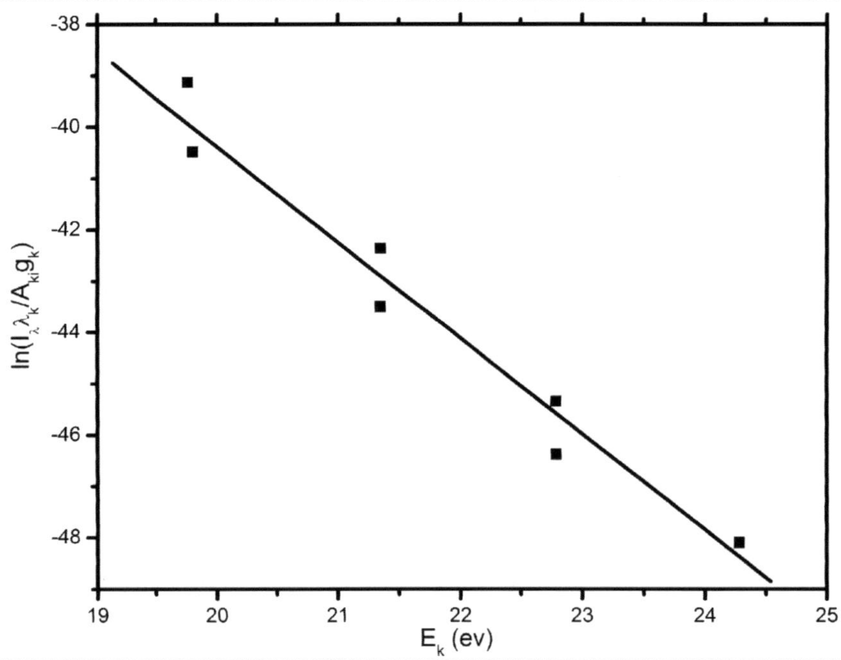

Figure 6. Typical Boltzmann's plot for the plasma jet produced by the plasma needle.

Journal of Physics: Conference Series **720** (2016) 012038 doi:10.1088/1742-6596/720/1/012038

4. Conclusion

A plasma needle have been developed by using a dc power source. The plasma jet produced by the plasma needle can easily interact with human body. Preliminary characterization of the plasma needle by the means of electrical discharge characteristics and OES are carried out.

Acknowledgments

Authors wishing to acknowledge the financial support of FONDECYT 11130048 and CONICYT-PIA grant ACT-1115.

References

[1] Stoffels E, Flikweert A J, Stoffels W W and Kroesen G M W 2002 Plasma Sources Sci. Technol. 11 383
[2] Sladek R E J, Stoffels E, Walraven R, Tielbeek P J A and Koolhoven R A 2004 IEEE Trans. Plasma Sci. 32 1540
[3] Kieft I E, Boers J L V, Caubet-Hilloutou V, Slaaf D W and Stoffels E 2004 Bioelectromagnetics 25 362
[4] I E Kieft, E P v d Laan and E Stoffels 2004 New Journal of Physics 6 149
[5] Lee YS, Bora B, Yap S L, Wong C S, Bhuyan H and Favre M 2012 Powder Technology 222 95
[6] Bora B, Wong C S, Bhuyan H, Lee Y S, Yap S L and Favre M 2013 J. Quant. Spectrosc. Radiat. Transfer. 117 1

Journal of Physics: Conference Series **720** (2016) 012039 doi:10.1088/1742-6596/720/1/012039

Plasma outflows from wire-based z-pinch experiments driven at currents of hundreds of kiloamperes.

Felipe Veloso[1], Gonzalo Muñoz-Cordovez, Luis Donoso-Tapia, Vicente Valenzuela-Villaseca, Mario Favre and Edmund Wyndham
Instituto de Física, Pontificia Universidad Católica de Chile, Av. Vicuña Mackenna 4860, Santiago, Chile

E-mail: fveloso@fis.puc.cl

Abstract. Preliminary results on the latest experiments regarding plasma outflows from different wire-based z-pinch configurations performed in the Llampudken generator (~350kA in ~350ns) are presented. These outflows are produced from three different experiments: cylindrical, conical and nested conical arrays. Our experiments show that it is indeed possible to produce plasma outflows from moderate size pulsed power drivers with currents of some hundreds of kiloamperes. Each one of the configurations studied here can produce a dense plasma outflow characterized by its own set of dimensionless parameters; such as Reynolds number, magnetic Reynolds number, amongst others. A dense magnetized, magneto-hydrodynamically unstable plasma outflow is produced using a modified cylindrical wire array, whereas strongly collimated jets are produced from the conical configurations. Moreover, it is possible to mimic the episodic emission of plasma outflow in a collimated jet by producing temporally separated implosions from the nested conical configuration. Finally, the characteristic and dynamics of each outflow are presented and discussed.

1. Introduction

Wire-based z-pinches are highly dynamic plasma configurations and an efficient technique to heat small amount of masses into extremely high temperatures. In this kind of configuration, the current flowing through thin metallic wires produces their ablation forming coronal plasma over their solid cores which act as mass reservoirs during the ablation stage [1]. In addition, this current also produces a global magnetic field as the summation of the local fields of each one of the wires. As the coronal plasma of each wire is driven by the $\mathbf{J}\mathrm{x}\mathbf{B}_{global}$ Lorentz force, the connection path of the wires between cathode and anode determinates the global plasma dynamics of the configuration. Amongst these wire-based z-pinch plasmas, several experiments can be included such as single-wires[2], crossed-wires[3], cylindrical[4], nested cylindrical [5], conical[6] or radial[7] wire arrangements. These wire- based plasmas are used as load of pulsed power generators, where fast rising electrical currents above 100's kA in 100's ns drive the experiments.

Nowadays, it is known that the wire- based z-pinches are efficient and powerful x-ray radiation sources. They are also used in a variety of research areas such as inertial confinement fusion,

[1] To whom any correspondence should be addressed.

laboratory astrophysics, radiation science and high energy density studies among others. An extensive review on the physics of z-pinch plasmas can be found in ref [8]. Depending on the geometry of the configuration, the global $\mathbf{J}\mathrm{x}\mathbf{B}_{global}$ force can produce plasma accumulation on the region where the magnetic field is zero, and it can be later expelled out of the configuration as a plasma outflow. These outflows provide useful information to study not only the basic plasma physics principles of wire-based z-pinch plasmas, but also a laboratory approach scenario for studying plasma outflows from different sources where magnetohydrodynamical (MHD) similarities can be found through comparing their dimensionless parameters. This paper presents some of the latest results on wire-based z-pinch plasmas driven at the Llampudken generator, where plasma outflows are produced. The preliminary results discussed on this paper shows that a moderate size current driver at hundreds of kiloamperes is capable of producing interesting plasma outflows for basic physics and applications studies.

2. Experimental setup

The experiments were performed using the Llampudken pulsed power generator [9] located at the Instituto de Física of the P. Universidad Católica de Chile. This generator consists of two Marx capacitor banks, each of 0.25μF, 240kV and 7.4kJ. Each Marx drives a series of transmission lines connected by line switches towards a couple section composed of an exponential line to the load section. In addition, an auxiliary exponential transmission line is coupled in parallel to the load in order to optimize the energy coupling from the transmission lines to a time-varying impedance load, as it is the case of z-pinch plasmas. A schematic diagram of the generator and the transmission lines currently in use can be seen on Figure 1. Further details on the design and construction aspects of the Llampudken generator can be found elsewhere [9]. At present operation, the generator delivers a peak current of ~350kA in ~350ns. A characteristic current trace is shown in Figure 2a. The wire arrays are mounted as load for the generator and their length and diameter can be adjusted on each experiment. The wire arrays are composed typically of either aluminium or tungsten wires having diameter of ~10s of μm diameter. The connection path of the wires between cathode and anode can be easily changed according to the experiment and it determinates the global magnetic field of the configuration and consequently, the dynamics of the wire-based z-pinch experiment under consideration.

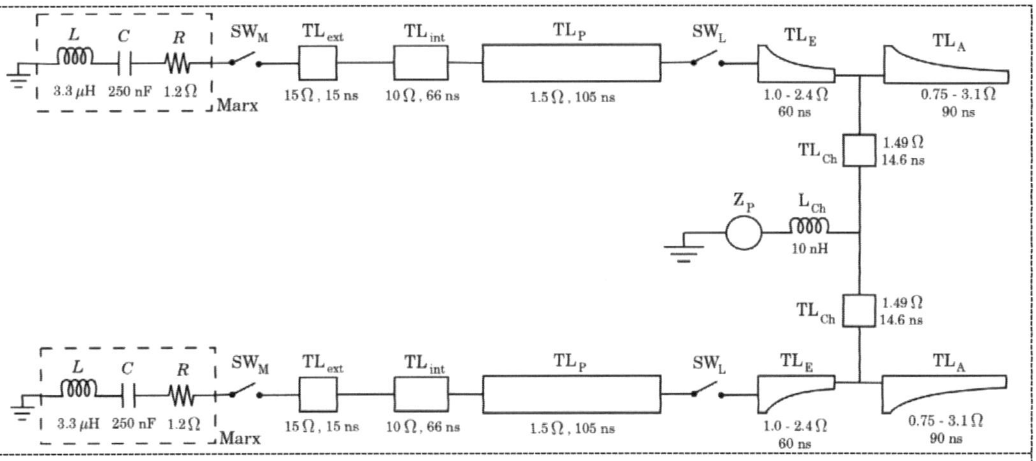

Figure 1. Schematic diagram of the transmission lines in the Llampudken pulsed power generator. *TL* represents transmission lines, *SW* are high voltage switches, whereas Z_p represents the plasma load.

The plasma dynamics were diagnosed using time-resolved imaging diagnostics in the side-on direction (i.e., perpendicular to the axis between electrodes). Extreme ultraviolet (XUV) self-emission imaging is performed using unfiltered pinhole imaging onto two 2-frame micro-channel plate cameras (MCP) with ~5ns temporal resolution. In addition, time-resolved laser shadowgraphy is performed using a frequency doubled Nd:YAG laser (532nm, 12ps) recorded using open shutter CMOS cameras with interference filters at the laser wavelength to avoid the recording of the plasma self-emission. In the conical wire array experiments described in section 3.2, time-integrated pinhole imaging is obtained in both end-on (i.e., collinear to the axis of the configuration) and side-on directions. The pinhole camera images are recorded on HP5 photographic paper, then scanned and contrast enhanced by an image processing program.

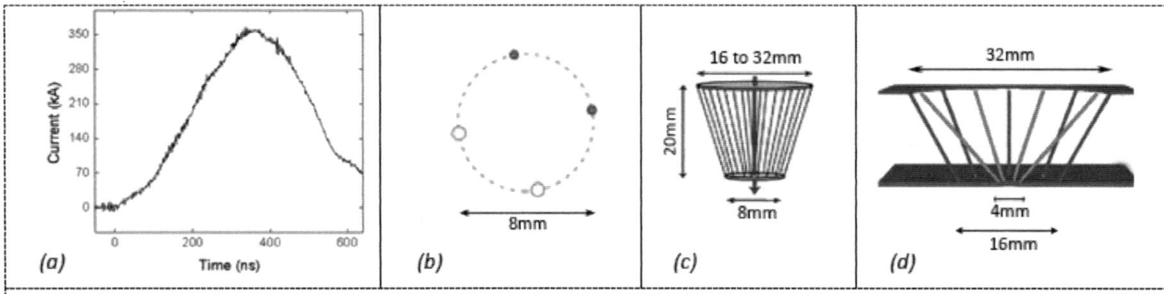

Figure 2. (color online) (a) Experimental current trace of the Llampudken generator (b-d) Schematic diagrams of the experiments (b) End-on view of 2x 10μm + 2x 25μm, (c) Side-on view of conical wire array, (d) Sliced side-on view of nested conical wire array.

In addition to the experimental equipment, some of the experiments are complemented with numerical simulations using the 3-dimensional resistive magnetohydrodynamical code GORGON [10]. This code has been developed at Imperial College of London, and it has proved to be a robust and accurate tool for modelling z-pinch plasmas. This code solves the set of equations that governs plasma dynamics such as the resistive-MHD, thermodynamics, ionization, radiation emission amongst others altogether in a single fluid approximation on a Cartesian grid. These numerical simulations are performed in a 128GB ram, 64-cores cluster located at the Instituto de Física, which is used exclusively on plasma experiments simulations. The input parameters of the code are the initial geometry of the experiment, current profile and the properties of the wire material (such as ionization levels, mass density, etc). In order to match experimental results with the numerical simulations, the initial temperature of the wires can be adjusted. For the experiments described here, a cell by cell random temperature perturbation along each wire length of the form $T=0.3^{(1+\chi)}eV$ (where χ is a random number between 0 and 1) is introduced. This randomly distributed temperature causes different rates of initial expansion along the wires, producing structures and results similar to those experimentally observed in different wire array configurations [11].

3. Results and discussion

3.1. Plasma outflows from cylindrical wire arrays under temporally variable B-field topology

Cylindrical wire array z-pinches are composed of several wires of a given material and diameter arranged in an axisymmetrical cylindrical configuration with the wires equally spaced among each other similar to a bird cage. Due to the azimuthal symmetry imposed on these arrays, the ablation flows from each wire radially converge toward the array axis, where plasma accumulates, forming a precursor plasma column. Even after most of the wires have been fully ablated and during the implosion phase as well, the null magnetic field region (and consequently, the precursor) remains at the array axis. However, the symmetry of the global magnetic field on this configuration can be modified by replacing a pair of consecutive wires with wires of a larger diameter. This modification leads to two separate effects; firstly, current is unevenly distributed between the wires and secondly the thicker wires take longer to fully ablate. Both of these effects impact the dynamics of the precursor plasma due to the production of non-axially symmetric temporal variations of the magnetic field

Journal of Physics: Conference Series **720** (2016) 012039 doi:10.1088/1742-6596/720/1/012039

topology. Since plasma accumulation on the precursor remains at the null magnetic field region, the variation on the global field topology also displaces the region where the magnetic field is zero, displacing the precursor accordingly. In order to produce a plasma outflow from the wire array precursor, the mass per unit length of the load must be kept similar (or below) those required for producing full ablation of the wires during the current rise of the generator. For the Llampudken generator and according to the 0D model for wire array implosions [8], it is equivalent to ~3μg/m in an 8mm diameter cylindrical array. For this reason, the experiments described here use four Al wires, 2x 10μm + 2x 25μm, in a cylindrical configuration of 23 mm in length and 8mm diameter. An schematic diagram of the array observed in end-on direction is shown in Figure 2b. These wire diameters meet both conditions: they produce temporal variation of the magnetic field topology and provide an appropriate mass per unit length for the production of plasma outflows.

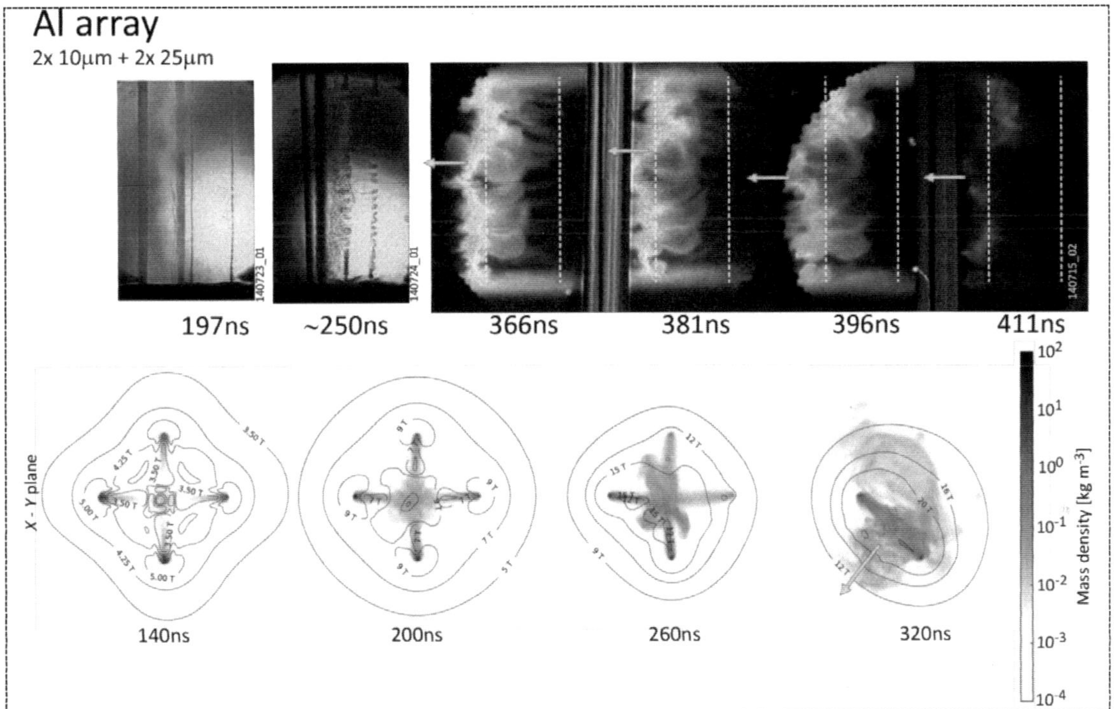

Figure 3. (color online) Experimental results of 2x 10mm + 2x 25mm. (top) Laser shadowgraphy and XUV imaging and (bottom) End-on view from numerical simulations together with isocontour of magnetic field. Dashed vertical yellow lines delimit the original array diameter. Green arrows indicate the direction of the plasma outflow

Figure 3 shows both laser probing and XUV self-emission of the plasma at different times of the discharge. It can be observed from early times in the discharge (~200ns) that the plasma ablated from the wires accumulates near the symmetry axis, but slightly shifted towards the side of the thicker wires (almost hidden by a thick wire). This indicates that the global magnetic field topology has been affected from a symmetric configuration, and the null magnetic field region is already located off-axis. This is ascribed to an uneven current distribution amongst thin and thick wires since early times of the discharge. However, it is possible to observe the precursor movement towards the outer region of the array after the appearance of breakages along the dense cores of the wires (at ~250ns). These discontinuities on the wires increase their equivalent impedance, producing the redistribution of the current towards the precursor and the remaining thicker wires. Since the precursor does not have a solid core or a fixed position in space, it gains momentum directed towards the gap produced in the inter-wires space increasing its mass from the ablation of the remaining thicker wires. These effects cause the precursor plasma to move outside the region delimited by the positions of the wires in the array as shown in the XUV images of Figure 3. From these results, a characteristic velocity of the precursor column as a plasma outflow from the wire array can be estimated. The precursor is observed to move from the axis to the array boundary during the time interval of ~190ns to 366ns, from which the average plasma velocity can be estimated as ~3 x 10^4 m/s. However, it is worthwhile mentioning that this characteristic velocity is not necessarily uniform along the precursor due to the development of instabilities (such as m=1 MHD mode observed in the shadowgraphy images) along the plasma column. Moreover, the numerical simulations of Figure 3 show how the global field topology is being temporally modified, with the consequent movement of the plasma accumulated originally on-axis toward the outer region of the array

In order to provide characterical parameters of these plasma outflows, some dimensionless parameters can be evaluated after some estimations on the plasma. Considering the velocity estimation of the plasma outflow and assuming a plasma temperature of 30eV, an average ionization state of 8, a characteristic length scale of L~1mm (i.e., of the order of precursor diameter), and considering transverse Spitzer resistivity, the Reynolds number has a high value (Re>10^5) and the magnetic Reynolds number is Re_m~1.5. This indicates that the movement of the plasma is more likely to produce turbulences and that both magnetic diffusion and advection are similarly important on the global magnetic field dynamics with characteristic timescales (L/v) on the order of ~30ns. Further details on the ablation dynamics in cylindrical wire arrays using temporal variations of the magnetic field topology can be found elsewhere [12]. Future experiments will include further detail on the dynamics of the plasma outflow from these arrays, together with their interaction with obstacles as a research tool for shock studies in high density plasmas.

3.2. Plasma outflows from conical wire array experiments

In conical wire arrays, the dynamics of the precursor plasma is intentionally modified by changing the angle of the wires with respect to the z- axis of the configuration, in contrast to the cylindrical wire array z-pinches previously discussed where the wires are placed parallel to it. Therefore, the diameter of the array at the cathode differs from its diameter at the anode, making a truncated cone-shaped array. Since plasma ablation from the wires is given by the global Lorentz force, the ablated plasma is driven perpendicular to them regardless of the array geometry. In the conical wire arrays, the plasma ablated from the wires reaches the array axis bottom first and then top, creating a zipper effect in the precursor. Hence, the precursor gains upwards momentum generating a collimated highly supersonic jet that is expelled at the top of the array. It has been studied that the collimated plasma outflows from these conical arrays could provide relevant information to the observations of astrophysical environments regarding the jet emission from young stellar objects, since both outflows can be compared using an appropriate set of dimensionless parameters [13-16]. In addition, numerical simulations have suggested that some of the properties of the plasma outflows from conical wire arrays, such as collimation and propagation velocities, depend on the current driver parameters [17]. However, most of the experiments have been conducted in megaampere devices having a large current rate; namely MAGPIE (1.4MA, ~5kA/ns) and COBRA (1MA, 10kA/ns). Therefore, there is a lack of experimental results in lower current rate devices, which is the case of Llampudken (350kA, 1kA/ns). The experiments described here use sixteen equally spaced tungsten wires of 25μm diameter each using 20mm cathode-anode separation. The array diameter at the cathode is 8mm, whereas the anode diameter is interchangeable between 16mm and 32mm. The anode has a central aperture to allow the propagation of the plasma out of the interelectrode region. An schematic diagram of the configuration is shown in Figure 2c.

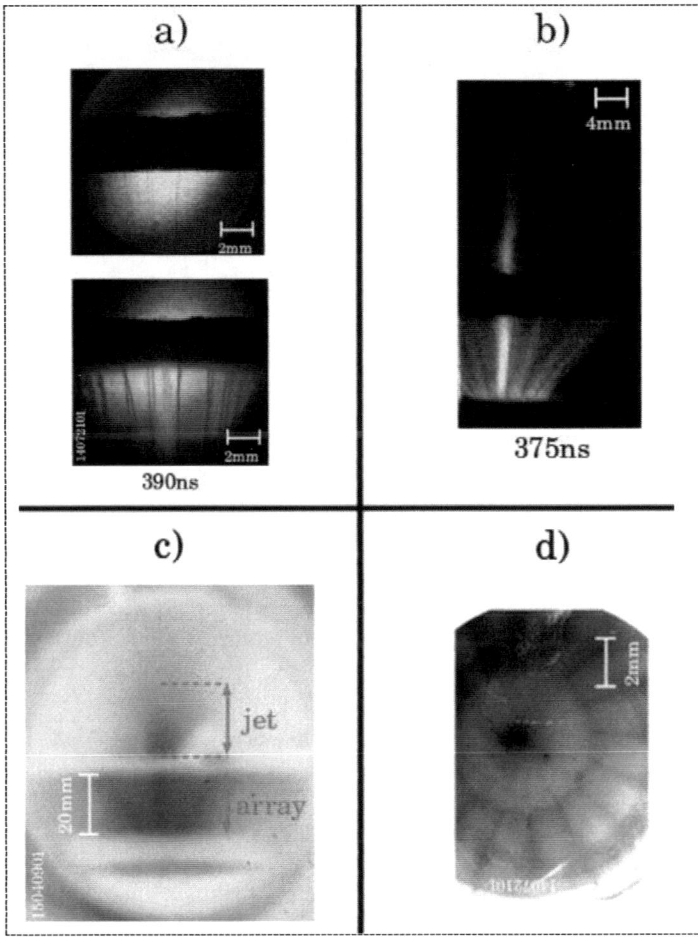

Figure 4. (color online) Experimental results from conical wire arrays. (a-b) Time-resolved side-on images (a) shadowgraphy of 16x 25μm W wires with 16mm anode diameter. (b) XUV image of 16x 25μm W wires with 32mm anode diameter. (c-d) Time-integrated visible images (c) Side-on of 16x 25μm W wires with 32mm anode diameter and (d) End-on of 16x 25μm W wires with 16mm anode diameter.

Figure 4a shows shadowgraphy imaging of the comparison between preshot array and precursor formation just before the jet is expelled out over the top of the upper electrode. It can be measured that the precursor plasma diameter of half height of the interelectrode space (lower part of the image) is 1.6 times wider than the top of it, showing the zipper-like formation of the precursor [6]. Later in time, this plasma is expelled as a jet which can be observed in the XUV images shown in Figure 4b. In order to measure a characteristic propagation velocity of the jet, the upper part of the jet for images of a single shot is considered providing a value of ~4 x 10^4 m/s [18]. It is worthwhile mentioning that this measurement provides an estimation of the velocity rather than an accurate value, since MCP cameras might have different gains for recording the self-emission of the plasma. In order to avoid this difficulty, it would be more adequate to follow a particular structure within the jet. However, there is no a clear structure which can be followed in the present results. Besides this, a highly collimated jet with aspect ratio (lenght/width) ~15 is clearly observed. The complete evolution of the plasma dynamics can be observed in the time-integrated images shown in Figure 4c-4d, showing that the initially collimated jet expands radially at later times than those observed in the time-resolved XUV images. From these results, and considering the jet as a 10eV, $10^{17\text{-}18}$ cm^{-3} plasma Z_{eff} = 5 and a magnetic field ~0.7T, the dimensionless parameters shown in Table 1 can be calculated. These dimensionless parameters agree with measurements of jet emerging from young stellar objects [15,16]. Future experiments will include improved diagnostics to measure the jet density, ionization level and temperature in order to avoid some of the estimations previously shown. Additionally, the interaction

Journal of Physics: Conference Series **720** (2016) 012039 doi:10.1088/1742-6596/720/1/012039

of the jet with a background atmosphere will be studied with the aims of comparing the possible shock structures to be formed with astrophysically relevant objects (such as Herbig-Haro observations [19]).

Table 1. Dimensionless parameters for plasma outflows.

	Aspect ratio (length/width)	Mach number (v_{jet}/c_s)	Collisionality (λ_{mfp}/ L)	Viscosity (Re)	Resistivity (Re_m)	Heat conduction (Pe)	Plasma beta (β)
Llampudken conical array	~ 15	~5	10^{-4}	>> 1 (10^7)	~20-30	>> 1 (10^3 - 10^4)	~8
Typical young stellar object [15,16]	10-100	> 1	<< 1	>>1 ($\gtrsim 10^8$)	>>1 ($\gtrsim 10^{15}$)	>>1 ($\gtrsim 10^7$)	> 1

3.3. Mimicking episodic plasma outflows using nested conical wire arrays.

Many astrophysical observations from the Hubble Space Telescope, and ALMA telescope, amongst many others have observed that the propagation of jet structures from young stellar objects present discontinuities on the plasma emissions propagating towards the interstellar medium. These structures are plasma jets emitted from a protostar (known as a Herbig-Haro object in astrophysics), where discontinuities are observed as variation on the luminosity along the jet which are often interpreted as perturbations to a relatively steady flow [16,19]. The origin of these discontinuities remains as an open issue in astronomy which can be addressed using laboratory experiments. A possible explanation has been provided by radial wire and foil array z-pinches as the formation of periodic episodic ejections of magnetic bubbles from the source naturally evolving into a heterogeneous jet propagating inside a channel made of self-collimated magnetic cavities [20]. In order to provide an alternative source of episodic emissions in the laboratory, a nested configuration of conical wire arrays is investigated using numerical simulations. These nested arrays have a common anode diameter, but differing on their cathode diameters. The difference on the base angle of these conical arrays produces two plasma outflows which are combined in the array axis and propagating as a single object. The simulations use eight 25μm diameter tungsten wires for each array, with base angles of 45° and 60° for 14mm height cones, and the Llampudken current parameters. A schematic diagram of the nested conical array is shown in Figure 2d.

Figure 5 shows an sliced 3D plot of the electron density, where close-ups of the jet at different times are included. As expected from a conical wire array, a plasma jet is emerging from the bottom of the configuration axially towards the open anode as a collimated plasma outflow. In addition, it can be observed that there is a reduction on the electron density which is propagating as part of the jet with a velocity ~4×10^4 m/s. This reduction in electron density is a discontinuity in the jet which can be seen as a gap similar to those described in the astronomical observations of the Herbig-Haro objects. This gap is formed over the anode between two regions of high density plasma at about 400ns, as shown in Figure 5. The gap between these high density regions is explained from the different times on convergence of the plasma streams coming from the different arrays towards the center of the configuration. When performing independent simulations of each array that compose the nested configuration separately, there is no observation of the gap in the plasma jet regardless the amount of current driving the wire array (either full or half of the Llampudken current); i.e., the gap is only observed in the nested configuration, which coincides with the experimental results discussed in section 3.2. It is worthwhile mentioning that the reduction in density of the nested configuration is sufficient to be experimentally observed on interferometry images of the jet on the future planned experiments to be performed. Besides this, the dimensionless parameters of the jet from the nested configuration are similar to those from a single conical array presented in Table 1. These observations suggest that the observed episodic emissions can also be produced by successive pulses of plasma outflows due to concentric sources imploding at different times. In the near future, experiments on the Llampudken generator will be carried out in order to complement these numerical simulations for providing a full description of the episodic emission phenomena from nested conical arrays.

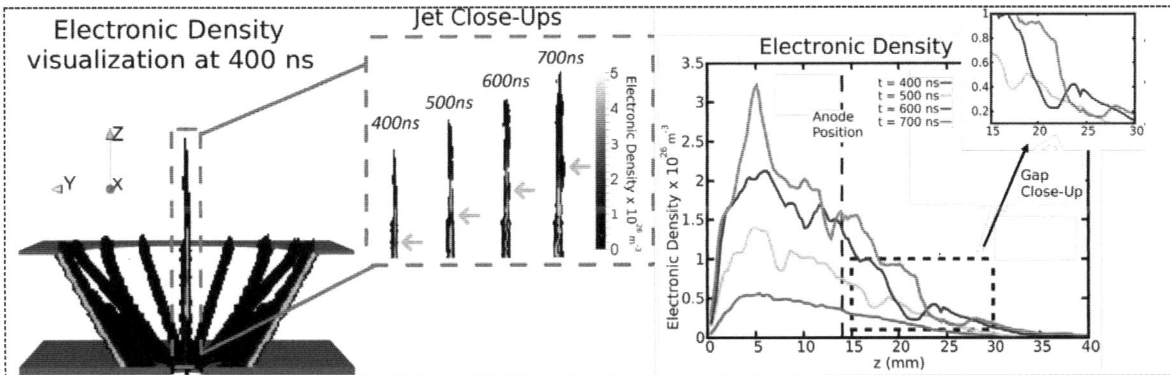

Figure 5. (color online) Numerical simulations of nested conical wire arrays driven at Llampudken. Yellow arrows in close-ups indicate the gap in electron density.

4. Final remarks and conclusions

The preliminary results described in this paper obtained from the Llampudken generator demonstrate that the production of dense transient plasma outflows is indeed possible in moderate size pulsed power generators driving currents of some hundreds of kiloamperes. These plasma outflows are not only useful for understanding the basic physics of wire- based z-pinch plasmas, but also they provide an optimal scenario where the plasma source of the outflow and the outflow itself can be studied together in detail. In cylindrical array using different wire diameters, the outflow is originally the dense precursor plasma which is driven out of the array due to temporal variations on the magnetic field topology. In contrast to the outflows produced in conical array experiments, the precursor outflow from cylindrical arrays carries an important fraction of the total current and therefore, there is a growing of an m=1 MHD instability while it moves through a diluted plasma background. This opens the possibility of studying different regimes of current-carrying plasmas, where the emission of high energy particles from the turbulent plasma motion or the growing of different kind of instabilities can be developing simultaneously. On the other hand, the plasma outflows from conical wire arrays are mainly hydrodynamic flows in nature once they come out of the array. It has been shown that these outflows produced in Llampudken provide an scenario for comparing them to astrophysically produced jets with the appropriate set of dimensionless parameters. It has also been shown that the episodic emission of outflows can be ascribed to temporally separate convergent sources of the same nature.

The experimental capability of easily performing modifications to the load parameters enables the modification of the plasma source in almost any sense for studying its impact on the plasma outflow, which is also beneficial for application studies. Further investigations on the three kinds of experimental configurations described here will be carried out in order to improve the simultaneous characterization of the source and outflow, which will be reported elsewhere.

The authors acknowledge Prof. J Chittenden for his collaboration and for providing us the GORGON code for the numerical simulations of the experiments. This work has been supported by Fondecyt

Iniciacion 11121621. G Muñoz-Cordovez and L Donoso-Tapia acknowledge financial support of CONICYT scholarship for graduate studies.

References

[1] S V Lebedev, F N Beg, S N Bland, J P Chittenden, A E Dangor, M G Haines, S A Pikuz and T A Shelkovenko 2000 *Phys Rev Lett* **85** 98
[2] F N Beg, A E Dangor, P Lee, M Tatarakis, S L Niffikeer and M G Haines 1997 *Plasma Phys Control Fusion* **39** 1
[3] I H Mitchell, J A Gomez, F A Suzuki, R Aliaga-Rossel, H Chuaqui, M Favre and E Wyndham 2005 *Plasma Sources Sci Technol* **14** 501
[4] F Veloso, F Suzuki-Vidal, F Molina, I H Mitchell, H Chuaqui, M Favre and E Wyndham 2012 *IEEE Trans Plasma Sci* **40** 3319
[5] C Deeney, M R Douglas, R B Spielman, T J Nash, D L Peterson, P L'Eplattenier, G A Chandler, J F Seamen and K W Struve 1998 *Phys Rev Lett* **81** 4883
[6] D J Ampleford, *et al* 2007 *Phys Plasmas* **14** 102704
[7] S V Lebedev, A Ciardi, D J Ampleford, S N Bland, S C Bott, J P Chittenden, G N Hall, J Rapley, C A Jennings, A Frank, E G Blackman, T Lery 2005 *Mon Not R Astron Soc* **361** 97
[8] M G Haines 2011 *Plasma Phys Control Fusion* **53** 093001
[9] H Chuaqui, E Wyndham, C Friedli and M Favre 1997 *Laser Part Beams* **15** 241
[10] J P Chittenden, S V Lebedev, C A Jennings, S N Bland and A Ciardi 2004 *Plasma Phys Control Fusion* **46** B457
[11] J P Chittenden and C A Jennings 2008 *Phys Rev Lett* **101** 055005
[12] F Veloso, G Muñoz-Cordovez, L Donoso-Tapia, V Valenzuela-Villaseca, F Suzuki-Vidal, G Swadling, J Chittenden, M Favre and E Wyndham, Ablation dynamics in wire array z-pinches under modifications on global magnetic field topology *submitted to Phys Plasmas*
[13] D D Ryutov, R P Drake and B A Remington 2000 *Astrophys J Suppl Series* **127** 465
[14] D D Ryutov, B A Remington, H F Robey and R P Drake 2001 *Phys Plasmas* **8** 1804
[15] D D Ryutov, R P Drake, J Kane, E Liang, B A Remington and W M Wood-Vasey 1999 *Astrophys J* **518** 821
[16] B Remington, R P Drake, D D Ryutov 2006 *Rev Mod Phys* **78** 755
[17] J P Chittenden, *private communication*
[18] G Muñoz, V Valenzuela, F Veloso, M Favre and E Wyndham 2014 *AIP Conf Proc* **1639** 92
[19] P Hartigan, A Frank, J M Foster, B H Wilde, M Douglas, P A Rosen, R F Coker, B E Blue and J F Hansen 2011 *Astrophys J* **736** 29
[20] F Suzuki-Vidal *et al* 2011 *Astrophys Space Sci* **336** 41

Journal of Physics: Conference Series **720** (2016) 012040 doi:10.1088/1742-6596/720/1/012040

Analytic Approximate for the Plasma Sheath Potential

Pablo Martin [1,2], Fernando Maass-Artigas [1,†], Luis Cortés-Vega[3]

1 Physics Department, Antofagasta University, Casilla 170, Antofagasta, Chile
2 Physics Department, Simón Bolivar University, Apartado 89000, Caracas 1083 A, Venezuela
3 Mathematics Department, Antofagasta University, Casilla 170, Antofagasta, Chile

E-mail: pmartin@usb.ve, pablo.martin@uantof.cl, fernando.maass@uantof.cl, luis.cortes@uantof.cl, luisvega@vtr.net

Abstract. Here a new analytic approximation for the Bhom Sheath Potential is presented, which is valid for any value of the characteristic parameter K, measuring the mean ion velocity. The procedure to obtain this approximation is different to those used by previous authors, because now, the characteristic exponential parameter λ, depends on the parameter K, as well as the wall potential ϕ_w. In previous works that parameter used to be function of K only, and all the approximation used to fail for $K \leq \frac{1}{2}$, which is not the case now.

1. Introduction

The plasmas in physics are characterized by the quasi-neutrality, which means that the number of ions and electrons are in nearly equal numbers. This property is lost near the walls containing the plasma. The classic treatment for this region, called Bhom Sheath model [1-3] leads to the idea that there are two regions a long one denoted as presheath, where the quasi-neutrality is still preserved, and the sheath region, where there are a few electrons, and the ions are those feeded by the plasma, whose velocity are determined by the wall potential and plasma density. There the electron density is determined by a Boltzman factor.

In this way the plasma potential near the wall is determined by Poisson equation, which can be written in dimessionaless variables as

$$\frac{\partial^2 \phi}{\partial y^2} = -\frac{1}{2}\left[e^{-\phi} - \frac{1}{\sqrt{1+\frac{\phi}{K}}}\right]. \tag{1}$$

here the dimesionless potential ϕ, is defined in terms of the wall potential φ as

$$\phi = \frac{e\varphi}{k_B T} \tag{2}$$

where $(-e)$ is the electron charge, k_B the Boltzmann constant and T_e the electron and ion temperature of the plasma. The demensionless distance to the wall y is measured in Debye length units λ_D, that is, $y = x/\lambda_D$; $\lambda_D^2 = \varepsilon_0 \, k_B \, T/2n \, e^2$, where x is the actual distance to the wall, and n is the ion and electron plasma density. The parameter K introduced in Eq. (1) is also a dimesionless quantity, measuring the characteristic ion plasma velocity v

$$K = \frac{\frac{1}{2} m v^2}{k_B T}$$

2. Previous Approximations to ϕ

The main problem with the Eq. (1) is that no analytic solution is known, and it has to be solved numerically for each value of K. Furthermore, from the equation is much easier, to obtain the distance x as a function of ϕ, than ϕ as function of x, which is usually needed it. For this reason several approximated solutions have been found for this equations [4,5].

The simplest and most usual one is

$$\phi_2(y) = \phi_4 \, exp\left[-\frac{1}{2}\sqrt{\frac{2K-1}{K}}\, y\right].$$

This is obtained by keeping the first term of the non-linear first order differential equation

$$\frac{d\phi}{dy} = -\sqrt{F(\phi)},$$

where

$$F(\phi) = 2K\sqrt{1+\frac{\phi}{K}} - 2K + e^{-\phi} - 1. \tag{3}$$

This equation is obtained by a first integration of Eq. (1), with the boundary condition that $\frac{d\phi}{dy}$ is zero at the end of the sheath, which is also the beginning of the presheath, that is, when ϕ is zero.

The Taylor expansion for $F(\phi)$ can be written as

$$F(\phi) = \left(\frac{2K-1}{4K}\right)\phi^2\left[1+\phi\left(\frac{3-4K^2}{12K^2-6K}\right)+\phi^2\left(\frac{18K^3-15}{96K^3-48K^2}\right)+\ldots\right]. \tag{4}$$

Now $F(\phi)$ is approached by

$$F(\phi) \simeq \lambda^2\phi^2\left(1+\frac{1}{2}\phi\right)^2; \quad \lambda^2 = \frac{2K-1}{4K}. \tag{5}$$

A better approximation for ϕ is obtained [5], in the following way

$$\tilde{\phi}(y) = \frac{\phi_w}{e^{\lambda y}\left(1+\frac{\alpha}{2}\phi_w\right) - \frac{\alpha}{2}\phi_w}, \tag{6}$$

with

$$\alpha = \frac{2}{\phi_w^2}\left[\sqrt{\frac{4K\left(2K\sqrt{1+\frac{\phi_w}{K}} + e^{-\phi_w} - (2K+1)\right)}{(2K-1)}} - \phi_w\right], \tag{7}$$

where α has been determined, imposing the condition that the slope at $y = 0$, must coincide with that of the exact function ϕ.

Journal of Physics: Conference Series **720** (2016) 012040 doi:10.1088/1742-6596/720/1/012040

3. New approximate solution

The previous approximate solutions have the problem that they fail for values of K equal to $\frac{1}{2}$, or nearly that value. Furthermore, in the case of the first approximation, the slope at the wall become independent of the potential wall and depend only of the value of K. However numerical integration of the main differential equation shows that the slope is also depending of ϕ_w. On the other hand, through the slope in the second approximation was chosen to coincide with the right one, however the coefficient of the exponential λ is the same than that in the first approximation. This means, that it is also independent of the potential ϕ_w. It seems that it will much better if that coefficient were also dependen of ϕ_w. With these ideas in mind we are presenting here, a way to obtain approximations, where all the parameters of the approximation will be depending of K and ϕ_w, that is, ion velocity and wall potential.

Here, the simplest approximation with those ideas in mind will be presented, and more elaborated ones will be also found in future works.

The simplest new approximation has the form

$$\tilde{\Phi}_1(y) = \phi_w e^{-\tilde{\lambda} y}. \tag{8}$$

Now, the parameter $\tilde{\lambda}$ is determined by the condition that the slope at $y = 0$ must be equal to the exact one, that is,

$$\frac{d\phi}{dy}\Big|_{y=0} = -\sqrt{F(\phi_w)}, \tag{9}$$

$$\frac{d\tilde{\Phi}_1(y)}{dy} = -\tilde{\lambda}\, \phi_w e^{-\tilde{\lambda} y}, \tag{10}$$

$$\frac{d\tilde{\Phi}_1(y)}{dy}\Big|_{y=0} = -\tilde{\lambda}\, \phi_w. \tag{11}$$

Thus

$$-\tilde{\lambda}\, \phi_w = -\sqrt{F(\phi_w)}. \tag{12}$$

Finally, it is obtained

$$\tilde{\lambda} = \frac{1}{\phi_w}\sqrt{F(\phi_w)} = \frac{1}{\phi_w}\sqrt{2K\sqrt{1+\frac{\phi_w}{K}} - 2K + e^{-\phi_w} - 1}. \tag{13}$$

Now the parameter in the exponential depends not only of K, but also of ϕ_w, as it should be. As it is shown in Figures 1 and 2, the accuracy of the approximation is only a little better than ϕ_2 and $\tilde{\phi}$, but the most important advantage is that, it is good for $K = \frac{1}{2}$, as well as, values of K smaller than $\frac{1}{2}$. This is very important, because of all previous approximations failed near to $\frac{1}{2}$ or lower values.

In the Figures 1a, 1b, 1c and 1d, the exact potential ϕ is compared with the usual exponential approximation ϕ_2, the most complicated approximation $\tilde{\phi}$ and the new approximation $\tilde{\Phi}_1$. In the figures, four values of K has been chosen: $K = 1$, 0.8, 0.6 and 0.51, which are shown respectively in Figures 1a, 1b, 1c and 1d. The Figure 2a, the absolute error of each approximation is presented.

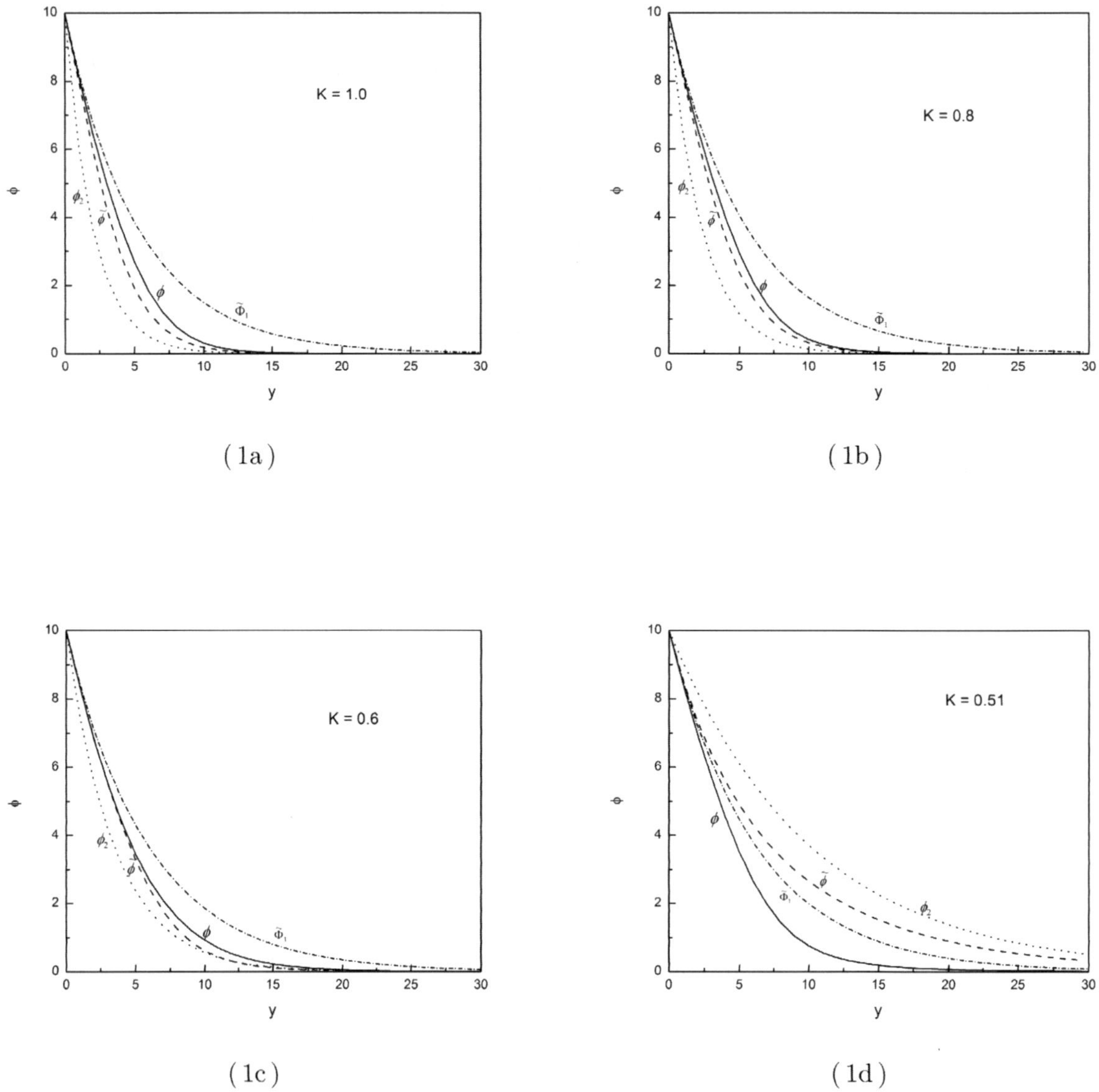

Figure 1. In the plots (1a), (1b), (1c) and (1d), the exact potential ϕ is compared with the usual exponential approximation ϕ_2, the best approximation in previous paper $\tilde{\phi}$, and the approximation here obtained $\tilde{\Phi}_1$, for different values of $K = 1.0$; 0.8; 0.6 and 0.51.

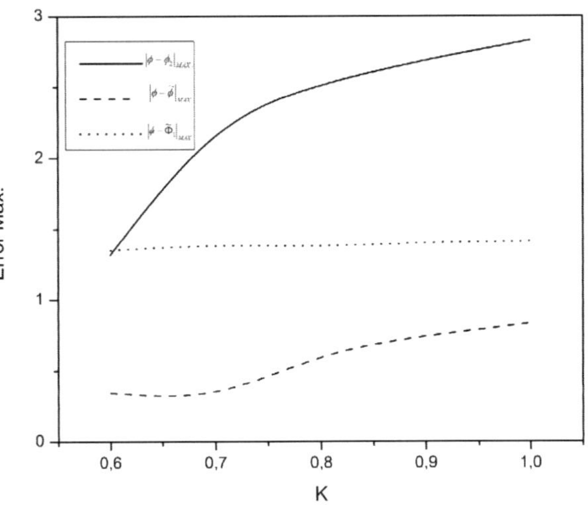

(2a)

Figure 2. In the figure (2a), the absolute error of each approximation is presented, $\Delta\phi_2$ (plain line), $\Delta\tilde{\phi}$ (point line) and $\Delta\tilde{\Phi}_1$ (dash line).

Conlusion

New analytic approximation for the Bohm sheath potential has been found. This new approximation, as previous one, allows the direct calculation of the sheath potential as a function of the distance to the wall. The errors of the new approximation and the previous one have been determined for four different values of K. Previous approximations use to fail for values smaller than $\frac{1}{2}$ or nearly to $\frac{1}{2}$. The new approximation is good also for any value of K, including $K = 1/2$, wich is a very important advantage compared with the previous ones. Furthermore it is also very simple, since it is an exponential as ϕ_2, notwithstanding that its accuracy is higher.

Acknowledgments

Work supported by: (1) Decanatura de la Facultad de Ciencias Básicas, Universidad de Antofagasta, Antofagasta, Chile; (2) Grant G-22 (P. Martin). Decanato de Investigaciones, Universidad Simón Bolivar, Caracas, Venezuela, and Grant FONDECYT (L. Cortés-Vega) $N°$ 1121103, Chile.

†The oral presentation was performed by F. Maass in SOCHIFI-2014 meeting.

References

[1] Bhom D. "The characteristics of electric discharges in magnetic fields", A.Guthrie, R Wakerling Ed. (McGraw Hill, New York, 1949) Chap. 3.
[2] Hazeltine R.D. and Waelbroeck F. "The framework of plasma physics", Perseus Books, Reading Massachuset., (1998), pp. 81-84.
[3] Stangeby P.C. and McCracken G.M., Nuclear Fusion, **30**, 1228 (1990).
[4] Martin P. and Cereceda C., 29th EPS Conference on Plasma Phys. and Contr. Fusion, Montreux, 17-21 June 2002 ECA Vol. 26B, P-4.009 (2002).
[5] Martin P., Cortés-Vega L. and Maass-Artigas F., "Precise approximate solution for the Bhom Sheath potential", J. Phys.: Conf. Ser. **574** (2015) 012107.

Journal of Physics: Conference Series **720** (2016) 012041 doi:10.1088/1742-6596/720/1/012041

Characterization of the ionization degree evolution of the PF-400J plasma sheath by means of time resolved optical spectroscopy

G. Avaria[1,2], O. Cuadrado[3], J. Moreno[1,2], C. Pavez[1,2] and L. Soto[1,2]

[1]Comisión Chilena de Energía Nuclear, Casilla 188-D, Santiago, Chile
[2]Center for Research and Application in Plasma Physics and Pulsed Power, P4, Chile
[3]Universidad Austral de Chile, Valdivia, Chile

E-mail: gavaria@cchen.cl

Abstract. Spectral measurements in the visible range of the plasma sheath ionization degree evolution on the plasma focus device PF-400J are presented. The measurements were done with temporal and spatial resolution in a plasma focus device of low stored energy: PF-400J (176-539 J, 880 nF, 20-35 kV, quarter period $\sim$ 300ns) [1]. An ICCD was attached to a 0.5 m focal length visible spectrometer, which enabled the acquisition of time resolved spectrum with 20 ns integration time throughout the whole current pulse evolution. The spatial resolution was attained using a set of lenses which allowed the focusing of a small volume of the plasma sheath in different positions of the inter-electrode space. Discharges were carried out in mixtures of Hydrogen with gases in different proportions: 5% Neon, 5% Krypton and 2% Nitrogen. Discharges using Neon as an impurity showed no ionization of the gas, just a very low intensity emission of Ne I at times much larger than the maximum current. Nitrogen, on the other hand, showed a high ionization reaching N V (N^{4+}) at the end of the axial phase, with a distinctive evolution of the ionization degree as the plasma sheath moved towards the end of the electrodes. A mixed result was found when using Krypton, since the ionization degree only reached levels around Kr II/III, even though it has an ionization potential lower than Neon.

1. Introduction

Plasma focus devices have recently become of interest to the international plasma community for its capabilities as a self scale plasma device [2], neutron source [1; 3–7], pulsed X-ray source [8], plasma shock [9], ultrasonic plasma jets [10], filamentation[11], scaled astrophysical experiments [9; 10] and a source for material testing for future fusion reactor walls [9; 12]. The interest in improving the yield of particles and radiation has lead to some studies in the influence that gas impurities could have on that production, showing that under the right conditions, an order of magnitude increase in neutron yield can be attained [13; 14]. Although this is an important finding, the observation of the ionization degree evolution of the gas impurity during the axial phase has been ignored. Speed and thickness characterization of the plasma sheath evolution for pure gases have been developed in detail in references [15; 16]. Few studies using visible spectroscopy have shown the ionization degree at the pinch for discharges in Argon [17] and the evolution of the plasma when it interacts with a target [18]. Results on the evolution of the plasma sheath ionization degree when using Neon, Nitrogen and Krypton as impurities during the axial phase of the discharge will be shown.

2. Experimental details

The experimental observations were developed at the low energy plasma focus device PF-400J (176-539 J, 880 nF, 20-35 kV, quarter period ~300 ns) [1]. The device was operated in three configurations: a mixture of 95% Hydrogen and 5% Neon; a mixture of 98% of Hydrogen and 2% of Nitrogen; a mixture of 95% Hydrogen and 5% Krypton. The mixture pressure was determined by maintaining the gas mass as close as possible to the equivalent mass used for pure Hydrogen shots. The device was operated at a repetitive scheme with a charging voltage of ~ 27 kV and the spectra was acquired with a 0.5m focal length Cerny-Turner spectrometer coupled with a 576x384 pixel ICCD detector. A 300 groove/mm diffraction grating was used due to the broad spectral region available and good response to the observed region. The plasma emission from two positions at the inter-electrode region (figure 1) was focused into a 19 circular-to-linear fiber optic bundle connected to the entrance of the spectrometer (see figure 2).The ICCD-Spark gap system was synchronized externally, allowing the acquisition of time resolved measurements with integration times of 20 ns.

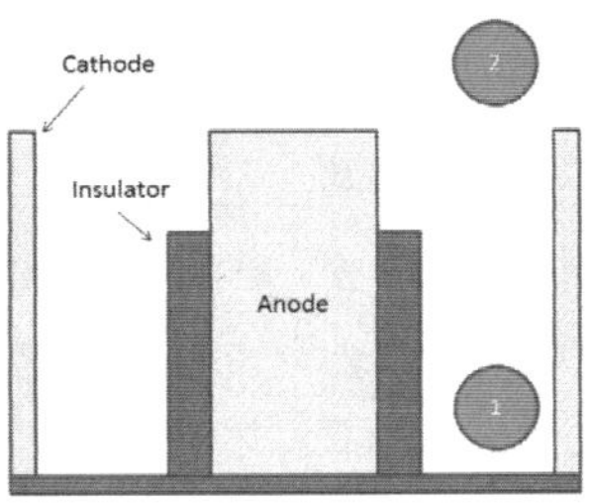

Figure 1. Plasma emission from two positions of the inter-electrode volume (green circle) was focused into the lens-to-fibre optic system.

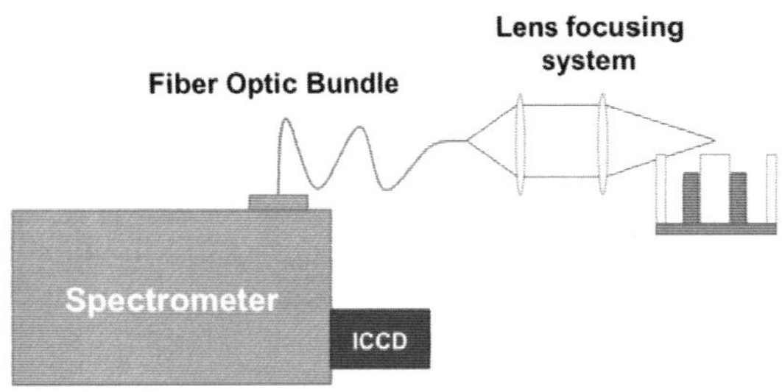

Figure 2. Scheme of the experimental setup for the spectroscopy measurements in both devices.

3. Experimental results

The electrical signals for a typical shot can be seen in figure 3. The timing of the images was obtained from the onset of the current derivative to the lowest point of the ICCD triggering signal labelled as (4). For the case of Hydrogen-Neon mixtures, the time was taken with respect to the moment of the maximum compression of the plasma column: a negative time indicates that the image was acquired before the pinch and a positive time, after the pinch. In order to observe the appearance of the H-alpha emission in the discharge chamber, a filtered photomultiplier tube was connected to an optical fibre pointed at the plasma volume, signal that is shown as (3) in the figure.

Figure 3 shows the evolution of the visible emission in the region around $\sim 625nm$ and $\sim 700nm$ for a mixture of Hydrogen and Neon. It can be seen that there is a broadening of the H_α transition at times close to the pinch (-16 ns), in comparison to the width at earlier times. This spectral region should show a larger emission at earlier times, since it includes the breakdown of the plasma sheath over the insulator, nevertheless this is not the case. In previous observations for a pure hydrogen discharge [19], the broadening of the H_α happened earlier in time. This might be related to the fact that for a H+Ne mixture, the plasma sheath is heavier and the breakdown conditions are modified. Also, the observation of excited Neon (Ne I) is only evident at times much larger than $1\mu s$ after the onset of the current. This might be related to either a rapid ionization of the Neon impurities, and thus the emission at different spectral regions; or the lack of any ionization due to the higher ionization potential of Neon compared

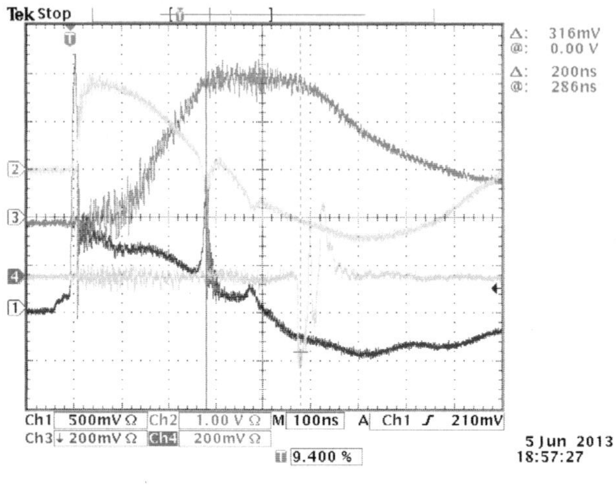

Figure 3. Electrical signals of: Voltage (1); Current Derivative, dI/dt (2); H-alpha filtered photomultiplier (3); ICCD pulser (4).

to Hydrogen. To be able to determine the emission from higher charge Neon ions, the use of a XUV spectrometer is under consideration.

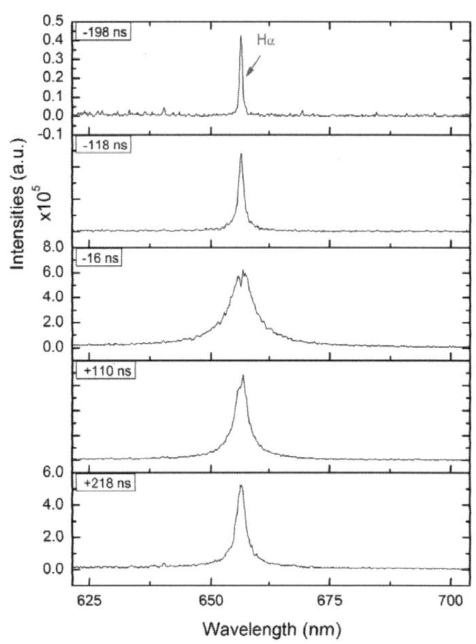

Figure 4. Evolution of the plasma sheath in time at the beginning of the axial phase. The discharge was operated at 9 mbar with a mixture of Neon (5%) and Hydrogen (95%) in the PF-400J discharge.

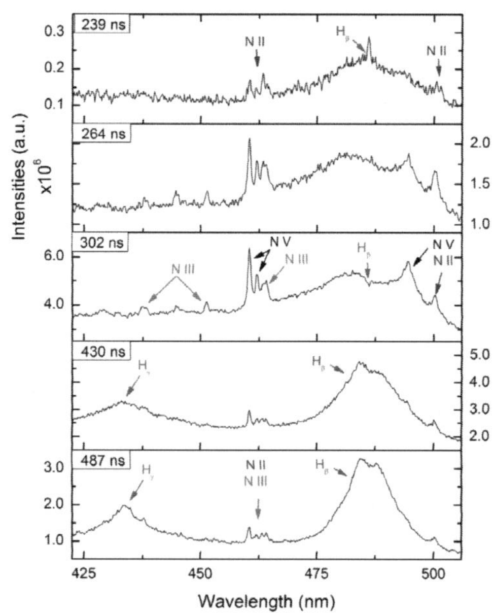

Figure 5. Temporal evolution of the visible emission of a 98% Hydrogen and 2% Nitrogen plasma, observed at the end of the axial phase in position 2 (Fig. 1).

In the case of the 98% Hydrogen and 2% Nitrogen mixtures, the acquired spectra can be seen in figure 5. The maximum current happens at $\sim 320ns$ and the pinch corresponds to $\sim 430ns$, times that are longer than expected compared to pure hydrogen (pinch at $\sim 280ns$). From the images it is possible to determine the evolution of the ionization degree of the Nitrogen impurity from NII(N^{1+}), $100ns$ before the pinch, to NV (N^{4+}) almost at pinch time. If the line emission intensity can be associated to the position of the plasma sheath, then the highest emission is seen at a time that should correspond to around $300ns$, which gives a difference of only $30ns$ with respect to the pinch time. This is a much shorter time than expected from observations in pure Hydrogen [15] ($\sim 85ns$), where the sheath should be lighter and faster than the one seen in a gas mixture. At this moment the authors do not have an explanation to this measurement.

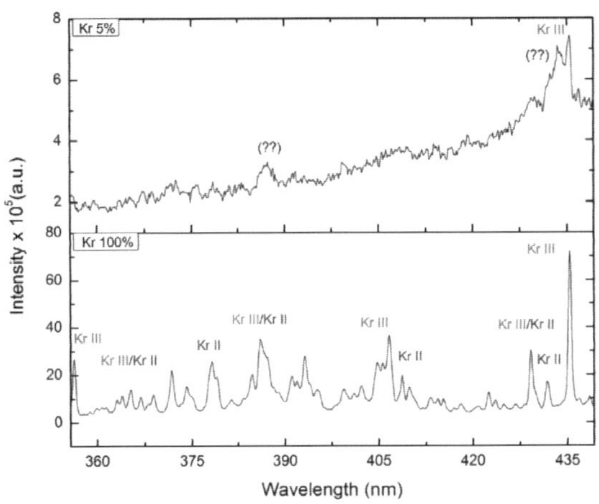

Figure 6. Spectral profiles for a 5% Krypton and a 100% Krypton discharge.

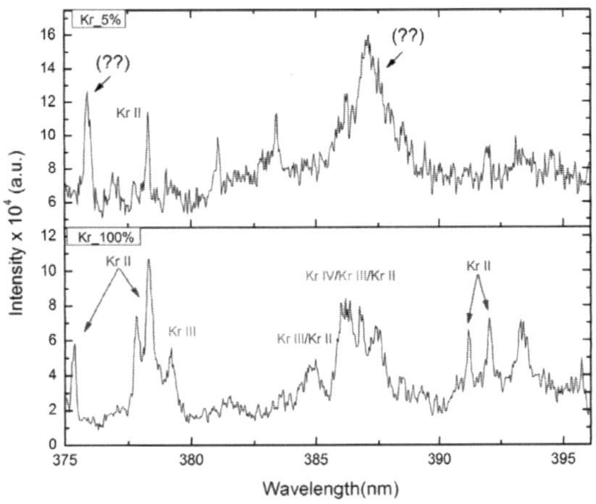

Figure 7. Detail of the spectral region of figure 6 in order to identify the broad peak at $\sim 387nm$

Several measurements were carried out using small percentages of Krypton to determine the behaviour of heavier impurities in the evolution of the plasma. Figure 6 shows two profiles, acquired at the same position as Nitrogen, where it is possible to observe that the use of a small concentration of Krypton does not allow the identification of line transitions. A 100% concentration of Krypton shows that the discharge is able to ionize the gas to low degrees of ionization (Kr II-III) but when the mixture is used, this ionization is not attained. This behaviour is not understood since the ionization potential of Krypton is even lower than the potential for Nitrogen. Further measurements using Krypton, in different spectral regions, are under development. In order to resolve the lines in the broad feature described before, spectra was acquired using the 1200 l/mm grating as seen in figure 7. From the spectra it is clear that there is no correspondence between the 5% and 100% concentration cases, an even some unresolved features can be seen.

4. Conclusion

The results presented above show the behaviour of the spectroscopic emission from the plasma sheath for different conditions of gas impurity in different positions of the plasma focus discharge. The addition of small concentrations of different gases directly impact the dynamics of the plasma sheath: It becomes slower, even when the mass is kept constant; and the ionization of the impurities is not related to the ionization potential (Nitrogen vs Krypton). The difference seen between the evolution of Nitrogen and Krypton impurities needs further studies, searching for higher ionization degrees in different spectral regions (UV-Visible/XUV).

Acknowledgments
The authors would like to acknowledge the financial support from CONICYT-PAI Inserción 791100020, FONDECYT Iniciación 11121587, CONICYT-PIA Anillo ACT1115, Bilateral Project Chile-Argentina: CONICYT ACE-01 / ANPCyT-PICT-2697 and FONDECYT Regular 1120801

References

[1] Silva P, Moreno J, Soto L, Birstein L, Mayer R E and Kies W 2003 *Applied Physics Letters* **83** 3269–3271

[2] Soto L, Pavez C, Tarifeno A, Moreno J and Veloso F 2010 *Plasma Sources Science and Technology* **19** 055017

[3] Soto L, Silva P, Moreno J, Zambra M, Kies W, Mayer R E, Clausse A, Altamirano L, Pavez C and Huerta L 2008 *Journal of Physics D: Applied Physics* **41** 205215

[4] Soto L, Pavez C, Moreno J, Barbaglia M and Clausse A 2009 *Plasma Sources Science and Technology* **18** 015007

[5] Soto L 2005 *Plasma Physics and Controlled Fusion* **47** A361

[6] Klir D and Soto L 2012 *Plasma Science, IEEE Transactions on* **40** 3273–3279

[7] Veloso F, Pavez C, Moreno J, Galaz V, Zambra M and Soto L 2012 *Journal of Fusion Energy* **31** 30–37

[8] Pavez C, Pedreros J, Zambra M, Veloso F, Moreno J, Ariel T S and Soto L 2012 *Plasma Physics and Controlled Fusion* **54** 105018

[9] Soto L, Pavez C, Moreno J, Inestrosa-Izurieta M J, Veloso F, Gutiérrez G, Vergara J, Clausse A, Bruzzone H, Castillo F and Delgado-Aparicio L F 2014 *Physics of Plasmas* **21** 122703

[10] Pavez C, Pedreros J, Tarifeño-Saldivia A and Soto L 2015 *Physics of Plasmas* **22** 040705

[11] Soto L, Pavez C, Castillo F, Veloso F, Moreno J and Auluck S K H 2014 *Physics of Plasmas* **21** 072702

[12] Inestrosa-Izurieta M, Ramos-Moore E and Soto L 2015 *Nuclear Fusion* **55** 093011

[13] Verma R, Lee P, Lee S, Springham S, Tan T, Rawat R and Krishnan M 2008 *Applied Physics Letters* **93** 101501–101501

[14] Bures B L, Krishnan M, Madden R E and Blobner F 2010 *Plasma Science, IEEE Transactions on* **38** 667–671

[15] Veloso F, Tarifeño-Saldivia A, Pavez C, Moreno J, Zambra M and Soto L 2012 *Plasma Physics and Controlled Fusion* **54** 095007

[16] Veloso F, Moreno J, Tarifeño-Saldivia A, Pavez C, Zambra M and Soto L 2012 *Measurement Science and Technology* **23** 087002

[17] Feugeas J and Grigioni G 1997 *Journal of Physics D: Applied Physics* **30** 2026

[18] Skladnik-Sadowska E, Malinowski K, Sadowski M, Kubkowska M, Jakubowska K, Paduch M, Scholz M, Garkusha I, Ladygina M and Tereshin V 2011 *Contributions to Plasma Physics* **51** 288–292

[19] Avaria G, Cuadrado O, Moreno J and Soto L 2015 *Journal of Physics: Conference Series* **591** 012024

Journal of Physics: Conference Series **720** (2016) 012042 doi:10.1088/1742-6596/720/1/012042

Ion beam measurement using Rogowski coils in a hundred of joules dense plasma focus device.

J. Jain[1,2,3,*], J. Moreno[2,3], C. Pavez[2,3], B. Bora[2,3], M. J. Inestrosa-Izurieta[2,3], G. Avaria[2,3] and L. Soto[2,3,*]

[1]Universidad de Talca, Chile
[2]Comsion Chilena de Energia Nuclear, Casilla 188-D, Santiago Chile
[3]Center for Research and Applications in Plasma Physics and Pulsed Power, P4

**Email:* jalajjain83@gmail.com , lsoto@cchen.cl

Abstract. In present work an effort has been made to measure the ion beams generated during experiment with PF-400J plasma focus device, using an array of two Rogowski coils with time of flight analysis. It was found that the coils measure the signals of beam for a particular range of operating pressure. The beam signals were recorded at 20, 15, 12, 10, 9,8,7,6 and 5 mbar filled pressure of hydrogen gas. The optimized pressure range for good plasma column formation for this device was found about 9 mbar. At 15 mbar no or very weak beam signals were observed by Rogowski coil which was kept relatively far from the top of the anode and at 20 mbar there were no beam signals observed in both of the coils. The calculated beam energy is found to have maximum value at 9 mbar of filled hydrogen gas pressure.

1. Introduction

Dense plasma focus devices (DPF) are well known to emit different kind of radiations as well as particles and particle beams [1-3]. Since the generated beams might have various applications, it is desirable to characterize the emitted beams. It is believed that the X-rays and beams emission from DPF devices takes place because of presence of the induced electric fields, generated due to different instabilities in pinch phase [4-6]. In fact the m = 0 instability is considered as one of the most strong candidate to induce the high electric field during plasma column formation. However the charged particle acceleration mechanism, which creates beam, is still under active investigation for more clear understanding. Intensive experimental investigations on beam generation from DPF devices have been carried out by many researchers [7-8].

The conventional diagnostic techniques (Faraday cup and/or Thomson parabola on CR-39 plastics) are active and charge particle interaction with the detector provide beam signal. Also pinholes, which usually have dimensions less than the plasma column, are used before the beam detectors. The beams may be generated at any place in plasma column and there might be difficulties to align pinhole with the beam originated place. Moreover, there are chances that high energetic charge particles which may arise during axial shock phase [9] mimic the beam signals by making interaction with Farady cup and leaving tracks on CR-39 plastics.

In this work we propose a passive non intrusive diagnostic technique using Rogowski coil in order to get information about beams. The choice of Rogowski coil is advantageous over the other diagnostic techniques in a way that it senses change in magnetic flux induced by beam current and does not perturb the measuring beam. Also the change in magnetic flux induced by the non-beam

Journal of Physics: Conference Series **720** (2016) 012042 doi:10.1088/1742-6596/720/1/012042

charge particles will not be able to produce a strong signal in Rogowski coil, thus the measured signal has the large chances to be a beam signal generated in plasma column. Also use of Rogowski coil does not demand pinhole arrangement and the beams which are generated at any place in plasma column can easily be detected. A more advantageous use of Rogowski coil is that the beam signal will not be lost and can be used further.

In following section 2 we will describe ion beam measurement using Rogowski coil and calculations for beam energy estimation. Results are discussed in section 3 and the work is concluded in section 4.

2. Experimental arrangement

In figure 1 a schematic sketch of experimental arrangement for PF-400J [10-11] is presented. The central electrode (anode) is made of stainless steel has diameter of 12 mm. The anode length covered by an insulator (alumina) is ~20.5 mm and the effective length was ~8.0 mm. The anode is surrounded by 8 copper cathode bars of diameter ~4.7 mm. Two Rogowski coils separated by 4.50 cm were placed in a drift tube at the top of the anode. The coils were designed by winding a thin copper wire on RG-58 co-axial cable. The coil nearer (designated as RC5) to the anode consist 45 turns and the coil farther (designated as RC1) from the anode has 20 turns. Both of the coils have similar inner diameter ~2.9 cm. The dissimilarity in number of turns will only make difference in the rate of increment or decrement in current derivative signal but the start time of signal in both the coils should not be affected by the number of turns.

The experiment was performed for different pressures (20, 15, 12, 10, 9, 8, 7, 6 and 5 mbar) of hydrogen gas. During the experiments 15 discharges for each pressure have been performed in dense plasma focus device PF-400J. The distance between the two coils (4.5 cm) is divided by the differences of time of appearing the current signals at two different positions in Rogowski coils (time of flight analysis). This estimation of beam speed is used to calculate the beam kinetic energy.

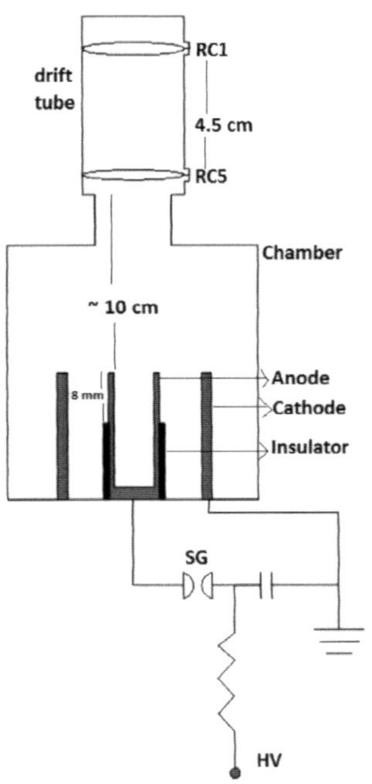

Figure 1. Schematic representation of plasma focus device PF-400J with Rogowski coil arrangement for beam measurements.

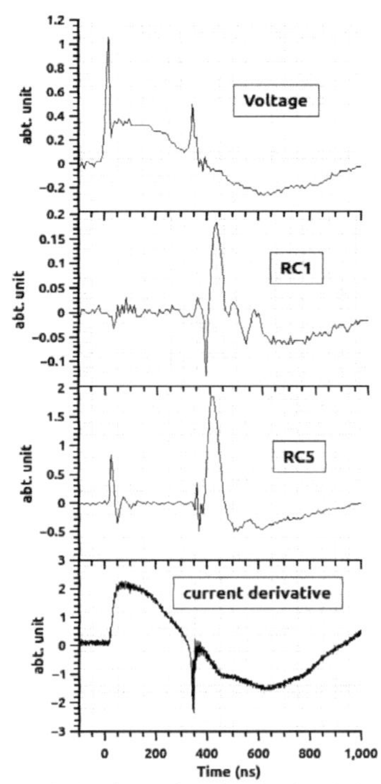

Figure 2.Various signals obtained during the experiment.

Journal of Physics: Conference Series **720** (2016) 012042 doi:10.1088/1742-6596/720/1/012042

3. Results and discussion

Typical plasma focus discharge signals obtained during the experiments with plasma focus device PF-400J are presented in figure 2. It is clearly visible from figure 2 that the ion beam and X-rays emission take place after plasma column formation, which is indicated by a dip in current derivative signal of plasma discharge. During the experiments more than one signal were detected by the Rogowski coils. In the present study the only signal (appears first) which is indicated by an arrow in figure 3 is analyzed. Here we made an assumption that this signal is produced by proton beam (filled gas is hydrogen). In true sense, apart from the proton beam signals, more than one signal can be produced by impurity ions, which may arise from the electrode material erosion during and after the plasma column formation. But due to their mass these signals should appear at later times than the proton beams. However further analysis is needed in order to understand the origin of the signals which appear at later time. The obtained beam signals (current derivatives in arbitrary units obtained from two Rogowski coils RC5 and RC1) at different pressures-12, 10, 9, 8, 7, 6 and 5 mbar of hydrogen gas are shown in figure 4 (we did not find the beam signals in RC1 at 15 mbar and in both the coils at 20 mbar). Two signals of the same beam in two different Rogowski coils, which are separated by a distance of 4.50 cm, are clearly visible. This distance between the coils is divided by the time (start of signal) taken by the beam to reach from first coil (RC5) to the second coil (Rc1) to estimate the beam speed and later this speed is used to estimate the kinetic energy of beam.

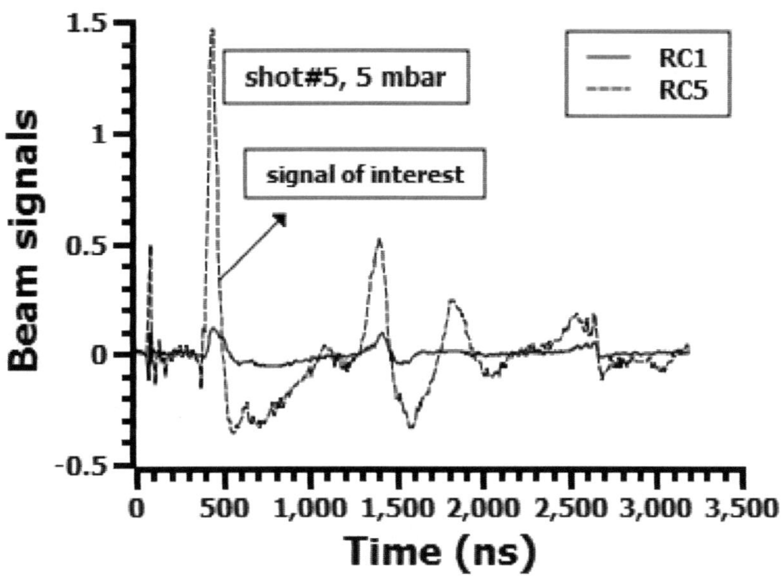

Figure 1 Experimentally observed beam signals using Rogowski coils RC5 and RC1. Only the first signal is taken in to account in order to calculate the proton beam energy.

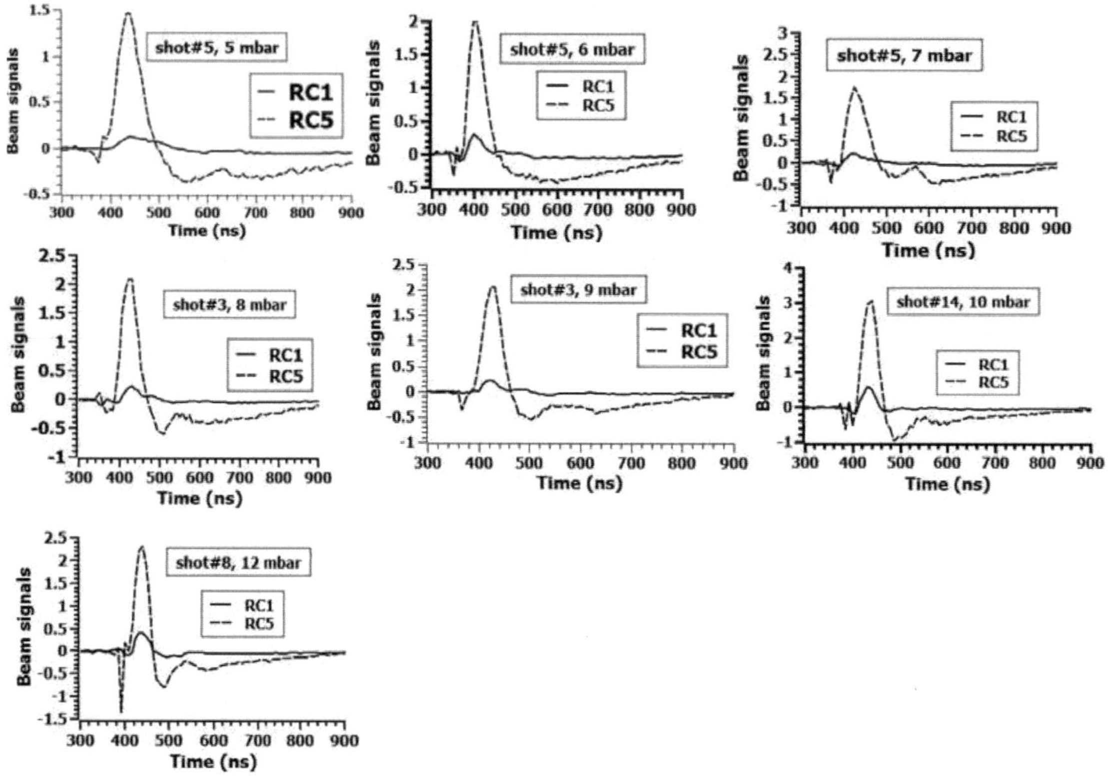

Figure 2 Beam signals at various pressures. The difference of appearing time in two Rogowski coils was used to calculate the beam speed. Later this speed is used to calculate the kinetic energy of beam.

Figure 5 shows the variation of kinetic energy of proton beam with pressure. The trend of this variation shows maximum kinetic energy of proton beams at 9 mbar. Nonetheless, the kinetic energy obtained at 8 mbar is out of trend. The plasma focus devices have an optimized range of working pressure for a good pinch, 7 – 12 mbar in case of PF-400J for hydrogen gas, there is a possibility that the maximum kinetic energy of proton beams may oscillate within this range if repeating the experiments several times. At this point we can only say that the energies of proton beam measured by the Rogowski coils are 28.55 ± 6.54, 40.81 ± 20.31, 79.10 ± 12.36, 48.0 ± 12.37, 98.42 ± 48.23, 53.32 ± 10.78, 35.27 ± 5.97 keV for the pressures 5, 6, 7, 8, 9, 10 and 12 mbar respectively, of hydrogen gas in PF-400J plasma focus device.

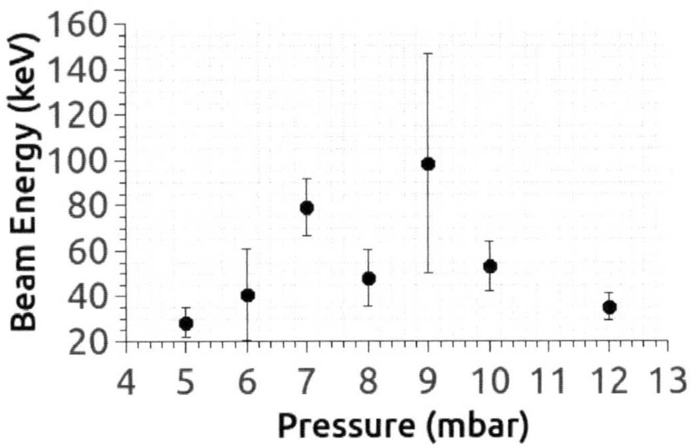

Figure 5 Variation of proton beam kinetic energy with pressure.

4. Conclusion

Proton beam measurements have been performed using an array of two Rogowski coils at the top of the anode in plasma focus device PF-400J, considering time of flight analysis. The experiments were performed at 20, 15, 12, 10 , 9, 8, 7, 6 and 5 mbar of filled hydrogen gas. The beam energy was estimated for the pressures 12, 10, 9, 8, 7, 6 and 5 mbar and at 15 mbar there was no or very weak signal was found in RC1 and at 20 mbar there were no signals found in both the coils. The energies of proton beams were found to have values ~30 ± 6 – 100 ± 50 keV for a pressure range from 5 to 12 mbar of hydrogen gas. The maximum kinetic energy was found at 9 mbar pressure.

Acknowledgement

The work is supported by grant ACT-1115, CNICYT, Chile.

References

[1] Soto L, 2005 Plasma Phys. and Control. Fusion **47** A361

[2] Silva P, Moreno J, Soto L, Birstein L, Mayer R E and Kies W 2003 *Appl. Phys.letter***83** 3269

[3] Soto L, Silva P, Moreno J, Zambra M, Kies W, Mayer R E Clausse A, Altamirano L, Pavez C and Huerta L 2008 *J. of Phys. D: Appl. Phys.***41** 205215

[4] L. Soto, C. Pavez, A. Tarifeño, J. Moreno and F. Veloso, 2010 *Plasma Sources Sci. and Technol.* **19** 055017

[5] Bernard A, Bruzzone H, Choi P, Chuaqui H, Gribkov V, Herrera J, Hirano K, Krejci A, Lee S, Luo C, Mezzetti F, Sadowski M, Schmidt H, Ware K, Wong C S, Zoita V 1998*J. Moscow Phys. Soc.* **8** 93

[6] Haines M G 1983*Nuclear Instruments and Methods***207** 179

[7] Stygar W, Gerdin G, Venneri F and Mandrekas J 1982 *Nucl. Fusion***22** 1161

[8] Bhuyan H, Chuqaqui H, Favre M, Mitchell I and Wyndham E 2005 *J. Phys. D: Appl. Phys.***38** 1164

[9] Soto L, Pavez C, Moreno J, Inestrosa-Izurieta M J, Veloso F, Gutierrez G, Vergara J, Clausse A, Bruzzone H, Castillo Fand Delgado-Aparicio L F 2014*Phys. Plasmas***21** 122703

[10] Soto L, Silva P, Moreno J, Silvester G, Zambra M, Pavez C, Altamirano L, Bruzzone H, Barbaglia M, Sidelnikov Y and Kies W 2004 *Braz. J. Phys.***34** 4b

Pavez C, Pedreros J, Zambra M, Veloso F, Moreno J, Ariel T S and Soto L 2012 *Plasma Phys. Control. Fusion* **54** 105018

Characterization of x-rays pulses from a hundred joules plasma focus to study its effects on cancer cells.

J. Jain[1,2,3,*], J. Moreno[1,2], G. Avaria[1,2], C. Pavez[1,2], B. Bora[1,2], M. J. Inestrosa Izurieta[1,2], D. Diez[4], O. Alvarez[4], J. Tapia[4], K. Marcelain[4], R. Armisen[4] and L. Soto[1,2,*]
[1]Comisión Chilena de Energía Nuclear, Casilla 188-D, Santiago Chile
[2]Center for Research and Applications in Plasma Physics and Pulsed Power, P4
[3]Universidad de Talca, Chile
[4]Instituto de Ciencias Biomédicas & Centro de Investigación y Tratamiento del Cáncer, Facultad de Medicina, Universidad de Chile, Chile

**E-mail:* *jalajjain83@gmail.com, lsoto@cchen.cl*

Abstract. With the purpose to study the effects of pulsed x-rays radiation on cancer cells, the total doses of x-ray pulses and the temporal duration of the pulse has been characterized in a hundred joules plasma focus device (PF-400J, 130 kA achieved in 300ns, 30 kV, 880 nF, 38 nH). TLD dosimeters were located outside of the discharge chamber, at 96 mm from the anode top. In addition, two photomultipliers with plastic scintillator were located in axial and radial directions. From the statistical analysis of the TLD and photomultiplier signals, was possible to estimate that a single shot has a total dose of the order of 30±15 μSv with a duration of the order of 12±3.6 ns at FWHM. Preliminary experiments using MCF7, a breast cancer cell line, were performed. Cells were irradiated at 96 mm from the anode top with 300 cumulative x-ray shots and cell proliferation was evaluated at 24, 48 and 72 hours later.

1. Introduction

Plasma focus devices (PF) have been paid attention due to their usefulness for fundamental as well applied studies [1, 2, 3]. On one hand, PF devices could be used for basic science [1, 2], including astrophysical plasma research [4, 5] and on the other hand they have various applications in material and biological sciences [6-11]. Indeed, during different phases: breakdown, rundown, compression, pinch and axial plasma shock of PF discharges; various phenomena take place [1, 3]. Pinch and axial plasma shock are considered as rich phases as a point of view of application as well as fundamental studies. Moreover, recently it has been discovered that the axial plasma shock phase [12] have application in the field of interaction of powerful plasma with the first wall of fusion reactor wall like Tokamaks and inertial fusion [13]. On the other hand, many groups have been studied the pinch phase [14] in order to understand the basic physical processes to make it more applied in nature: magnetic field reconnection [15], filaments [16, 17], plasma jets [18]. During the pinch phase there is emission of soft and hard x-rays [18] as well as neutrons if the filled gas is deuterium. The emission of various types of radiation during pinch phase makes PF devices applicable to material [19] and biological sciences [7]. In this work the effects of pulsed x-rays radiation on cancer cells, using a table-top plasma focus device is explored.

Radiation is used for the treatment of cancers. In radiation therapy, tumors are exposed to high doses of radiation from a continuous source. The amount of dose delivered to cancerous tumor depends upon the type and the stage of the cancer. For example, solid epithelial tumors can be treated with dose range 60 – 80 Gy while lymphomas can be treated with 20 – 40 Gy. Total radiation dose can be fractionated over a long time. Fractionate dose can be kept as 1.5 – 2.0 Gy per day, five days a week [20]. Fractioning the total dose allows healthy cells (which are also exposed to the radiation along with cancer cells) to recover themselves. Also fractionation may help to re-oxygenate the chronically or acutely hypoxic cells, which are more radio-resistance.

The effect of high dose exposure on cancerous cells is well characterized. Nonetheless, low doses effects cannot be predicted by extrapolating linearly from higher to lower dose effects. Given that decreasing radiation dose would have lower side effects on healthy cells. Now days many groups are trying to study the behavior of cancer cells under the action of low doses. N.V. Litvyakovet. al. [21] used pulsed periodic x-rays at pulse frequency 10 – 16 Hz on cultured cells from tumors of different histogenesis. They claimed that the pulse periodic x-rays exposure to the cells induces apoptosis in 20 – 40 % of cells. In their work the total absorbed doses were 30 – 120 mGy. Since the study done in [12] showed positive results of low dose irradiation on cancer cell, devices that can produce such low doses need to be studied.

Low energy plasma focus devices may be a good choice considering radiation emission. The plasma focus devices emit radiation (x-rays, neutrons) in pulse form. The advantage to choose plasma focus over other devices for this study is that they can produce the pulses of x-rays and neutrons depending on the filled gas. For example if the filled gas is hydrogen only x-rays are produced and if the deuterium is used as a filled gas, x-rays as well as neutron emission takes place. In addition, the size and cost of low energy plasma focus devices are lower than that of conventional radiation sources like-linear accelerators (LINAC) [22]. Therefore low energy plasma focus devices may provide an option of portable radiation emitting source for low dose radiation therapy applicable to cancers. Although, a lot of scientific efforts are needed in order to transfer pulse radiation technology of plasma focus devices to clinical trials.

In section 2 we will describe the experimental setup. The results are discussed in section 3 and the work is concluded in section 4.

2. Experimental setup

2a. Pulsed x-ray characterization and equivalent dose

Pulsed radiation of X-rays was produced using dense plasma focus device PF400J (C_0 = 880 nF, L_0 = 38 nH, 30 kV charging voltage, 130 kA of peak current achieved in 300 ns) [23]. A schematic of plasma focus PF-400J is shown in figure 1. The device PF400J consists of 8 copper cathodes and a central stainless steel hollow anode, which has an effective length of 8 mm. Rest of the length of anode is covered by an insulator (alumina, shown in figure 1). Diameter of anode is 12 mm. Hydrogen gas at 9 mbar of pressure was used in order to get maximum x-rays signal from the device. The experiment was performed with TLD-100 dosimeters array, which contained 21 dosimeters, in order to estimate the doses. Reason behind to use TLD-100 dosimeters for this study is that the material of these dosimeters have similar response to radiation as human body. The dosimeters array, as shown in figure 1, was kept on a specific holder which was specially designed for biological cell experiment. The holder was mounted outside the PF400J in the axial position. Three MCF-7 cells containing dishes are kept at the top of the holder before dosimeter array. The distance between the top of the anode and the bottom culture dish was 96 mm (95 mm from top of the anode to bottom of the holder +

1 mm thickness of holder). Height of each culture dish is about 13.0 mm so the distance between top of the anode and dosimeters array was about 135.00 mm. Two photomultiplier tubes (PMT), namely FM1, FM3 for hard x-rays detection were aligned in the axial and radial positions. The PMT which was in axial position was separated by 96 cm from top of the anode and radial PMT was at a distance about 131 cm from the center of the anode. In a first stage, the PMTs are used as referential devices. Later, the PMT signal will be used directly as doses measurements, considering the crossed calibration with the TLD dosimeters. The experiment was performed for different number of PF discharges, but in this work only the results for 300 discharges with x-rays are presented.

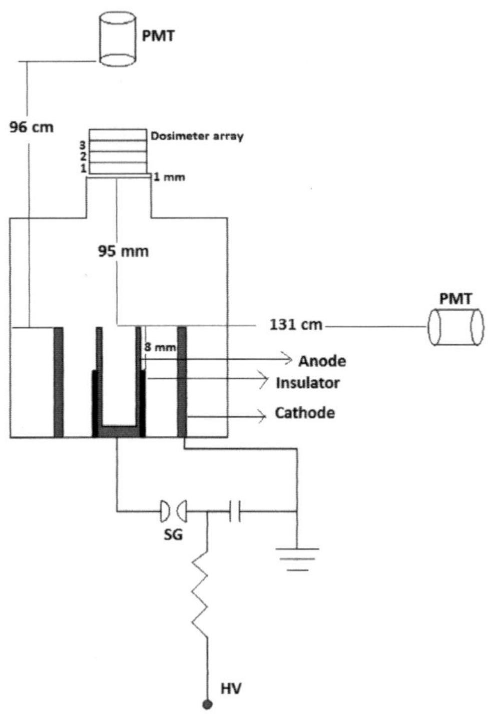

Figure 1. Schematic of PF-400J for cancer cell irradiation.

Figure 2 shows the electrical signals obtained during the experiments with plasma focus device PF400J. A dip in current derivative signal (obtained using Rogowski coil) is a signature of plasma column formation. In fact, during plasma column formation there is an increment in total impedance and hence reduction in current. We can clearly see the emission of pulse x-ray in figure 2. For this experiment maximum current was obtained about ~130 kA and the x-ray signal per shot, which is mean of area under curve of all the 300 x-ray signals, obtained using PMT in axial direction is about $(6.55 \pm 3.16) \times 10^{-8}$ V.s. Total accumulated signal of x-ray (sum of area under curve of all 300 x-ray signals) in PMT in axial direction was found about 2.00×10^{-5} V.s. These data were obtained for 300 discharges with x-rays signal. The total accumulated doses for 300 x-rays signal were found about tens of mSv. The dose per pulse was obtained about tens of μSv. The average x-ray pulse duration was ~ 10 ns.

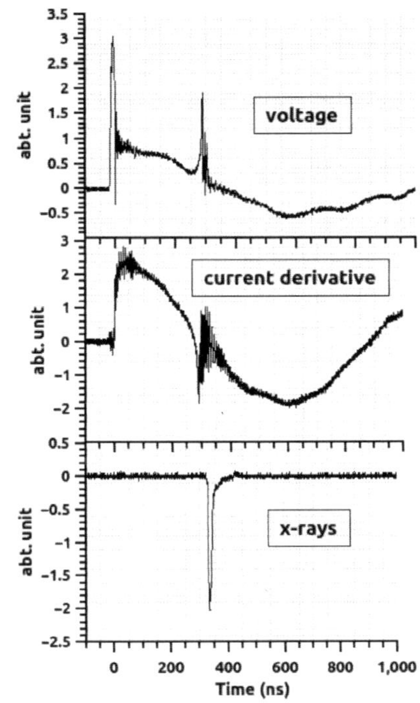

Figure 2. Various electrical signals obtained during the experiment with plasma focus device PF-400J.

2b. Cell culture preparation

Breast adenocarcinoma derived cells, MCF7, were maintained in Dulbecco's modified medium (DMEM) supplemented with 10% bovine fetal serum, in an atmosphere of 5 % CO_2 at 37 ^{0}C. The day before irradiation, cells were cultured in five 35mm dishes: one control dish was kept in the incubator, one control was mock irradiated and three plates were irradiated. After irradiation,every cell culture was seeded in 96-well plates, at a density of 2×10^3 cell per well in 100 μl complete medium.

2c. Cell Proliferation assay (MTS)

Cell proliferation was assessed at 24, 48 and 72 hours after irradiation by using a MTS (3-(4,5-dimethylthiazol-2-yl)-5-(3-carboxymethoxyphenyl)-2-(4-sulfophenyl)-2H-tetrazolium) assay. 20 μl MTS solution was added to each well of the 96 well plates and incubated at 37 ^{0}C. Live cells interact with MTS solution to produce formazan, which absorb light at 570 nm. Proliferation rate was estimated as the absorbance (abs) of every time point normalized by the abs at time 0.

3. Results and discussion

In figure 3, proliferation rate of breast cancer cells, MCF-7,exposed to 300 shots of pulsed x-ray is shown. Control incubator represent the MCF-7cells that were not exposed to x-rays and were kept in control conditions. Mock control corresponds to cells that although were not irradiated, they had the same manipulation as irradiated cells.There were three cell culture dishes numbered 1, 2 and 3 as shown in figure 1. Proliferation rate for the cells kept first on the holder (at top of the anode) (culture1) and cells which were kept at last (culture 3) are shown. As it can be seen from figure 3, 48 hours after irradiation, the proliferation rate seems to be higher for the irradiated cells (cultures 1 and 3) than for control cells. While 72 hours after irradiation, the effect was reversed for culture 3, which seems to show a decrease in proliferation rate. However, important is to notice that the present study is limited to only one experiment and biological samples, especially cell lines, have high intrinsic variance, therefore, more experiments are required to state confident conclusions. With this preliminary study the dose per shot was found about tens of μSv in ~ 10 ns pulse duration time, which is about mSv for total 300 shots. There is a possibility to enhance the emission of x-rays from PF-400J either by pre-ionizing mechanism [24] or by inserting another material (lead/copper) in hollow anode [25].

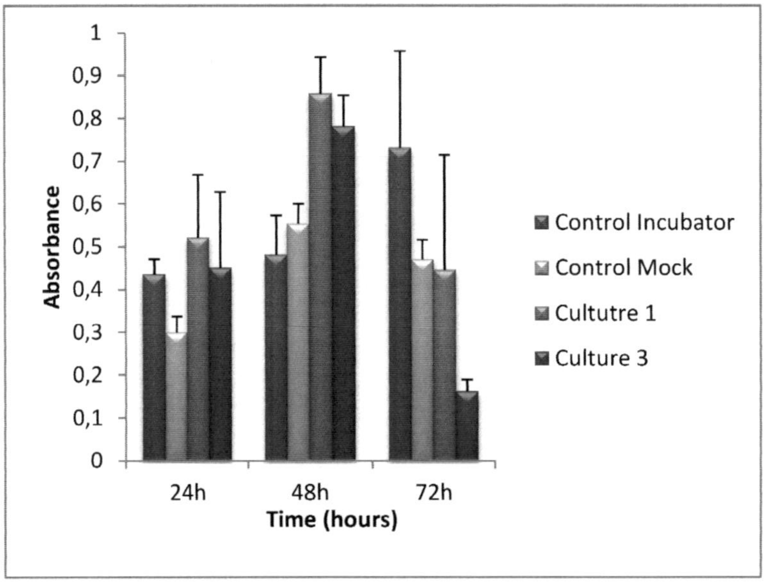

Figure 3.Proliferation rate of cancer cell line MCF-7 at 24, 48 and 72 hours after irradiation. The cells were exposed to 300 shots of x-ray pulses.

4. Conclusion

Preliminary experiments with plasma focus device PF-400J has been carried out in order to irradiate the breast cancer cells MCF-7, by pulsed x-rays. Based on the obtained results it can be concluded that the x-rays radiation emitted from plasma focus device affect the cancer cells. However in the present work the studies performed on cancerous cells are limited only for one experiment with 300 x-ray pulses. In order to make a trustable conclusion a repetitive behavior of both the studies-emission of radiation (x-rays, neutrons) from PF400J and response of cancerous cells to the irradiation should appear with some standard deviation. Due to this reason further studies by keeping in mind the statistical approach will be carried out.

Acknowledgement
The work is supported by grant ACT-1115, CONICYT, Chile.

References

[1] Soto L, 2005 *Plasma Phys. and Control. Fusion* **47**, A361
[2] L. Soto, C. Pavez, A. Tarifeño, J. Moreno and F. Veloso, 2010 *Plasma Sources Sci. and Technol.* **19**, 055017
[3] Bernard A, Bruzzone H, Choi P, Chuaqui H, Gribkov V, Herrera J, Hirano K, Krejci A, Lee S, Luo C, Mezzetti F, Sadowski M, Schmidt H, Ware K, Wong C S, Zoita V 1998 *J. Moscow Phys. Soc.***8** 93
[4] Pouzo J O and Milanese M M 2003 *IEEE Trans. Plasma Sci.* **31**, 1237

[5] Pavez C, Pedreros J, Tarifeño-Saldivia A and Soto L, 2015 *Physics of Plasmas* **22**, 040705

[6] Dubrovsky V, Gazaryan I G, Gribkov V A, Ivanov Yu P, Kost O A, Orlova M A, and Troshina N N, 2003 *Journal of Russian Laser Research* **24**, 289

[7] Orlova M A, Kost OA, Gribkov V A, Gazaryan I G, Dubrovsky A V, Egorov V A, and Troshina N N 2000 *Applied Biochemistry and Biotechnology* **88**243

[8] Stevens C V 1988 *EIR Science and Technology* **15**, 22

[9] Moreno C, Vénere M, Barbuzza R, Fresno M D, Ramos R, Bruzzone H, González F P J and Clausse A, 2002 *Braz. J. Phys.* **32**, 20

[10] Hussein E and Waller E, 1998 *Radiat. Meas.* **29,** 581.

[11] Lee S , Kudryashov V, Lee P, Zhang G, Serban A, Liu M, Feng X, Springham S,Wong T and Selvam C, 1998 *ICPP and 25th EPS Conf. on Controlled Fusion and Plasma Physics* vol 22C, 2591.

[12] Soto L, Pavez C, Moreno J, Inestrosa-Izurieta M J, Veloso F, Gutierrez G, Vergara J, Clausse A, Bruzzone H, Castillo Fand Delgado-Aparicio L F, 2014 *Physics of Plasmas* **21**, 122703

[13] Inestrosa-Izurieta M J, Ramos-Moore E and Soto L, 2015 *Nuclear Fusion* **55**, 093011

[14] Haines M G, 2011 *Plasma Phys. Control. Fusion* **53**, 093001

[15] Kubes P, Klir D, Kravarik J, Rezac K, Kortanek J, Krauz V,Mitrofanov K, Paduch M, Scholz M, Pisarczyk T, Chodukowski T, Kalinowska Z, Karpinski L and Zielinska E, 2013 *Plasma Phys. Control. Fusion* 55, 035011

[16] Soto L, Pavez C, Castillo F, Veloso F, Moreno J, and Auluck S K H, 2014 *Physics of Plasmas* **21**, 072702

[17] Kubes P, Paduch M, Cikhardt J, Kortanek J, Cikhardtova B, Rezac K, Klir D, Kravarik J and Zielinska E, 2014 *Phys. Plasmas* **21**, 122706

[18] Pavez C, Pedreros J, Zambra M, Veloso F, Moreno J, Tarifeño-Saldivia A and Soto L, 2012 *Plasma Phys. and Control. Fusion* **54**, 105018

[19] Rawat R S, 2013 *IEEE Trans. on Plasma Sci.* **41**, 701

[20] http://en.wikipedia.org/wiki/Radiation_therapy

[21] Litvyakov N V, Rostov V V, Bol'shakov M A, Buldakov M A, Afanas'ev K V, Astapenko A N, Kutenkov O P and Cherdyntseva N V, 2005 *Biophycs* **50**, S51

[22] Barker Jr J L, Garden A S, Ang K K, O'Daniel J C, Wang H, Court L E, Morrison W H, Rosenthal D I, Clifford Chao K S, Tucker S L, Mohan R, Dong L, 2004 *International Journal of Radiation Oncology BiologyPhysics* **59**, 960

[23] Silva P, Moreno J, Soto L, Birstein L, Mayer R, Kies W, 2003, *Applied Physics Letters* **83**, 3269

[24] Ahmad S, Sadiq M, Shafiq M, Waheed A, Lee P and Zajaullah M, 2006 *Plasma Sources Sci. Technol.* **15**, 314

[25] Hussain S, Shafiq M, Ahmad R, Waheed Aand Zakaullah M, 2005*Plasma Sources Sci. Technol.* **14**, 61

Collision dynamics of laser produced carbon plasma plumes

M. Favre[1], H. M. Ruiz[2], D. Cortés[1] F. Merello[1], H. Bhuyan[1], F. Veloso[1] and E. Wyndham[1]

[1] Instituto de Física, Pontificia Universidad Católica de Chile, Av. Vicuña Mackenna 4860, Santiago, Chile

[2] Departamento de Física, Universidad Técnica Federico Santa María, Av. España 1680, Valparaíso, Chile

E-mail: mfavre@fis.puc.cl

Abstract. We present preliminary experimental observations of the collision processes between two orthogonal laser produced plasmas in a low pressure neutral gas background. A Nd:YAG laser, 340 mJ, 3.5 ns, at 1.06 μm, operating at 10 Hz, is used in the experiments. The main laser beam is divided in two beams by a 50% beam splitter, and then focused over two rotating graphite targets, with characteristic fluence 3.5 J/cm^2. Experiments are conducted in a range from a base pressure of 0.3 mTorr, up to 50 mTorr argon. The dynamics of the laser plasmas is characterized by time resolved and time integrated optical emission spectroscopy (OES), with 20 ns and 10 ms time resolution, and 50 ns time resolved plasma imaging of visible plasma emission. Clear effects of the neutral gas background on the post-collision plasma dynamics are identified. The overall dynamics of the post-collision plasma is found to be consistent with high collisionality of the carbon plasma plumes, which results in full stagnation on collisioning.

1. Introduction

There is a renewed interest in investigating the dynamics of laser produced plasma plumes expanding in neutral background gas environments [1–3]. These investigations have been motivated by the study of collisional processes between laser plasma and neutral gas species, which play a crucial role in plasma assisted deposition [4], nanoparticle formation [5] and high harmonic generation [6], among others. In this context, particular attention has been given to the study of collisioning laser plasma plumes [7–11]. Here the main motivations are pulsed laser deposition (PLD) [19], composite nanoparticle formation [13] and simulation of astrophysical processes [14]. As a continuation of previous studies of the dynamics of single laser carbon plasma plumes expanding in argon background [15], we report now on preliminary studies of plasma features in orthogonal laser Carbon plasma plumes collisions in low pressure Argon background.

2. Experimental set-up

In the experiments we use a Nd:YAG laser, 340 mJ, 3.5 ns, at 1.06 μm, operating at 10 Hz. The main laser beam is divided in two beams by a 50% beam splitter, which are then focused over two rotating targets, with characteristic fluence 3.5 J/cm^2. The orthogonal laser plasma plumes intersect at 1 cm from the target surfaces. Graphite is used as target material. Base pressure is 0.3 mTorr, and argon gas background is used, at pressures from 1 to 50 mTorr. The laser plasmas are characterized by time resolved and time integrated optical emission spectroscopy (OES), with 20 ns and 10 ms time resolution, and 50 ns time resolved plasma imaging of visible plasma emission. Further details on the experimental set-up and associated plasma diagnostics have been published elsewhere [15].

3. Results and discussion

Figure 1 shows characteristic images of colliding laser plasma plumes, at two different argon background pressures. The light intensity has been converted linearly to a color scale, in order to enhance features of the plasma expansion. Before colliding, the laser plasma plumes behave as single plasma plumes, with characteristic initial axial expansion velocities of the order of 10^6 cm/s, depending on background gas pressure. In general, the plasma expansion dynamics of laser carbon plasma plumes is complex, with several plasma fronts of different characteristic velocities, as reported elsewhere [15, 16]. The fast component of the laser plasma plume is well characterized by the shock model, whereas the slower component is better described by the drag model [17]. This is particularly evident at higher background gas pressures, as shown by the bow like feature in the expanding plasma front at 50 mTorr argon, in figure 1-b. Other noticeable effects of increasing the argon background pressure are the lateral expansion seen at the head of the propagating plasma plume at higher pressure and the localization of the stronger light emitting region close to the collision region at low pressure, as opposite to the displacement along the expanding plasma plume observed at higher pressure.

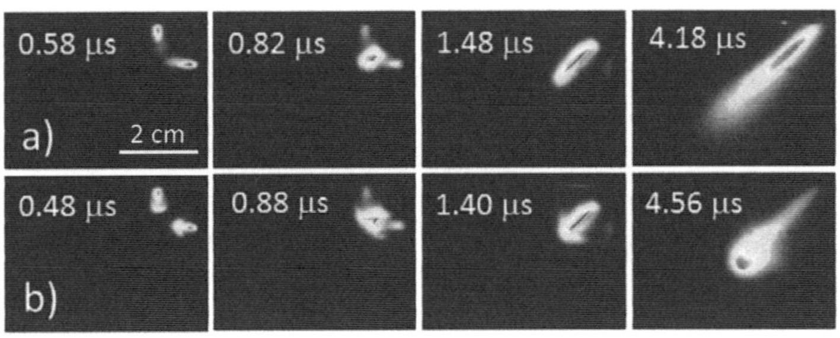

Figure 1. Time resolved images of colliding laser plasma plumes. a) 1 mTorr, and b) 50 mTorr argon background pressures.

After colliding, a single plasma plume is seen to expand along the diagonal, at 45° with respect to the original plasma plumes trajectory. In order to highlight the main features of the post collision plasma expansion dynamics, we have plot the light emission profile long the plasma plume axis, as a function of time. This is shown in figure 2, for three different pressures. The axial distance is measured along the diagonal, from the point where both graphite target planes intersect. It is seen that the plasma plumes expand in both directions, backwards and forward along the diagonal. The characteristic forward expansion velocity is much higher, decreasing from $6.0 \cdot 10^5$ cm/s at base pressure, to $3.8 \cdot 10^5$ cm/s at 50 mTorr argon background pressure. This characteristic velocity range is almost an order of magnitude below that for a single, non-colliding carbon plasma plume, as stated above. This overall behavior is and indication of a high collisionality of the carbon plasma plumes and consistent with the fact that the expansion velocity is predominantly axial, with a minor radial component. On inspecting the light emission profiles it can be seen that at base pressure the plasma plume expands monotonically in both directions. However, as the argon background pressure increases, the light emission profile in the forward direction steeps at the plasma boundary, thus suggesting that the carbon plasma density increases due a drag effect on the background gas. This dragging effect might also explain the slowing down of the expansion velocity, as pressure increases.

Journal of Physics: Conference Series **720** (2016) 012044 doi:10.1088/1742-6596/720/1/012044

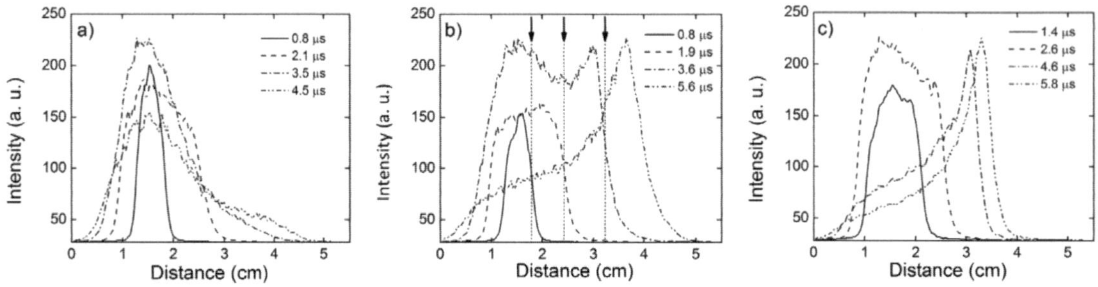

Figure 2. Axial profiles of visible light emission from the laser plasma plumes after collision. a) base pressure, b) 25 mTorr, and c) 50 mTorr argon background pressures.

In figure 3 we present a characteristic visible emission spectrum of a single plasma plume, obtained with 10 ms time resolution, and averaged over 10 laser shots. Light emission is collected from a ~3 mm size volume, at 2.0 cm form the graphite target. The spectrum shows emission corresponds to vibrational states of the $d^3\Pi_g \rightarrow a^3\Pi_u$ Swan band, of the neutral C_2 molecule and, to a lesser degree, vibrational states from the $a^1\Pi_u \rightarrow X^1\Sigma_g^+$ Swings band, of the neutral C_3 molecule. Weak single ionized carbon emission is seen to dominate only at base pressure [4].

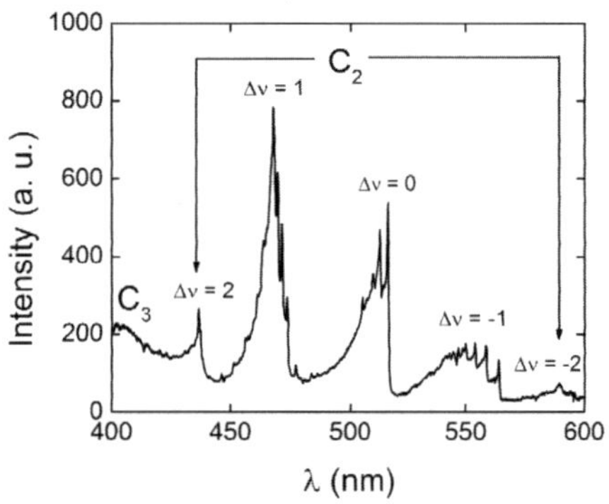

Figure 3. Carbon plasma emission, at 25 mTorr argon background pressure, collected from a ~3 mm size region in a single plasma plume, 2.0 cm from the graphite target surface. The labels identify emission from Argon C_2 and C_3 molecular bands.

Figure 4 shows emission spectra, with 25 mTorr argon background, corresponding to the $\Delta\nu = 1$ vibrational band of the C_2 molecule, obtained with 20 ns time resolution. Light is collected from three ~3 mm characteristic size locations along the plasma plume, which are identified by the arrows and dotted lines in figure 2-b. The time scale for emission growth and decay is similar, of around $\sim 3\mu s$, at the three different places along the plasma plume. However, the emission intensity is seen to diminish along the plume, which might be caused by plasma density decrease due to radial expansion of the propagating plasma plume, as seen in figure 1.

Although most of the experimental observations reported on colliding laser plasmas correspond to parallel plasma plumes which collide as a result of radial expansion [9–11], there have been some reports on collisions between orthogonal laser plasmas [7, 8, 18]. The investigations on perpendicular colliding plasmas have focused mainly on characterizing the formation of a stagnation layer in the interaction region. In particular, experimental observations on orthogonal calcium laser plasma plumes produced in high vacuum, at 300 J/cm^2 fluence show formation of a well defined stagnation region with almost no interpenetration [7]. In another recent series of experiments, the collision dynamics of two orthogonal laser produced plasmas obtained by line focusing at 5 J/cm^2 in high vacuum, showed that high-Z target (tungsten and molybdenum) plasma ions interpenetrate each other, while low-Z (Carbon and Aluminium) plasmas stagnate at the collision plane [9]. The effect of neutral gas background has been studied in PLD with orthogonal dissimilar targets, Platinum-Gold and platinum-ruthenium, in helium at pressures in the range of a few Torr [19]. The experiments highlighted the effect of neutral gas background on both, deposition rate and spatial homogeneity of the deposited films. In this context, our preliminary experiments focus on the effect of low pressure gas background in the collision dynamics of two orthogonal laser plasma plumes. The 45° trajectory observed after collision over the argon background pressure range investigated indicates that the plasma plumes stagnate into each other, without evidence of interpenetration. It is also evident that due to the identical laser fluence deposited on each orthogonal graphite target, the carbon plasma mass and velocity should be also nearly identical, which through momentum conservation explains the 45° trajectory after collision.

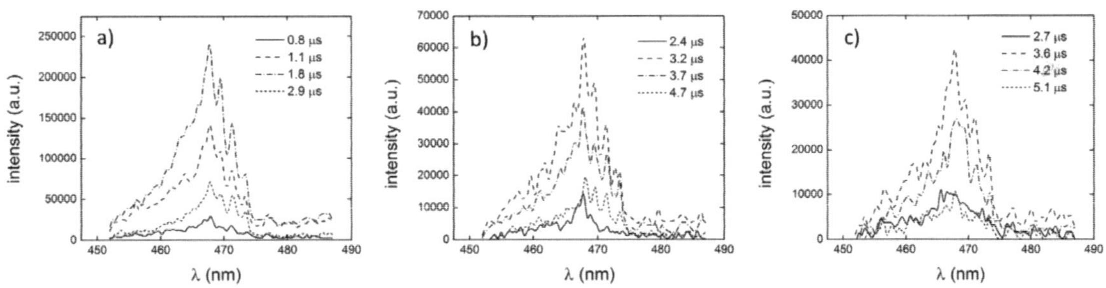

Figure 4. Time resolved emission spectra, with 25 mTorr argon background pressure, corresponding to the $\Delta\nu = 1$ of the C_2 molecule. Time resolution is 20 ns. Light is collected from three axial positions along the plasma plume: a) 1.8 cm, b) 2.5 cm, and c) 3.2 cm.

4. Conclusions

Our preliminary results show that the post-collision dynamics of two orthogonal laser carbon plasma plumes of equal mass density and characteristic kinetic energy is consistent with high a collisionality of the carbon plasma plumes, which results in full stagnation on collisioning. Although, in the background pressure investigated, the presence of a low pressure argon background does not modify the the collision process,it affects the post-collision dynamics, mainly through drag effects at the plasma-neutral gas boundary. Further experiments are being conducted to improve our characterization of the collision processes and post-collision plasma dynamics, which will be reported elsewhere.

Acknowledgments

This work has been funded by project FONDECYT 1141119.

References

[1] Farid N, Harilal S S, Ding H, and Hassanein A 2014 *J. Appl. Phys.* **115** 033107
[2] Ma Q, Motto-Ros V, Bai X, and Yu J 2013 *Appl. Phys. Lett.* **103** 204101

Journal of Physics: Conference Series **720** (2016) 012044 doi:10.1088/1742-6596/720/1/012044

[3] Singh R P, Gupta S L and Thareja R K 2013 *Phys. Plasmas* **20** 123509
[4] Guzmán F, Favre M, Ruiz H M, Hevia S, Caballero L S, Wyndham E S, Bhuyan H, Flores M and Mändl S 2013 *J. Phys. D: Appl. Phys.* **46** 215202
[5] Deng Z, Chu L, Ding X, Qin A, Fu G and Wang Y 2014 *J. Nanoparticles* **2014** 760712
[6] Ganeev R A, Baba M, Suzuki M and Kuroda H 2014 *J. Phys. B: At. Mol. Opt. Phys.* **47** 135401
[7] Luna H, Kavanagh K D and Costello J T 2007 *J. Appl. Phys.* **101** 033302
[8] DardisJ and Costello J T 2010 *Spectrochimica Acta B* **65** 627
[9] Al-Shboul K F, Harilal S S, Hassan S M, Hassanein A, Costello J T, Yabuuchi T, Tanaka K A and Hirooka Y 2014 *Phys. Plasmas* **21** 013502
[10] Singh R P, Gupta S L and Thareja R K 2014 *J. Appl. Phys.* **116** 183301
[11] Gupta S L, Pandey P K and Thareja R K 2013 *Phys. Plasmas* **20** 013511
[12] Irissou E, Vidal F, Johnston T, Chaker M, Guay D and N. Ryabinin A N 2006 *J. Appl. Phys.* **99** 034904
[13] L. Gupta S L and Thareja R K 2013 *J. Appl. Phys.* **113** 143308
[14] Mascalia D, Tudiscoa S, A. Bonannoa A, Gambinoa N, Ivanovskic S, Anzalonea A, Gamminoa S, Miracolia R and Musumecia F 2010 *Radiat. Effects Defects Solids* **165** 730
[15] Ruiz H M, Guzmán F, Favre M, Bhuyan H, Chuaqui and Wyndham E S 2012 *Plasma Sources Sci. Technol.* **21** 034014
[16] Al-Shboul K F, Harilal S S, Hassanein A and Polek M 2011 *J. Appl. Phys.* **109** 053302
[17] Geohegan D B 1992 *Appl. Phys. Lett.* **60** 2732
[18] Sánchez-Aké C, Bolaños M and C.Z. Ramírez C Z 2009 *Spectrochimica Acta B* **64** 857
[19] Irissou E, Vidal F,Johnston T, Chaker M and Guay D 2006 *J. Appl. Phys.* **99** 034904

Deposition of materials using a plasma focus of tens of joules

M. J. Inestrosa-Izurieta[1,2]*, P. Jauregui[3] and L. Soto[1,2]*

[1] Comisión Chilena de Energía Nuclear, Casilla 188-D, Santiago Chile.
[2] Center for Research and Applications in Plasma Physics and Pulsed Power, P^4.
[3] Facultad de Ciencia, Universidad de Santiago de Chile, Santiago, Chile.

**Email:* mjinestrosa@cchen.cl , lsoto@cchen.cl

Abstract. Physical properties of transient plasmas, energetic ions and electrons, as produced in plasma focus (PF) discharges are substantially different than the conventional plasma devices used for plasma nanofabrication. In particular, PF discharges provide new and unique opportunities in processing and synthesis of new materials. Since PF discharges have very short duration and produce plasmas of high ion density, the anode is exposed to a high energy density causing its pulverization and generating a vapour of material that allows a fast deposit. In this paper a table top plasma focus of tens of joules, PF-50J, was used to produce material deposition. First deposits obtained from detached anode material (steel) or a metallic insert (titanium) from the plasma ejected after the pinch in the axial direction are presented.

1. Introduction

In the last decade, there is evidence of the increasing interest in the study of small and compact plasma focus (PF) devices [1] and their various applications [2]. Specially they have attracted great interest due to the possible applications in different technological fields where you can take advantage of the fact that these devices are not only a source of high density and temperature plasmas, but also a rich source of high-energy ions usable for modification of surface properties of thin films deposition, doping of semiconductors and production/coating of nanocomposites [3-15].

The material deposition through a pulsed plasma device, such as a PF, has some special features in comparison with other methods which makes them very appropriate for the generation and treatment of technological materials. These special features include high deposition rate; an energetic deposition process; the ability to operate at high frequency; and the low operating costs. Nevertheless, these methods still require further studies to improve the quality of the resulting surfaces [7].

The dynamic of a PF allows the formation of a column of plasma of high temperature and high density at the end of the of radial collapse phase. This plasma column is then disintegrated due to plasma instabilities, generating a plasma burst of high ion density and beams of relativistic energy ions and electrons that are accelerated in the direction of the axis of the plasma column in both directions. Then, this plasma burst and energy charged particle beams, are responsible for both a possible ablation of the anode and its subsequent material deposition on a substrate, and a possible direct chemical deposition on a substrate placed at some height on the same axis.

The present study focuses on the use of the PF-50J [17, 18] device for material deposition, as well as the study of the first deposits which can be obtained from material removed from a steel hollow anode or from a titanium insert, through the plasma ejected in the axial direction. Deposits at different distances measured from the tip of the anode were obtained and characterized. Morphological and

Journal of Physics: Conference Series **720** (2016) 012045 doi:10.1088/1742-6596/720/1/012045

elemental characterization is performed by means of Scanning Electron Microscopy (SEM) with Energy-dispersive X-ray Spectroscopy (EDX), and the structural characterization through X-ray Diffraction (XRD).

2. Experimental setup and diagnostics

The experiments presented here were performed in a low-energy plasma focus designed and built in our group. This device operates at energy levels of tens of Joules (32-72J of energy stored in the capacitor bank), 160nF, 38nH, 20-30kV, maximum current of 40-60kA reached in 150ns [18].

The main vacuum chamber of the PF-50J was fitted with a sample holder and a shutter, both externally manipulated, to control the position (height from the anode, "d") of the sample and the pulses of deposition respectively (See Figure 1). Discharges were conducted at a pressure of 6 mbar of hydrogen gas and with a charging voltage of 28 kV. The PF-50J was operated with hollow steel anodes of different depth and inner diameter. The discharges for depositions were made at a variable distance between the anode and substrate and accumulating 100 continuous discharges in each case.

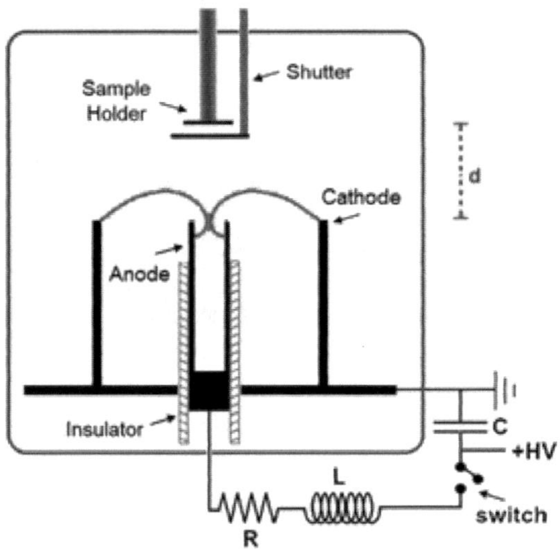

Figure 1: Schematic diagram of the improved PF-50J showing the moment of the pinch.

As standard in-situ electrical diagnostics during plasma pulses (see Figure 2), the voltage *V(t)* was measured with a fast resistive divider, and the current derivative *dI(t)/dt* with a calibrated Rogowski coil. A photomultiplier was implemented to detect x-rays emitted from the pinch compression, and monocrystalline silicon wafers were placed within the chamber, as substrates to deposit the generated material.

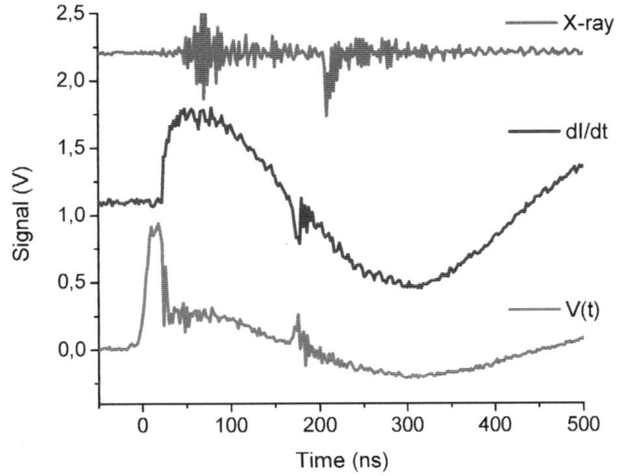

Figure 2: In-situ electrical diagnostics during plasma pulses.

3. First results and discussion

3.1. Deposition with a steel hollow anode

Four silicon substrates were exposed to plasmas generated using a steel hollow anode. The deposits are shown in Figure 3 and their experimental conditions are detailed in Table 1.

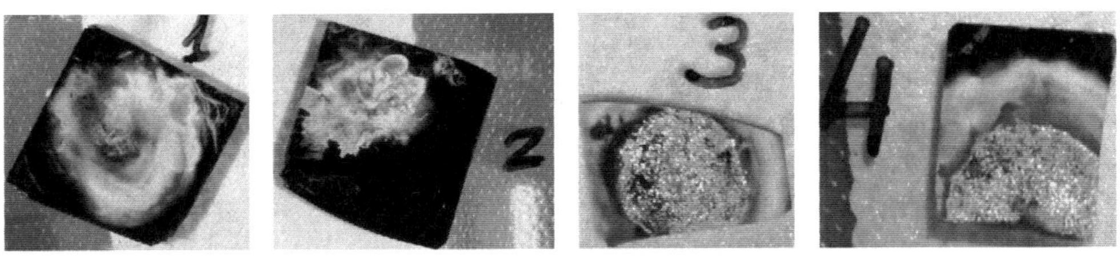

Figure 3: Images of the deposited samples. Dimensions: (1) 12.8x12.3; (2) 12.9x14.7; (3) 9.3x6.3; (4) 6.6x6.6 mm.

Table 1: Experimental differences of samples deposited with hollow anode.

N°	Sample height (mm)	Anode inner diameter (mm)	Anode depth (mm)
1	16	5.2	10.9
2	21	5.2	10.9
3	14	4.6	18.7
4	14	5.2	10.9

Observing the foregoing images some differences in color and amount of deposited material can be clearly seen. Samples (1) and (2) for example, are those with less material: their appearance is rather amorphous, seems smooth and the colors that stand out of the deposit range from light brown to some blue with touches of green. While samples (3) and (4) clearly have a larger amount of material on the substrate, their appearance is more roughened and a dark metallic color is present, whose edges exhibit similar colors to those of samples (1) and (2). With these observations is possible to determine an increase in the obtained amount of material when the substrate height is decreased, as well as larger increase with the decrease in the anode inner diameter.

Surface morphology of the deposited material with different anode geometry was imaged with a TESCAN Vega 3 LMU Scanning electron microscopy. The morphological features of the samples are shown in Figure 4. The images from sample (3) and (4) present some cracks, revealing that the samples placed closer to the anode also receive a larger proportion of the plasma shock attack, as shown in previous articles [19, 20].

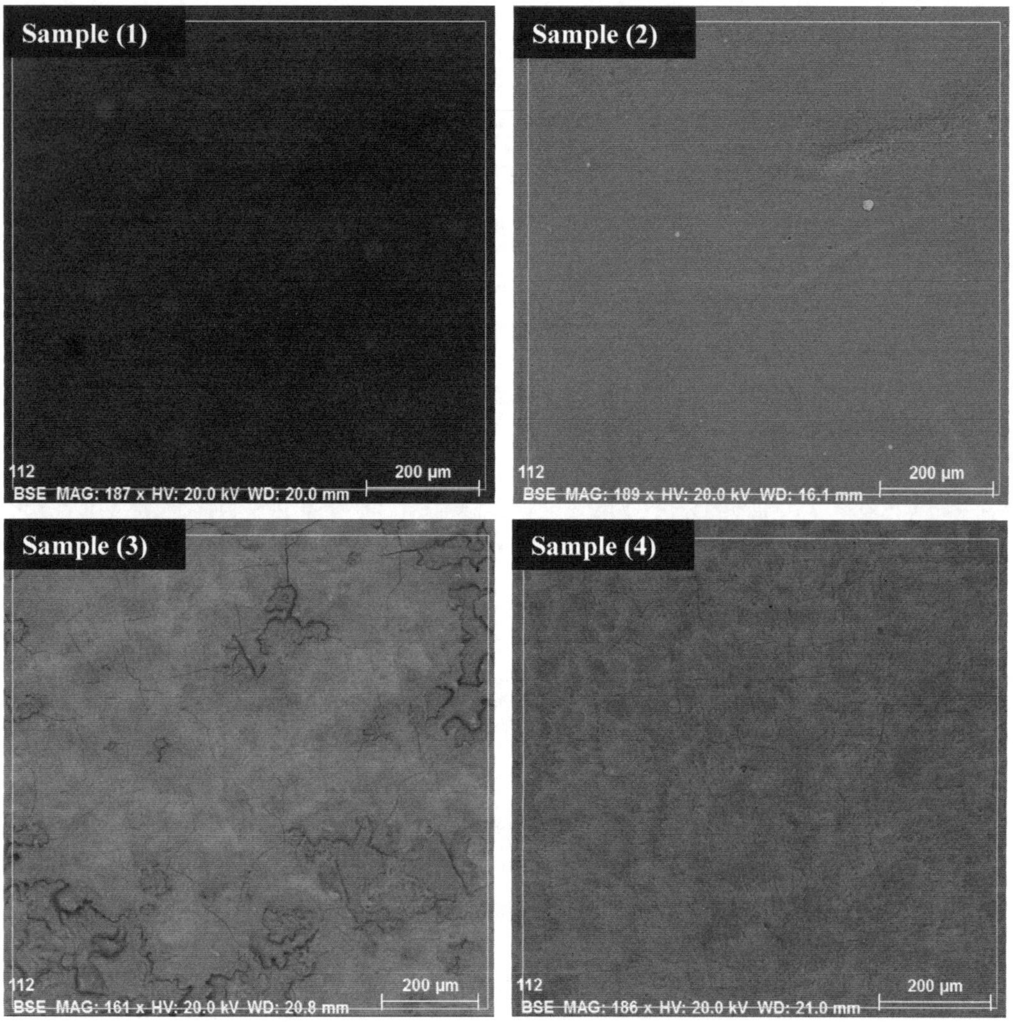

Figure 4: SEM images of the center of the deposited samples from a hollow anode PF.

EDS was used to investigate the elements that are present in the deposited material. The spectra obtained for all the samples are similar in nature presenting only elements that correspond to the steel of the anode. The observation confirms the hypothesis of ablation of the anode and deposition on the substrate surface. The spectra obtained from the samples reveal that between 6% and 14% of material (in standardized atomic concentration) corresponds approximately to the elements of steel (~ 70% Fe, 20% Cr and 10% Ni) and its concentration decreases near the edges of the deposition.

Structural characterization of the deposited material with different anode geometry was made by a SIEMENS D5000 X-ray Diffractometer. The patterns obtained exhibit the diffraction peaks related to different compounds from Fe-Cr-Ni-Si which not only confirm the EDS result about the ablation of the anode, but also the ablation of the Si wafer surface by energetic ions and their re-solidification on the substrate surface, as presented in previous work [21].

3.2. Deposition with a titanium insert

By implementing a hollow anode (inner diameter of 5.2 mm and 10.9 mm depth with a solid insert of titanium (Ti) of about 5 mm diameter) three other deposits were made. See images in Figure 5 and experimental differences in Table 2.

Figure 5: Images of deposited samples. Dimensions: (3) 13.3x8.1; (4) 11.3x8.6; (5) 6.7x8.3 mm.

Table 2: Experimental differences of samples deposited with Ti insert.

N° of sample	Sample height (mm)
3	12
4	17
5	14

The above images show clear differences of the deposit obtained compared with the samples in Figure 3. The deposit presents a bright metallic silvered color with a rough appearance in the center and a smoother appearance around the edges, without the brown and blue colors seen in the previous case (just hollow anode). It seems that those colors are possibly due to the anode material of thinner layers (stainless steel). The silvered color can be distinctively associated with the titanium insert since for the samples with the expected steel deposition, the metallic appearance is darker than in the samples with the expected titanium deposition.

The morphological features of the samples imaged by SEM are shown in Figure 6. The images from sample (3) y (5) present some cracks, reproducing the same effect as in the above samples (Figure 4): closer to the anode the samples receive a good proportion of the plasma shock attack, independently of the material that is deposited or if there is an insert on the anode. Also on the images of samples (3) and (5) the melting of surface material due to the plasma attack can be distinguished.

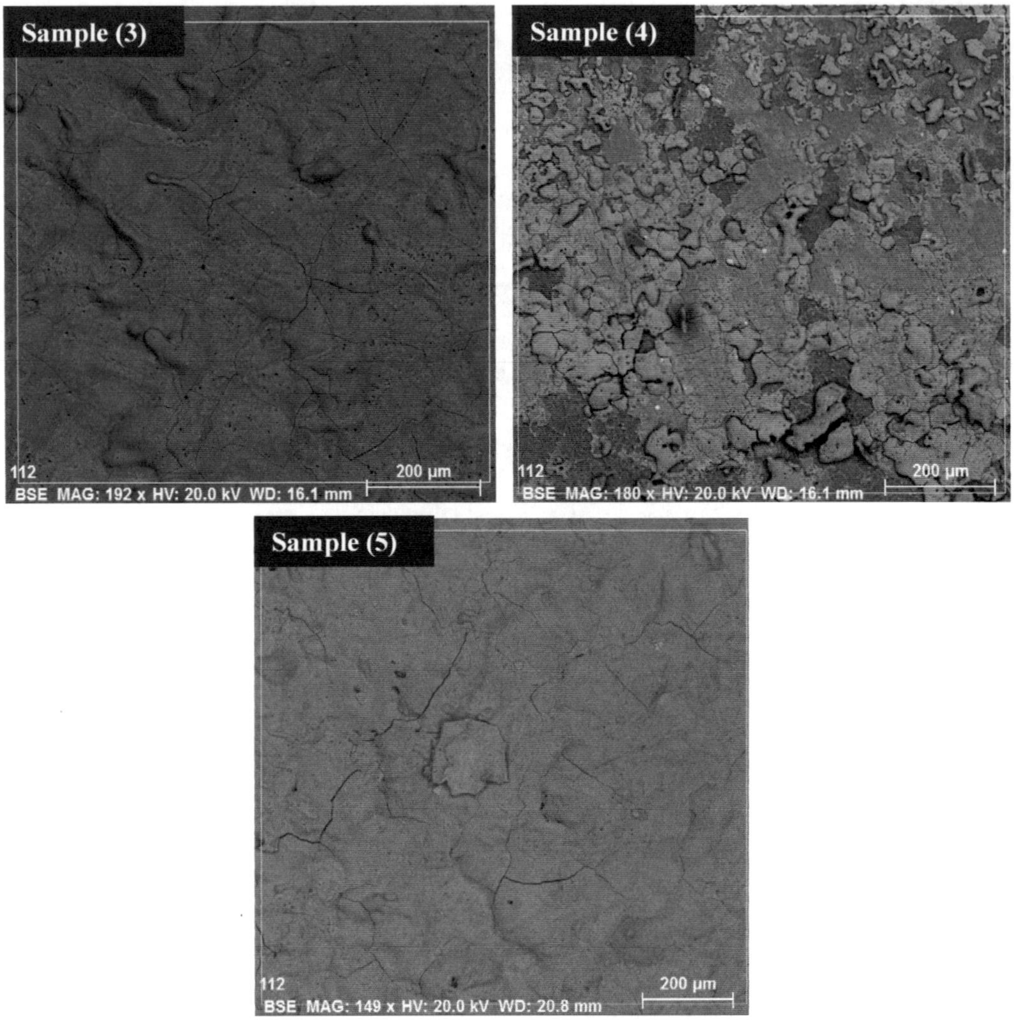

Figure 6: SEM images of the center of the deposited samples from a hollow anode PF with Ti insert.

The spectra obtained with EDS for all the samples are similar presenting only elements that corresponds to titanium from the insert and a bit of the steel from the anode. This observation confirms the hypothesis of ablation of the anode and deposition on the substrate surface as in the previous case. The spectra obtained for the samples reveal (Figure 6): (3) The center of the sample contains around 7% of material that correspond to Ti and 18% of Ti and 5% of elements from steel (Fe & Cr) around the edges; (4) Near the center of the sample it contains around 15% of material that correspond to Ti and around the edges there is 28% of Ti and 5% of elements from steel (Fe & Cr); (5) The center of the sample contains around 12% of material that correspond to Ti and around the edges contains 34% of Ti and 3% of elements from steel (Fe & Cr). Combining these data it can be determined that the deposition of the anode edges (in this case steel) is much less than the one that comes from de center (Ti insert). In addition, considering that the discussed melting of the surface material due to the attack promotes a smoother surface allowing more penetration of response in the EDS spectra and thus more silicon partial percentage, we can conclude that higher relative percentage of titanium is deposited when the sample is placed closer to the anode. The latter could be explained after observing how the titanium does not seem to have a great expansion as we move the sample away from the anode. In the case of steel it is clearly spreading which means that the material comes to the sample with less energy than the titanium. Considering that the ejected plasma and particles are mainly in the axial direction, if we place a sample near the anode, the trajectory of the material is shorter which means that it has less probability of interaction with the gas until deposition and can go directly to the silicon wafer. If the sample is placed far from the anode, the material collected could be less because it has the possibility to expand radially.

The XRD patterns obtained for these samples exhibit diffraction peaks mainly related to Ti (not sample (3)) or different Ti-Si (for example $TiSi_2$ or Ti_5Si_3), which again confirms the ablation of the Si wafer surface by energetic ions and its re-solidification on the substrate surface generating an interlayer and even presenting a possible limit in which the deposition of pure Ti over the interlayer starts.

4. Conclusions

We demonstrate that thin films of Ti, different Ti-Si composites (for example TiSi2 or Ti5Si3) or compounds from Fe-Cr-Ni-Si can be deposited using a table top plasma focus device of only tens of joules. From our knowledge, it is the first time that this kind of plasma focus device, with low operating costs and the ability to operate at high frequency, has been used for thin film deposition. Nevertheless, this method still require further studies to improve the quality of the resulting surfaces.

It was observed an increase in the deposited amount of material when the height between anode and sample is decreased, as well as an increase in the deposited amount when a smaller inner diameter of the hollow anode was used. Furthermore, the samples placed closer to the anode receive a larger amount of plasma shock attack, presenting some cracks or even melting of the surface material independently if they were deposited from a hollow anode or from an insert. The ablation of the anode and its deposition on the substrate was confirmed. Also at the time of the deposition there is an ablation of the Si wafer surface by energetic ions that permits its re-solidification on the substrate surface and can form compounds generating an interlayer. However, it was observed that the deposition of the anode edges (steel) is much less and comes to the sample with less energy than the one that comes from de center (Ti insert) implying that the material coming from edges of the anode is removed by a different mechanism than the material of the center.

Acknowledgements
This project has been supported by PAI-CONICYT 79130026 and PIA- CONICYT ACT-1115.

References
[1] L. Soto, "New Trends and Future Perspectives on Plasma Focus Research", Plasma Physics and Controlled Fusion vol. 47, A361 (2005).
[2] M. Krishnan, "The Dense Plasma Focus: A Versatile Dense Pinch for Diverse Applications," IEEE Transactions on Plasma Science, vol. 40, no. 12, pp. 3189–3221, 2012.
[3] R. S. Rawat, "High energy density pinch plasma - A unique non-conventional tool for plasma nanotechnology", R. S. Rawat, IEEE Trans. on Plasma Sci. vol. 41 , pp. 701- 715, 2013.
[4] Lee, P., Feng, X., Zhang, G. X., Liu, M. H. & Lee, S. "Electron lithography using a compact plasma focus", Plasma Sources Science and Technology, vol. 6, pp. 343–348, 1997.
[5] S. K. Ngoi, S. L. Yap, B. T. Goh, R. Ritikos, S. a. Rahman, and C. S. Wong, "Formation of Nano-Crystalline Phase in Hydrogenated Amorphous Silicon Thin Film by Plasma Focus Ion Beam Irradiation," Journal of Fusion Energy, vol. 31, no. 1, pp. 96–103, 2011.
[6] C. R. Kant, M. P. Srivastava, and R. S. Rawat, "Thin carbon film deposition using energetic ions of a dense plasma focus," Physics Letters A, vol. 226, no. 3–4, pp. 212–216, 1997.
[7] H. Bhuyan, M. Favre, E. Valderrama, G. Avaria, F. Guzman, H. Chuaqui, I. Mitchell, E.

Wyndham, R. Saavedra, and M. Paulraj, "Effect of high energy ion irradiation on silicon substrate in a pulsed plasma device", Applied Surface Science, vol. 254, pp. 197–200, 2007.
[8] G. Murtaza, S. S. Hussain, M. Sadiq, and M. Zakaullah, "Plasma focus assisted carburizing of aluminium," Thin Solid Films, vol. 517, no. 24, pp. 6777–6783, 2009.
[9] T. Hussain, R. Ahmad, N. Khalid, Z. a. Umar, and a. Hussnain, "Synthesis of nanostructured multiphase (Ti,Al)N/a-Si3N4 thin films using dense plasma focus device," Nuclear Instruments and Methods in Physics Research Section B: Beam Interactions with Materials and Atoms, vol. 269, no. 18, pp. 1951–1955, 2011.
[10] Z. a. Umar, R. S. Rawat, K. S. Tan, a. K. Kumar, R. Ahmad, T. Hussain, C. Kloc, Z. Chen, L. Shen, and Z. Zhang, "Hard TiCx/SiC/a-C:H nanocomposite thin films using pulsed high energy density plasma focus device," Nuclear Instruments and Methods in Physics Research Section B: Beam Interactions with Materials and Atoms, vol. 301, pp. 53–61, 2013.
[11] T. Hussain, R. Ahmad, N. Khalid, Z. A. Umar, and A. Hussnain, "Synthesis of nanostructured multiphase (Ti,Al)N/a-Si3N4 thin films using dense plasma focus device," Nuclear Instruments and Methods in Physics Research Section B: Beam Interactions with Materials and Atoms, vol. 269, no. 18, pp. 1951–1955, 2011.
[12] T. Hussain, R. Ahmad, I. A. Khan, J. Siddiqui, N. Khalid, A. S. Bhatti, and S. Naseem, "Deposition of titanium nitride on Si(100) wafers using plasma focus," Nuclear Instruments and Methods in Physics Research Section B: Beam Interactions with Materials and Atoms, vol. 267, no. 5, pp. 768–772, 2009.
[13] R. S. Rawat, P. Lee, T. White, L. Ying, and S. Lee, "Room temperature deposition of titanium carbide thin films using dense plasma focus device," Surface and Coatings Technology, vol. 138, no. 2–3, pp. 159–165, 2001.
[14] R. S. Rawat, T. Zhang, K. S. T. Gan, P. Lee, and R. V. Ramanujan, "Nano-structured Fe thin film deposition using plasma focus device," Applied Surface Science, vol. 253, no. 3, pp. 1611–1615, 2006.
[15] R. S. Rawat, V. Aggarwal, M. Hassan, P. Lee, S. V. Springham, T. L. Tan, and S. Lee, "Nano-phase titanium dioxide thin film deposited by repetitive plasma focus: Ion irradiation and annealing based phase transformation and agglomeration," Applied Surface Science, vol. 255, no. 5, pp. 2932–2941, 2008.
[16] G. R. Etaati, M. T. Hosseinnejad, M. Ghoranneviss, M. Habibi, and M. Shirazi, "Deposition of tungsten nitride on stainless steel substrates using plasma focus device," Nuclear Instruments and Methods in Physics Research Section B: Beam Interactions with Materials and Atoms, vol. 269, no. 10, pp. 1058–1062, 2011.
[17] P. Silva, L. Soto, W. Kies, and J. Moreno, Plasma Sources Sci. and Technol. 13, 329 (2004).
[18] Leopoldo Soto, Patricio Silva, José Moreno Moreno, Marcelo Zambra, Walter Kies, Roberto E. Mayer, Alejandro Clausse, Luis Altamirano, Cristian Pavez, and Luis Huerta. J. Phys. D: App. Phys. vol. 41, 205215 (2008).
[19] L. Soto, C. Pavez, J. Moreno, M. J. Inestrosa-Izurieta, F. Veloso, G. Gutiérrez, J. Vergara, A. Clausse, H. Bruzzone, F. Castillo, and L. F. Delgado-Aparicio, "Characterization of the axial plasma shock in a table top plasma focus after the pinch and its possible application to testing materials for fusion reactors," *Phys. Plasmas*, vol. 21, no. 122703, pp. 1–6, 2014.
[20] M. J. Inestrosa-Izurieta, E. Ramos-Moore, and L. Soto, "Morphological and structural effects on tungsten targets produced by fusion plasma pulses from a table top plasma focus," *Nucl. Fusion*, vol. 55, no. 9, p. 093011, 2015.
[21] T. Hussain, R. Ahmad, N. Khalid, Z. a. Umar, and A. Hussnain, "Synthesis of nanostructured multiphase (Ti,Al)N/a-Si3N4 thin films using dense plasma focus device," *Nucl. Instruments Methods Phys. Res. Sect. B Beam Interact. with Mater. Atoms*, vol. 269, no. 18, pp. 1951–1955, Sep. 2011.

Parallel processing implementation for the coupled transport of photons and electrons using OpenMP

Edgardo Doerner

Institute of Physics, Pontificia Universidad Católica de Chile, Vicuña Mackenna 4860, Santiago 7820436, Chile

E-mail: edoerner@fis.puc.cl

Abstract. In this work the use of OpenMP to implement the parallel processing of the Monte Carlo (MC) simulation of the coupled transport for photons and electrons is presented. This implementation was carried out using a modified EGSnrc platform which enables the use of the Microsoft Visual Studio 2013 (VS2013) environment, together with the developing tools available in the Intel Parallel Studio XE 2015 (XE2015). The performance study of this new implementation was carried out in a desktop PC with a multi-core CPU, taking as a reference the performance of the original platform. The results were satisfactory, both in terms of scalability as parallelization efficiency.

1. Introduction

The EGSnrc (**E**lectron-**G**amma-**S**hower) corresponds to a system designed for the MC simulation of coupled transport for photons and electrons and therefore allowing the study of many problems in radiotherapy and dosimetry [1]. EGSnrc may be executed in parallel through a Batch-Queuing-System (BQS) [2]. However, this system presents several disadvantages: (a) any code must be extensively modified in order to support this parallelization, (b) the BQS configuration process is not a trivial task and (c) currently the parallel processing process through a BQS is available only on GNU/Linux operative systems (OS). For these reasons is presented a new parallel computing implementation for EGSnrc through the OpenMP API [3], named OMP_EGS.

2. Materials and methods

The different OpenMP directives were added to the MC code using a modified EGSnrc platform [4], allowing the use of XE2015 integrated in VS2013. The use of these programming tools makes possible an effective and safe implementation of the parallel processing in the OMP_EGS platform. The OMP_EGS platform was developed with the aid of the Intel Parallel Advisor XE component of XE2015, which reveals the potential costs and benefits of parallelism by modeling this behavior in the serial program. Then, it was tuned using the Intel Vtune Amplifier XE profiler.

The performance of the OMP_EGS platform was tested in a 8-core AMD FX8350 desktop PC running under GNU-Linux Debian 7 Wheezy. OMP_EGS was compiled with GNU Fortran 4.7.2 (GFORTRAN), part of the GNU Compiler Collection (GCC) and the Intel Fortran Compiler 15.0 (IFORT), inside XE2015. This implementations were compared against the original EGSnrc BQS parallel processing solution in terms of parallel scalability.

To study the parallel scalability different work load partition schemes were considered. Weak scalability reflects the ability to solve larger problems with more resources whereas strong scalability considers a fixed amount of work while increasing the processing resources. For the first scalability test $5x10^4$ particle histories per processing core were considered. For the later, a total amount of $3x10^5$ initial particles were simulated while increasing the number of used processing cores.

3. Results and Discussion

The results for the weak scalability test are shown in Figure 1. Perfect scalability is achieved when the simulation time remains constant. It can be seen the parallel overhead costs of the parallel processing in both implementations. However, the results show that the OMP_EGS platform is faster than the EGSnrc BQS solution for both compilers. The strong scalability study is depicted in Figure 2. Again, the elapsed times are lower for the OMP_EGS platform. As shown by the results, the parallel scalability of the OMP_EGS platform is satisfying, even surpassing the EGSnrc BQS results.

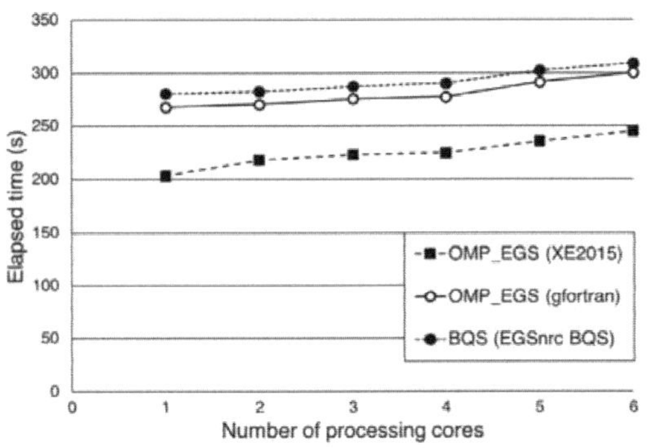

Figure 1. Results for the weak scalability test.

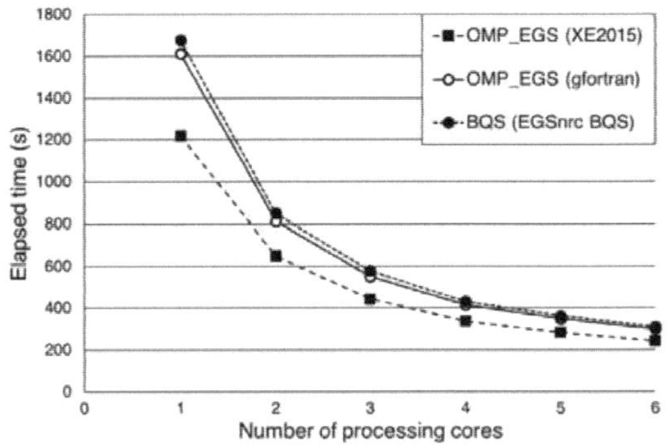

Figure 2. Results for the strong scalability test.

4. Conclusion

OMP_EGS is a portable and compact solution that can be implemented with any compiler that supports OpenMP, between them GFORTRAN and IFORT. Therefore the support of parallel processing depends on the compiler and not on the availability of an external system like a BQS. Thanks to the use of OpenMP most of the serial code is preserved, and moreover, it is possible to recover the serial implementation by turning off the compiler's OpenMP support. This is of great value for debugging purposes. Finally, the performance results have shown that OMP_EGS is a good alternative to the EGSnrc BQS implementation. OMP_EGS obtained better runtimes for all the studied cases.

References

[1] Kawrakow I, Mainegra-Hing E, Rogers D, Tessier F and Walters B 2013 *NRCC Report PIRS-701*
[2] Kawrakow I, Mainegra-Hing E and Rogers D W O 2006 *NRCC Report PIRS-877*
[3] OpenMP Architecture Review Board 2013 Openmp application program interface, version 4.0 URL `http://omp.org/wp/openmp-specifications/`
[4] Doerner E and Hartmann G 2012 *Phys. Med. Biol* **57** 3093

Characteristic variogram for land use in Multispectral Images

E. Mera[1], A. Condal [2], C. Ríos [2] & L. Da Silva [3]

[1] Universidad Tecnológica Metropolitana, Physics Department, Box 9845, Santiago, Chile.
[2] Université Laval, Département des Sciences Géomatiques, G1K 7P4, Canadá.
[3] Universidad Tecnica Federico Santa Maria, Box 110-V, Valparaiso.

E-mail: emera@utem.cl

Abstract. In remote sensing is the concept of spectral signature in multispectral imagery to recognize different land uses in the area; This study proposes the existence of a characteristic variogram for land use in multispectral images. To test this idea we proceeded to work with a sector of a scene image of multispectral Landsat 7 ETM +, in 6 of their bands (1- 450nm to 520nm, 2 - 520nm to 600nm, 3 - 630nm to 690nm, 4 - 760nm to 900nm 5 - over 1550nm to 1.750nm and 7 - 2.080nm to 2.350nm), corresponding to two uses of urban land and agricultural, the omnidirectional variogram for each band was analyzed and modal variogram for each land use was established in the stripe set. Of the analyzed claims data for each land use is a model characteristic and modal cross variogram how their wavelengths.

1. Introduction

Considering that there is the concept of spectral signature on satellite images [1] for land use in multispectral images, the idea that there should be the characteristic variogram [2][3] for land use in multispectral images is proposed. To test this idea we proceeded to:

- Study two uses of Agricultural soil and urban land use.
- Proceeded to analyze the omnidirectional variogram in each of the bands
- Finally the modal variogram a established in each band.

Figure 1. Landsat 7 ETM+ Satellite Image Agricultural Soil

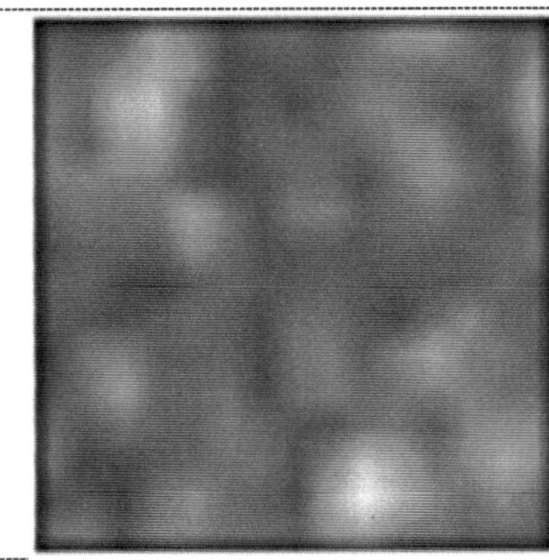

Figure 2. Landsat 7 ETM+ Satellite Image Urban Land Use.

The image used was Landsat 7 ETM +, in 6 of their bands (1,2,3,4,5 and 7).

2. Results

Based on the above steps, the data obtained:

*2.1. Variograms and variograms models obtained in the a*gricultural soil*:*

The first step we proceeded to calculate the histogram image in all his bands:

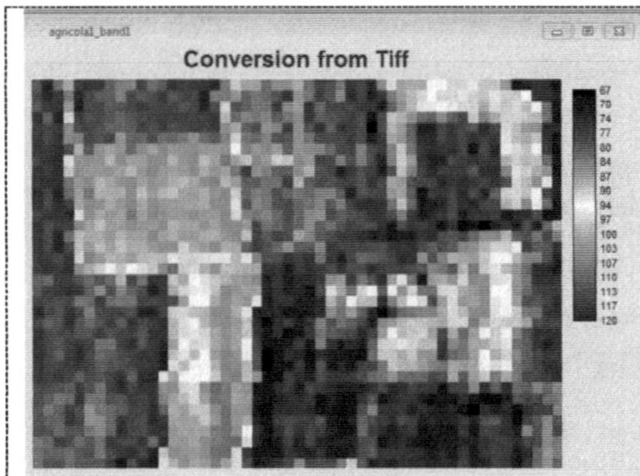

Figure 1. Band 1 - Landsat 7 ETM+ Satellite Image Agricultural Soil – in digital level

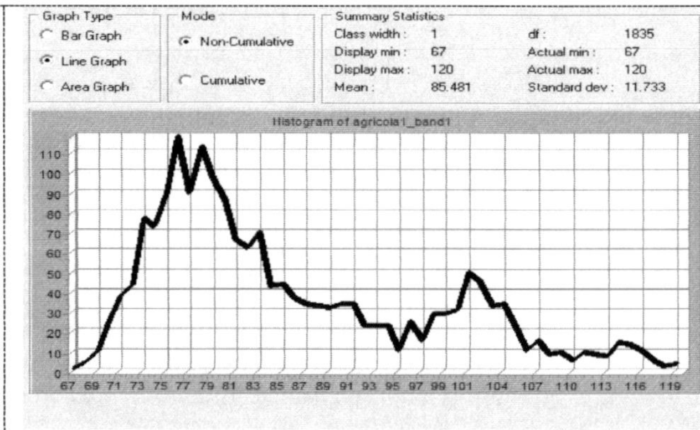

Figure 2. Band 1 - Landsat 7 ETM+ Satellite Image Agricultural Soil – Histogram

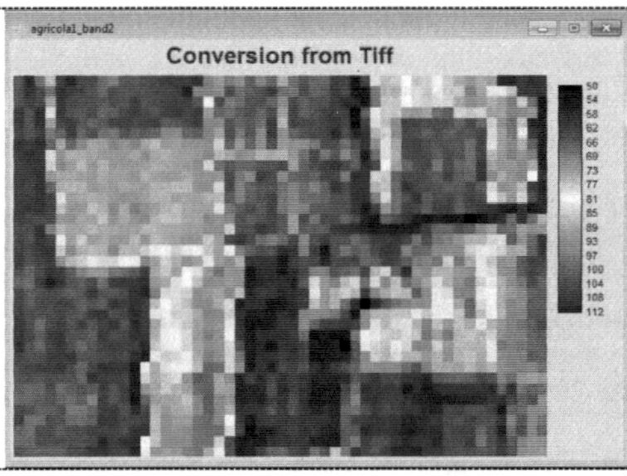

Figure 3. Band 2 - Landsat 7 ETM+ Satellite Image Agricultural Soil – in digital level

Figure 4. Band 2 - Landsat 7 ETM+ Satellite Image Agricultural Soil – Histogram

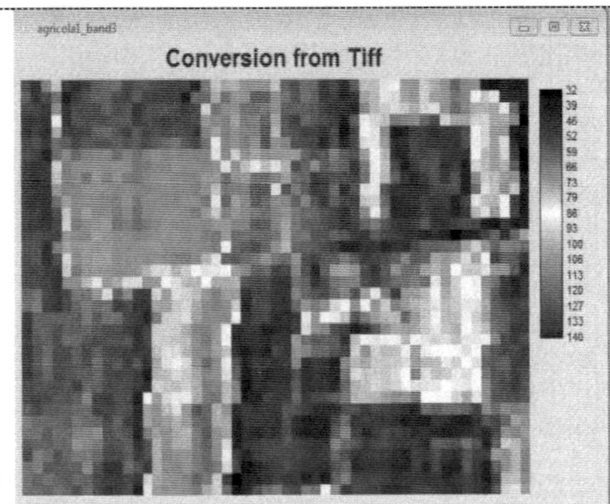

Figure 5. Band 3 - Landsat 7 ETM+ Satellite Image Agricultural Soil – in digital level

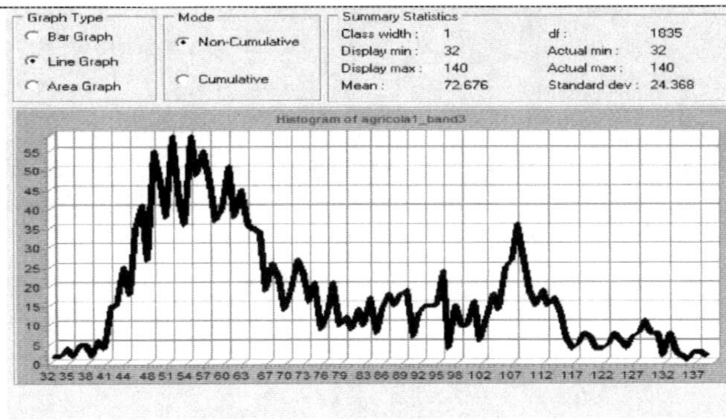

Figure 6. Band 3 - Landsat 7 ETM+ Satellite Image Agricultural Soil – Histogram

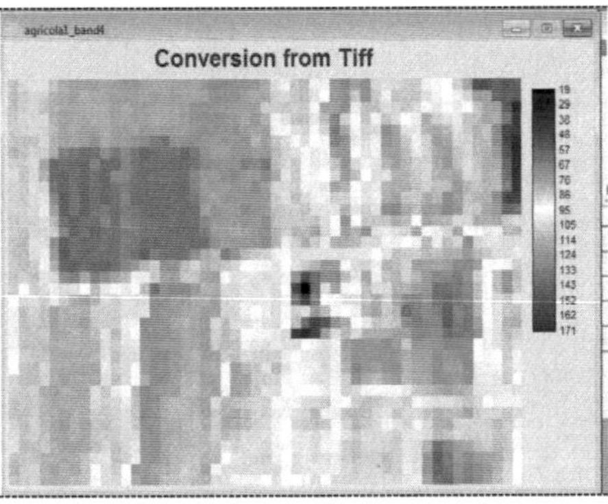

Figure 7. Band 4 - Landsat 7 ETM+ Satellite Image Agricultural Soil – in digital level

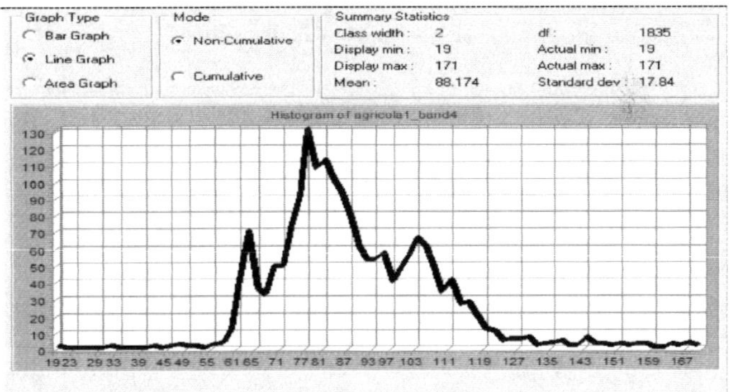

Figure 8. Band 4 - Landsat 7 ETM+ Satellite Image Agricultural Soil – Histogram

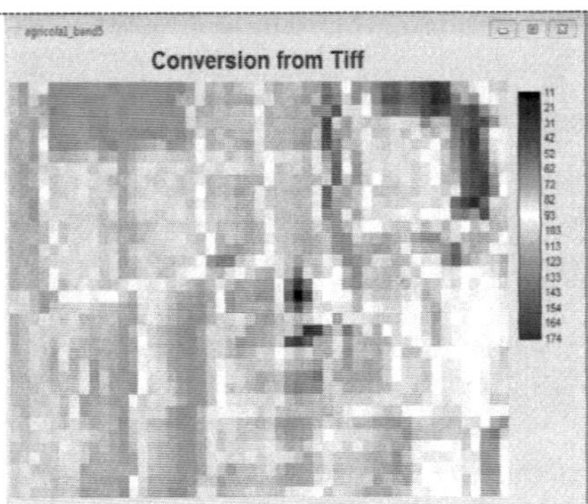

Figure 9. Band 5 - Landsat 7 ETM+ Satellite Image Agricultural Soil – in digital level

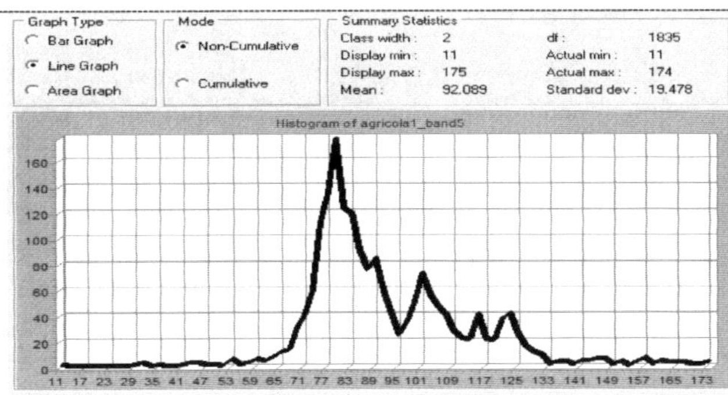

Figure 10. Band 5 - Landsat 7 ETM+ Satellite Image Agricultural Soil – Histogram

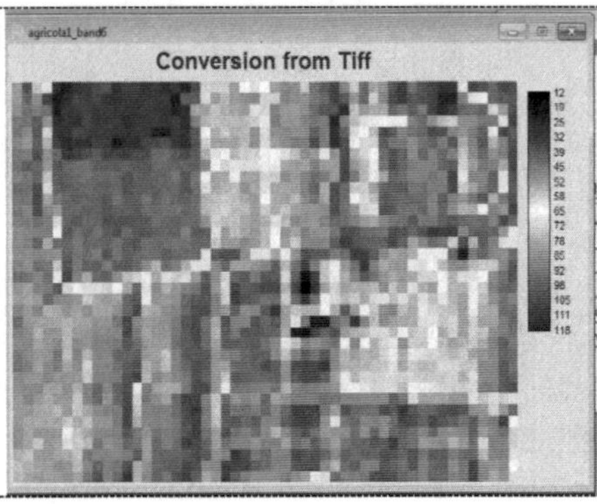

Figure 11. Band 4 - Landsat 7 ETM+ Satellite Image Agricultural Soil – in digital level

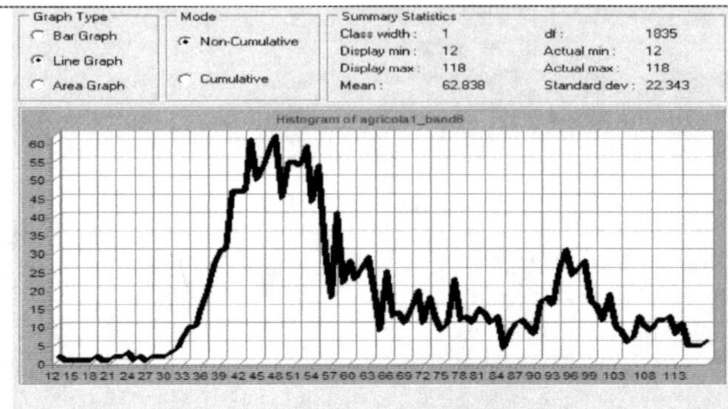

Figure 12. Band 4 - Landsat 7 ETM+ Satellite Image Agricultural Soil – Histogram

Subsequently proceeded in each band to calculate the experimental variogram (black line) and the best variographic model (blue line) was adjusted:

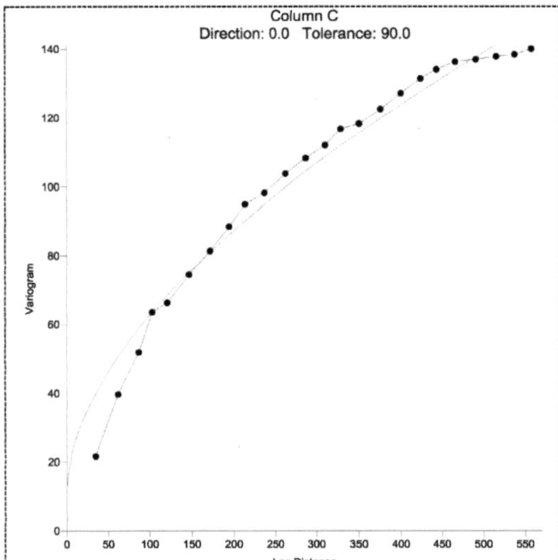

Figure 13. Band 1 - Best variographic model fitted is Power model with Nugget effect

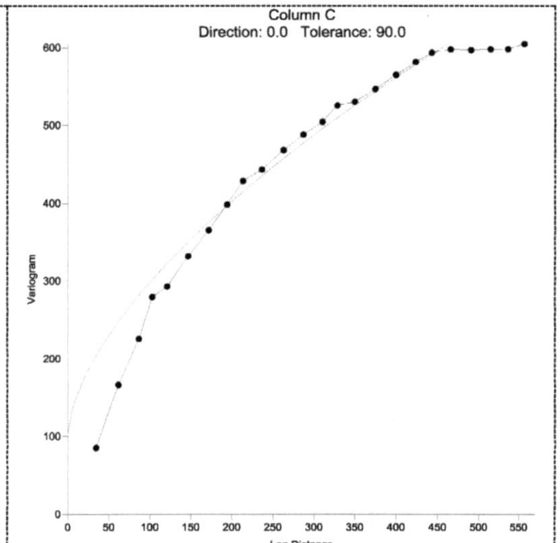

Figure 14. Band 1 - Best variographic model fitted is Power model with Nugget effect

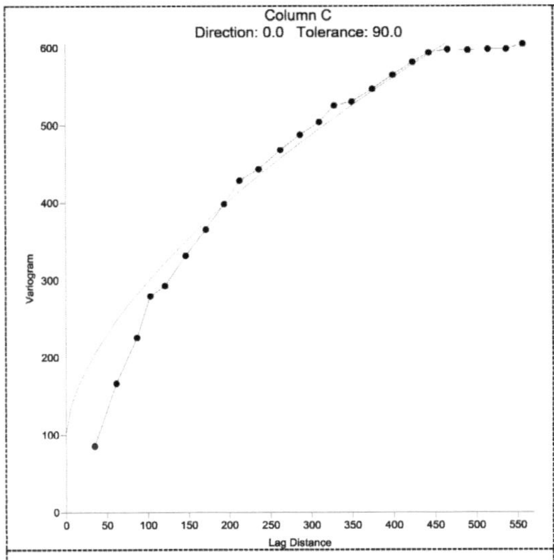

Figure 15. Band 3 - Best variographic model fitted is Power model with Nugget effect

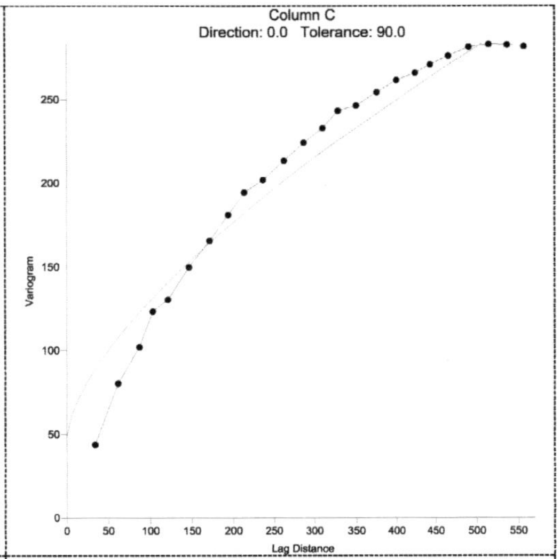

Figure 16. Band 4 - Best variographic model fitted is Power model with Nugget effect

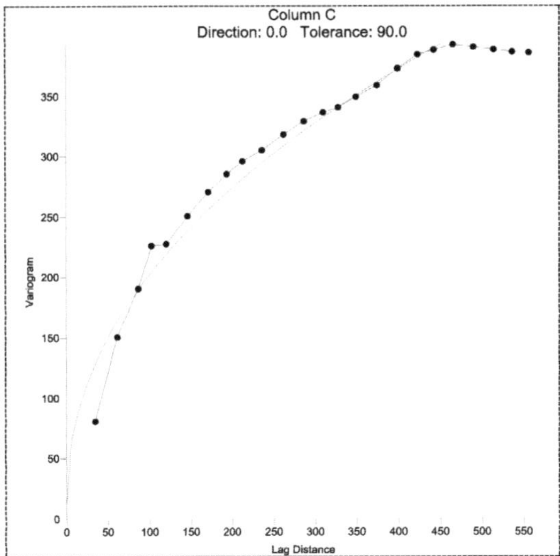

Figure 17. Band 5 - Best variographic model fitted is Power model with Nugget effect.

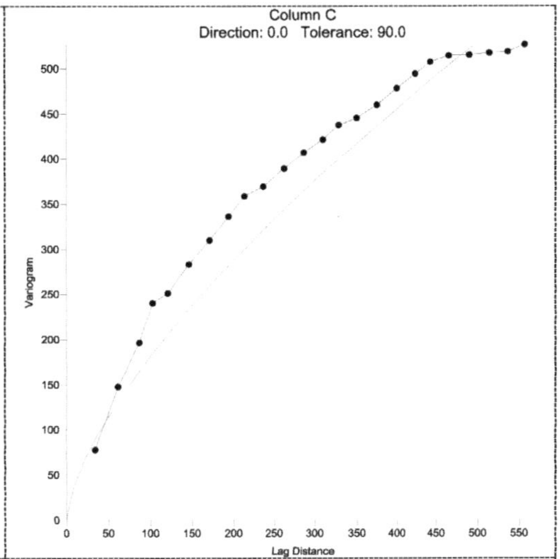

Figure 18. Band 7 - Best variographic model fitted is Power model with Nugget effect.

The analysis of the graphs above it is concluded that the characteristic variographic model for the urban case is the Power model with nugget effect.

*2.2. Variograms and variograms models obtained in the u*rban land use*:*

The first step we proceeded to calculate the histogram image in all his bands:

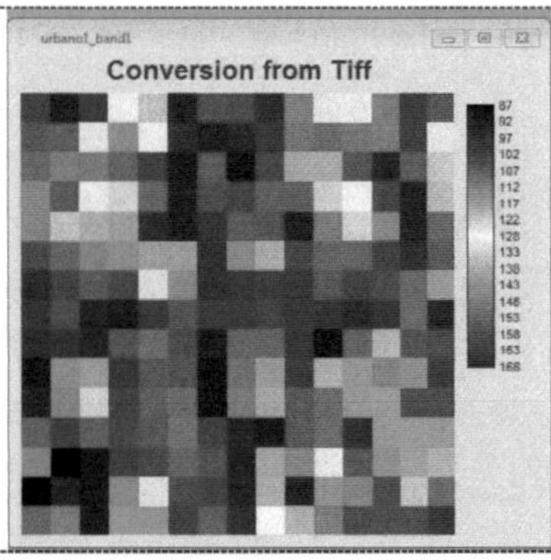

Figure 19. Band 1 - Landsat 7 ETM+ Satellite Image Urban Land Use – in digital level

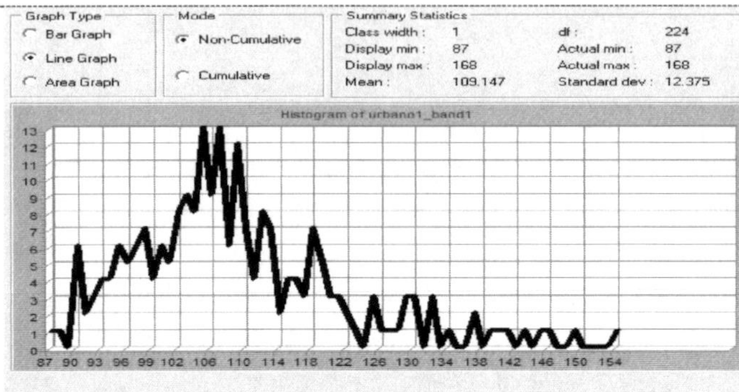

Figure 20. Band 1 - Landsat 7 ETM+ Satellite Image Urban Land Use – Histogram

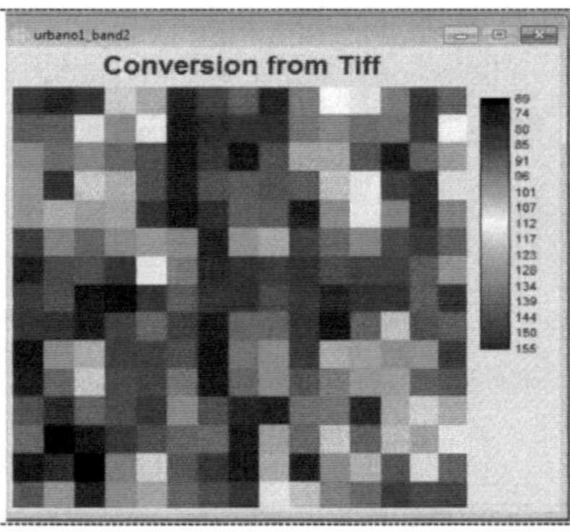

Figure 21. Band 2 - Landsat 7 ETM+ Satellite Image Urban Land Use – in digital level

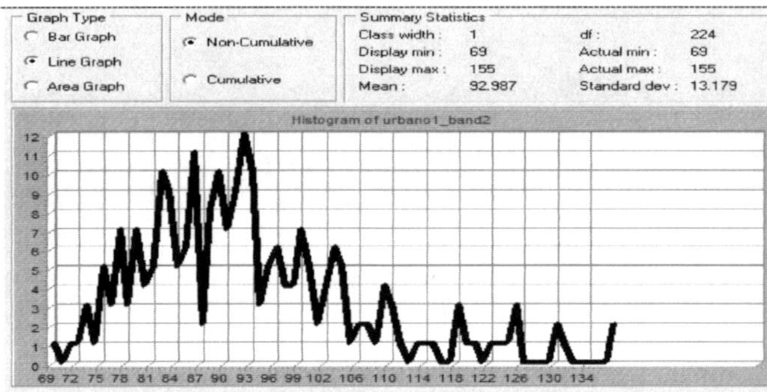

Figure 22. Band 2 - Landsat 7 ETM+ Satellite Image Urban Land Use – Histogram

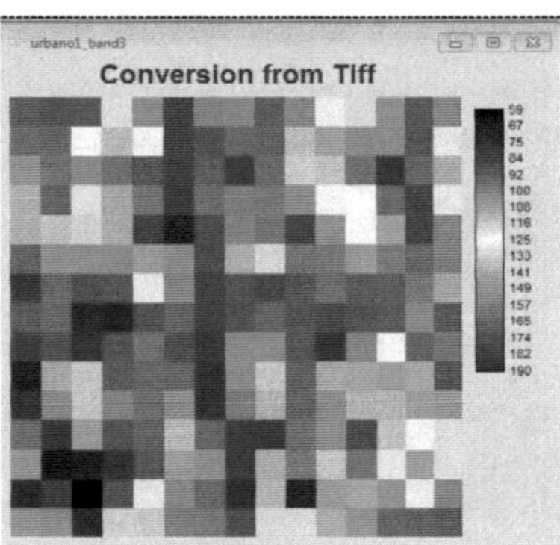

Figure 23. Band 3 - Landsat 7 ETM+ Satellite Image Urban Land Use – in digital level

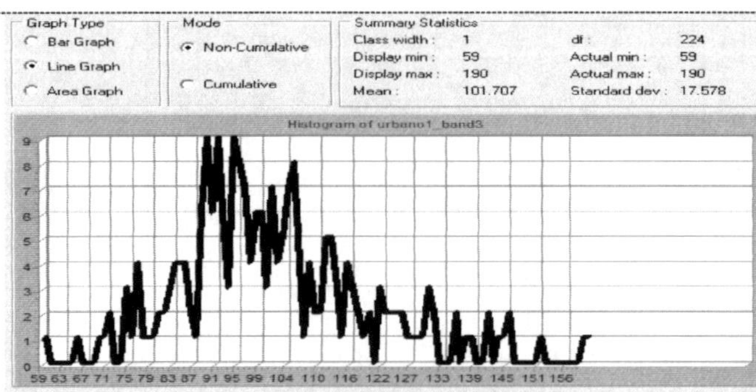

Figure 24. Band 3 - Landsat 7 ETM+ Satellite Image Urban Land Use – Histogram

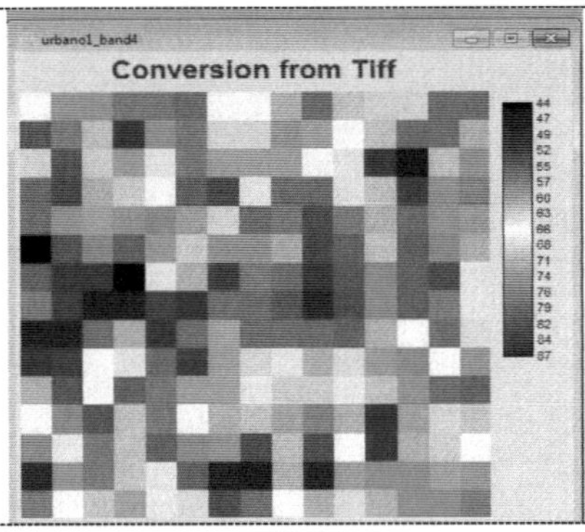

Figure 25. Band 4 - Landsat 7 ETM+ Satellite Image Urban Land Use – in digital level

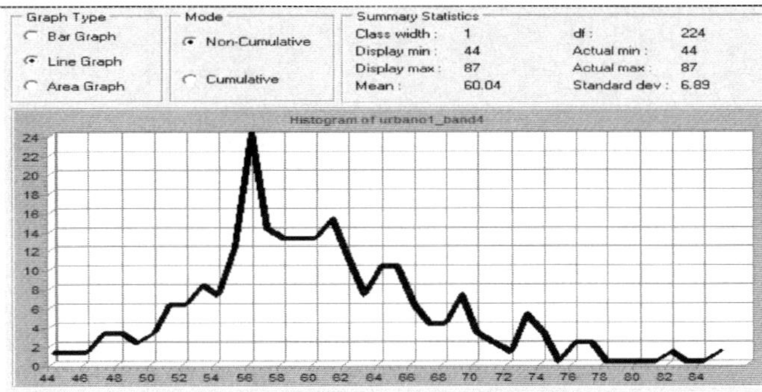

Figure 26. Band 4 - Landsat 7 ETM+ Satellite Image Urban Land Use – Histogram

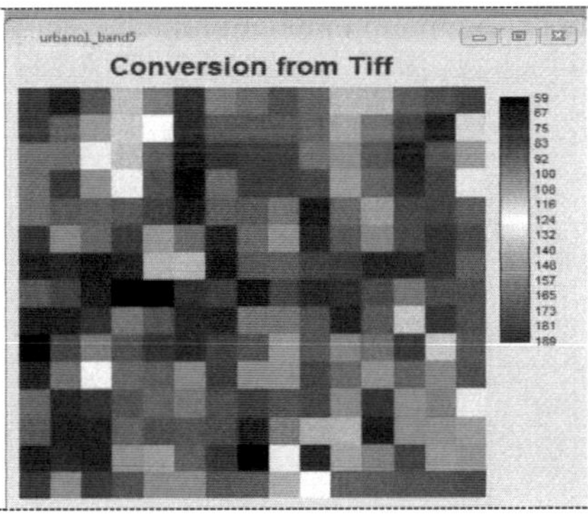

Figure 27. Band 5 - Landsat 7 ETM+ Satellite Image Urban Land Use – in digital level

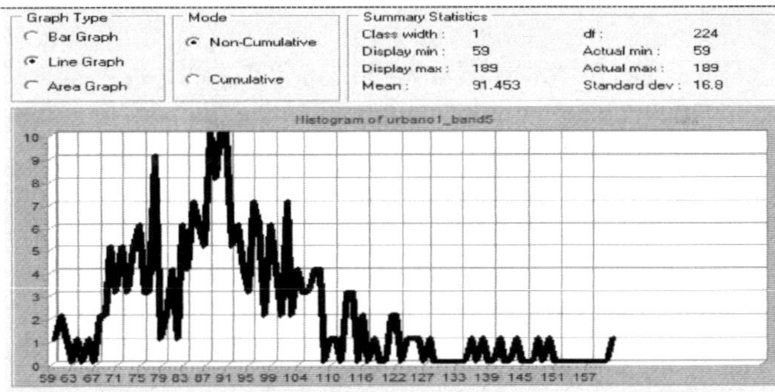

Figure 28. Band 5 - Landsat 7 ETM+ Satellite Image Urban Land Use – Histogram

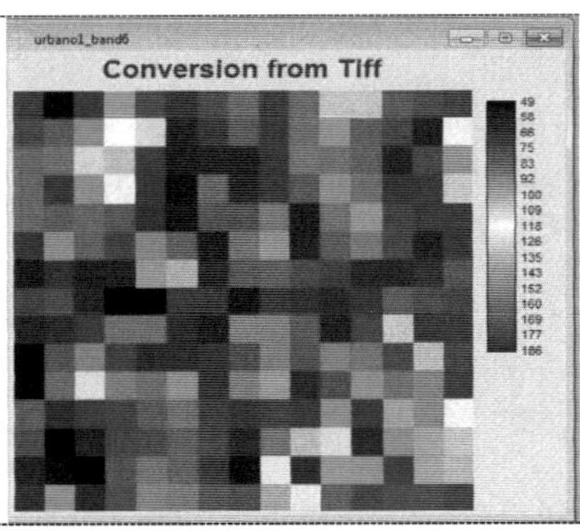

Figure 29. Band 7 - Landsat 7 ETM+ Satellite Image Urban Land Use – in digital level

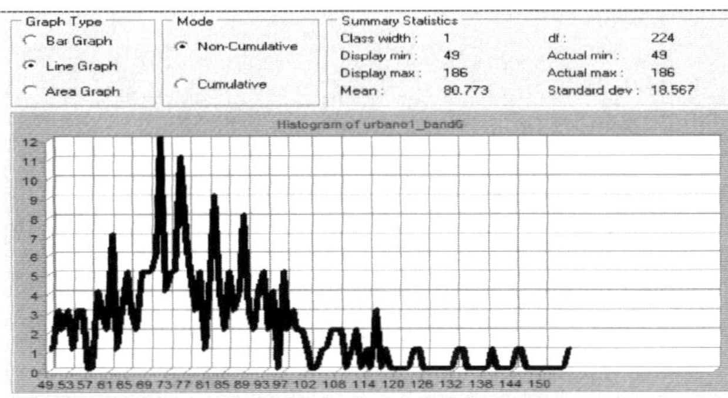

Figure 30. Band 7 - Landsat 7 ETM+ Satellite Image Urban Land Use – Histogram

Subsequently proceeded in each band to calculate the experimental variogram (black line) and the best variographic model (blue line) was adjusted:

Journal of Physics: Conference Series **720** (2016) 012047 doi:10.1088/1742-6596/720/1/012047

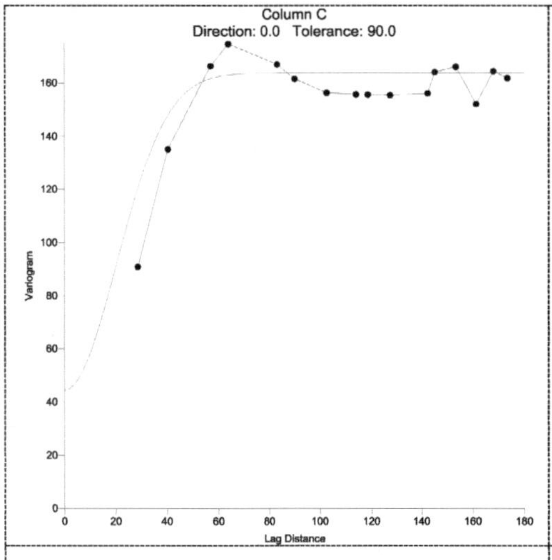

Figure 31. Band 1 - Best variographic model fitted is Gaussian model with Nugget effect

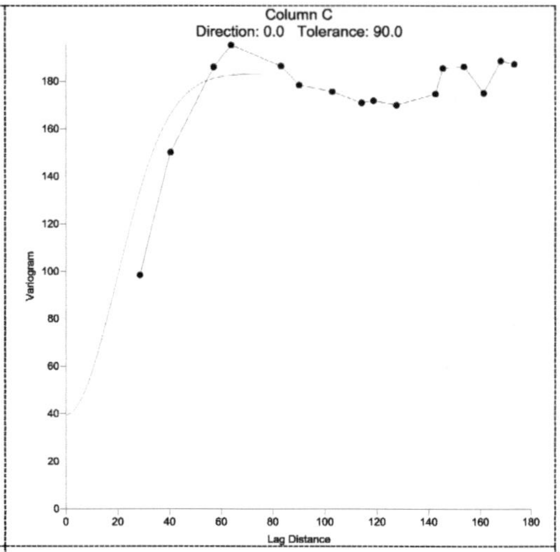

Figure 32. Band 1 - Best variographic model fitted is Gaussian model with Nugget effect

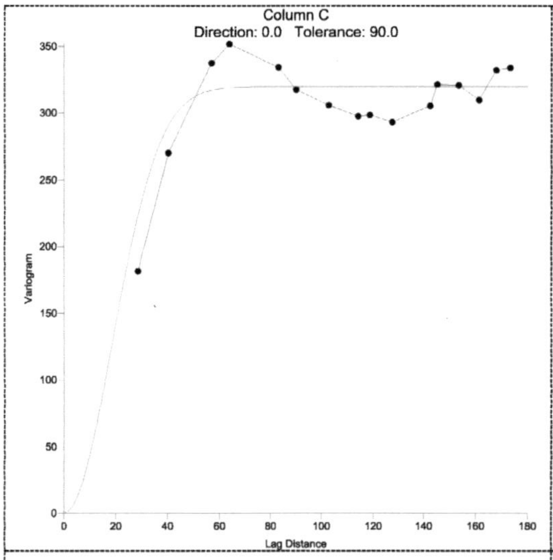

Figure 33. Band 3 - Best variographic model fitted is Gaussian model with Nugget effect

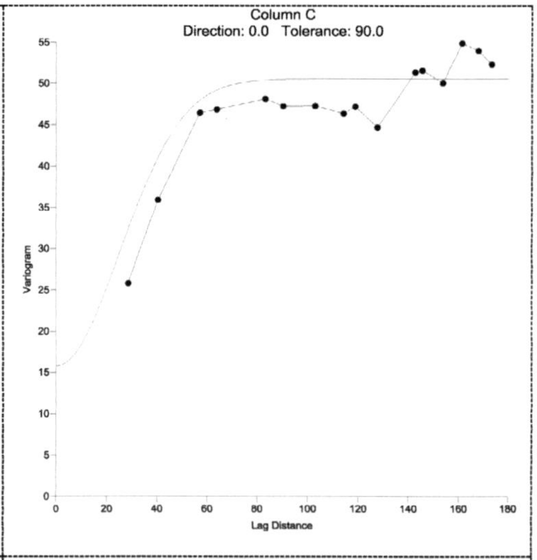

Figure 34. Band 4 - Best variographic model fitted is Gaussian model with Nugget effect

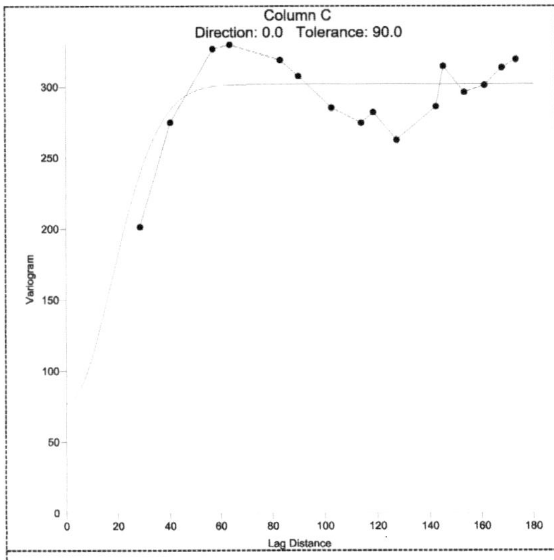

Figure 35. Band 5 - Best variographic model fitted is Gaussian model with Nugget effect.

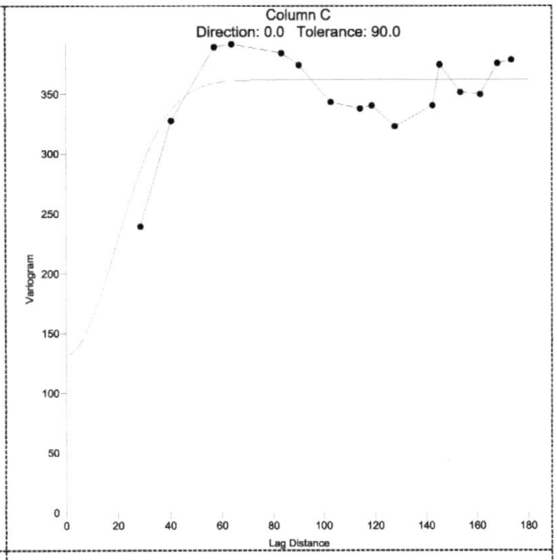

Figure 36. Band 7 - Best variographic model fitted is Gaussian model with Nugget effect.

The analysis of the graphs above it is concluded that the characteristic variographic model for the urban case is the Gaussian model with nugget effect.

3. Discussion

The above analysis indicates that for each land use is a characteristic shape and variographic cross-modal how their wavelengths.

References

[1] Jensen, J. R. Introductory Digital Image Processing: A Remote Sensing Perspective Ed. Prentice Hall, Upper Saddle, NJ, 1996.

[2] Krige, D. G.Two dimensional weighted moving average trend surfaces for ore valuation.- Proc. Symp. on math . statist. and computer applications in ore valuation, J. Johannesburg: 13-38. 1966.

[3] Matheron, G. Principles of Geostatistics.- Econ. Geol. 58: 1246-1266. 1963.

Numerical modeling of solar irradiance on earth's surface

E. Mera [1], **L. Gutierez** [2], **L. Da Silva** [2] **& E. Miranda** [1]

[1] Universidad Tecnológica Metropolitana, Physics Department, Box 9845, Santiago, Chile.
[2] Universidad Tecnica Federico Santa Maria, Box 110-V, Valparaiso.

E-mail: emera@utem.cl

Abstract. Modeling studies and estimation of solar radiation in base area, touch from the problems of estimating equation of time, distance equation solar space, solar declination, calculation of surface irradiance, considering that there are a lot of studies you reported the inability of these theoretical equations to be accurate estimates of radiation, many authors have proceeded to make corrections through calibrations with Pyranometers field (solarimeters) or the use of satellites, this being very poor technique last because there a differentiation between radiation and radiant kinetic effects. Because of the above and considering that there is a weather station properly calibrated ground in the Susques Salar in the Jujuy Province, Republic of Argentina, proceeded to make the following modeling of the variable in question, it proceeded to perform the following process: 1. Theoretical Modeling, 2. graphic study of the theoretical and actual data, 3. Adjust primary calibration data through data segmentation on an hourly basis, through horizontal and adding asymptotic constant, 4. Analysis of scatter plot and contrast series. Based on the above steps, the modeling data obtained: Step One: Theoretical data were generated, Step Two: The theoretical data moved 5 hours, Step Three: an asymptote of all negative emissivity values applied, Solve Excel algorithm was applied to least squares minimization between actual and modeled values, obtaining new values of asymptotes with the corresponding theoretical reformulation of data. Add a constant value by month, over time range set (4:00 pm to 6:00 pm). Step Four: The modeling equation coefficients had monthly correlation between actual and theoretical data ranging from 0.7 to 0.9.

1. Introduction

Modeling studies and estimation of solar radiation in base area, touch from the problems of estimating equation of time, distance equation solar space, solar declination, calculation of surface irradiance, considering that there are a lot of studies you reported the inability of these theoretical equations to be accurate estimates of radiation, many authors have proceeded to make corrections through calibrations with Pyranometers field (solarimeters) or the use of satellites, this being very poor technique last because there a differentiation between radiation and radiant kinetic effects.
Because of the above and taking into account the existence of a weather station Vaisala, it proceeded to perform the following process:

1. Theoretical Modeling

Journal of Physics: Conference Series **720** (2016) 012048 doi:10.1088/1742-6596/720/1/012048

2. Graphic study of the theoretical and actual data
3. Adjust primary calibration data through data segmentation on an hourly basis, through horizontal and adding asymptotes constant.
4. Analysis of scatter plot and contrast series.

2. Results

Based on the above steps, the modeling data obtained:

2.1. Theoretical Model Generation:

For the calculation of the equations Spencer, Cooper, Perrin, Lamm, Walreaven with changes Michalsky [1][3], where parameters of distance, decline, daylight radiation and assessments of surface graphics and will generate equations were calculated in this study they are occupied They were:

Tau =2*3.14*(Dia-1)/365

Equation 1. Daily conversion factor to Tau Expression Angle

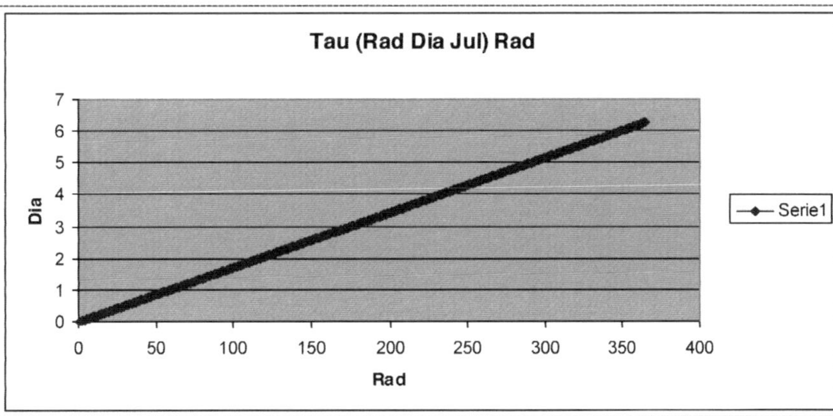

Figure 1. Graphic expression Tau

Solar declination = 0.006918 - 0.399912 * Cos(Tau) + 0.070257 * Seno(Tau) - 0.006758 * Cos(2*Tau) + 0.000907* Seno(2*Tau)-0.002697* Cos(3*Tau) + 0.00148 * Seno(3*Tau))

Equation 2. Solar declination angle multiple Fourier terms.

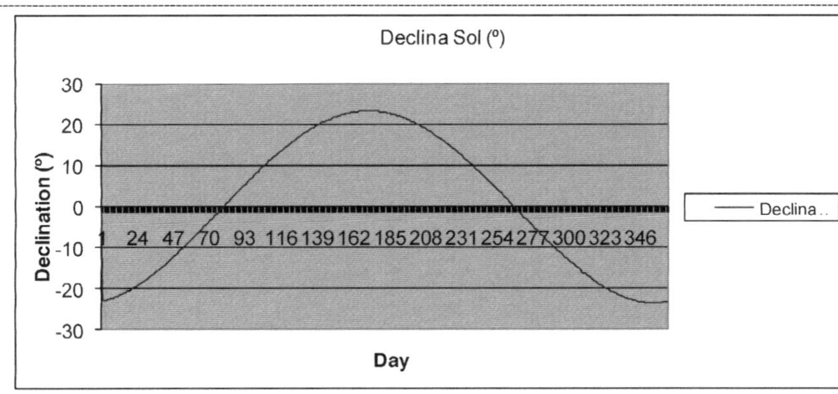

Figure 2. Graphic solar declination

Journal of Physics: Conference Series **720** (2016) 012048 doi:10.1088/1742-6596/720/1/012048

Eo =1.00011 + 0.034221*cos(Tau) +0.00128*Seno(Tau)+ 0.000719*COS(2*Tau)+ 0.000077*Seno(2*Tau)

Equation 3. Earth Sun Factor Correction

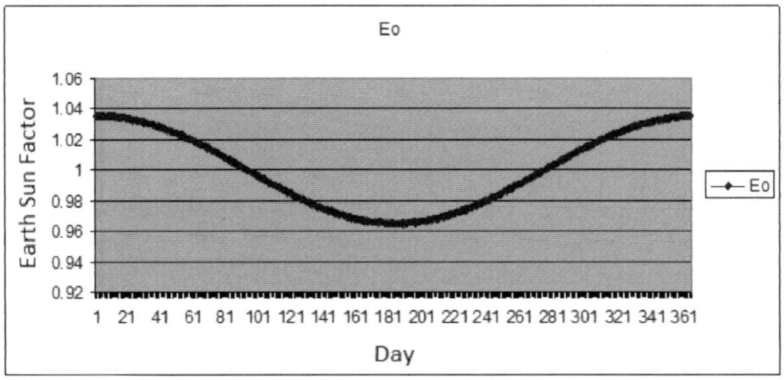

Figure 3. Graphic Earth Sun Factor Correction

Then the calculation model of solar radiation on the surface, took the following formula:

Surface Solar Irradiance =Io*Eo*(((Seno (Solar declination) *Seno(Slope) +Cos(Solar declination) * Cos (Hour Angle)) * Cos(Latitude)) * Cos(Slope) + Seno(Slope) * Seno (ACos ((Seno (Solar declination) *Seno(Latitude) + Cos(Solar declination) * Cos(Hour Angle)) *Cos(Latitude))) *(Seno(Latitude)- Seno(Eo))/ Cos(Latitude))

Equation 4. Surface irradiance

The result given in terms of surface radiant flux (W / m2), because the above equation depends on the hour angle, you can calculate the irradiance, every hour of the day of the year (Io is the Solar constant).

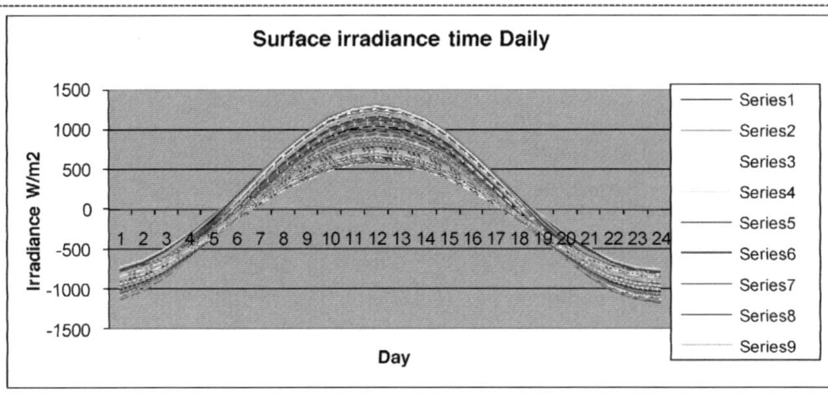

Figure 4. Surface irradiance Time Daily

Journal of Physics: Conference Series **720** (2016) 012048 doi:10.1088/1742-6596/720/1/012048

2.2. Graphic study of the theoretical and actual data:

He proceeded to take the theoretical data modeling seed, proceeded to contrast with the actual data, it was found a strong time lag and amplitude variation in the data.

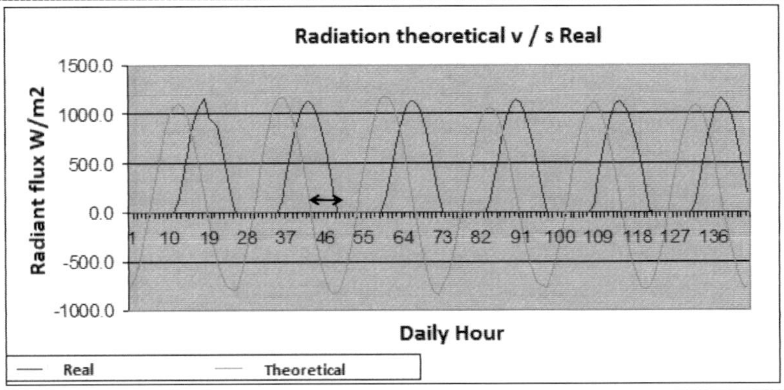

Figure 5. Temporary difference of n hours

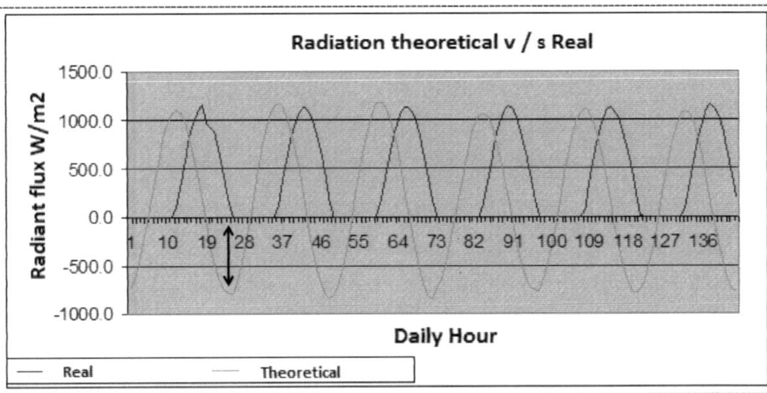

Figure 6. flux amplitude difference of n hours

Which indicated that the theoretical models, do not come close to solving the local reality because they can not only specify latitude and longitude, the above made you decide to make a temporary displacement and based on asymptotic model for radiation.

2.3. Adjust primary calibration data through data segmentation on an hourly basis, through horizontal and adding asymptotes constant:

Generic steps were:

2.3.1. Displacement t-n hours:

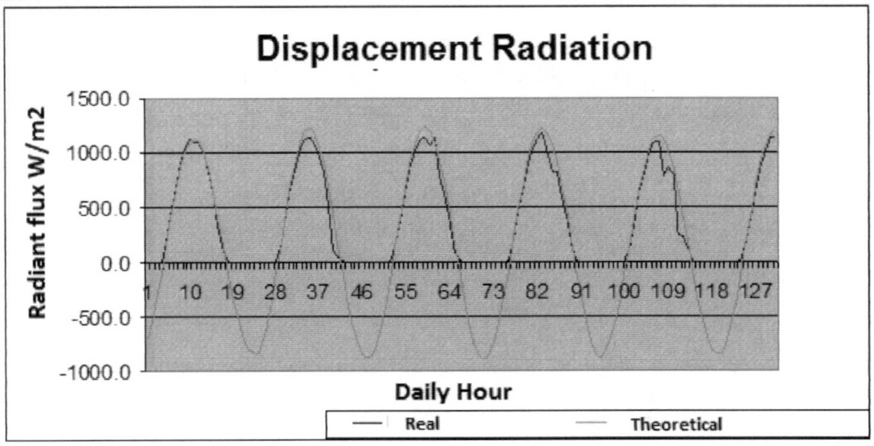

Figure 7. Temporary displacement of n hours

It follows that the displacement was 5 hours, such as data modeling in December 20:00 hr, it was 15:00 hr.

Table 1. Temporal displacement.

Month	Feb	Jan	Mar	Apr	May	Jun	Jul	Agu	Sept	Oct	Nov	Dec
Hours	5	5	5	5	5	5	5	5	5	5	5	5

2.3.2 Asymptotic curve fitting process and first least-squares adjustment

Must be the actual values regarding the theoretical, show asymptotic gap type, so their values radical, as seen in the figure:

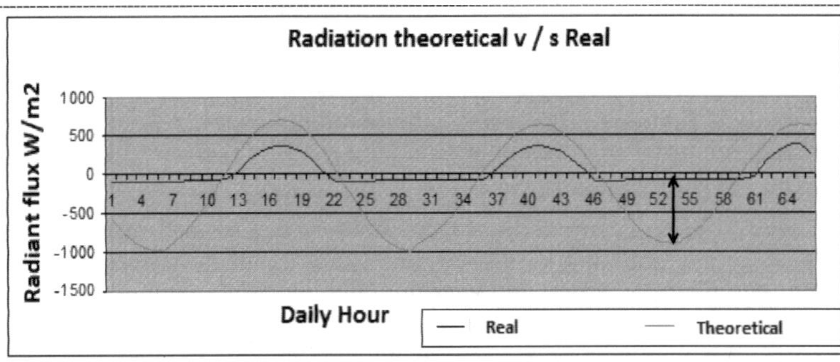

Figure 8. Asymptotic gap

It is noted that segmenting the feature temporarily, it was indicated that all the radiation they had negative flow the following values:

Table 2. Asymptotic initial value.

Month	Feb	Jan	Mar	Apr	May	Jun	Jul	Agu	Sept	Oct	Nov	Dec
Hours	-150	-150	-150	-150	-150	-150	-150	-150	-150	-150	-150	-150

Subsequently the total series underwent error minimization square minimum, where the values obtained were:

Table 3. Asymptotic fit value.

Month	**Feb**	**Jan**	**Mar**	**Apr**	**May**	**Jun**
Hours	-9.4	-23	-29.9	-57.5	-71.3	-85.5
Month	**Jul**	**Agu**	**Sept**	**Oct**	**Nov**	**Dec**
Hours	-108.6	-91.3	-92.1	-108.4	-16.2	-16.1

Where the values of error minimization are obtained:

Table 4. Optimization parameters.

Month	**Feb**	**Jan**	**Mar**	**Apr**	**May**	**Jun**
a.	0.66	0.8	0.76	0.6	0.52	0.44
b.	90.9	96.32	85.3	32.26	6.63	-18.99
Month	**Jul**	**Agu**	**Sept**	**Oct**	**Nov**	**Dec**
a.	0.53	0.49	0.49	0.53	0.83	0.75
b.	-29.03	-18.44	-18.25	-28.52	109.13	95.04

The error minimization model is linear type:

Optimized Radiation value = Asset Value Restricted by Asymptotic * a + b

Equation 5. Equation Theoretical Parameter Settings

2.3.3 Displacement amplitude of the curve

Figure 8 is observed in different months amplitude differences between the theoretical model and the actual data, maintaining similarity in curved shape, so it is assumed, it could be corrected by an additive process. It must at this point as a correction minimum squares had, it must

F (t) = Equation 5 + C (time restricted domain)

Equation 6. Parameters reformulation

He proceeded to add constant values in different months in different regimes of time, as indicated by the table:

Table 5. Addition constant value.

Month	**Feb**	**Jan**	**Mar**	**Apr**	**May**	**Jun**	**Jul**	**Agu**	**Sept**	**Oct**	**Nov**	**Dec**
Hours	4 to 6 pm											
Value	250	50	50	50	50	80	25	80	55	50	50	200

It has to curve made with the above processes, is:

Journal of Physics: Conference Series **720** (2016) 012048 doi:10.1088/1742-6596/720/1/012048

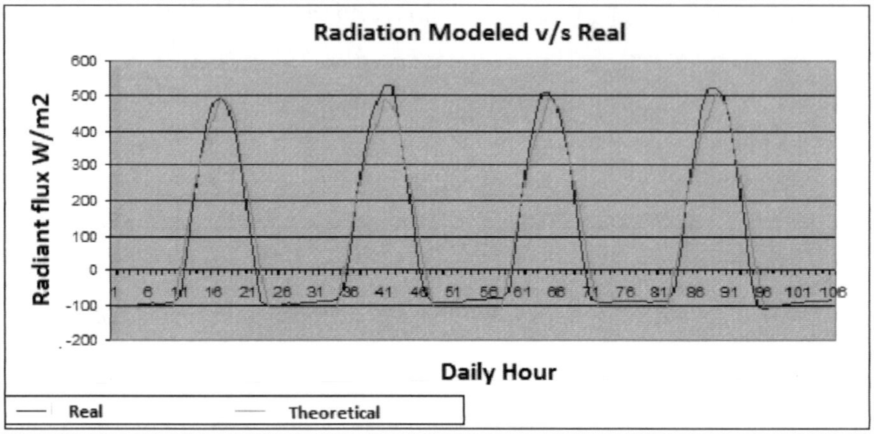

Figure 9. Adjusted curve additively

The modeling equation is as follows, depending on the month and time in hours:

$$Rad(t)(\frac{W}{m^2}) = (Theorial\_Rad(t-n)\_with\_flux < 150\_como\_-150) * a + b + Cte\_of\_4pm\_at\_6pm$$

Equation 7. Model

2.4. Analysis of scatter plot and contrast series:

After all regenerative processes, facts on the theoretical series, to achieve similarity to the actual data, in order to achieve environmental demonstration event observed, the values to be had, as exemplified June:

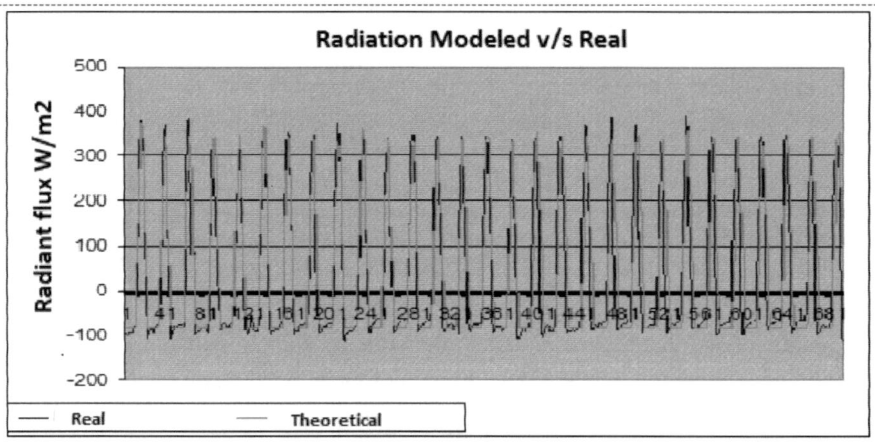

Figure 10. Contrast Series

In the analysis of the dispersion obtained following correlations:

Journal of Physics: Conference Series **720** (2016) 012048 doi:10.1088/1742-6596/720/1/012048

Table 6. Correlation value.

Month	**Jun**	**Jul**	**Agu**	**Sept**	**Oct**	**Nov**	**Dec**	**Jan**
Value	0.77	0.93	0.77	0.74	0.86	0.93	0.908	0.79

3. Discussion

It is noted that:

Radiation Theoretical values (t-5), Table 1 (This is because the problem of parallax).

Values of "a" and "b" in Table 4 (error minimization theoretical real value adjusted).

Constant values 4-6 pm in Table 5 (Setting Base).

It means that for example for the month of September, if you want to calculate the radiation at 6 pm, radiation at 1 pm is sought, if the flow is greater than zero this value is left then is multiplied by 0.48 and that adds value and how this value -18.25 in the time range between 4pm and 6pm, adds 55.

References

[1] Boland, J. Modeling Solar Radiation at the Earth's Surface, chapter 11. Springer Verlag Berlin Heidelberg, 2008

[2] Díaz, F., Montero, G., Escobar, J.M., Rodríguez, E., y Montenegro, R. An Adaptive Solar Radiation Numerical Model. Journal of Computational and Applied Mathematics, 236(18):4611–4622, 2012b.

[3] Holton J.R. "Introducción a la meteorología dinámica", Ed. Instituto Nacional de Meteorología, 1990.

[4] Janjai, S. Modeling Solar Radiation at the Earth's Surface, chapter 19. Springer Verlag Berlin Heidelberg, 2008.

[5] Muneer, T. Solar Radiation and Daylight Models. Elsevier Butterworth Heinemann, 2 edición, 2004.

Multi-V-type and Λ-type electromagnetically induced transparency experiments in rubidium atoms with low-power low-cost free running single mode diode lasers

S. Lavín Varela, J. A. León Suazo, J. Gutierrez González, J. Vargas Roco, T. Buberl[2] and J. G. Aguirre Gómez[1,3]

[1] Departamento de Física, Facultad de Ciencias Físicas y Matemáticas, Universidad de Concepción, Av. Esteban Iturra s/n, Barrio Universitario, Concepción - Chile
[2] Technische Uiversität München, Munich, Germany
[3] Center for Optics and Photonics, Universidad de Concepción, Casilla 4012, Concepción, Chile

E-mail: jaguirre@udec.cl

Abstract. In this work we present the experimental realization of electromagnetically induced transparency (EIT) in Λ-type and multi-V-type configurations in a sample of rubidium atoms inside a vapor cell at room temperature. Typical EIT windows are clearly visible in the Doppler-broadened absorption signal of the weak probe beam. The coherent optical pump and probe fields are produced by two tunable low-cost, low-power, continuous-wave (cw), free-running and single mode operated diode laser systems, temperature stabilized and current controlled, tuned to the D2 line of rubidium atoms at 780.2 nm wavelength. The continuum wave and single mode operation of our laser systems are confirmed by direct and saturated absorption spectroscopy techniques. Among other applications, these simple experiments can be used as a low-cost undergraduate laboratory in atomic physics, laser physics, coherent light-atom interaction, and high resolution atomic spectroscopy.

1. Introduction

Today, high-precision optical techniques allow bi-directional control of light and atoms; atoms can be manipulated with light, and light can be controlled by atoms [1]; in a quantum network a quantum bit must be stored in a quantum memory, which in turn re-emit a photon that must be transmitted between two distant quantum memories [2]; the experimental realization of super-radiance and of slow-light effects, which also can be used as quantum memories [3]; and the theoretical study of atomic ensembles in Λ-configuration, also good candidates for quantum memories [4], have been object of recent study. Although in most cases the proposed scheme are based in ultra-cold atomic samples, quantum memories schemes at the level of one-photon at room temperature, have also been studied [5], and in the last years long-live quantum optics memories have been also proposed [6]. In those applications, at some stage in its experimental realization, the requirement of an optical pump source, emitting light tuned with the atomic transitions, i.e., for the production of photons where the information is encoded, or for the cooling and trapping of the atoms where the information must be stored, becomes an unavoidable task.

Our experimental group, at the Department of Physic of the University of Concepción (Chile), started its experimental activities in 2012 with the aim to conduct experimental research studies in coherent light-atom interaction such as that stated in the previous paragraph, in particular, on the interaction of laser light beams tuned to atomics transitions of rubidium atoms. Due to small budget for operation we have built our own coherent light sources based on semiconductor laser chips (diode lasers). The central wavelength and the range of tunability, as well as, the operation temperature range and the driven current of diode lasers are given by the manufacturer. Once the diode laser is chose for the specific atomic specimen, it can be tuned to the specific atomic transition changing its temperature, its current [9], or by means of an external cavity setup, in which a diffraction grating in a Littrow or LittmanMetcalf configuration, act as a wavelength selective element. The tunable properties of diode lasers have been used largely in atomic physics experiments [7]. Diode lasers systems with optical feedback, known as external cavity diode lasers (ECDL), have been used in saturated absorption spectroscopy of cesium and rubidium atoms [8] and in laser cooling and trapping experiments for undergraduate laboratories [10]. In the present paper we present our simple laser diode system built to produce a beam of laser light tuned to specifics transitions of rubidium atoms, in particular, to the D2 line at 780.2 nm wavelength. The beam of these sources are used to show experimental signal of electromagnetically induced transparency (EIT) in Λ-type and multi-V-type configurations. The present paper is structured as follow: After the introduction §1, in section §2 we describe our tunable freerunning, low-cost, low power light source: Continuous wave and single mode operation of the light source are presented through direct and saturated absorption spectroscopy setups, respectively (see subsection §2.1). In section §3 we present the use of our source of light in electromagnetically induced transparency experiments. This section is divided in two subsections: In subsection §3.1 we present the main results for Λ-type configuration EIT, while section§3.2 is devoted to multi-V-type configuration. Finally, we present our conclusions in section §4.

2. Tunable laser light source

Our low-cost, low-power laser light system is a free running diode laser stabilized in temperature and controlled in current. A tunable infrared diode laser 780 nm, 80 mW laser diode with PD DL-7140-211N (SANYO) [11] is placed inside a diode laser collimation tube with optic LT220P-B (Thorlabs), which in turn is placed inside a home-made aluminum case for temperature stabilization. A piece of aluminum about three times in size compared to the home-made case is used as heat reservoir. The diode laser case and the reservoir are sandwiched with a thermo-electric cooler TEC3-6 (Thorlabs) in between for temperature stabilization and a thermistor TH10k (Thorlabs) is placed in a drilled hole in the case for temperature monitoring. Both the temperature stabilization and the driven current for the diode laser are taken from a commercial laser diode and temperature controller ITC102 (Thorlabs), current driven by an AC-DC linear regulated power supply (Acopian).

Because the typical wavelength of our tunable diode laser is 785 nm and due to the fact that we are not using it in an external cavity setup, we achieve single mode operation at the desired wavelength only by stabilizing the temperature and controlling the injected current to the system. So, for a work temperature $\sim 20^0$C and an injected current $\sim$ 120 mA our system is able to emit $\approx$ 60 mW of power in a single-mode continuum-wave (cw) at 780 nm wavelength, resonant with the D2 line of rubidium atoms. Two systems with similar characteristics are developed and their spectral emission are tested with simple pass and saturated absorption techniques in a rubidium vapor cell at room temperature. Another similar system, tuned to

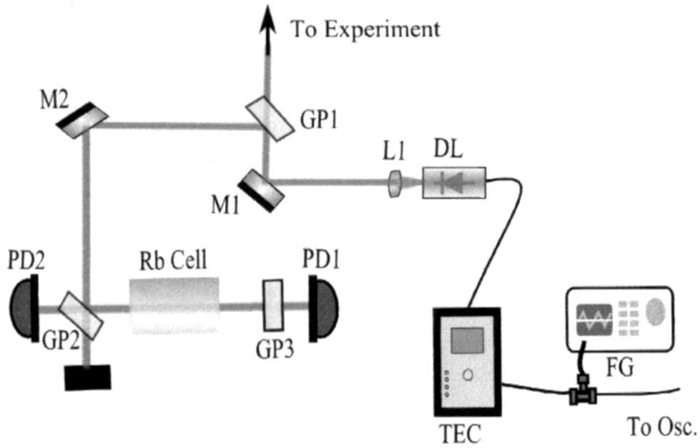

Figure 1. Tunable laser light source and direct and saturated absorption spectroscopy setup. DL: diode laser; TEC: Diode laser controllers; FG: Function Generator; GP: glass plate; L1: aspheric lens; M: Dielectric mirrors; PD2: saturated absorption signal photo-detector; PD1: direct absorption signal photo-detector.

the D1 line of rubidium atoms at 795 nm wavelength, is used in a simple experiment of second harmonic generation at 397.5 nm wavelength inside a PPKTP, Type-I, nonlinear crystal [12].

2.1. Direct and saturated absorption spectroscopy

The single-mode continuum-wave operation of our laser light system is tested extracting a small portion of the beam to a direct and saturated absorption spectroscopy, schematically showed in Fig.1: The setup consists of two 25 mm diameter common glass plates, a 25 mm diameter and 75 mm long cell filled with natural ^{85}Rb and ^{87}Rb isotopes composition of rubidium atoms (Triad Technology) at room temperature, a pair of broad-band photo-detectors, and an oscilloscope (not showed in the figure). The frequency of the diode laser is scanned around the D2 resonance line for both ^{85}Rb and ^{87}Rb isotopes, injecting a small modulating voltage from a commercial Tektronix AFG3021B function generator via the modulation injection port available in the current driver controllers system. Direct and saturated absorption signals are registered by photo-detectors PD1 and PD2, respectively, and their photo-currents are visualized in an oscilloscope. Typical signals from this setup are shown in Fig.2.

The hyperfine atomic structure of the D2 resonance line of rubidium atoms at 780 nm wavelength of both ^{85}Rb and ^{87}Rb isotopes are clearly visible in Figure 2. Figure 2.a) [2.b)] show the $5^2S_{1/2} \rightarrow 5^2P_{3/2}$ transition between the fundamental $F = 2[F = 3]$ and the excited $F' = 1, 2, 3$ $[F' = 2, 3, 4]$ hyperfine levels of the ^{85}Rb isotope. The separation in frequency between the two fundamental hyperfine levels $F = 2$ and $F = 3$ for ^{85}Rb is 3.035 GHz. Figure 2.c) [2.d)] show the $5^2S_{1/2} \rightarrow 5^2P_{3/2}$ transition between the fundamental $F = 1$ $[F = 2]$ and the excited $F' = 0, 1, 2$ $[F' = 1, 2, 3]$ hyperfine levels of the ^{87}Rb isotope. The splitting in frequency between the two fundamental hyperfine levels $F = 1$ and $F = 2$ for ^{87}Rb is 6.834 GHz. These curves are taken with a small amplitude modulation signal, a few millivolts peak-to-peak, injected in the driven current of the diode laser. When the amplitude of the modulation signal is increased, we are able to see these four curves in only one screen without mode-hope, that means, we are able to scan the frequency of the diode laser system in more than 7 GHz. This is a clear evidence of a single-mode free-running continuum-wave operation of our laser system. Furthermore, the ability to resolve the hyperfine structures of both ^{85}Rb and ^{87}Rb isotopes

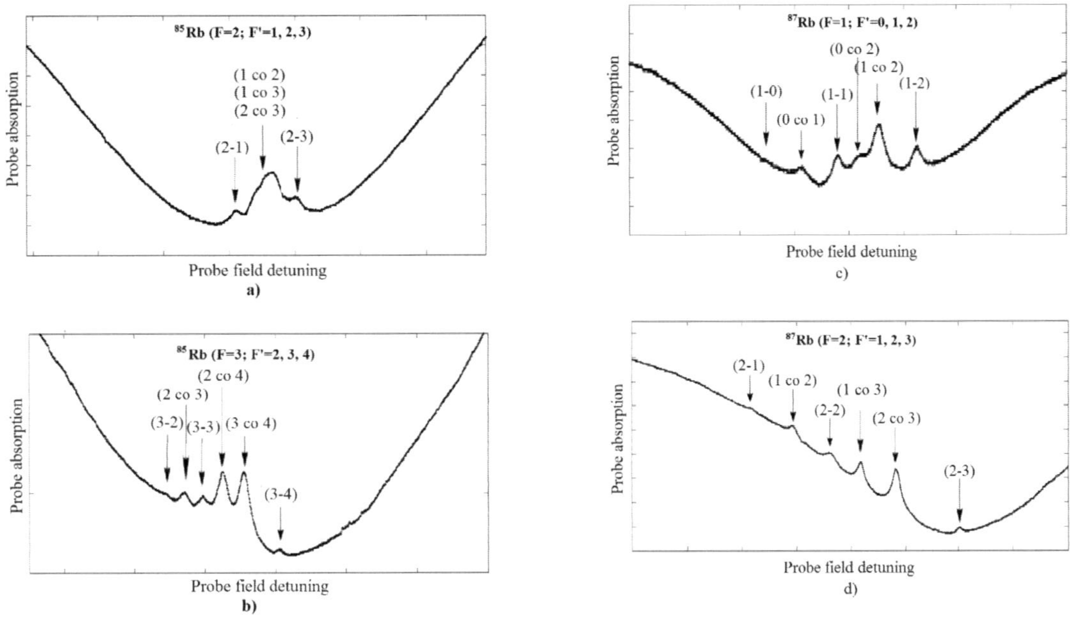

Figure 2. Saturated absorption signals for the $5^2S_{1/2} \rightarrow 5^2P_{3/2}$ transitions of both ^{85}Rb [a) and b) curves)]) and ^{87}Rb [c) and d) curves] isotopes. In all plots "co" means "cross-over".

revels us that the emission line-width of our system is, at least, less than the natural line-width of the transition, $\Gamma \sim 2\pi \times 6 \times 10^6$ Hz [13, 14] for the D2 line at 780 nm wavelength.

3. Electromagnetically induced transparency

The term EIT, acronym of Electromagnetically Induced Transparency, was introduced by Harris *et al* [15] in 1990 to show that a three-state atomic system exhibit a significant reduction in the absorption of a beam from a weak probe laser field in the presence of a strong laser field which couples an atomic transition that shares a common state with the transition coupled for the probe beam. The first experimental evidence of EIT was presented in 1991 [16] using high-power pulsed laser to produce the coupling and probe fields resonant with atomic transitions of strontium vapor. From that experiment, EIT has gained great attention in both theoretical and experimental aspects [17]. Although Λ-type EIT is the most common configuration used, multi-V-type [18] and ladder-type EIT [19] have been also reported.

Electromagnetically induced transparency involve three electronics transitions, commonly labeled $|1\rangle$, $|2\rangle$ and $|3\rangle$, and two laser beams: The strong pumping, driving or coupling beam of angular frequency ω_c is resonant with the $|2\rangle \rightarrow |3\rangle$ transition; and the weak probe beam (non absorbed by the medium) of angular frequency ω_p being scanned around the $|1\rangle \rightarrow |2\rangle$ transition. In the present paper we focus our attention in the experimental realization of Λ-type and multi-V-type EIT configurations using rubidium atom as the atomic medium and the diode laser system, described in the previous section, as the sources for the coupling and probe beams. The three energy states for these configurations are sketched in Fig.3 where the state $|2\rangle$ is used as the common state shared for the two EIT types.

The experimental setup, used for both multi-V and Λ-type EIT, is shown in Fig.4. Two diode laser systems equipped with their respective saturated absorption setups (for reference in

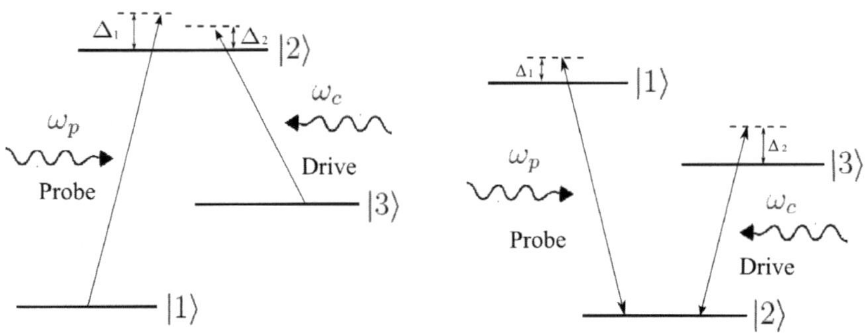

Figure 3. Three energy-levels diagrams for Λ-type (Left) and V-type (Right) EIT. In the two diagrams (ω_p, Δ_1) and (ω_c, Δ_2) denotes the angular frequency and the de-tuning from the probe and coupling laser beams, respectively.

frequency) produces the driving or coupling (Drive Laser) and the probe (Probe Laser) beams. Half-wave plates (HWP) and a polarizing-beam splitter cube PBS1 are used to control the polarization and the power of the beams interacting inside the vapor cell (Rb cell) filled with ^{85}Rb and ^{87}Rb isotopes at room temperature. At the polarizing-beam splitter cube PBS1 the two beams are recombined and sent to the rubidium cell: The probe laser beam is transmitted through the cube with horizontal polarization, while the driving laser beam is reflected with vertical polarization in the same cube. After their interaction with the atoms in the cell both the probe and the driving beam are separated using another polarizing-beam splitter cube PBS2: The transmitted probe beam is detected using a photo-detector and the photo-current signal is sent to an oscilloscope; the reflected driving beam is blocked. Two apertures, one at the input face and the other at the output face of the rubidium cell (not shown in the figure) are used to control the cross-sectional area of the beams and their co-propagation into the atomic medium.

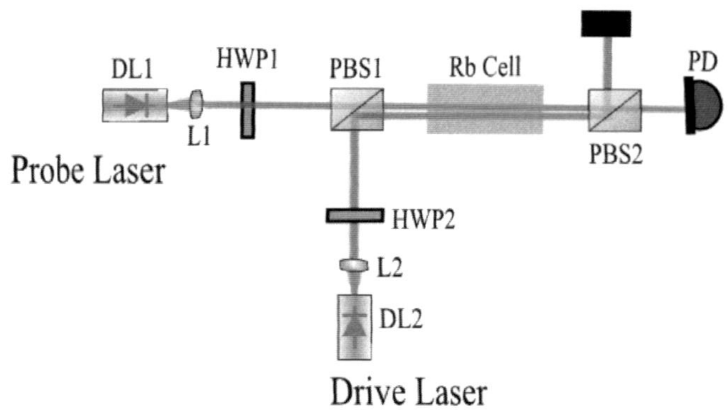

Figure 4. Experimental setup for Λ-type and multi-V-type electromagnetically induced transparency. HWP: half-wave plate; PBS: polarizing beam-splitter cube.

Armed with this experimental setup we proceed to record the typical EIT and saturated absorption signals for the probe laser beam. The results are presented in the following two subsections.

3.1. Λ-type EIT

A general scheme for Λ-type EIT is presented in Fig.3 (Left): The driving laser beam of frequency ω_c is tuned to the $|3\rangle \rightarrow |2\rangle$ transition, while the probe laser beam is scanned around the $|1\rangle \rightarrow |2\rangle$ transition. As mentioned above, in this EIT configuration the two field couples transitions from different ground levels to a common one high-energy level. In Fig.5 (Left) we show a typical probe laser beam absorption signal from the PD3 photo-detector (see Fig.4) with a clear EIT peak (red curves) together with the saturated absorption signal of the same beam, from which the hyperfine splitting of the $5^2P_{3/2}$ state is clearly visible (blue curves); Fig.5 (Right) show the corresponding energy-level diagram for this EIT configuration. In this particular case the driving laser beam is resonant ($\Delta_2 \approx 0$) with the $5^2S_{1/2}$ $(F = 3) \equiv |3\rangle \rightarrow 5^2S_{1/2}$ $(F' = 3) \equiv |2\rangle$ transition, with the probe laser beam being scanned around the $5^2S_{1/2}$ $(F = 2) \equiv |1\rangle \rightarrow 5^2S^{1/2}$ $(F' = 1, 2, 3) \equiv |2\rangle$ transitions for ^{85}Rb. From those curves we see a non-absorption peak in the Doppler-broadened absorption spectra of the probe beam through the rubidium atoms cell "Rb cell" (see Fig.4) exactly at the $5^2P_{3/2}$ $(F' = 3)$ hyperfine level of ^{85}Rb, the same level coupled by the driving field resonant with the $5^2S_{1/2}$ $(F = 3) \equiv |3\rangle \rightarrow 5^2S_{1/2}$ $(F' = 3) \equiv |2\rangle$. The power of the probe and coupling fields at the input face of the rubidium cell are 93 μm and 1.37 mW, respectively.

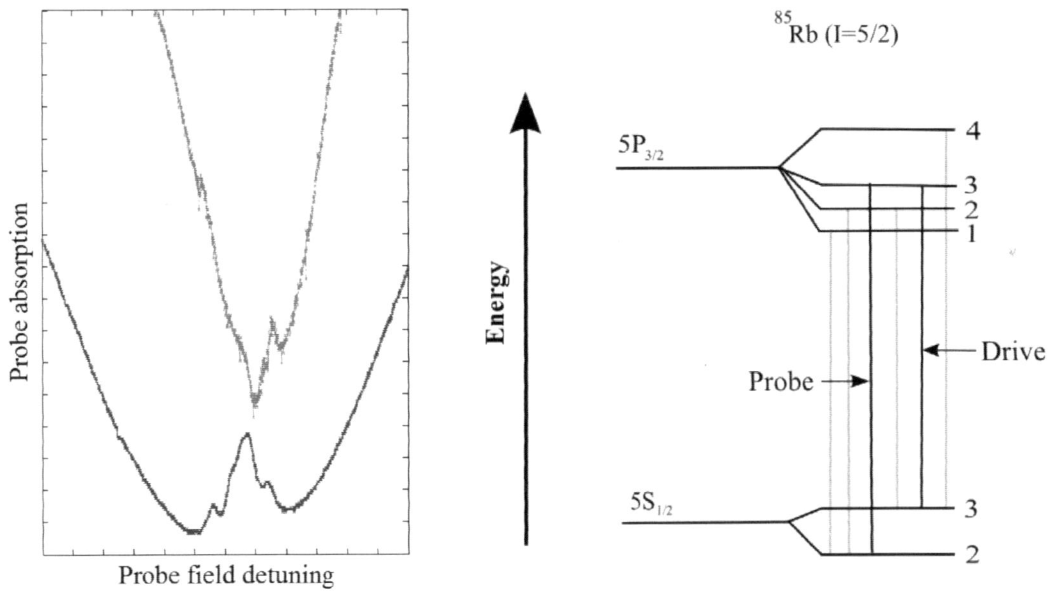

Figure 5. Probe laser beam absorption versus probe frequency. Left: Doppler-broadened probe absorption signal with EIT peak (red curves) and saturated probe absorption signal (blue curves). Right: Energy-levels diagram for ^{85}Rb with bold vertical lines showing the corresponding transitions coupled for the fields.

3.2. Multiple-V-type EIT

Electromagnetically induced transparency in V-configuration or V-type EIT has been observed experimentally and theoretically modeled in sodium atomic beam [20] showing a decreasing fluorescence and absorption related to the detuning of the probe and coupling fields. After that,

EIT in an open V-type Na_2 molecular system, in both co- and counter-propagating arrangements for the direction of propagation of the driving and probe fields, was also experimentally observed and simulated [21]. Cesium atoms have been used in experimental observation of absorption and dispersion in V-type configuration [18] and in multi-V-type EIT and hyperfine structure of atomic levels measurement [22]. Using rubidium atoms as the atomic medium, V-type mismatched EIT of a probe laser field tuned to the $5^2S_{1/2} \rightarrow 6^2P_{3/2}$ transition at 420 nm wavelength, using a low power near infrared driving field at 780 nm wavelength, has also been experimentally observed and theoretically modeled [23], and a detailed experimental investigation and theoretical analysis of a V-type EIT using ^{85}Rb isotope, showing seven transparency windows- with a double-peak-structure included- have been also reported [24].

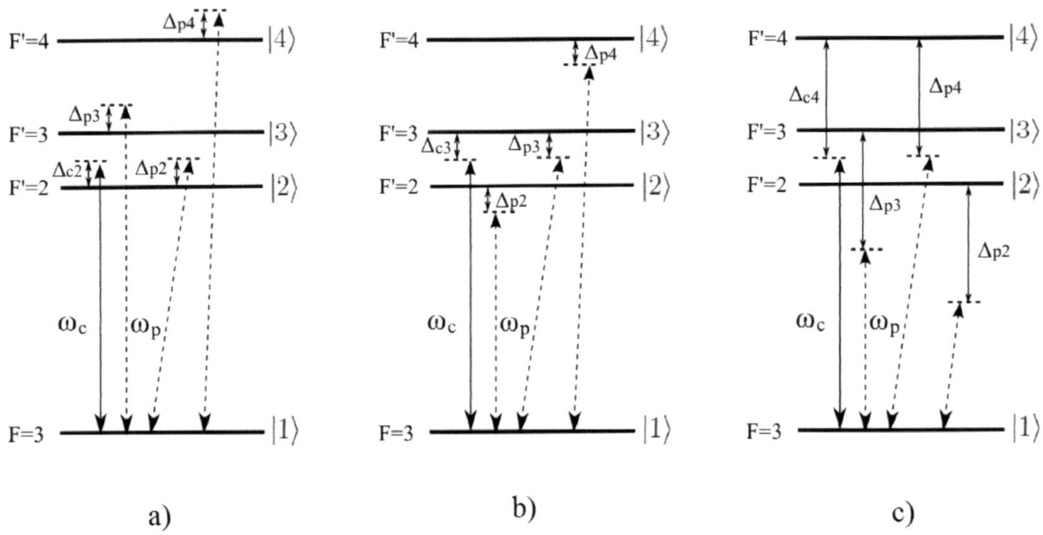

Figure 6. Schematic energy-levels structure of a multi-V-type EIT configuration. ω_c and ω_p denotes the frequency of the coupling and probe beams, respectively. Δp_i and Δc_j (with i, $j = 2, 3, 4$) denotes detuning of the probe and coupling fields from the $|i\rangle$, $|j\rangle$ levels.

A schematic diagram of the energy levels involved in a multi-V-type EIT configuration, adapted from Ref.[24], is shown in Fig.6. In Fig.6.a) the coupling laser beam of frequency ω_c is seen blue shifted by an amount Δc_2 from the transition $|1\rangle \rightarrow |2\rangle$ and the probe laser beam, being swept in frequency around the $|1\rangle \rightarrow |2\rangle$, $|3\rangle$, $|4\rangle$ transitions, can be seen blue shifted by the amounts Δp_2, Δp_3 and Δp_4 from the $|1\rangle \rightarrow |2\rangle$, $|1\rangle \rightarrow |3\rangle$ and $|1\rangle \rightarrow |4\rangle$ transitions, respectively. Figures 6.b) and 6.c) represent similar situations where the coupling field is seen red shifted by the amount Δc_3 and Δc_4 from the $|1\rangle \rightarrow |3\rangle$ and $|1\rangle \rightarrow |4\rangle$ transitions, respectively. The EIT structure will appear when the two-photon resonance condition $\Delta p_i = \Delta c_j$ (with i, $j = 2, 3, 4$) be satisfied by the sweep in frequency of the probe field. As pointed in the reference, this is the origin of the multi-EIT-windows for this configuration.

In this paper we do not make any theoretical analysis nor a detailed experimental investigation of the multi-V-type EIT, but we show that our low cost low power diode laser system also can be used for this type of research. In Fig.7 we show a Doppler-broadened probe laser beam absorption signal (red curves) for a multi-V-type EIT experiment at the D2 line of ^{85}Rb atoms at 780 nm wavelength. Also in the same figure we show a typical saturated absorption signal for the same probe beam (blue curves). In this particular case we set the frequency of the coupling laser beam resonant with the frequency of the $5^2S_{1/2}$ $(F = 3) \rightarrow 5^2P_{3/2}$ $(F' = 2)$ transition, being the

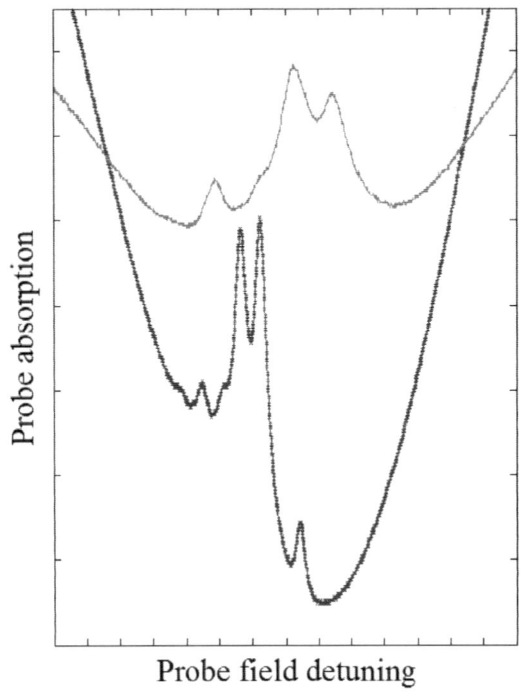

Figure 7. Probe laser beam absorption versus probe detuning. Doppler-broadened probe laser beam absorption signal with EIT peaks (red curves) and saturated probe absorption signal (blue curves) of ^{85}Rb for the transition $5^2S_{1/2}$ $(F = 3) \rightarrow 5^2P_{3/2}$ $(F' = 2, 3, 4)$ of the D2 line at 780 nm wavelength.

probe laser beam scanned around the $5^2S_{1/2}$ $(F = 3) \rightarrow 5^2P_{3/2}$ $(F' = 2, 3, 4)$ transitions. The Doppler-broadened curve show a non-absorption peak for the probe laser beam exactly at the $5^2S_{1/2}$ $(F = 3) \rightarrow 5^2P_{3/2}$ $(F' = 2)$ transition, the same transition coupled for the driving laser field (typical signature of the EIT phenomenon), together with a very weak peak followed by a double-peak-structure. Our experimental result agree partially with those reported in Ref.[24] and we attribute this to different experimental conditions, such as, coupling and probe field powers, line-width of our diode laser systems and the temperature of the rubidium sample.

4. Conclusions

In conclusion, we have reported the experimental implementation of electromagnetically induced transparency in Λ-type and multi-V-type configurations in a cell of rubidium atoms at room temperature. For the former, our results are in agreement with the ones previously reported in the literature, showing the clear EIT window for a weak probe laser beam interacting with an atomic sample pumped by another coherent laser field. For the latter, our results are in partial agreement with result recently reported in the literature and we attribute this to different experimental conditions, such as, coupling and probe laser beam powers, line-width of our diode laser systems and the temperature of the rubidium sample. For the realization of these experiments we developed low-cost, low-power diode laser systems in a continuum wave (cw) and single mode operation, stabilized in temperature and current controlled, tuned to the D2 line of rubidium atoms at 780.2 nm, whose spectral characteristics were also showed by standard direct and saturated absorption spectroscopy techniques. Among others applications, these simple experiments can be used as a low-cost and suitable ones in undergraduate laboratory in atomic physics, laser physic, coherent light-atom interaction, and high resolution atomic spectroscopy.

5. Acknowledgment

The present work was partially supported from PIA-CONICYT PFB0824, to which we give our thanks.

References

[1] M. D. Lukin, 2003 *Rev. Mod. Phys.* **75** 475
[2] T. Chaneliere, D. N. Matsukevich, S. D. Jenkins, S. Y. Lan, T. A. B. Kennedy and A. Kuzmich 2005 *Nature Letters* **438** 457
[3] A. Walther, A. Amari, S. Kröll and A. Kalachev 2009 *Phys. Rev. A* **80** 0121037
[4] E. Seuthen, A. Grodecka-Grad and A. Sorensen 2011 *Phys. Rev. A* **84**, 043838
[5] K. F. Reim, P. Michelberger, K. C. Lee, J. Nunn, N. K. Langford and I. A. Walmsley 2011 *Phys. Rev. Lett.* **107** 053603
[6] R. Zhao, Y. O. Dudin, S. Jenkins, C, Campbell, D.N. Matsukevich, T. A.B. Kennedy and A. Kuzmich 2009 *Nature Physics* **5**, 100
[7] A reference for the characteristics of diode lasers emission an the improvement of their spectral properties for atomic physics is Carl E. Wieman and Leo Hollberg 1991 *Rev. Sci. Inst.* **62** 1
[8] K. B. MacAdam, A. Steinbach and C. Wieman 1992 *Am. J. Phys.* **60** 1098
[9] A good reference for diode lasers system without grating feedback is G. N. Rao, M. N. Reddy, and E. Hecht 1998 *Am. J. Phys.* **66** 702
[10] C. Wieman, G. Flowers, and S. Gilbert 1995 *Am. J. Phys.* **63** 317
[11] We buy 50 pieces by 59.90(1.20 per piece) at http://www.aliexpress.com/store/product/50pcs-New-import-780nm-80mW-Laser-Diode-with-PD-Sanyo-DL-7140-211N/802826 1012778766.html
[12] José Vargas Roco, Sebastián Lavín Varela, Javier León Suazo, Javier Gutiérrez González, and José G. Aguirre Gómez 2015, submitted for revision at Am. J. Phys. ID submission number 11149.
[13] Daniel Adam Steck, *Rubidium 85 Line Data* 2013
[14] Daniel Adam Steck, *Rubidium 87 Line Data* 2013
[15] S. E. Harris, J. E. Field, and A. Immamoglu 1990 *Phys. Rev. Lett.* **64** 1107
[16] K. J. Boller, A. Imamoglu, and S. Harris 1991 *Phys. Rev. Lett.* **66** 2593
[17] Good references for the review of the EIT topic are: J. P. Marangos 1998 *J. Mod. Opt.* **45** 471 and M. Fleischhauer, A. Imamoglu, and J. Marangos 2005 *Rev. Mod. Phys.* **77** 633
[18] Jianming Zhao, Lirong Wang, Liantuan Xiao, Yanting Zhao, Wangbao Yin, and Suotang Jia 2002 *Opt. Commun.* **206** 341
[19] J. Gea-Banacloche, Yong-qing Li, Shao-zhen Li, and Min Xiao 1995 *Phys. Rev. A* **51** 576 and Abraham J. Olson and Shannon K. Mayer 2009 *Am. J. Phys* **77** 116
[20] George R. Welch, G. G. Padmabandu, Edward S. Fry. Mikhail D. Lukin, Dmitri E. Nikonov, Frank Sander, Marlan O. Scully, Anton Weis, and Frank K. Tittel 1997 *Found. Phys.* **28** 621
[21] A. Lazoudis, T. Kirova, E. H. Ahmed, P. Qi, L. Huennekens, and A. M. Lyyra 2011 *Phys. Rev. A* **83** 063419
[22] Zhao Jian-Ming, Yin Wang-Bao, Wang Li-Rong, Xiao Lian-Tuan, and Jia Suo-Tang 2001 *Chin. Phys.* **11** 0241
[23] Silvije Vdović, Ticijana Ban, Damir Aumiler, and Goran Pichler 2007 *Opt. Commun.* **272** 407
[24] Kang Ying, Yueping Niu, Yihong Qi, Dijun Chen, Haiwen Cai, Ronghui Qu, and Shangqing Gong 2014 *J. Mod. Opt.* **61** 635

Generation of stable cnoidal waves in an erbium doped fiber laser system

Mario Wilson[1], Andrés González-García[2], and Vicente Aboites[3]

[1] Departamento de Física, Facultad de Ciencias Físicas y Matemáticas, Universidad de Chile, Blanco Encalada 2008, Casilla 487-3, Santiago, Chile.

[2] Coordinación de Mecatrónica, Instituto Tecnológico Superior de Guanajuato, carretera estatal Guanajuato-Puentecillas km. 10.5, Predio el Carmen, 36262, Guanajuato, México.

[3] Centro de Investigaciones en Óptica A.C., Loma del Bosque 115, Lomas del Campestre, C.P. 37150, León, México.

E-mail: mario.wilson@ing.uchile.cl

Abstract. In this work an all-fiber laser system able to produce stable cnoidal waves in both limits, i.e. in sinusoidal and soliton limits, is presented. The system is constructed using an erbium highly–doped fiber as an active media and an electrooptic modulator as the core of the control tool. This control tool is able to modulate the cavity losses. The direct modulation allows to control the pulses properties, shape, width and intensity. The proposed system is well described by a three level laser model based on the Statz–de Mars rate equations. Numerical and Experimental results are presented.

1. Introduction

A cnoidal wave, in one of their limits, can be identified as a soliton pulse. The first solitary wave was reported in 1834 [1], nevertheless, it was not until the late 19th century when the first attempt to explanain this phenomenon was done[2]; the subject was left to rest until the 1960's [3]. One rough definition of a soliton is a solitary wave that maintains its shape while it travels through a nonlinear dispersive medium with constant speed. A balancing mechanism between nonlinearity and dispersion is responsible for this phenomenon [4]. Optical solitons can be either spatial [5] or temporal [6, 7]. Hasegawa and Tappert [7] suggested that the balance between self-phase modulation and anomalous dispersion could explain soliton formation in optical fibers. Many exactly solvable models have soliton solutions, including the Korteweg–de Vries, the nonlinear Schrödinger, the Sine-Gordon and the Manakov equations; actually this represents a very active field of mathematical and physical research. The main soliton practical feature, that makes it a good carrier for reliable optical communications, is its capacity to maintain its energy and frequency (shape) while propagating along an optical fiber. However, the soliton production in a laser system its normally expensive [8, 9, 10]. The presented proposal reduces the cost for this kind of pulses. In their other limit, a cnoidal wave is transformed into a sinusoidal wave that can be useful in the study of the nonlinear propagation inside plasmas [11, 12]. In this work we propose an all-fiber system composed by a semiconductor laser acting as a pump source, an erbium doped fiber, that is used as an active medium, and an Electrooptic modulator, placed inside the cavity, as a control tool; the active medium acts also as a saturable absorber for this system. The cavity losses are controlled via the Electrooptic modulator through

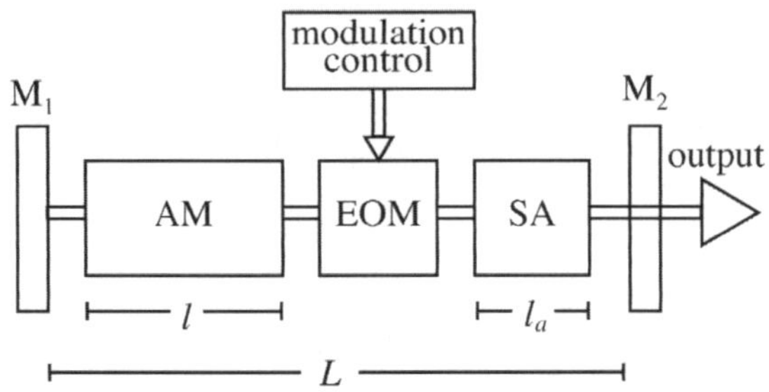

Figure 1. Proposed scheme for cnoidal wave generation. M_1 and M_2 are total reflection and semi-transparent mirrors, AM and SA are active medium and saturable absorber, and EOM is an electrooptic modulator

a functions generator. We demonstrate in this work that such a system can act as a cnoidal wave generator in a wide frequency range at a quite reasonable cost.

2. Theoretical model

The laser system depicted in Fig. 1 can be described by the Statz–de Mars equations [13]. A modified set of Statz–de Mars equations for a three-level laser with a saturable absorber (SA) without modulation control [14, 15] is written as follows:

$$\begin{aligned} \frac{dS}{dt} &= \Gamma\nu\sigma NS - \Gamma\nu\frac{l_a}{l}k_a S - \frac{1}{T}S, \\ \frac{dN}{dt} &= -\beta\frac{\sigma}{\hbar w}NS + \frac{N_0 - N}{\tau}, \\ \frac{dk_a}{dt} &= -\frac{2\sigma_a k_a S}{\hbar w} + \frac{k_{0a} - k_a}{\tau_a}, \end{aligned} \tag{1}$$

where S is the emitted photon density, N is the population inversion of the active medium, and k_a is the resonant absorption of the saturable absorber. Γ, ν, σ, and T stand, respectively, for cavity filling coefficient, optical frequency, active medium cross-section, and photon lifetime in the cavity, β is the coefficient which accounts for the difference in population inversion coursed by lasing, l and l_a are, respectively, the active medium and the SA lengths, k_{0a} is the linear resonant SA absorption coefficient without lasing, σ_a is the SA cross-section, N_0 is the population inversion in the active medium without radiation, τ and τ_a stand for relaxation time in the active medium and in the SA, respectively, and finally $\hbar w$ is the photon energy.

Equations (1) can be rewritten in an adimensional form by introducing new parameters and variables defined as $t' = t/\tau$, $G = \tau/T$, $\delta = \tau/\tau_a$, $\rho = 2\sigma_a/\beta\sigma$, $\alpha = \Gamma\nu\sigma TN$, $\alpha_a = -\Gamma\nu Tk_{0a}(l_a/l)$, $n(t') = \Gamma\nu\sigma TN(t')$, $n_a(t') = -\Gamma\nu(l_a/l)Tk_a(t')$, and $m(t') = 2\pi\beta\sigma\tau S(t')/\hbar w$, and including the normalized harmonic modulation control $(1 + cos(\omega t))/2$ to α_a in the third equation which describes the SA:

$$\begin{aligned} \frac{dm}{dt'} &= Gm(n + n_a - 1), \\ \frac{dn}{dt'} &= \alpha - n(m+1), \\ \frac{dn_a}{dt'} &= \delta\alpha_a\frac{1 + cos(\omega t)}{2} - n_a(\rho m + \delta). \end{aligned} \tag{2}$$

Journal of Physics: Conference Series **720** (2016) 012050 doi:10.1088/1742-6596/720/1/012050

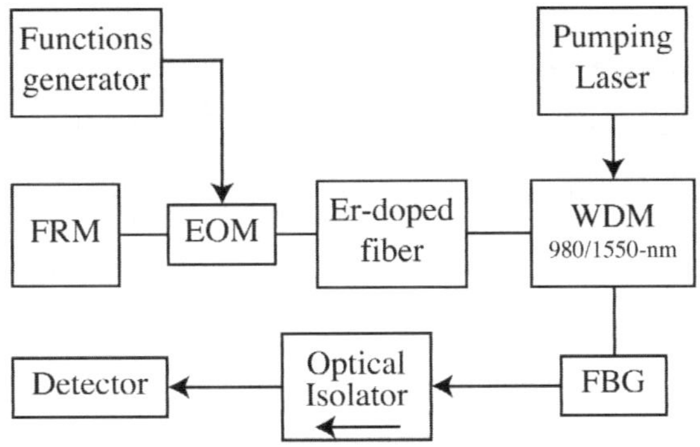

Figure 2. Experimental all-fiber set-up for cnoidal wave generation. FRM is a Faraday rotating mirror, EOM stands for electrooptic modulator, WDM represents a wavelength-division multiplexer and FBG is a fiber Bragg grating.

While an external signal injected into an optical cavity has often been used to modify the shape of the output signal [15, 16, 17], in this work it can be injected into the SA through the electrooptic modulator (EOM) in order to modify the system dynamics and generate stable cnoidal waves. In this context, the EOM becomes an active device since the laser output is regulated by it through the cavity losses.

3. Experimental results

For the experiment shown in Fig. 2, we used a 1550-nm erbium-doped fiber laser (EDFL) pumped with a semiconductor laser operating at 980-nm. The active media is given by a section of excited rare-earth erbium-doped fiber (EDF). The resonant cavity has a Fabry-Perot structure. The EDF consists in a 90-cm span of a SCL 110-01 actively doped fiber (with an Er_2O_3 concentration of 2300 ppm), this fiber was pumped by a 980-nm laser diode through a 980/1550-nm wavelength-division multiplexer (WDM) coupler. In order to form the resonator, a Faraday rotating mirror (FRM), and a fiber Bragg grating (FBG) centered at 1550 nm with a 100 pm FWHM, having, respectively, 100 and 88% reflectivities at the laser wavelength were used; the FRM was used in order to avoid polarization mode beating. An electrooptic modulator (EOM), modulated through a function generator, was placed inside the cavity. The fiber output, after passing through the WDM, was recorded with a high-speed photodetector and analyzed with an 8GHz oscilloscope. An opto-isolator was placed at the output in order to avoid any feedback that could cause instabilities in the laser dynamics. In this experiment the pumping-laser current was fixed at 63 mA, which resulted in a pumping power of 15.3 mW corresponding to about 10% above the laser threshold of 14 mW. This value was chosen in order to have a cw-laser operation. The EDFL's relaxation frequency was around 28 kHz due to the applied pumping power. As previously demonstrated [16, 17], depending on the relaxation frequency, certain modulation losses can be applied to the cavity through the EOM to produce cnoidal waves.

Figure 3 shows the obtained cnoidal waves for the presented configuration. The laser's output power registered for the soliton-limit was 2.5 mW with a repetition rate of 8.28MHz, corresponding to 90% of the injected modulation. These pulses are generated when the frequency applied to the EOM is around 9.1 MHz, which corresponds to 325 times the laser's relaxation frequency. The frequency needed to generate cnoidal waves depends on the length of the EDFL; the highest repetition rate is given by the maximum control modulation frequency injected into the EOM and its matching erbium-doped fiber length, meaning that it is limited by

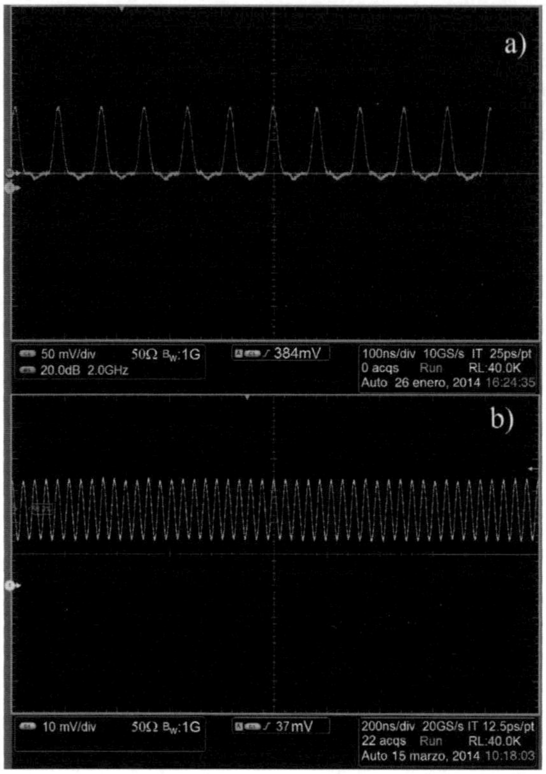

Figure 3. cnoidal wave generation. description.

the functions generator. Figure 4 shows the relationship between different erbium-doped fiber lengths and the control frequency needed to produce the cnoidal waves in their bounds. As it can be observed from Fig.4, the relationship between frequency and length becomes almost linear in the soliton limit (solid squares) with two jumps at lengths of 0.15 and 0.9 m. However, at the sinusoidal limit (solid circles) the growth presents jumps and plateaus at different fiber lengths than the soliton-limit. The optical spectrum remains fixed centered at 1552.2-nm. Figure 5 shows the temporal dynamics of the laser output, while the modulation frequency is increased for two different EDFL lengths. It is clear from Figs. 5(b) and 5(d) that the pulses shift from a soliton to a sinusoidal shape, having a strong relationship with the control frequency. Following our results, the cnoidal waves are bounded between two values of the control parameter (this parameter corresponds to the control modulation frequency). In this case, at a low frequency, we can observe that the cnoidal waves are transformed into a *sech* shape and, as the frequency increases, the wave changes into a *cos* shape. Note that the waves produced are asymptotically stable due to the quadratic nonlinearities of the erbium (acts as a saturable absorber and active medium).

4. Conclusions

It is demonstrated that linearly stable cnoidal waves can be produced in an erbium-doped fiber laser assisted by an EOM, which modulates the erbium absorbent centers. Both cnoidal wave limits were achieved, i.e. sinusoidal and soliton limits showing that it is possible to obtain such kind of pulses for different pairs of control frequency and doped fiber length. The dependence between the erbium-doped fiber length and modulation frequency of the EOM

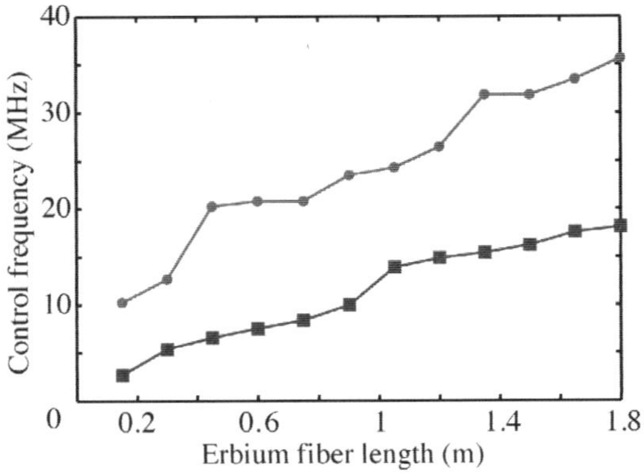

Figure 4. Relationship between fiber length and control frequency. Solid blue squares represent the soliton limit and solid green circles the sinusoidal one.

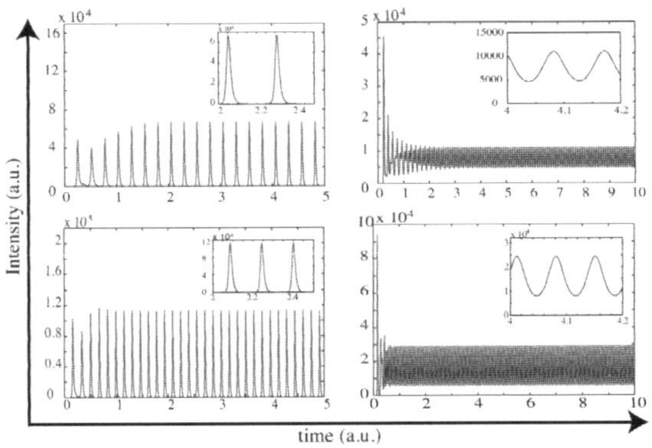

Figure 5. Numerical results for cnoidal waves limits. a) *sech* limit for 0.9m, b) *cos* limit for 0.9m, c) *sech* limit for 1.7m d) *cos* limit for 1.7m.

was experimentally demonstrated. The presented numerical results show the same qualitative behavior as the obtained experimental results. Furthermore, the proposed method is reliable and simple for the creation of soliton-shaped pulses which can be used for optical communication purposes and for sinusoidal-shaped pulses that can be useful in the study of the nonlinear propagation of pulses in different media.

Acknowledgement
The authors acknowledge partial financial support from CONACYT. M. W. acknowledges FONDECYT project 3140387. A. G. G. acknowledges CONCyTEG project 27 for young scientist 2014.

References
[1] J. S. Russell, Report on waves, Fourteenth Meeting of the British Association for the Advancement of Science (1844).
[2] L. Rayleigh, On waves, Philos. Mag. **1**, 257–279 (1876).
[3] N. J. Zabusky and M. D. Kruskal, Interaction of solitons in a collisionless plasma and the recurrence of initial states, Phys. Rev. Lett. **15**(6), 240–243 (1965).

[4] P. G. Drazin and R. S. Johnson, Solitons: An Introduction, 2nd ed. (Cambridge University Press, 1989).
[5] J. E. Bjorkholm and A. A. Ashkin, Cw self-focusing and self-trapping of light in sodium vapor, Phys. Rev. Lett. **2**(4), 129–132 (1974).
[6] A. Hasegawa and F. Tappert, Transmission of stationary nonlinear optical physics in dispersive dielectric fibers I: Anomalous dispersion, Appl. Phys. Lett. **23**(3), 142–144 (1973).
[7] A. Hasegawa and F. Tappert, Transmission of stationary nonlinear optical physics in dispersive dielectric fibers II: Normal dispersion, Appl. Phys. Lett. **23**(4), 171–172 (1973).
[8] F. Grme, P. Dupriez, J. Clowes, J. C. Knight, and W. J. Wadsworth, High power tunable femtosecond soliton source using hollow-core photonic bandgap fiber, and its use for frequency doubling, Opt. Express **16**(4), 2381–2386 (2008).
[9] R. Herda and O. G. Okhotnikov, All-fiber soliton source tunable over 500 nm, in Conference on Lasers and Electro-Optics/Quantum Electronics and Laser Science and Photonic Applications Systems Technologies, Technical Digest (CD) (Optical Society of America, 2005), paper JWB39.
[10] S. Chouli and P. Grelu, Rains of solitons in a fiber laser, Opt. Express **17**(14), 11776–11781 (2009).
[11] A. K. Upadhyay, G. Raj, R. K. Mishra, A. Malviya, and P. Jha, Phys. Plasmas **14**(9), 093107 (2007).
[12] A. K. Upadhyay, G. Raj, R. K. Mishra, and P. Jha, Phys Plasmas **14**(11), 113105 (2007).
[13] C. L. Tang, H. Statz, and G. de Mars, J. Appl. Phys. **34** 2289 (1963).
[14] L. Tarassov, Physique des Processus dans les Générateurs de Rayonnement Optique Cohérent (Éditions MIR, 1981).
[15] M. Wilson, V. Aboites, A. N. Pisarchik, V. Pinto and Y. Barmenkov, Rev. Mex. Fis **57**(3), 250-254 (2011).
[16] M. Wilson, V. Aboites, A. N. Pisarchik, V. Pinto and M. Taki, Opt. Express **19**(15), 14210 (2011).
[17] M. Wilson, P. Rangel-Fonseca, and A. González-García, Rev. Mex. Fis **61**(1), 14-17 (2015).

Design of a program in Matlab environment for gamma spectrum analysis of geological samples

M Rojas[1,2] and R Correa[1]
[1]Physics Departament, Technological Metropolitan University (UTEM), Santiago, Chile
[2]Invited Researcher, Universidad Nacional San Antonio Abad del Cusco (UNSAAC), Cusco, Perú

E-mail: mrojasgamarra@gmail.com

Abstract. In this work we present the analysis of gamma ray spectra Ammonites found in different places. One of the fossils was found near the city of Cusco (Perú) and the other in "Cajón del Maipo" in Santiago (Chile). Spectra were taken with a hyperpure germanium detector (HPGe) in an environment cooled with liquid nitrogen, with the technique of high-resolution gamma spectroscopy. A program for automatic detection and classifying of the samples was developed in Matlab. It program has the advantage of being able to make direct interventions or generalize it even more, or make it automate for specific spectra and make comparison between them. For example it can calibrate the spectrum automatically, only by giving the calibration spectrum, without the necessity of putting them. Finally, it also erases the external noise.

1. Introduction.

The study of fossils is important in different areas of science such as geology and paleontology. Development of automatic procedures of the analysis of gamma spectrums is important for correct datation of samples. There are many ways of analyzing fossil, the most common is based on the gamma radiation emitted by radioactive nuclei present in the fossil, gamma-ray spectroscopy. Unfortunately, the existing programs do not allow any interference. Neither it is possible to plot the spectra in an easier way in order to get a visualization of radioactive series present. For these reason we had to develop our software from scratch.

In this work we obtained and analysed the spectra of two samples of fossilized ammonite. One of the fossils was found near the city of Cusco and the other in the Cajón del Maipo in Santiago. Gamma-ray spectroscopy of high resolution was done at the Center for Experimental Physics, University of Chile.

The article is organized as follows: in section 2 we will describe the experiment, the spectrum analysis; we have plotted, detected and classified in radioactive series the present elements. Section 3 contains conclusions.

2. Gamma-ray spectroscopy in application to the Ammonite.

The spectra of the two Ammonite samples were taken with a detector of hyperpure Germanium (HPGe) cooled with liquid nitrogen, with the technique of Gamma-ray spectroscopy with high resolution.

The spectral data were collected 24 hours for the fossil of Cusco and 50 hours for the fossil of Santiago. In both cases, the detector was calibrated with NA22. Figure 2 and 3 shows both spectra.

For the correct identification of elements it was necessary to compare our spectrum with calibration curves, plot the calibrated spectrum and correctly identify the pecks. This identification is based on a data base of the transitions which help to find the radioactive nucleus and series to which they belong to. Figure 2 contains the summary of algorithms we used. Figures 3 and 4 contain the results of the spectral analysis. It is possible to can clearly visualize the radioactive series, mainly Th-232, U-238 and U-235, the deleted extreme points, Background and peck.

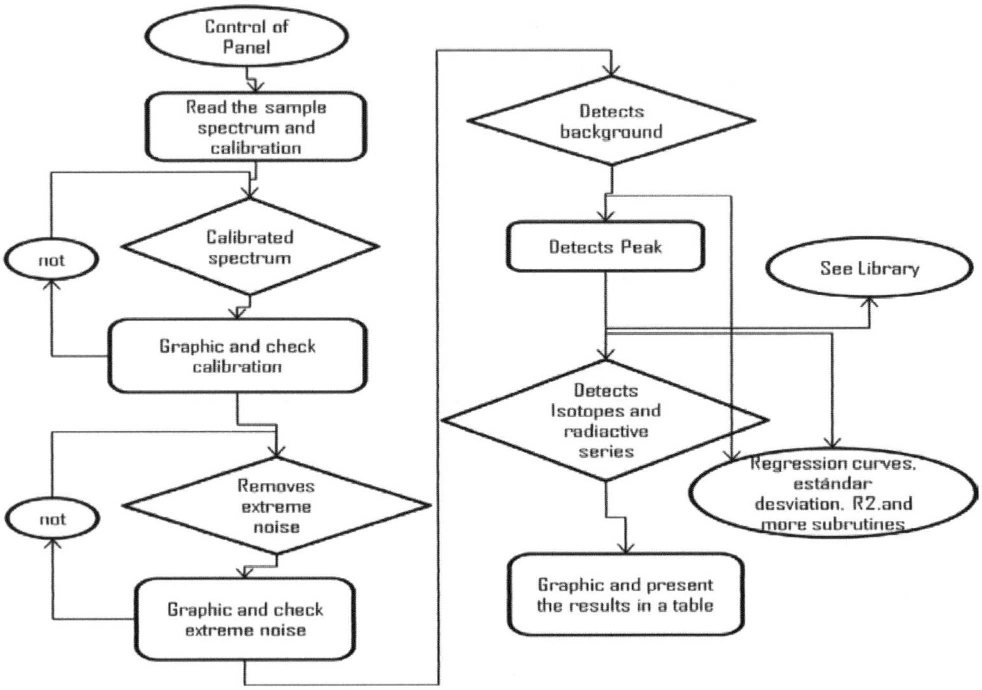

Figure 1. Main algorithms and subroutines.

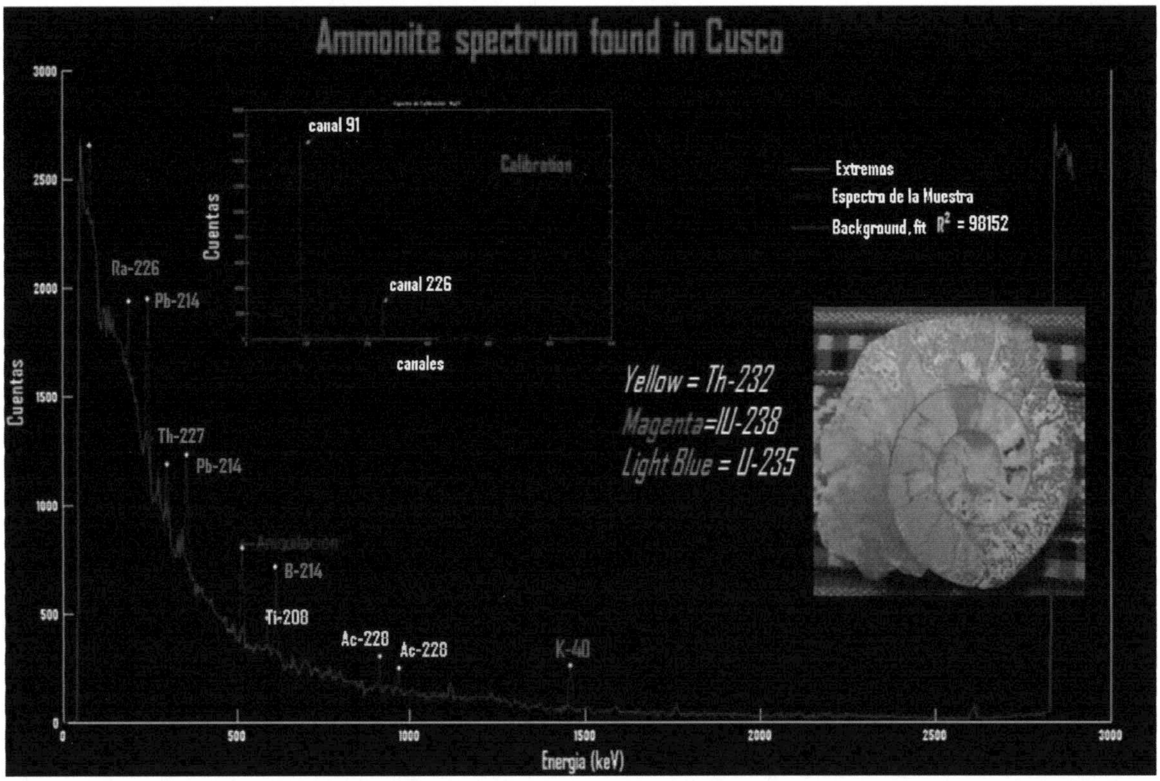

Figure.2. Graphic spectrum Fossil Cusco.

For the fossil of Cusco and the Cajon del Maipo we can observe the radioactive series of U-238 in magenta, light-blue U-235 and Th-232 in yellow. Both are the peak of annihilation and Potassium 40. The peaks of the fossil of Cajon del Maipo were higher as well as their areas because their data were taken with more time. Another thing to notice is that the areas of the peaks are in simple counts and not in energy counts.

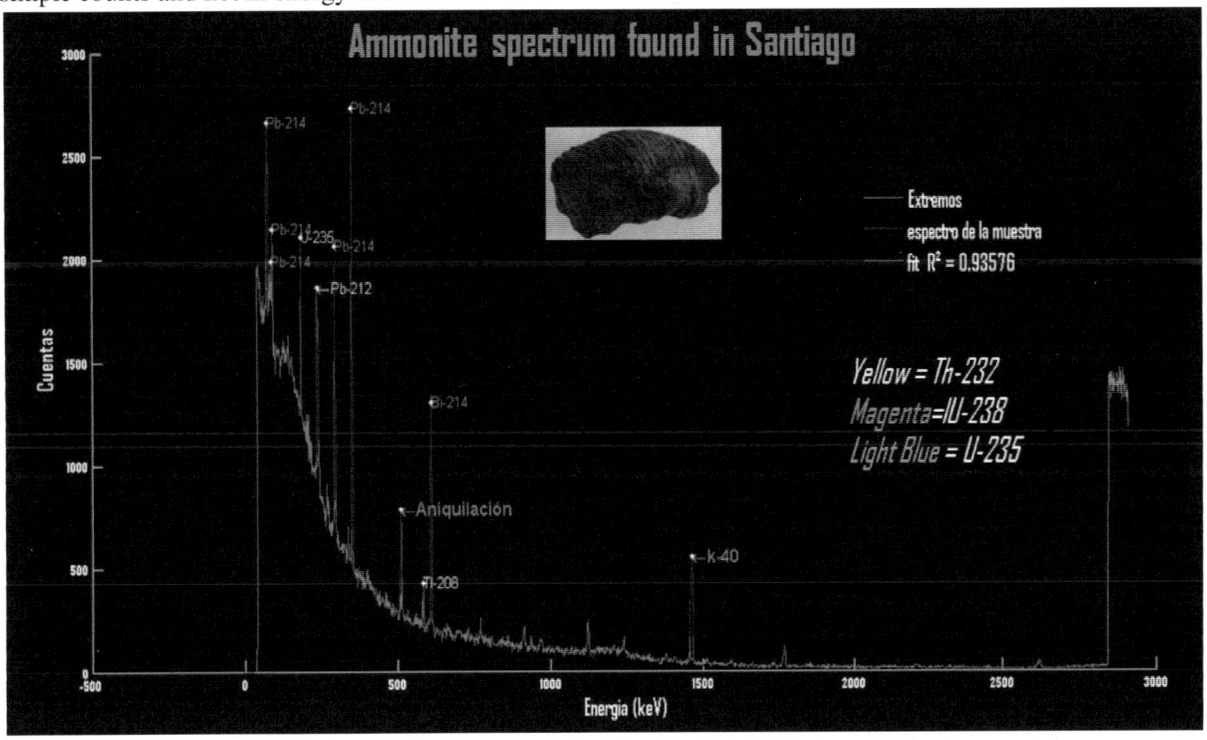

Fig.3. Graphic spectrum Fossil of Cajón del Maipo.

The results of the gamma-ray spectrum, the Ammonite Cusco and Santiago are presented in Table 1. In order: The Peaks (abscissa and ordinate), FWHM, resolution, area, standard deviation, Failure to detect the isotope compared with the values of library (less than 5% error. which can be changed).

Ammonite of Cusco												Ammonite of Cajon del Maipo											
Peak	1	2	3	4	5	6	7	8	9	10	11	1	2	3	4	5	6	7	8	9	10	11	
Energ del máx (keV)	78.4	188.6	243	299	355	513.6	584.5	610.6	912.6	969.1	1458.2	74.5	86.5	91.5	185.1	239.5	295	351.9	511.1	584.4	609.7	1463	E_0_v
FWHM(keV)	6.65	0.00	6.64	6.66	6.65	6.64	0.00	6.51	0.00	0.00	6.65	8.81	0.00	1.68	5.07	9.59	3.92	3.62	5.14	1.68	4.94	8.66	FWHM
Resolución (en %)	8.5	0.0	2.7	2.2	1.9	1.3	0.0	1.1	0.0	0.0	6.7	11.8	0.0	1.8	2.7	4.0	1.3	1.0	1.0	0.3	0.8	0.6	R_E_v
Area (Cuentas)	455	329	983	341	850	641	119	515	133	98	365	2319	187	606	1600	3053	2408	4652	1128	317	3157	2300	Area_v
Desv std (keV)	2.8	0.0	2.8	2.8	2.8	2.8	0.0	2.8	0.0	0.0	2.8	3.7	0.0	0.7	2.2	4.1	1.7	1.5	0.2	0.7	2.1	3.7	sigma_v
Error(%)	1.7	1.3	0.2	0.4	1.0	0.5	0.2	0.2	0.2	0.0	0.1	0.4	0.9	1.2	0.3	0.4	0.1	0.0	0.0	0.2	0.1	0.1	Error_v
Elemento		Ra-226	Pb-21	Th-22	Pb-2:	Aniq	Tl-208	Bi-214	Ac-228	Ac-228	K-40	Pb-21	Pb-21	Pb-21	U-235	Pb-21:	Pb-2	Pb-21	Aniq	Tl-208	Bi-214	K-40	

Table 1. Results spectrum Fossil Ammonite Cusco and Cajón del Maipo.

3. Conclusion.

Our studies samples of Ammonite fossils were done with a hyperpure Germanium (HPGe) detector, in an environment cooled with liquid nitrogen, and with the technique of Gamma-ray spectroscopy of high resolution. It is shown that both Ammonite fossils contain the same radioactive chemical elements on two, depite the fact that they were found 3000km from one to the other. These results could be useful for Paleontological research.

Journal of Physics: Conference Series **720** (2016) 012051 doi:10.1088/1742-6596/720/1/012051

To do this study we developed a program in Matlab environment, which can show the spectrum in a graphic, detects the peaks, and detects the radioactive series present.

The program is free for anyone who requires it. It can be freely manipulated, whether to make it even more general, or to automatize it in the analysis of new particular spectrum. It doesn´t need the values for calibration pecks, for it detects them by itself.

In the future, we could generalize the program even more, making it capable of analyzing more than one spectrum at a time. This would be useful for comparing the different samples.

4. Acknowledgment.

To the Center for Experimental Nuclear Physics, University of Chile. The experiment was done under the supervision and collaboration of Dr. Roberto Morales.
Dr Stepanova M. and Ms Vargas C. for the wording.

References

[1] Krane, K. (1988). *Introductory Nuclear Physics.* Oregon: John Wiley and Sons.
[2] Kennth R., L. (1974). *Astrophysical Formulae.* Springer Verlag.

Theoretical prediction of gold vein location in deposits originated by a wall magma intrusion

Abstract. The isotherm time-evolution resulting from the intrusion of a hot dike in a cold rock is analized considering the general case of nonvertical walls. This is applied to the theoretical prediction of the gold veins location due to isothermal evolution. As in previous treatments earth surface effects are considered and the gold veins are determined by the envelope of the isotherms. The locations of the gold veins in the Callao mines of Venezuela are now well predicted. The new treatment is now more elaborated and complex that in the case of vertical walls, performed in previous papers, but it is more adequated to the real cases as the one in El Callao, where the wall is not vertical.

Pablo Martin [1,2,†], **Fernando Maass-Artigas**[2], **Luis Cortés-Vega**[3]

1 Physics Department, Simón Bolivar University, Apartado 89000, Caracas 1080A, Venezuela
2 Physics Department, Antofagasta University, Casilla 170, Antofagasta, Chile
3 Mathematics Department, Antofagasta University, Casilla 170, Antofagasta, Chile

E-mail: `pmartin@usb.ve`, `pablo.martin@uantof.cl`, `fernando.maass@uantof.cl`, `luis.cortes@uantof.cl`, `luisvega@vtr.net`

1. Introduction

In a previous paper [1], calculations of temperatures in and around an intrusive dike was performed, assuming that the contact surface between the magma and the cold host rocks, was a vertical plane. The real case in El Callao gold mine in Venezuela is that this wall is inclined in 15° with respect to the vertical. In the present paper a theoretical treatment is presented considering this inclination. In this sense, a little better approximation of the location of the gold veins is given, compared with that given previously [1]. Here, the 400°C and 500°C isotherms are studied in detail, because of theirs importance in the determination of the gold vein sites. As it is usual, we consider that the magma heat propagates normal to the surface separating magma and cold rocks [2]. Our treatment will be 2-D, because the earth magma isotherm is considered, instead of the simplified 1-D diffusion problem considered by other authors [3]. Here it is assumed that the magma fill at the beginning all the semi-infinite space on one side of the wall separating magma and cold rock. This is like in the paper by Aldaz and Martin, but different to a most recient paper on this problem [4].

In this paper the following procedure will be followed:

(i) First, the basis of the theoretical description and the main assumptions will be presented.

(ii) Second, the solution and the mathematical procedures will be described in a general way.

(iii) Numerical calculations and comparison with the real gold vein sites in El Callao, Venezuela, will be followed.

(iv) Finally this work will finish with a short summary and conclusion.

2. Theoretical treatment and main assumptions

The assumptions in this work are: (1) The temperature of the magma at the moment of the intrusion ($t = 0$) is uniform and equal to T_m; (2) The temperature T_0 of the cold rock is also uniform at time $t = 0$; (3) The earth surface is a plane of uniform temperature T_0 at any time; (4) The evolution of the cold rock temperature is studied at any time. The magma is brought suddenly in contact with the cold rock at time $t = 0$, and the changes in temperature are due to cooling and solidification of the magma; (5) A unique melting point temperature T_m is assumed and the degree of overheating is considered small, assumption that seems well justified [5]. All the temperatures here will be given in centigrades degrees.

Besides our previuos considerations, the following quantities will be considered well defined. Latent heat of fusion L; density of the country rocks ρ; its specific heat c, and thermal conductivity k. In the present treatment convection in the magma will not taken into account, neither heat transport due to hydrotermal currents or other fluids.

The starting point in our treatment is the heat conduction equation

$$\frac{\partial T}{\partial t} = \kappa \nabla^2 T, \tag{1}$$

where T is the local temperature of the country rock and κ is the thermal diffusivity, given by $\kappa = k/\rho c$.

The main equation (1) is derived from the heat flow $\vec{q}$ per unit area and unit time, due to temperature gradient

$$\vec{q} = k \nabla T, \tag{2}$$

and the energy balance, that shows, that the heat changes in the medium must be compensated by flux diverging from the system

$$\nabla \bullet \vec{q} + c\rho \frac{\partial T}{\partial t} = 0. \tag{3}$$

Now, dimenssionless temperature is introduced as

$$\theta = \frac{T - T_0}{T_m - T_0} \tag{4}$$

given the new equation

$$\frac{\partial \theta}{\partial t} = \kappa \nabla^2 \theta. \tag{5}$$

Independent dimensionless similarity variables are also defined as

$$\eta = \frac{x}{\sqrt{2\kappa t}}, \qquad \varsigma = \frac{y}{\sqrt{2\kappa t}}, \tag{6}$$

where x is measured along the earth surface on the line perpendicular to the intersection line at $t = 0$, between magma and country rock, and y is measured along the down vertical line starting on the earth surface. In this way, the diffusion equation takes the form

$$\eta \frac{\partial \theta}{\partial \eta} + \varsigma \frac{\partial \theta}{\partial \varsigma} = -\frac{1}{2}\left[\frac{\partial^2 \theta}{\partial \eta^2} + \frac{\partial^2 \theta}{\partial \varsigma^2}\right] \tag{7}$$

The solutions of (7) now have to acomplish with the boundary conditions determined by the dike, which means that θ should be one in the magma contact surface. For $t = 0$, this surface is a plane forming an angle α, with the a vertical plane (x, y). As in previuos works [1,3,4] symetry conditions along the z axis are assumed. This is the reason that a 2-D treatment is performed.

Journal of Physics: Conference Series **720** (2016) 012052 doi:10.1088/1742-6596/720/1/012052

If this condition is not verified, 3-D treatment has to be performed, which becomes much more complicated.

It is important to point out that this boundary condition is the main one, because the temperatures are mainly determined by the solidification of the dike. There are also the condition that the earth is a isotherm also, but the influence of this boundary is no so important and decisive as the previous one.

From here on, the mathematical analysis becomes much more complex and difficult than in previous works [1,4].

3. Mathematical treatment

Several changes of variables have to be performed in order to obtain analytic solution acomplishing the boundary conditions.

First, let us consider a non-orthogonal reference frame, such that the "$\tilde{y}$" axis of coordinates form an angle $2\tilde{\alpha}$ with the vertical line or y-axis. The new coordinates will be $\tilde{x}$ and $\tilde{y}$, such that

$$\tilde{x} = x - y\,tg(2\tilde{\alpha}), \qquad \tilde{y} = y\,sec(2\tilde{\alpha}) \tag{8}$$

$$x = \tilde{x} + \tilde{y}\,sin(2\tilde{\alpha}), \qquad y = \tilde{y}\,cos(2\tilde{\alpha}) \tag{9}$$

It is important to point out that $\tilde{\alpha}$ is different of α, though its values are nearby.

In similarity variables, the previuos change of variables are written as

$$\tilde{\eta} = \frac{\tilde{x}}{\sqrt{2\kappa t}}, \quad \tilde{\varsigma} = \frac{\tilde{y}}{\sqrt{2\kappa t}}, \tag{10}$$

$$\tilde{\eta} = \eta - \varsigma\,tg(2\tilde{\alpha}), \qquad \tilde{\varsigma} = \varsigma\,sec(2\tilde{\alpha}) \tag{11}$$

$$\eta = \tilde{\eta} + \tilde{\varsigma}\,sin(2\tilde{\alpha}), \qquad \varsigma = \tilde{\varsigma}\,cos(2\tilde{\alpha}). \tag{12}$$

The difussion equation is now written as

$$\tilde{\eta}\frac{\partial\tilde{\theta}}{\partial\tilde{\eta}} + \tilde{\varsigma}\frac{\partial\tilde{\theta}}{\partial\tilde{\varsigma}} = -\frac{1}{2}\left[sec^2(2\tilde{\alpha})\frac{\partial^2\tilde{\theta}}{\partial\tilde{\eta}^2} - 2tg(2\tilde{\alpha})sec(2\tilde{\alpha})\frac{\partial^2\tilde{\theta}}{\partial\tilde{\eta}\partial\tilde{\varsigma}} + sec^2(2\tilde{\alpha})\frac{\partial^2\tilde{\theta}}{\partial\tilde{\varsigma}^2}\right] \tag{13}$$

with $\theta(\eta,\varsigma) = \tilde{\theta}(\tilde{\eta},\tilde{\varsigma})$. Unfortunatly in the equation (13), the term $\frac{\partial^2\tilde{\theta}}{\partial\tilde{\eta}\partial\tilde{\varsigma}}$ hinder the posibility of making an expansion in eingenfunctions as it was done in previous tratments [1,4]. In order to avoid this term a new change of variables must be done, as follows,

$$u = \tilde{\eta}\,cos(\tilde{\alpha}) + \tilde{\varsigma}\,sin(\tilde{\alpha}), \quad v = \tilde{\eta}\,sin(\tilde{\alpha}) + \tilde{\varsigma}\,cos(\tilde{\alpha}), \tag{14}$$

and

$$\tilde{\eta} = \frac{cos(\tilde{\alpha})}{cos(2\tilde{\alpha})}u - \frac{sin(\tilde{\alpha})}{cos(2\tilde{\alpha})}v, \quad \tilde{\varsigma} = -\frac{sin(\tilde{\alpha})}{cos(2\tilde{\alpha})}u + \frac{cos(\tilde{\alpha})}{cos(2\tilde{\alpha})}v \tag{15}$$

So, with $\tilde{\theta}(\tilde{\eta}.\tilde{\varsigma}) = \Theta(u,v)$, one appropiate partial differential equation appears:

$$\frac{\partial^2\Theta}{\partial u^2} + \frac{\partial^2\Theta}{\partial v^2} = -2u\frac{\partial\Theta}{\partial u} - 2v\frac{\partial\Theta}{\partial v} \tag{16}$$

Now, it is possible to proceed to separate variables writting $\Theta(u,v) = \phi(u)\psi(v)$, and from here it is obtained

$$\frac{1}{2\phi}\frac{d^2\phi}{du^2} + \frac{u}{\phi}\frac{d\phi}{du} = -\left(\frac{1}{2\psi}\frac{d^2\psi}{dv^2} + \frac{v}{\phi}\frac{d\psi}{dv}\right) = \frac{\beta}{2}, \tag{17}$$

where β is a constant. The same analysis performed at the infinite in the paper by Aldaz and Martin [1] leads to that β must be zero, and the solution for Θ is

$$\Theta(u,v) = \frac{erfc(u)\; erf(v)}{erfc(-\lambda_1^{'})\; erf(\lambda_2^{'})},$$

where the integration constants are now replaced by $\lambda_1^{'}$ and $\lambda_2^{'}$, which will be determined by the following process

$$\lambda_1^{'} = cos(\tilde{\alpha})\,\tilde{\lambda}_1 + sin(\tilde{\alpha})\,\tilde{\lambda}_2, \quad \lambda_2^{'} = sin(\tilde{\alpha})\,\tilde{\lambda}_1 + cos(\tilde{\alpha})\,\tilde{\lambda}_2, \tag{18}$$

$$\tilde{\lambda}_1 = \lambda_1 - \lambda_2\, tg(2\tilde{\alpha}), \qquad \tilde{\lambda}_2 = \lambda_2\, sec(2\tilde{\alpha}), \tag{19}$$

where λ_1 and λ_2 are calculated as it was shown in Aldaz-Martin paper [1], by the equations

$$\frac{L}{c}\frac{\sqrt{\pi}}{T_m - T_0} = \frac{e^{-\lambda_1^2}}{\lambda_1 erfc(-\lambda_1)} \tag{20}$$

and

$$\frac{L}{c}\frac{\sqrt{\pi}}{T_m - T_0} = \frac{e^{-\lambda_2^2}}{\lambda_2 erf(\lambda_2)}, \tag{21}$$

respectively.

The last unknown is the angle $\tilde{\alpha}$, which has to be determined by the condition that the surface $u = 0$, must be coincident with the magma-cold rock contact of the surface dike at time $t = 0$. This means that $x/y = tg(\alpha) = \eta/\varsigma$. Now, $u = 0$ leads to

$$0 = \tilde{\eta} cos(\tilde{\alpha}) + \tilde{\varsigma} sin(\tilde{\alpha}) = \eta cos(\tilde{\alpha}) + [sec(2\tilde{\alpha}) sin(\tilde{\alpha}) - tg(2\tilde{\alpha}) cos(\tilde{\alpha})]\,\varsigma. \tag{22}$$

From here the following equation for $\tilde{\alpha}$ is obtained

$$tg(2\tilde{\alpha}) - sec(2\tilde{\alpha}) tg(\tilde{\alpha}) = tg(\alpha) \tag{23}$$

If α is measured in radians, then for α small the value of $\tilde{\alpha}$ approachs to α plus terms of order α^2. In the case of El Callao mines, the angle α is 15°, and solving the equation (23), $\tilde{\alpha}$ is obtained as $\tilde{\alpha} = 13.72°$, the difference with α is about 10%.

4. Isotherm envelope

Let us consider the envelope of the isotherms for a given temperature T_1 (e.g. $T = 400°C$, $500°C$, etc.), the curve tangent to all the isotherms at this temperature will be the envelope of these isotherms. The procedure is similar to that previously followed in the previous works [1,4]. The result is that the isotherm envelope is a straight line defined by the equation :

$$x = m_1\, y = tg(\gamma)\, y, \tag{24}$$

where

$$tg(\gamma) = m_1 = tg(2\tilde{\alpha}) + sec(2\tilde{\alpha}) \left[\frac{v_1 sin(\tilde{\alpha}) - u_1 cos(\tilde{\alpha})}{u_1 sin(\tilde{\alpha}) - v_1 cos(\tilde{\alpha})}\right], \tag{25}$$

and u_1 and v_1 are given by the equations (26) and (27), given below, that is:

$$u_1\, e^{-u_1^2}\, erf(v_1) = v_1\, e^{-v_1^2}\, erfc(u_1) \tag{26}$$

and

$$T_1 - T_0 = \alpha_0^{'}\, erfc(u_1) erf(v_1), \tag{27}$$

Journal of Physics: Conference Series **720** (2016) 012052 doi:10.1088/1742-6596/720/1/012052

with

$$\alpha_0' = \frac{T_m - T_0}{erfc(-\lambda_1')\, erf(\lambda_2)}.$$

The graph of the isotherms and envelope are show in Figure 1 for the case of $T_1 = 400°C$ and $\alpha = 15°$.

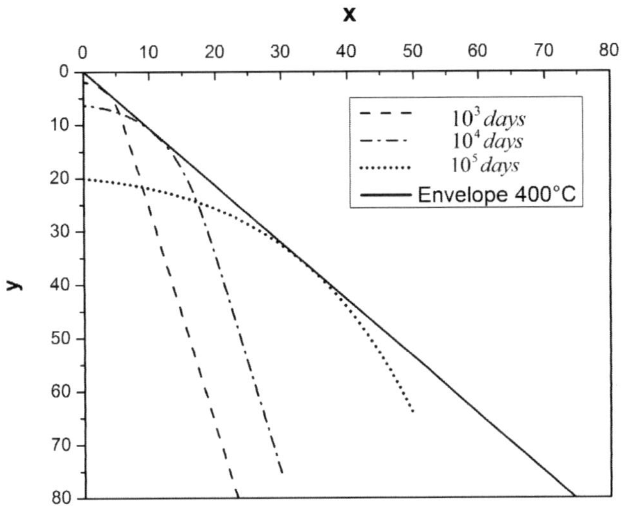

Figure 1. In this figure, the cases $T_1 = 400°C$ and $\alpha = 15°$ are presented.

In the next graph it is shown the location of the gold mine galleries in El Callao, Venezuela, compared with the theoretical predictions.

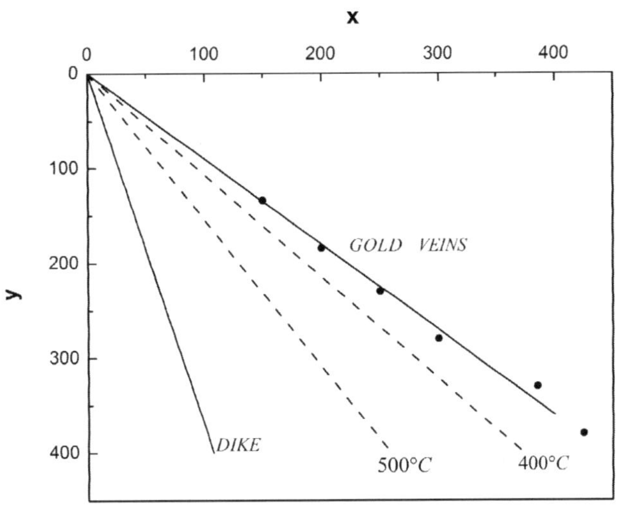

Figure 2. In this figure, the location of the gold mine galleries in El Callao, Venezuela, vs. theoretical predictions are presented.

The theoretical results compare very well with the actual location of the gold mine galleries (Colombia mine) in El Callao, Venezuela, operated by Minerven Company. The results are in a little better agreement than previous results [1], and are justified in a much better way, because now there is not need to add the angle α of the dike to the theoretical results, as it was done on this previous work [1]. Here everything is coming out in a natural way, and the results are justified in a much better way. Now, the angle γ is 43° compared with $48^\circ = (33^\circ + 15^\circ)$, which was the actual angle in the Colombia mine in El Callao, Venezuela. The previous angle was 27°, which becomes 42°, once the 15° are added [1]. The difference is only one degree in the right direction, but the important point is that now the result is coming directely from the equations.

5. Conclusion

The analysis here presented of the heat difussion due to the intrussion of a hot non-vertical dike in a cold rock, leads to a prediction of gold vein location almost in according with those of actual Colombia mine in El Callao, Venezuela. The analysis here done now is very general, and it does not need the assumption of vertical dikes. Our results almost coincide with those previously published [1] in the case of vertical dikes. In our treatment the earth is also considered as a isotherm, that is, it is a 2-D analysis. However, our analysis requires more elaborated mathematical tools than all the previous ones. The isotherm considered by the gold deposition, are those between $400^\circ C$ and $500^\circ C$. However, the isotherm does not give the gold vein location, but instead of that, the envelope of the isotherm are the important ones.

Acknowledgments

Work supported by: (1) Decanatura de la Facultad de Ciencias Básicas, Universidad de Antofagasta, Antofagasta, Chile; (2) Grant, Project $N^\circ 3125$, "Determinación de las vetas auríferas en las minas de Minerven, El Callao", PEI, FONACIT, Venezuela; (3) Grant Proyect FONDECYT (L. Cortés-Vega) N° 1121103, Chile.

We also thank to Eng. Abigail Morales from Minerven for the facilities given in the visit (P. Martin) to the galleries of Colombia mines, Venezuela.

†The oral presentation was performed by P. Martin in SOCHIFI-2014 meeting.

References

[1] R. Aldaz and P. Martin, Geophysics. **68**, 1159-1168 (2002).
[2] H. S. Carslaw and J. C. Jaeger, Clarendon Press. p. 78, Ch. I. p. 52. Appendix II (1967).
[3] D. L. Turcotte and J. Schubert, Geodynamics: John Wiley & Sons 134-159, 172-174 (1982).
[4] P. Martin, E. Castro and R. Aldaz, J. Earth & Planetary Studies. **2**, 53-64 (2012).
[5] R. Kerride and W. S. Fife, Chem. Geol. **33**, 256-294 (1981).

Journal of Physics: Conference Series **720** (2016) 012053 doi:10.1088/1742-6596/720/1/012053

Background considerations in the analysis of PIXE spectra by Artificial Neural Systems.

R. Correa[1], **J.R. Morales**[2], **I. Requena**[3], **J. Miranda**[4] **and V.A. Barrera**[4]

[1] Universidad Tecnológica Metropolitana, Departamento de Física. Av. José Pedro Alessandri 1242 Ñuñoa, Santiago, Chile.

[2] Universidad de Chile, Facultad de Ciencias, Departamento de Física. Las Palmeras 3425, Ñuñoa, Santiago, Chile.

[3] Universidad de Granada, Departamento de Ciencias de la Computación e Inteligencia Artificial. Daniel Saucedo Aranda s/n, 18071Granada, España.

[4] Universidad Nacional Autónoma de México. Instituto de Física. Ap. Postal 20-364, Mexico. D. F. 010000. Mexico.

E-mail: rcorrea@utem.cl

Abstract. In order to study the importance of background in PIXE spectra to determine elemental concentrations in atmospheric aerosols using artificial neural systems ANS, two independently trained ANS were constructed, one which considered as input the net number of counts in the peak, and another which included the background. In the training and validation phases thirty eight spectra of aerosols collected in Santiago, Chile, were used. In both cases the elemental concentration values were similar. This fact was due to the intrinsic characteristic of ANS operating with normalized values of the net and total number of counts under the peaks, something that was verified in the analysis of 172 spectra obtained from aerosols collected in Mexico city. Therefore, networks operating under the mode which include background can reduce time and cost when dealing with large number of samples.

1. Introduction.

Since implementation of the PIXE method by Johansson et al [1] it became one of the most used accelerator-based analytical method. Its capability to detect many elements simultaneously, roughly from Na to Pb, in concentration in the order of tens of parts per million, ppm, without significant damage to the sample, make it specially adequate to analyze low mass samples. One successful application has been the analyses of atmospheric aerosols. In environmental studies it is important to identify the most abundant elements in the sample and their concentrations. This knowledge permits the identification of natural and anthropogenic sources of aerosols, thus contributing to evaluate the air quality of the site where the samples were collected. Some countries have implemented systematic collection campaigns in cities and non urban sites. In these studies, a rather large amount of samples is collected for later irradiation by a proton beam in a PIXE laboratory. At least one energy spectrum is obtained for each sample requiring a careful analysis to identify the emitting atom and to determine its concentration. Usually, an analytical function fits the spectrum which is composed by peaks associated to each element detected in the sample, and by a continuous background covering the energy range accepted by the X-ray detector. The net number of counts in each peak, which is related to the elemental concentration, is calculated by integration around the peak and subtraction of background. In general, elemental concentrations are obtained by comparison with spectra given by targets of reference material with known concentrations. Some computational codes like AXIL, GUPIX, and others are used in the analysis of the spectra.

In previous works [2,3] we tested the application of artificial neural networks (ANN), to the analysis of this type of spectrum from materials like atmospheric aerosols and organic tissues. Also, it has been shown that a configuration based on a system of networks presents some advantages with respect to the use of single networks. In this work an Artificial Neural System (ANS) was constructed with several feed-forward ANN of the similar topology and operating in parallel inside the ANS. Following the training phase, elemental concentrations are calculated in very short time, in the order of microseconds.

The determination of the net number of counts in each peak, require the subtraction of the counts in the background, implying an additional mathematical operation. Here, we show that when applied to the analysis of PIXE spectra, trained ANNs working with the total number of counts including background, or with the net number of counts in each peak, give statistically similar results of the elemental concentration.

Our results are based on the analyses of a large number of fine atmospheric aerosols samples, thirty eight collected in Santiago, Chile, and one hundred seventy two collected in the metropolitan area of Mexico city [4].

2.- Structure of the ANS

The construction of the ANS, followed the same pattern used in a previous application [3] with several ANS operating in parallel, one for each chemical element detected in the sample. In Figure 1 the neural system schema is shown. Several feed-forward ANN, of similar topology, operate in parallel inside each ANS trained with the well known back propagation algorithm in an independent process. The only difference is in the training process, so that the initial weights are different in each ANN, i.e., several ANN with the same topology and training set, are running at the same time. The ANS obtain the capacity to learn and apply this knowledge to unknown cases not included in the training phase.

The input and output parameters in the training phase are the same for all ANN in the ANS. Each ANS output value is obtained by combining the ANN outputs in the decision system. The advantages in the use of a system of networks rather than single networks have been already shown [3].

For each one of the selected element a map with four input parameters was generated in an ANN, with three layers in the ANS, in order to provide a single output parameter, i.e. the elemental concentration *C(Z)*. The four input parameters were: i) m, the deposited mass in the filter; ii) Q, the total electric charge transported by the ions while crossing through the filter during irradiation; iii) S, the size of the filter surface irradiated, and iv) *N(Z)*, the total number of counts in peak.

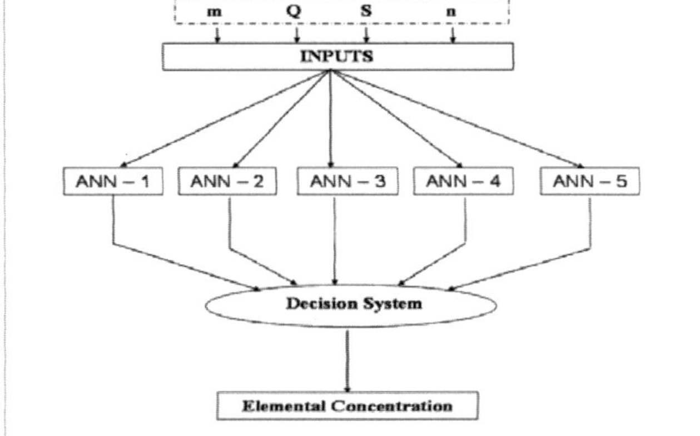

Figure 1. Schematic for ANS spectrum analysis for aerosol samples.

In the training process the weights of the connections between each neuron in the input layer with the hidden layer and between this one and the output layer are adjusted through an iterative process that minimize the error by a function considering the approximation between the network output and the expected answer. Now, if one decides to take as an input the net number of counts instead of the total number of counts, in the training process the network will learn to work under that condition, and the unknown concentration will be calculated accordingly.

Less computer work is required by taking the total number of counts rather than the net number of counts in the peak.

The architecture of the ANN has the topology shown in Figure 2, where a feed forward neural network has I neurons in the input layer, H, neurons in the hidden layer and one neuron in the output layer. The aggregation function Σ_j in each neuron h_j in the hidden layer, j = 1,....H, represents the scalar product between the input vector $X = (X_1,........,X_I)$ and the associated weight vector $W_j = (W_{jl},........W_{jI})$ where W_{ji} is the value of the connection between the neuron X_i in the input layer and the neuron H_j in the hidden layer. The activation function $f_j = f(\sum_j)$ in each hidden neuron is a sigmoid function of type

$$f_j = (1 + \exp(-\Sigma_j XW_j))^{-1} \quad (1)$$

representing the output value of each neuron in the hidden layer. In the output layer $\sum_o$ represents the aggregation function of the outputs from the hidden layer and corresponds to the scalar product of the output vector in the hidden layer $f = (f_1,......f_H)$ and the weight vector $W_o = (W_1,........W_H)$ associated to the value of the connections between each neuron H_j of the hidden layer with the neuron Y in the output layer.

The output value O is determined by a linear activation function depending on $\sum_o$.

$$O = \sum_o = \sum f\,W_o \quad (2)$$

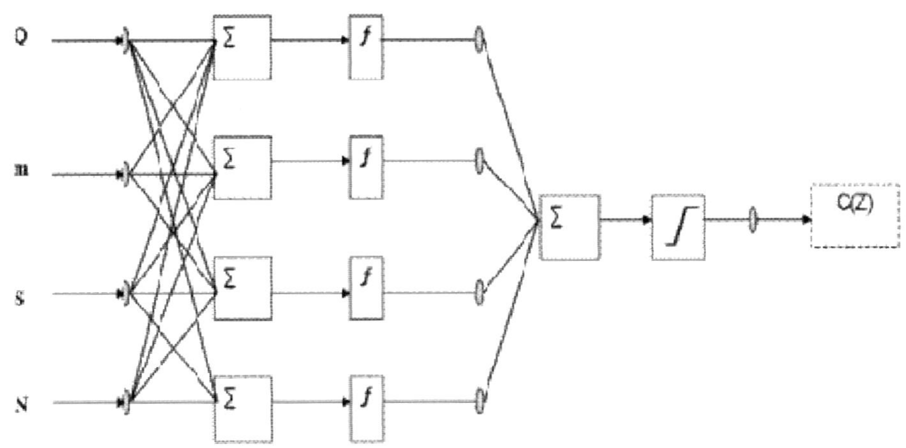

Figure 2. Architecture of ANN used inside ANS.

The neural networks used architecture with three layers having four input neurons, four hidden neurons and one output neuron. The input parameters form the vector (m, S, Q, N) and the network output is the concentration *C(Z)* of the element detected in the sample using the

Journal of Physics: Conference Series **720** (2016) 012053 doi:10.1088/1742-6596/720/1/012053

proper values of the inputs. Input parameters were normalized to the electric charge passing across the filter and collected at the Faraday cup during irradiation.

3. Background considerations.

In a typical PIXE spectrum the peaks lie on a continuum radiation background generated by the interactions of the incoming ions with electrons belonging to the atoms in the target, beam collimators, Faraday cup and eventually in the chamber walls. Effects like secondary electron bremsstrahlung, quasi free electron bremsstrahlung, atomic bremsstrahlung and ion bremsstrahlung produce a continuous radiation from zero to energies higher than those from K, L, or M transitions. Several authors have studied and calculated the probabilities associated to these effects. [5,6,7,8,9]. Prediction of peak to background ratio in gamma ray spectra have been performed using an artificial neural network [10].

In general, there are two contributions to the background under the peak, one is due to the interaction of the ions in the beam with collimators, Faraday cup, target holder, and the substrate to which the material under analysis is fixed, and secondly that produced by the interaction with the material itself. With an adequate experimental setup, the first contribution can be reduced to a minimum while the second would fluctuate with the mass deposited on the filter. This fact has been observed in the spectra considered in this work and is illustrated in Figure 3 for silicon and potassium, where the ratio between the number of counts in the background, *b*, to the total number of counts in the region of interest around the peak, *N*, for each spectrum, remains almost constant within a small statistical deviation. Defining this ratio as $B(Z) = b(Z)/N(Z)$, the mean values of the ratios for the major elements detected in the set and their corresponding standard deviation, σ,are presented in Table 1.

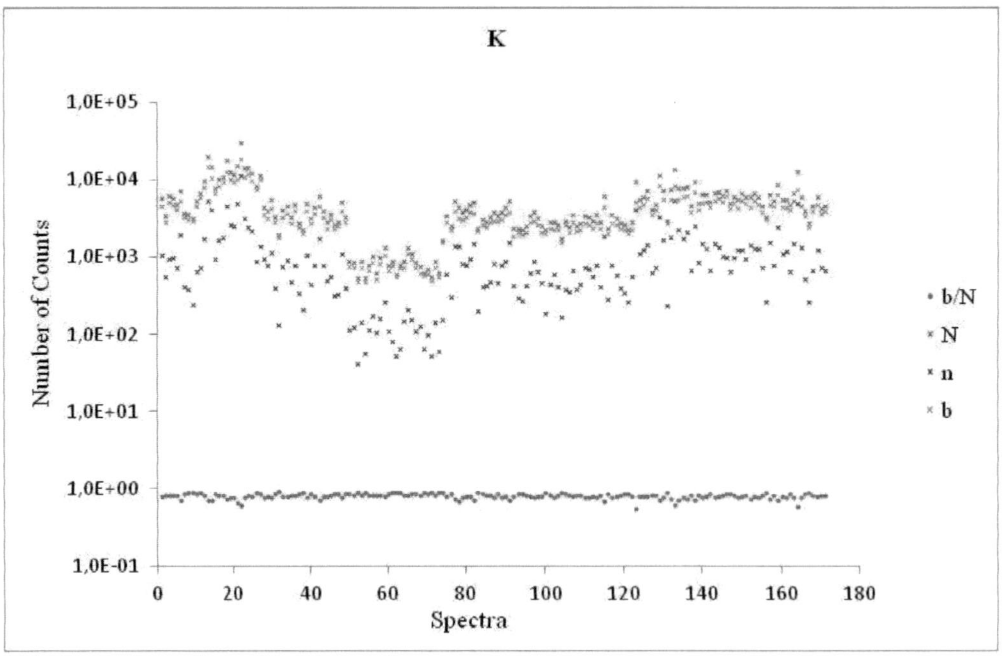

(a)

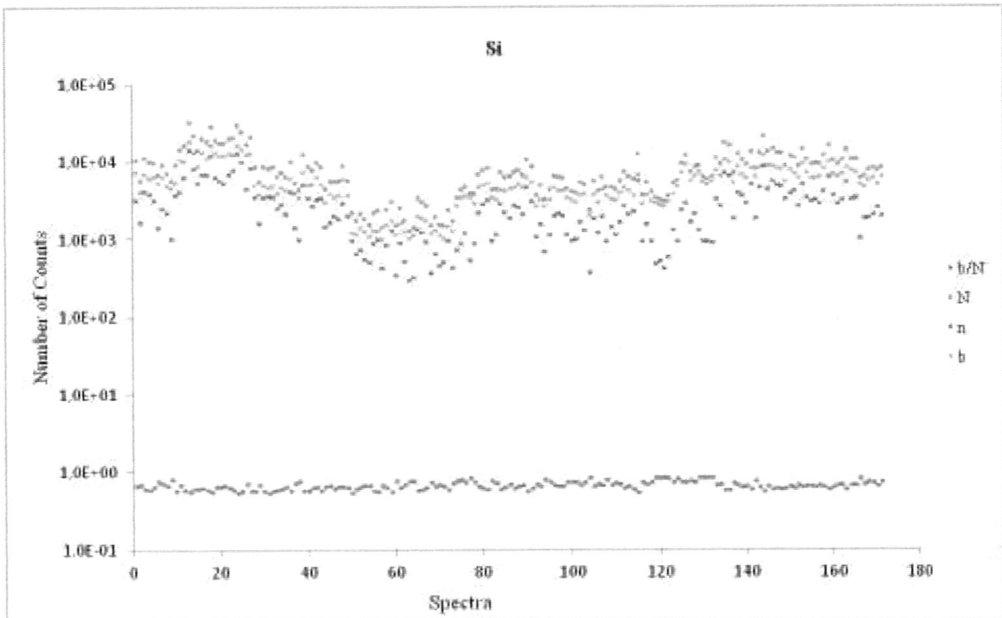

(b)

Figure 3 Ilustrates ratio between the number of counts in the background, *b*, to the total number of counts in the region of interest around the peak, *N*, for all 172 spectra collected in the metropolitan area of Mexico city analyzed for (a) K and (b) Si

Table 1: shows the mean values $B(Z)_{mean}$ of the ratios for the major elements detected in the set and their corresponding standard deviation,

	$B(Z)_{mean}$	σ
Si	0.686	5.8×10^{-3}
S	0.692	6.0×10^{-3}
K	0.822	4.8×10^{-3}
Ca	0.574	6.0×10^{-3}
Fe	0.310	4.7×10^{-3}

In general, the total number of counts in the region of interest around a characteristic peak of each detected element, of atomic number Z, can be expressed as

Journal of Physics: Conference Series **720** (2016) 012053 doi:10.1088/1742-6596/720/1/012053

$$N_i(Z) = n_i(Z) + b_i(Z) \tag{3}$$

where the subscript i denotes one of the spectrum in the set, $N_i(Z)$ corresponds to the total counts, $n_i(Z)$ represents the net number of counts and $b_i(Z)$ is the number of counts due to the background. In the evaluation of the elemental concentration, $C_i(Z)$, by standard methods, it is assumed that this value is proportional to the net number of counts in the peak.
Since ANS operates with values normalized, i.e. between 0 and 1, all inputs are normalized to the maximum value of the set. Then, we define the normalized values $N_i'(Z)$ and $n_i'(Z)$ as

$$N_i'(Z) = N_i(Z) / N(Z)_{max} \tag{4}$$

and

$$n_i'(Z) = n_i(Z) / n(Z)_{max} \tag{5}$$

Since it has been observed that for each element, $Bi(Z)$ fluctuates in a narrow range around the mean value $B(Z)_{mean}$, it can be expressed approximately as

$$B_i(Z) = B(Z)_{mean} +/- \sigma(Z) \tag{6}$$

From equations (6), $b_i(Z)$ can be expressed *as*

$$b_i(Z) \approx N_i(Z)(B(Z)_{mean} +/- \sigma(Z)) \tag{7}$$

and from equation (3)

$$n_i(Z) \approx N_i(Z)(1 - B(Z)_{mean} +/- \sigma(Z)) \tag{8}$$

where

$$k(Z) = 1 - B(Z)_{mean} \tag{9}$$

is a constant and then

$$n_i (Z) \approx N_i(Z) (k(Z) +/- \sigma(Z)) \tag{10}$$

So, the values n_{max} and N_{max} become related by a constant

$$n_{max}(Z) \approx N_{max}(Z)(k +/- \sigma(Z)) \tag{11}$$

where

$$(k +/- \sigma(Z)) \approx n_{max}(Z)/ N_{max}(Z) \tag{12}$$

Combining with (10) it is found that the normalized values $n_i'(Z)$ and $N_i'(Z)$ are alike

$$n_i'(Z) \approx n_i(Z) / n(Z)_{max} \approx N_i(Z) / N(Z)_{max} \approx N_i'(Z) \tag{13}$$

The similarity between $n_i'(Z)$ and $N_i'(Z)$ was verified for the most abundant elements in 172 spectra of aerosols collected in Mexico city. The correlations between $n_i'(Z)$ and $N_i'(Z)$ for Si and K are shown in Figure 4, and for other elments are given in Table 2.

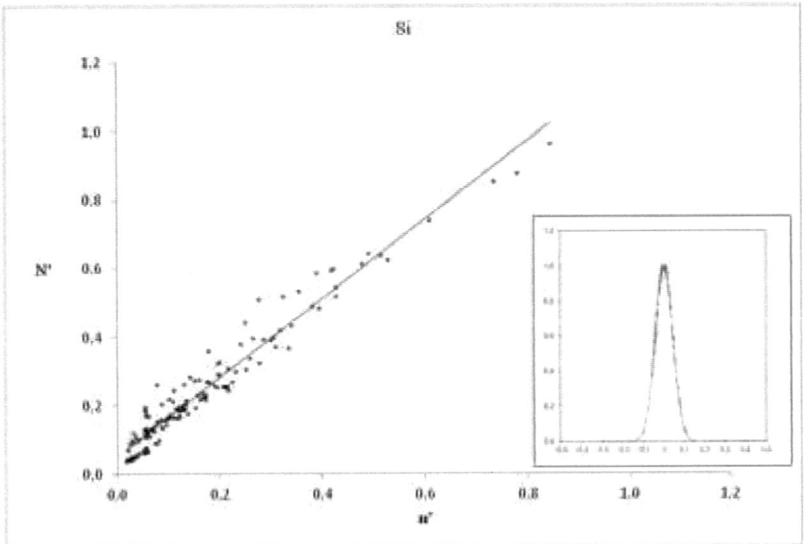

(a)

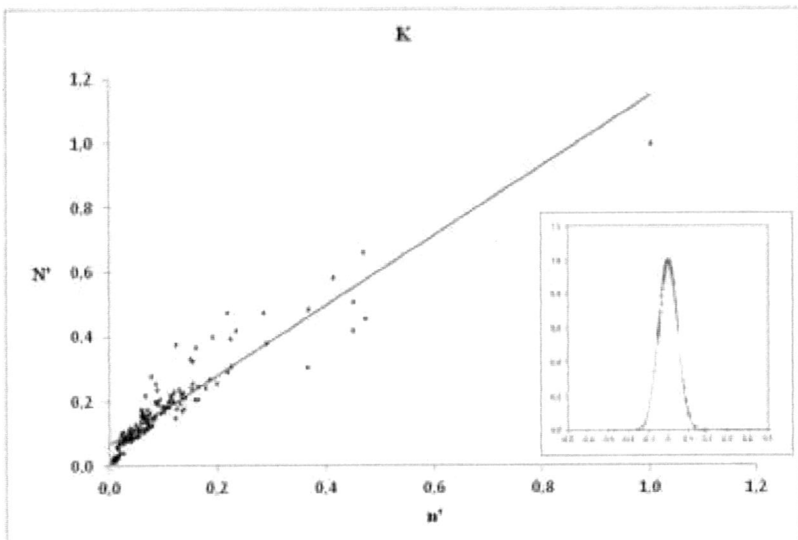

(b)

Figure 4 Correlation between the normalized total number of counts N' and the normalized net number of counts n' for Si and K are presented in (a) and (b) respectively. The residual distributions are shown in the inserted squares. A total of 172 spectra collected in the metropolitan area of Mexico city were considered.

In Table2 the linear correlation factors r^2 and their standard errors σ for Si, S, K, Ca and Fe are given. This result supports the assumption that the neural network system provides similar elemental concentration values operating with either one of the normalized values. Values of

r^2 and σ show strong statistics dependence and relation linear between and N' and n' for each element analyzed.

Table 2: Presents the linear correlation factors r^2 and their standard errors σ for Si, S, K, Ca,Fe

	r^2	σ
Si	0.95	4.07×10^{-2}
S	0.93	5.48×10^{-2}
K	0.87	4.84×10^{-2}
Ca	0.97	3.58×10^{-2}
Fe	0.99	3.19×10^{-2}

3. Results.-

In order to verify the above mentioned assumption, a different set of 38 aerosol PIXE spectra with known elemental concentrations, obtained by a standard method using the computer code AXIL [11] were taken as expected concentrations for ANS $C\ (Z)_{st}$ were reanalyzed by an ANS working either with the net number of counts or with the total number of counts. The comparison between $C\ (Z)_{st}$, and concentrations obtained by ANS using n'(Z) or using N'(Z), gave linear correlation factors of 0.88 and 0.94 respectively (See Figure 5). The correlation between normalized values n'(Z) and N'(Z), was verified with a set of 172 PIXE spectra of aerosols from Mexico city.

It is worth to mention that both sets of filters, those from Mexico city and those from Santiago, were irradiated in the same laboratory and under similar experimental conditions.

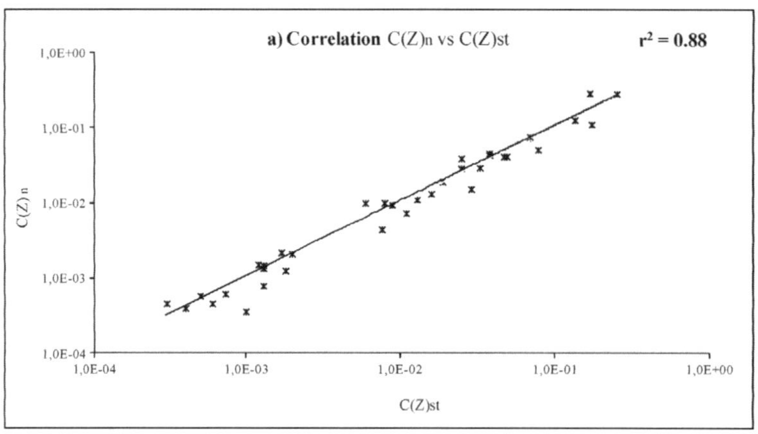

(a)

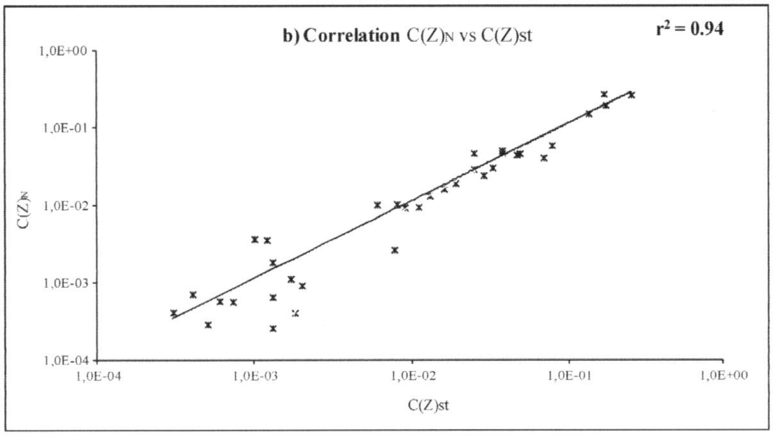

(b)

Figure 5 Concentrations of five elements, Si, S, K, Ca and Fe obtained by ANS using a) the net number of counts in the peak, b) and the total number of counts in the peak.

4. Conclusions.

Peaks associated to detected elements in aerosol PIXE spectra have been analyzed by artificial neuronal systems trained to operate with either the net number of counts or, with the integral number of counts. The elemental concentrations obtained in both cases turned out to agree quite well with concentration obtained by standard methods. This result is consistent with the fact that a constant added to the value of an input parameter in an artificial neuronal system do not affect the final results.

Networks operating under this mode can reduce time and cost when dealing with large number of samples.

Acknowledgments.

Partial support from the Centre for Experimental Physics, University of Chile, is acknowledged.

References

[1] Johansson, S.A.E., Campbell J.L., 1988. PIXE: A Novel Technique for Elemental Analysis. John Wiley and Sons, Chichester.

[2] Correa, R., Chesta, M.A, Morales, J.R., Dinator, M.I., Requena I., Vila, I., 2006.. Nuclear Instruments and Methods in Physics Research Section B: Beam Interactions with Materials and Atoms, Volume 248, Issue 2, pp. 324-328

[3] Correa, R., Dinator, M.I., Morales, J.R., Miranda, P.A., Cancino, S.A., Vila, I., I. Requena, 2008 International Journal of PIXE, Volume 18, Issue 3/4, 147-155.

[4] Barrera,V.A., Miranda, J., Espinosa, A. A., Meinguer, J, Martínez, J.N., Cerón, E., Morales, J.R., Miranda, P.A., Dias, J.F., 2012..International Journal of Environmental Research, Volume 6 Issue 3: 597-612.

[5] Folkmann,F., Gaarde, C., Huus T., Kemp, K., 1974. Nuclear Instruments and Methods, Volume 116, Issue 3, pp. 487-499.

[6] Ishii, K., Morita, S., 1976.. Physics Review A, Volume 13, pp131-138

[7] Ishii, K., 2006. Radiation Physics And Chemistry, Volume 75, Issue 10, pp. 1135-1163.

[8] Murozono, K., Ishii, K.,Yamazaki, H., Matsuyama, S., Iwasaki, S., 1999.. Nuclear Instruments and Methods in Physics Research. B 150, pp. 76-82

[9] Potiriadis, C., Karydas, A.G., Zarkadas, C., Clouvas, A., 2005.. X-Ray Spectrometry, Volume 34, Issue 4, pp. 335-340.

[10] Dragovic, S., Onjia, A. Applied Radiation and Isotopes Volume 63, Issue 3 (2005) 363–366

[11] Morata, D., Polvé, M., Valdés, A., Belmar, M., Dinator, M.I., Silva, M., Leiva, M.A., Algouy, T, Morales, J.R., 2008. Environmental Geology, Volume 56, Number 1, pp-81-95.

Journal of Physics: Conference Series **720** (2016) 012054 doi:10.1088/1742-6596/720/1/012054

Radiative effects and the missing energy paradox in the ideal two capacitors problem

Gilberto A Urzúa, Omar Jiménez, Fernando Maass and Álvaro Restuccia

Departamento de Física, Universidad de Antofagasta, Casilla 170, Antofagasta, Chile

E-mail: gilberto.urzua@uantof.cl, omar.jimenez@uantof.cl, fernando.maass@uantof.cl, alvaro.restuccia@uantof.cl

Abstract. Starting from the Poynting theorem, which arises from the exact Maxwell equations, we establish the balance of energy for the radiating ideal two capacitors problem. This balance of energy results in a nonlinear differential equation governing the time evolution of the voltage V. Boykin, Hite and Singh, following an approach not based on first principles, were the first to obtain this nonlinear differential equation and proposed an exponentially decaying voltage as a unique solution for it. We claim that the space of solutions for this differential equation is much richer. In fact, besides the exponentially decaying solution just mentioned there exist solutions with a sudden death behavior. The radiative effect introduced by Boykin, Hite and Singh, complemented with our analysis based on the exact Maxwell equations and the characterization of the more general space of solution of the nonlinear differential equation, explain the missing energy paradox in the ideal two capacitors problem.

1. Introduction

We consider a system of two capacitors in a vacuum, initially C_1 charged and C_2 uncharged, that can be connected with wires through a switch. When the switch is closed the initial charge redistributes itself and the final electrostatic energy becomes lower than the initial one. The appearance of a transient current I causes radiation, in such a way that a fraction of the lacking energy within the capacitors is emitted as radiation and the complement is dissipated as heat energy by Joule effect or, for short, Joule heating. In fact, the decrease of electromagnetic energy within some volume containing the capacitors is equal to the energy radiated through the boundaries of that volume plus the energy dissipated by Joule heating.

The role of radiation for this system was first considered by Boykin, Hite and Singh [1]. They model the radiating system by a lumped element whose resistance changes with current, and such that its dissipative power equals the dissipative power of a magnetic dipole. Using Kirchhoff voltage rule without resistance R, nor selfinductance L, they obtain a third order non-linear differential equation for the voltage time evolution. Being a third order differential equation, it accepts three independent initial data. They found a particular analytical solution with the voltage decaying exponentially with time. In Boykin solution only the initial voltage is an independent datum, while its initial first and second derivatives are determined by the initial voltage itself.

We address the problem from a slightly different point of view. We consider the conservation of energy as stated in the Poynting theorem. Neglecting the system resistance R and its selfinductance L, and modeling the system radiation by a magnetic dipole, we arrive to an integral equation observing that, under certain restrictions, the third order differential equation mentioned is a differential version of this integral equation. We analyze the solutions of the third order non linear differential equation for three independent initial data $V|_0$, $dV/dt|_0$ and $d^2V/dt^2|_0$ on a three dimensional sector of R^3. We claim that solutions with independent initial data within this sector of R^3 present a "sudden death" behavior which is a new effect characterizing the radiation problem.

2. Conservation of energy and Maxwell equations

The balance of energy within an spherical region of radius r and volume v containing the system, for any time $t > 0$, is given by

$$\int_0^t dt' \int_v d^3x \left(\frac{\partial u}{\partial t'} + \vec{J} \cdot \vec{E}\ d^3x + \vec{\nabla} \cdot \vec{S} \right) = 0, \tag{1}$$

where u is the electromagnetic field energy per unit volume, $\vec{J} \cdot \vec{E}$ is the rate of work done by the field on each unit volume of matter, and the magnitude of $\vec{S}$ gives the flow of field energy per unit time across a unit area of the spherical region.

Explicit expressions for u and $\vec{S}$ can be obtained after the Poynting theorem:

$$u = u_e + u_m = \frac{1}{2}\epsilon_0 E^2 + \frac{1}{2}\epsilon_0 c^2 B^2, \tag{2}$$

$$\vec{S} = \epsilon_0 c^2 \vec{E} \times \vec{B}, \tag{3}$$

where $\vec{S}$ is the Poynting vector. The derivation of these expressions can be found in Feynman [2]. The volume integral appearing in Eq. (1) is considered over the sphere of radius r, and its first term gives the rate of change of the field energy U within this sphere, its second term gives the rate of work done by the field on matter inside the sphere and its third term gives the rate of the field energy flow across the sphere. The integral over time gives the total values of these physical quantities from $t = 0$ up to some time t, while still there is current in the system. Finally we consider the integral until reaching the stationary state at $t = \infty$.

3. The RC model with radiation

We assume that the current I is a function of the time only and not of the spatial coordinates. This assumption seems to be reasonable in order to give a qualitative analysis of the system energy. The field energy U has electric and magnetic components: $U = U_e + U_m$, where

$$U_e = \int_v u_e d^3x = \int_v \frac{1}{2}\epsilon_0 E^2\ d^3x, \tag{4}$$

$$U_m = \int_v u_m d^3x = \int_v \frac{1}{2}\epsilon_0 c^2 B^2\ d^3x. \tag{5}$$

Also, for this physical system

$$\int_v \vec{J} \cdot \vec{E}\ d^3x = I^2 R. \tag{6}$$

3.1. The magnetic dipole model for the energy radiated in the two capacitors problem
The system radiated power $P_{rad}(t) \equiv dU_{rad}(t)/dt$ is the outgoing flux of the Poynting vector $\vec{S}$, and is calculated through the spherical surface Σ of radius r,

$$\frac{dU_{rad}(t)}{dt} = \int_{\Sigma} \vec{S} \cdot d\vec{a} = \int_{v} \vec{\nabla} \cdot \vec{S}\, d^3x, \tag{7}$$

where the last equality is Gauss's theorem. For this purpose the system can be modeled by a loop of radius b and current I, which is a magnetic dipole. The radiated power due to this magnetic dipole has been calculated in the paper of Boykin [1], its result being

$$\frac{dU_{rad}(t)}{dt} = \frac{\pi b^4}{6\epsilon_0 c^5}\left(\frac{d^2I}{dt^2}(t - r/c)\right)^2 \quad \text{if } t \geq r/c, \tag{8}$$

and $dU_{rad}(t)/dt = 0$ if $t < r/c$. In Eq. (8) the quantity $t - r/c$ is the retarded time. For other models see the paper of Choy [3]. Within this model the field energy radiated up to time t becomes

$$U_{rad}(t) = \int_{r/c}^{t} K\left(\frac{d^2I}{dt'^2}(t' - r/c)\right)^2 dt', \tag{9}$$

where $K = \pi b^4/6\epsilon_0 c^5$. After the change of variable $\tau = t' - r/c$, we get

$$U_{rad}(t) = \int_0^{t-r/c} K\left(\frac{d^2I(\tau)}{d\tau^2}\right)^2 d\tau \tag{10}$$

$$U_{rad}(t) = \int_0^{t} K\left(\frac{d^2I}{d\tau^2}\right)^2 d\tau - \int_{t-r/c}^{t} K\left(\frac{d^2I}{d\tau^2}\right)^2 d\tau. \tag{11}$$

After replacing Eqs. (11), (6), (5) and (4) back into Eq. (1), we obtain

$$\int_0^t dt'\left(\frac{d}{dt'}(U_e + U_m) + I^2R + K\left(\frac{d^2I}{dt'^2}\right)^2\right) - \int_{t-r/c}^{t} dt' K\left(\frac{d^2I}{dt'^2}\right)^2 = 0. \tag{12}$$

We can split the electric energy as $U_e = U_e^{in} + U_e^{out}$, where U_e^{in} is the energy of the electric field in the capacitors including border effects, and U_e^{out} is the energy of the electric field in the complementary space. If we define a physical quantity V as

$$V = \frac{Q_1}{C_1} - \frac{Q_2}{C_2}, \tag{13}$$

we get the relationship

$$I = -C_{eq}\frac{dV}{dt}, \tag{14}$$

where

$$\frac{1}{C_{eq}} = \frac{1}{C_1} + \frac{1}{C_2}. \tag{15}$$

Plugging Eq. (14) back in Eq. (12) we obtain

$$\int_0^t \left(K\left(\frac{d^3V}{dt'^3}\right)^2 + R\left(\frac{dV}{dt'}\right)^2 + \frac{1}{C_{eq}}V\frac{dV}{dt'}\right) dt' =$$

$$- \frac{1}{C_{eq}^2} \left[U_e^{out}(t') + U_m(t') \right] \Big|_0^t + \int_{t-r/c}^{t} K \left(\frac{d^3V}{dt'^3} \right)^2 dt'. \quad (16)$$

This equation is an expression of the conservation of the energy for any time t during the process. Taking the time derivative on both sides we have

$$K \left(\frac{d^3V}{dt^3} \right)^2 + R \left(\frac{dV}{dt} \right)^2 + \frac{1}{C_{eq}} V \frac{dV}{dt} =$$

$$- \frac{1}{C_{eq}^2} \frac{d}{dt} \left[U_e^{out}(t) + U_m(t) \right] + K \left(\frac{d^3V}{dt'^3} \right)^2 \Bigg|_{t-r/c}^{t}. \quad (17)$$

For the case where the time t is much larger than r/c and $\left(-1/C_{eq}^2\right) d \left[U_e^{out}(t) + U_m(t) \right] / dt \approx 0$, we can neglect the right hand side of Eq. (17) and obtain

$$K \left(\frac{d^3V}{dt^3} \right)^2 + R \left(\frac{dV}{dt} \right)^2 + \frac{1}{C_{eq}} V \frac{dV}{dt} = 0. \quad (18)$$

Under these restrictions, the solutions of Eq. (18) will be a good approximation to the behavior of the system. When we compare Eq. (18) with the well known equation of the RC model without radiation which is

$$R \left(\frac{dV}{dt} \right) + \frac{1}{C_{eq}} V = 0, \quad (19)$$

we see that, on the one hand, Eq. (19) is a first order linear differential equation for V, whose solution depends just upon one initial datum, which is $V|_0$. From its solution one gets that $dV/dt|_0 = -\left(Q_1^0/C_1\right)\left(1/RC_{eq}\right)$ which, in the limit when R tends to zero, diverges to minus infinity. On the other hand, Eq. (18) is a third order non-linear differential equation for V, whose solution depends on three initial data, which are $V|_0$, $dV/dt|_0$ and $d^2V/dt^2|_0$. In this case $dV/dt|_0$ is a finite initial datum which may be chosen to be independent of the value of R and, therefore, in the limit when R tends to zero, it remains unaffected. Actually, when R tends to zero, dV/dt remains bounded for all t, and the second term in Eq. (18) can be neglected.

4. The model with just capacitance and radiation

In what follows we restrict the discussion of Eq. (18) to the case $R = 0$. Then the equation for V becomes

$$\left(\frac{d^3V}{dt^3} \right)^2 + \frac{1}{KC_{eq}} V \frac{dV}{dt} = 0. \quad (20)$$

This third order non-linear differential equation for V was first derived by Boykin, Hite and Singh [1]. We notice that Eq. (20) arises from the more general Eq. (17), after neglecting the terms we have already mentioned. Equation (17) is a consequence of first principles, and it is a proof of the validity of the differential equation in reference [1].

A particular explicit solution for Eq. (20) was found by Boykin, Hite and Singh [1], given by

$$V(t) = V(0) \exp(-\alpha^{-1/5} t), \quad (21)$$

where $\alpha = KC_{eq} > 0$. (Just for the sake of clarity, in Boykin *et al.* is used $V(t) \equiv Q_2(t)/C_2 - Q_1(t)/C_1 \Rightarrow V(0) = -Q_1^0/C_1$; whereas in Eq. (13) we have defined $V(t) \equiv Q_1(t)/C_1 - Q_2(t)/C_2 \Rightarrow V(0) = Q_1^0/C_1$.)

4.1. The space of solutions for V

The solution given in Eq. (21) is obtained by taking the initial data of Eq. (20) to depend just upon one parameter: $V(0)$, which we also denote as $V|_0$. In fact, for this solution $dV/dt|_0 = -\alpha^{-1/5}\,V|_0$ and $d^2V/dt^2|_0 = \alpha^{-2/5}\,V|_0$. However Eq. (20) has a much larger space of solutions because its initial data in general depend upon three independent parameters: $V|_0$, $dV/dt|_0$ and $d^2V/dt^2|_0$.

In the problem that we consider here $V|_0 > 0$, and Eq. (20) implies $dV/dt|_0 < 0$, while $d^2V/dt^2|_0$ can takes negative or positive values. The sector of R^3 we are going to consider is the following: $V|_0 > 0, dV/dt|_0 < 0$ and $d^2V/dt^2|_0 < 0$. We claim that any solution V with these initial conditions reaches the value zero at a finite time T, and that its behavior near T is given by

$$V(t) \approx G(t-T), \tag{22}$$

where G has a negative value. We say that these solutions have a sudden death behavior. The model given by Eq. (20) needs to be upgraded for t near T by taking into account the selfinductance of the system. The effect of the selfinductance is to smooth out the behavior of V when it approaches zero. We will give an explicit proof of these claims elsewhere.

5. Conclusions

We have analyzed the ideal two-capacitor problem. Starting from Maxwell´s equations and using the conservation of energy as given by the Poynting theorem, we derived a third order non-linear differential equation for the physical quantity V. After neglecting some terms in the equation we arrived at Boykin *et al.* equation which was obtained using a different approach.

There exist real solutions for this differential equation depending on the initial data $V|_0$, $dV/dt|_0$, and $d^2V/dt^2|_0$. Without loss of generality, we have taken $V|_0$ strictly positive, i. e. $V|_0 > 0$. Real solutions exist if and only if $dV/dt|_0 \leq 0$. We claim that for any $d^2V/dt^2|_0 \leq 0$ there exists a unique solution satisfying the initial data at $t = 0$. The solution satisfying $dV/dt|_0 = 0$, $d^2V/dt^2|_0 = 0$ is the constant V solution which corresponds to the uninteresting case when the switch remains open. In all other cases we expect that the solution will present the "sudden death" behavior. That is, it will decay to $V = 0$ in a finite time T.

It is a very interesting property of the nonlinear differential equation that it admits two qualitatively different kind of solutions: the exponential decay and the "sudden death" decay solutions. The appearance of the "sudden death" solutions in the radiated system has not been noticed before.

Acknowledgments

G. U. wants to acknowledge Julio Yáñez for bringing to our attention the paper of T. B. Boykin et al. and enlightening comments. A. R. is partially supported by Fondecyt 1121103.

References

[1] Boykin T B, Hite D and Singh N 2002 *Am. J. Phys.* **70** 415–420
[2] Feynman R P, Leighton R B and Sands M 1965 *The Feynman Lectures on Physics* vol II (Reading, Massachusetts: Addison-Wesley) chapter 27
[3] Choy T C 2004 *Am. J. Phys.* **72** 662 – 670

Journal of Physics: Conference Series **720** (2016) 012055 doi:10.1088/1742-6596/720/1/012055

Small oscillations of a 3D electric dipole in the presence of a uniform magnetic field

L A del Pino[1], B Atenas[1], S Curilef[1]

1 Departamento de Física, Universidad Católica del Norte, Avenida Angamos 0610, Antofagasta, Chile

E-mail: scurilef@ucn.cl

Abstract. The classical behavior of a 3D electric dipole in the presence of a uniform magnetic field is studied in the small oscillations approximation. Using the Lagrangian formulation, the equations of motion are obtained, as well as their solutions and constants of motion. Normal modes of oscillation and their corresponding normal coordinates are obtained. Furthermore, the existence of a type of bound states without turning points, so-called *trapped states* conjectured by Troncoso and Curilef [Eur. J. Phys **27** (2006) 1315-1322], is investigated.

1. Introduction

Nowadays, a lot of applications are being developed in molecular-scale technologies because of the effort made to build *nanobots* (real intelligent nanostructures) by several specialists. Although such systems are not easy to model, we may do it through electric dipoles into external magnetic (and/or electric) fields[1, 2, 3].

In the study of this kind of systems, an atypical class of states was previously proposed[4] for a rigid electric dipole rotating around an axis into a plane perpendicular to the magnetic field, in such model the state corresponds to a bound state without turning points. In a different context[5], where the direction of axis of rotation of the electric dipole is supposed to be coincident with the magnetic field[5]. Through a prior analysis is conjectured the existence of a kind of bound states without turning points, that are so-called *trapped states*[5]. Recently, the motion and the relationship between the constants of motion was obtained. The trapped states [5] are shown for the model and they are completely described[6, 7]. The solutions are expressed in terms of Jacobi elliptic functions[6, 7].

In this work, the primary aim is to describe, from the perspective of the small oscillations approximation, the 3D motion of an electric dipole in the presence of an external magnetic field, without any other restriction. Particles immersed in magnetic field have been studied by several authors from different perspectives and imposing certain constraints on the motion of the center of mass and the relative coordinate[4, 5, 6, 7, 8, 9]. However, the study of small oscillations of this class of systems has not been yet addressed, and it is crucial to the creating real nanobots. The study of normal modes is a useful tool that allows classifying of molecules and studying of forces according with its symmetry.

The study of small oscillations is a relevant perspective to introduce widespread physical applications in optics, acoustics, molecular spectra, mechanical and electrical vibrations. We expect that these solutions significantly impact on the future of the applications to molecular motors and their constructions, as they describe the overall behavior of a dipole from a classical perspective.

This paper is organized as follows: In Section 2, we present the model and its theoretical basis, deriving the equations and constants of motions. In Section 3, the solution of the equation of motion is obtained in the small oscillations approximation, including its normal modes and coordinates. In Section 4 we investigate the possibility of existence of trapped states as conjectured before [5]. Finally, in Section 5, we offer some concluding remarks.

2. The model

Let us consider an electric dipole composed by two identical and opposite charges ($e_1 = -e_2$), separated by a fixed length a and whose masses are equal, in a uniform magnetic field $\vec{B} = B\hat{e}_{iii}$. The Lagrangian function is given by

$$L(\vec{r}_1, \vec{r}_2; \dot{\vec{r}}_1, \dot{\vec{r}}_2) = \frac{1}{2}m_1\dot{\vec{r}}_1^2 + \frac{1}{2}m_2\dot{\vec{r}}_2^2 - \frac{e_1}{c}\vec{A}(\vec{r}_1)\cdot\dot{\vec{r}}_1 - \frac{e_2}{c}\vec{A}(\vec{r}_2)\cdot\dot{\vec{r}}_2 - \frac{e_1 e_2}{\kappa|\vec{r}_2 - \vec{r}_1|}, \tag{1}$$

where $r_{1,2}$, $m_{1,2}$ are the position of the particles (1,2) and their corresponding masses, respectively. We additionally use the symmetric gauge for the vector potential $\vec{A}(\vec{r}_i) = \frac{1}{2}\vec{B}\times\vec{r}_i$ for $i = 1, 2$ and by using the coordinate of the center of mass $\vec{R} = \frac{m_1\vec{r}_1 + m_2\vec{r}_2}{M}$ and the relative coordinate $\vec{r} = \vec{r}_2 - \vec{r}_1$, the Lagrangian takes the following shape[5, 6, 7]:

$$L(\vec{R}, \vec{r}; \dot{\vec{R}}, \dot{\vec{r}}) = \frac{1}{2}M\dot{\vec{R}}^2 + \frac{1}{2}\mu\dot{\vec{r}}^2 + \frac{e^2}{\kappa a} + \frac{e}{2c}\left[\vec{B}\times\vec{R}\cdot\dot{\vec{r}} + \vec{B}\times\vec{r}\cdot\dot{\vec{R}}\right], \tag{2}$$

where $M = m_1 + m_2$ is the total mass, $\mu = \frac{m_1 m_2}{M}$ is the reduce mass and the charge $e_1 = e$.

The so-called *pseudomomentum* is a constant of motion[5, 9, 10] given by

$$\vec{C} = \vec{P}_R + q\vec{A}_R + e_c\vec{A}_r, \tag{3}$$

where $q = e_1 + e_2$ is the total charge and $e_c = e_1\frac{m_2}{M} - e_2\frac{m_1}{M}$ is the coupling charge, $\vec{A}_R = \frac{1}{2c}\vec{B}\times\vec{R}$, $\vec{A}_r = \frac{1}{2c}\vec{B}\times\vec{r}$ and $\vec{P}_R = \frac{\partial L}{\partial\dot{\vec{R}}} = M\dot{\vec{R}} + \frac{e}{2c}\vec{B}\times\vec{r}$ is the conjugate momentum of the center of mass. Thus, the relation between the linear momentum of the center of mass and the relative coordinates, for $q = 0$ is given by

$$M\dot{\vec{R}} = \vec{C} - \frac{e}{c}\vec{B}\times\vec{r} \tag{4}$$

Additionally, we define a constant vector that we call *transverse* pseudomomentum $\vec{\gamma}$.

$$\vec{\gamma} = \vec{C}\times\hat{e}_{iii} = C_{ii}\hat{e}_i - C_i\hat{e}_{ii}, \tag{5}$$

where C_i and C_{ii} are the components of $\vec{C}$ in a plane perpendicular to $\vec{B}$. Combining the Eq.(5) with (4), we arrive at the following relation:

$$M\dot{\vec{R}} = (-\gamma_{ii} + \frac{eB}{c}r_{ii})\hat{e}_i + (\gamma_i - \frac{eB}{c}r_i)\hat{e}_{ii} + C_{iii}\hat{e}_{iii} \tag{6}$$

Certainly, other constant of motion is the energy of the system

$$\mathcal{E}(\vec{R}, \vec{r}; \dot{\vec{R}}, \dot{\vec{r}}) = \vec{P}_{\vec{R}}\cdot\dot{\vec{R}} + \vec{p}_{\vec{r}}\cdot\dot{\vec{r}} - L(\vec{R}, \vec{r}; \dot{\vec{R}}, \dot{\vec{r}}), \tag{7}$$

where $\vec{p}_r = \frac{\partial L}{\partial\dot{\vec{r}}}$ is the relative conjugate momentum. Consequently, the energy fulfills:

$$\mathcal{E}(\vec{R}, \vec{r}; \dot{\vec{R}}, \dot{\vec{r}}) = \frac{1}{2}M\dot{\vec{R}}^2 + \frac{1}{2}\mu\dot{\vec{r}}^2 - \frac{e^2}{\kappa a}. \tag{8}$$

Substituting Eq.(6) into (8), the energy may be expressed in terms of the relative variable:

$$\mathcal{E} = \frac{1}{2}\mu\dot{\vec{r}}^2 + \frac{1}{2M}\left\{C^2 + \frac{e^2B^2}{c^2}(r_i^2 + r_{ii}^2) - 2\vec{\gamma}\cdot\vec{r}\right\} - \frac{e^2}{ka}. \tag{9}$$

Defining a set of dimensionless variables: $\rho = \frac{r}{a}$, $\xi = \frac{R}{a}$, $\frac{d}{dt} = \omega_c \cdot \frac{d}{d\tau}$ and $\omega_c = \frac{eB}{Mc}$ (the cyclotron frequency). The pseudomomentum and its corresponding transverse component are also defined as dimensionless constants as follows: $\vec{C} = \vec{C}/M\omega_c a$, $\vec{\gamma} = \vec{\gamma}/M\omega_c a$.

The Eqs.(9) and(6) in terms of the dimensionless variables are given by

$$\dot{\vec{\xi}} = (-\gamma_{ii} + \rho_{ii})\hat{e}_i + (\gamma_i - \rho_i)\hat{e}_{ii} + c_{iii}\hat{e}_{iii} \tag{10}$$

$$\epsilon = \frac{1}{\alpha}\dot{\vec{\rho}}^2 + \{c^2 - 2\vec{\gamma}\cdot\vec{\rho} + \rho_1^2 + \rho_2^2\} - \epsilon_c. \tag{11}$$

with $\alpha = \frac{M}{\mu}$, $\epsilon_c = \frac{2e^2}{\omega_c^2 M\kappa a^3}$ and $\epsilon = \frac{2\mathcal{E}}{\omega_c^2 a^2 M}$. Now, using spherical coordinates:

$$\begin{aligned}\rho_i &= \cos\psi\sin\theta,\\ \rho_{ii} &= \sin\psi\cos\theta,\\ \rho_{iii} &= \cos\theta,\end{aligned} \tag{12}$$

where ψ stands for the polar angle and θ stands for the azimuthal angle. Substituting the above equations in Eq.(11), we get:

$$\epsilon = \frac{1}{\alpha}\left(\dot{\theta}^2 + \sin^2\theta\dot{\phi}^2\right) + c^2 - 2\gamma\cos\phi\sin\theta + \sin^2\theta - \epsilon_c. \tag{13}$$

Next, we take $\phi = \psi - \psi_0$.$\gamma = |\vec{\gamma}|$, $c = |\vec{C}|$. The angle subtended between the polar axis and the direction of the vector $\vec{\gamma}$ is represented by ψ_0.

3. Small oscillations

When oscillations of systems occur about positions of equilibrium is interesting to consider the small oscillations approximation. We are interested in the motion within the immediate neighborhood of a configuration of stable equilibrium. If the deviation from equilibrium is too small, the kinetic and potential energies can be expanded in Taylor series about the equilibrium and retaining only the quadratic terms in generalized coordinates and velocities. From de Eq.(13) is easy to see that:

$$\begin{aligned}T(\dot{\theta},\dot{\psi},\theta,\psi) &= \frac{1}{\alpha}\left(\dot{\theta}^2 + \sin^2\theta\dot{\phi}^2\right)\\ V(\theta,\psi) &= c^2 - 2\gamma\cos\phi\sin\theta + \sin^2\theta - \epsilon_c,\end{aligned} \tag{14}$$

where $T(\dot{\theta},\dot{\psi},\theta,\psi)$ represents the kinetic energy and $V(\theta,\psi)$ the potential function. The potential $V(\theta,\psi)$ has three stable configurations of equilibrium, depending on the value of the constant of motion γ:

$$\begin{aligned}\phi_0 &= 0, \quad \theta_0 = \arcsin\gamma,\\ \phi_0 &= 0, \quad \theta_0 = \pi - \arcsin\gamma, \quad \gamma \leq 1\\ \phi_0 &= 0, \quad \theta_0 = \frac{\pi}{2}, \quad \gamma > 1\end{aligned} \tag{15}$$

In Figures 1 and 2, we depict the potential as a function of the angles θ and ϕ. Other parameters are defined to take the following values: $\gamma = 0.6$, $c = 1$.

Journal of Physics: Conference Series **720** (2016) 012055 doi:10.1088/1742-6596/720/1/012055

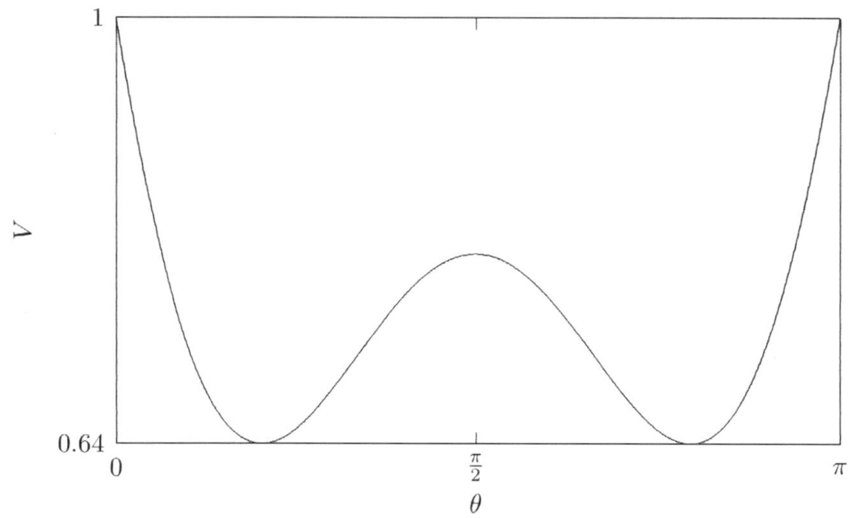

Figure 1. The potential V is depicted as a function of θ for the following values of parameters: $\gamma = 0.6$, $c = 1$ and $\phi = 0$.

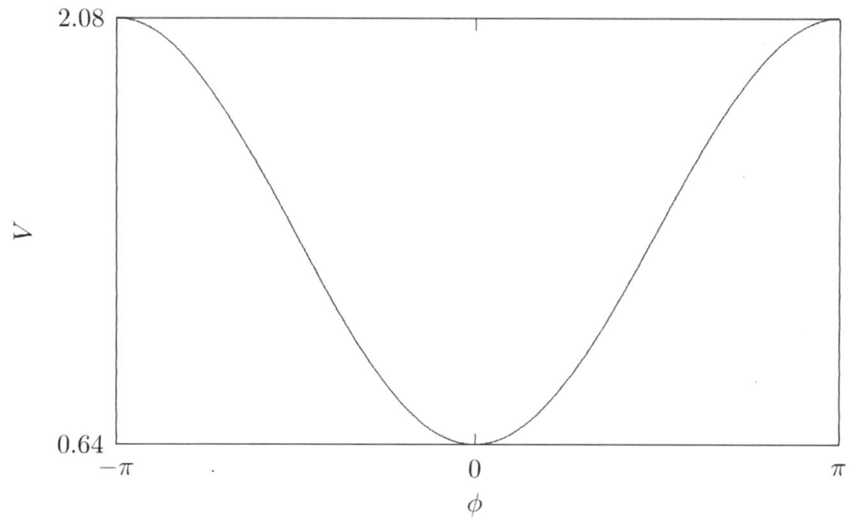

Figure 2. The potential V is depicted as a function of ϕ for $\gamma = 0.6$, $c = 1$ and $\theta = \arcsin\gamma$

For the first two configurations, we have:

$$\begin{aligned} V &\approx c^2 - \gamma^2 + (1-\gamma^2)(\theta-\theta_0)^2 + \gamma^2\phi^2 \\ T &\approx \frac{1}{\alpha}\left(\dot{\theta}^2 + \gamma^2\dot{\phi}^2\right) \\ \epsilon &\approx \frac{1}{\alpha}\left(\dot{\theta}^2 + \gamma^2\dot{\phi}^2\right) + c^2 - \gamma^2 + (1-\gamma^2)(\theta-\theta_0)^2 + \gamma^2\phi^2 - \epsilon_c \end{aligned} \tag{16}$$

and for the last one:

$$V \approx c^2 - 2\gamma + 1 + \gamma\phi^2 + (\gamma - 1)\left(\theta - \frac{\pi}{2}\right)^2$$

$$\begin{aligned} T &\approx \frac{1}{\alpha}(\dot{\theta}^2+\dot{\phi}^2) \\ \epsilon &\approx \frac{1}{\alpha}(\dot{\theta}^2+\dot{\phi}^2)+\varsigma^2-2\gamma+1+\gamma\phi^2+(\gamma-1)\left(\theta-\frac{\pi}{2}\right)^2-\epsilon_c \end{aligned} \tag{17}$$

For $\gamma < 1$, we make the following change of variables:

$$\begin{aligned} \theta &= \theta_0+\sqrt{\frac{\alpha}{2}}\eta_1 \\ \phi &= \sqrt{\frac{\alpha}{2\gamma^2}}\eta_2 \end{aligned} \tag{18}$$

If we substitute Eq.(18) in (16), we obtain the following energy function:

$$\epsilon = \frac{1}{2}\dot{\eta}_1^2+\frac{1}{2}\dot{\eta}_2^2+\varsigma^2-\gamma^2+\frac{\alpha(1-\gamma^2)}{2}\eta_1^2+\frac{\alpha}{2}\eta_2^2-\epsilon_c \tag{19}$$

The above equation represents the energy of two uncoupled harmonic oscillators, thus η_1 and η_2 are the normal coordinates and the normal modes of oscillation are:

$$\begin{aligned} \phi \equiv 0, &\qquad \omega_1=\sqrt{\alpha(1-\gamma^2)} \\ \theta \equiv \arcsin\gamma, &\qquad \omega_2=\sqrt{\alpha} \end{aligned} \tag{20}$$

In Figures 3 and 4, we represent graphically the oscillation modes. As seen in mode 1 ($\phi \equiv 0$), the dipole oscillates about the equilibrium direction $\vec{e_q}=\vec{\gamma}+\sqrt{1-\gamma^2}\hat{e}_{iii}$ in a plane parallel to the magnetic field and in mode 2 ($\theta = \arcsin\gamma$), the swing is in a plane perpendicular to the magnetic field.

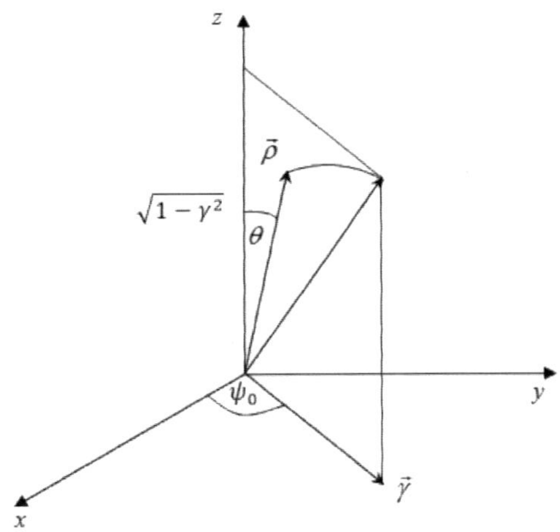

Figure 3. Typical directions of mode 1 are represented.

The general solution of the Eq.(19) is:

$$\begin{aligned} \eta_1 &= A_1\sin(\omega_1 t+\Delta_1) \\ \eta_2 &= A_2\sin(\omega_2 t+\Delta_2) \end{aligned} \tag{21}$$

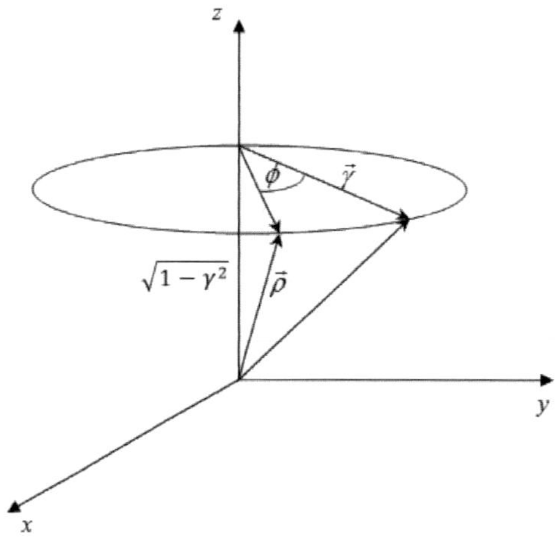

Figure 4. Typical directions of mode 2 are represented.

$A_{1,2}$ and $\Delta_{1,2}$ are determined from the initial conditions.

For $\gamma > 1$, the normal coordinates are:

$$\eta_1^{'} = \sqrt{\frac{\alpha}{2}}(\theta - \frac{\pi}{2}) \quad (22)$$

$$\eta_2^{'} = \sqrt{\frac{\alpha}{2}}\phi$$

(24)

and the normal modes of oscillation are:

$$\phi \equiv 0, \qquad \omega_1^{'} = \sqrt{\alpha(\gamma - 1)} \quad (25)$$

$$\theta \equiv \frac{\pi}{2}, \qquad \omega'_2 = \sqrt{\alpha\gamma}$$

(26)

We have noticed that the normal modes of oscillation do not satisfy the property of isochronism, i.e. the normal frequencies of oscillation depend on the initial conditions.

It is also noteworthy that exact solutions, for the equations of motion (11), have been obtained for all values of γ whether $\theta = \theta_0$ (mode 2). Hence, for $\theta = \theta_0$ the Eq.(13) is:

$$\epsilon = \frac{\gamma^2}{\alpha}\dot{\phi}^2 + c^2 - 2\gamma^2\cos\phi + \gamma^2 - \epsilon_c, \quad \gamma < 1, \quad (27)$$

$$\epsilon = \frac{1}{\alpha}\dot{\phi}^2 + c^2 - 2\gamma\cos\phi + 1 - \epsilon_c, \qquad \gamma > 1, \quad (28)$$

which correspond to the equations of a nonlinear pendulum, whose exact solutions are written in terms of the elliptic Jacobi function[6, 7, 11].

If $\gamma = 0$ and $\phi = 0$ (mode 1), the Eq.(11) also admits exact solution in terms of Jacobi functions. Further, $\dot{\phi} \equiv 0$ and $\theta = \varphi/2$, then Eq.(11) yields:

$$\epsilon = \frac{1}{4\alpha}\dot{\varphi}^2 + c^2 + \frac{1}{2}(1 - \cos\varphi) - \epsilon_c \quad (29)$$

Journal of Physics: Conference Series **720** (2016) 012055 doi:10.1088/1742-6596/720/1/012055

We have again obtained the equation of the nonlinear pendulum.

For $\gamma = 1$, the potential V is completely inharmonious in the generalized coordinate θ and the discussion of a case as this goes beyond the scope of the present article.

4. Trapped states

Trapped states exhibit the property of having a mean velocity equal to *zero*[5, 4]. For the variable of the center of mass, the Eq.(11), means:

$$\gamma_i = \langle \rho_i \rangle \qquad (30)$$

$$\gamma_{ii} = \langle \rho_{ii} \rangle$$

$$c_{iii} = 0, \qquad (31)$$

where $\langle \rho_i \rangle$, $\langle \rho_{ii} \rangle$, means the time average value of the $\rho_{i,ii}$. The set of Eqs.(31) is satisfied only if the third component of the pseudomomentum is zero, then, a necessary condition for the existence of trapped states is that the center of mass moves in the plane perpendicular to the magnetic field and $c = \gamma$.

Hereafter, we consider $c_{iii} = 0$. For $\gamma < 1$ and $\phi \equiv 0$ (mode 1), the expansions in Taylor series about the equilibrium of the spherical coordinates of Eqs.(12) are:

$$\rho_i \approx \gamma_i\left(1 - \frac{\sqrt{1-\gamma^2}}{\gamma}(\theta - \theta_0) - \frac{1}{2}(\theta - \theta_0)^2\right) \qquad (32)$$

$$\rho_{ii} \approx \gamma_{ii}\left(1 - \frac{\sqrt{1-\gamma^2}}{\gamma}(\theta - \theta_0) - \frac{1}{2}(\theta - \theta_0)^2\right)$$

Taken into account the Eqs.(18,21):

$$\langle \rho_i \rangle \approx \gamma_i \left(1 - \frac{A_1^2 \alpha}{4}\right) \qquad (33)$$

$$\langle \rho_{ii} \rangle \approx \gamma_{ii} \left(1 - \frac{A_1^2 \alpha}{4}\right)$$

From the above equations, it follows that the set of Eqs.(31) only have the trivial solution $A_1 \equiv 0$, i.e. the relative variable remains at rest. The same result is reached if $\theta \equiv \theta_0$ (mode 2).

If $\gamma = 0$, the Eqs.(31) are immediately satisfied because of the periodicity of the Jacobi functions and the center of mass is trapped by the magnetic field. For $\gamma > 1$, a similar analysis, leads to the conclusion of the absence of trapped states. Similarly, this result demonstrates the absence of trapped states in the neighborhood of the stable equilibrium, except for the particular case $\gamma = 0$.

Furthermore, it is intuitively presumed that small oscillations approximation is not valid for energies larger than the energy of the unstable equilibrium (see Figure 1,2), the challenge of finding the relationship between the constants of motion (γ and ϵ), which allows the existence of trapped states, still remains[4].

5. Concluding remarks

In the present study, we obtain the normal coordinates and normal modes of oscillation of a rigid dipole in the presence of a uniform magnetic field and investigates the possibility of existence of trapped states to the coordinate of the center of mass in the small oscillations approximation.

The normal mode of oscillation depends strongly on the initial conditions via the parameter γ and violates the isochronous property, inherent to the potentials that depend quadratically on the generalized coordinates. Besides, it is shown in the energy range where the small oscillations approximation is applicable, that the coordinate of the center of mass can only be trapped by the magnetic field whether $\gamma = 0$.

We think the system has the following special properties: two degrees of freedom with several stable equilibrium points that depend on the initial conditions, the violation of the property of isochronism present in its normal modes of oscillation and the possibility of existence of trapped states of the center of mass, represents a good challenge to follow studying in the future.

Acknowledgements

We acknowledge partial financial support from CONICYT-UCN PSD-065.

References

[1] V. V. Volchkov, M. N. Khimich, M. Ya. Melnikov, and B. M. Uzhinov, "*A Fluorescence Study of the Excited State Dynamics of Boron Dipyrrin Molecular Rotors*, Hihg Energy Chemistry" **47** (2013) 224-229.

[2] Y Yoshida, Y Shimizu, T Yajima,G Maruta, S Takeda, Y Nakano, T Hiramatsu,H Kageyama, H Yamochi and G Saito, "*Molecular Rotors of Coronene in Charge-Transfer Solids*" Chemistry-A European Journal **19** (2013) 12313-12324.

[3] N Ogiwara, T Yanagibashia, Y Hikichia, M Nishikawaa, J Kamiyaa, K Wadab, "*Development of a turbo-molecular pump with a magnetic shield function*", Vacuum **98** (2013) 1821.

[4] Pursey D L, Sveshnikov N A, Shirokov A M "*Electric dipole in a magnetic field: Bound states without classical turning points*" Theoretical and Mathematical Physics **117**, (1998) 1262-1273

[5] Troncoso P, Curilef S., "*Bound and trapped states of an electric dipole in a magnetic field*" European Journal of Physics **27** (2006) 1315-1322.

[6] Atenas B. ,del Pino L. A. and Curilef S. "*Classical states of an electric dipole in an external magnetic field: Complete solution for the center of mass and trapped states*" Annals of Physics. **350** 605-614 (2014)

[7] Atenas B. ,del Pino L. A. and Curilef S. "*Anomalous states from a classical Hamiltonian: Electric dipole in a magnetic field*, AIP Conference Proceedings **1648** (2015) 080004.

[8] Curilef S. and Claro F. "*Dynamics of two interacting particles in a magnetic field in two dimensions*" American Journal of Physics **65** (1997) 244-250.

[9] Escobar- Ruiz M. A., Turbiner A.V. "*Two charges on a plane in magnetic field: special trajectories*" Journal of Mathematical Physics **54** (2013) 022901.

[10] Gorkov L. P. and Dzyaloshinskii I. E. "*Contribution to the theory of the Mott excition in a strong magnetic field*" Zh. Eksp. Teor. Fiz. **53** 717-722 (1967)

[11] Karlheinz O. "*A comprehensive analytical solution of the nonlinear pendulum*" European Journal of Physics **32** (2011) 479-490.

[12] Hebert Bristol Dwight, "*Table of integrals and other mathematical data*" (New York: Chicago, 1957)

[13] P. F. Byrd and M. D. Friedmann "*Handbook of Elliptical Integrals for Engineers and Scientists*" (New York: Springer, 1971)
M. Abramowitz and I. A. Stegun, "*Handbook of Mathematical Functions with Formulas Graphs and Mathematical Tables*" (New York: Dover, 1972)
H. Hancock, " *Elliptic Integrals*" (New York: Dover, 1958)

Study of connectivity in student teams by observation of their learning processes

We can´t solve problems
by using the same kind of thinking
we used when we created them.
Albert Einstein

Patricio Pacheco H, Rafael Correa D.

J.P.Alesandri 1242, Ñuñoa,Santiago
Universidad Tecnológica Metropolitana, Santiago de Chile

patricio.pacheco@utem.cl
rcorrea@utem.cl

Abstract. A registration procedure based data tracking classroom activities students formed into teams, which are immersed in basic learning processes, particularly physical sciences is presented. For the analysis of the data various mathematical tools to deliver results in numerical indicators linking their learning, performance, quality of relational nexus to transformation their emotions. The range of variables under observation and further study, which is influenced by the evolution of the emotions of the different teams of students, it also covers the traditional approach to information delivery from outside (teaching in lecture) or from inside each team (abilities of pupils) to instructional materials that enhance learning inquiry and persuasion.

1. Introduction

This study is centered on the interaction processes at the level of students inside the classroom. Since through these is one of the essential pillars of the learning process and achievement of the same goals and that is connectivity.
A significant proportion of human activities is built, in first approximation, driven by certain horizon of expectations fed back by margins of predefined prediction, generating a "world of homogeneity" [1, 2], linear reductionism of life and actions of people that is very suitable and functional in Economy and its definition of quality [3, 4]. But its application in Education is valid in a restricted domain, it is not true that more infrastructure buildings, more computers, more information technology and communication [5, 6], more Internet [7], etc., provoke, in direct proportion, most significant learning [8], since they depend on the relational forms among people [9, 10, 11]. The educational system is highly complex [12, 13, 14] because of the fundamental variables that characterize it, especially in learning are interrelated [15, 16, 17], which is the basis of all real human system.
Behaviors displayed by teams of students learning processes are observed, studying codifiable dimensionalities formed by their emotions, inquiries skills, persuasive, search for internal and external information. In its analysis, nonlinear dynamic elements that reveal how they influence and how they

Journal of Physics: Conference Series **720** (2016) 012056 doi:10.1088/1742-6596/720/1/012056

can get to promote or restrict the process towards achieving meaningful learning teams students are employed. For this purpose the following objectives:

- Specify a set of significant variables that account for the dimensionalities involved in performing a teaching track and record them in real time.
- Validate measurement instruments, according to the universal sample of study, statistical distribution for t – student, sample correlation coefficient of Pearson [18] and index Cronbach [19]
- Use initial conditions for the induction of emotional transformation.
- Apply mathematical methods of analysis databases: Time series of Fourier, Fast Fourier Transformation (FFT) and Cross Correlation Function, non linear dynamics [20]
- Build graphical representations of the variables in both two-dimensional and three-dimensional form.
- To characterize the transitions between dynamic and numerically obtain connectivity giving relational bonds and emotional transformation of students in each team which directly affects the quality of their learning.

2. Emotion

The set of emotions or moods that holds a person or a group of them and that allows the interactions among them by the atmosphere or environment to which the person is incorporated, we will call emotional field, like a field of physical forces, whose name explained by the fact that emotions have the capacity to mobilize people. Emotions play a critical role in learning [21, 22, 23] that students acquire and the actions to be taken for it. This follows the statement by Humberto Maturana, when he says that [21]: When we talk about emotions, they are different domains of possible actions on people and animals, and the various bodily dispositions that constitute and perform. (...) There is no human action without emotions that melts as such and makes possible an act. (...) Is not the reason that leads us to action but emotion.

3. Characterization of each dimension is measured by teams inside the classroom

Conduct likely observational follow [11, 12, 24, 25, 26] were coded.

Variable Y is essential in this work which corresponds to emotions, has been divided into: Positivity (POS.) when positive acts as support and understanding in the team and Negativity (NEG.) manifest if it manifests in the form of negative acts such as disapproval, sarcasm or cynicism.

Variable X has been divided into: Inquiry (INQ) that relates to use didactic elements [27] of the experimental activity and Persuasion (PER) if discussion components are used in favor of the point of view of someone on the team getting others to understand and take that view.

Variable Z is composed of External Information (EI) which is linked to whether the information is directed outward or from outside the team and classroom. Internal Information (II) refers to the (s) person (s) dealing topics appealing to resources within the team: students with mathematical skills, another one in physics, etc..

4. Research Methodology

The sample is composed of 160 students (40% female and 60% male) Risk Prevention Engineering and Industrial Processes Engineering at Inacap IP (Professional Institute) Campus Santiago Sur. These students are a segment who finished high school with no prerequisite university selection test and a significant proportion of them, make up a human group of first generation family in accessing higher education.

They are divided into five courses of physical mechanical issue, leaving one the courses as a control, but all observational monitoring. Each class has an average of 32 students who are divided, for specific laboratory sessions, in two of 16 students working conforming 4 teams with 4 students each, planning to exploit the potential of cooperative learning (and also the parity gender). The temporal tracking matrix team of students is divided into intervals of 9 minutes (attentional optimization criterion) to complete the 90-minute class [23]. Emotions in the micro-scale (micro human emotions and micro expressions), whereas Gottman [24] uses times of the order of 1 second in scale coding specific affections of 20 categories to explore possible emotions that can come to express a couple during a conversation, is beyond the range of this study.

Each dimensionality involved in the study (X or Y or Z) is recorded by one person, psychologist or trained professional. To illustrate the dual polarity scale for the variable Y (t) (= Positivity / Negativity (t)) is shown, and subsequently graph:

Scheme N°1. Points out the domain dual polarity and scale coding, in its dual polarity, Positivity.

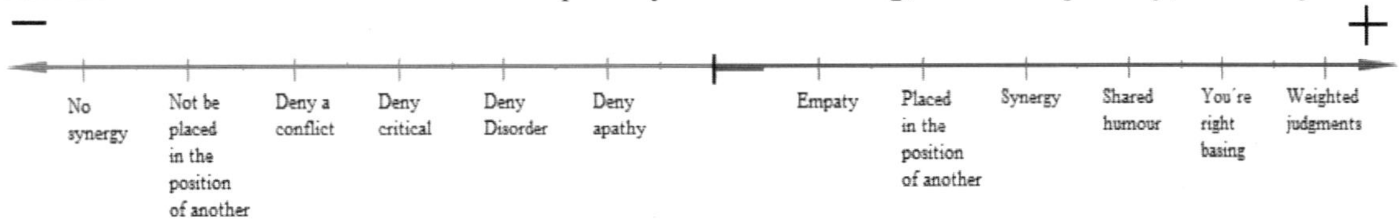

Table N°1 presents the behavior to be followed in time and taxonomy of coding.

	coding	t (min)	positivity behavior
6	weighted judgments	0	
5	You're right basing	9	
4	synergy	18	
3	shared humour	27	
2	placed in the position of another	36	
1	empathy	45	
0	neutral	54	
-1	deny apathy	63	
-2	deny disorder	72	
-3	deny critical	81	
-4	deny a conflict	90	
-5	not be placed in the position of another		
-6	no synergy		
total: 13			

Figure 1. It represents the behavior in time, t, of the variable Y (defined as quotient Positivity / Negativity). The first maximum of the curve is produced by applying the initial condition at 36 minutes:

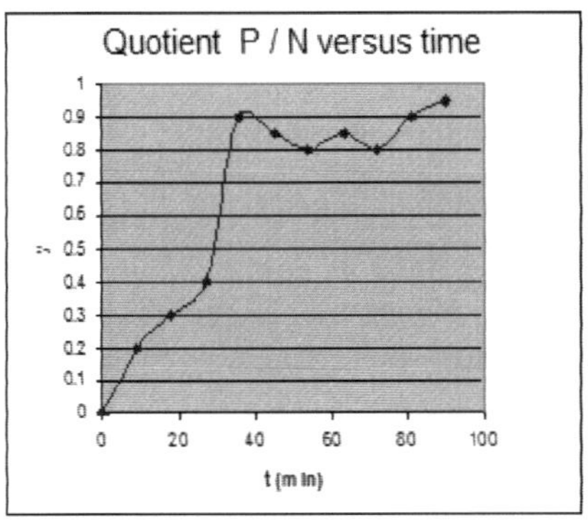

Figure 1. Quotient Positivity / Negativity versus time (chaotic case).

The same treatment is performed to the X and Z allowing ultimately build the other Tables of Values and their respective graphical representations variables. By transforming the data into time Fourier series [28], the observed interaction between team members detected processes of mutual influence or interlaced between the team members behavior. This is the environment of the control parameter, r, and essential in the learning process of arduous monitoring and quantification, based on behaviors within teams of students, enhancing or impeding progress towards the achievement of an educational objective [29, 30]. Each temporal Fourier series [28], consequent with the non-linear mathematical model, has the algebraic form:

$$\frac{dX}{dt} = a_0 + \sum_{i=1}^{\infty}(a_i \cos i\omega t + b_i \sin i\omega t); \frac{dY}{dt} = b_0 + \sum_{i=1}^{\infty}(c_i \cos i\omega t + d_i \sin i\omega t); \frac{dZ}{dt} = c_0 + \sum_{i=1}^{\infty}(e_i \cos i\omega t + f_i \sin i\omega t)$$

By adapting a series of continuous time Fourier experimental curve of the form Y (t) versus t, dY (t) / dt versus t (similarly for X (t) versus t, Z (t) versus t, dX (t) / dt versus t and dZ (t) / dt versus t) graphs are obtained:

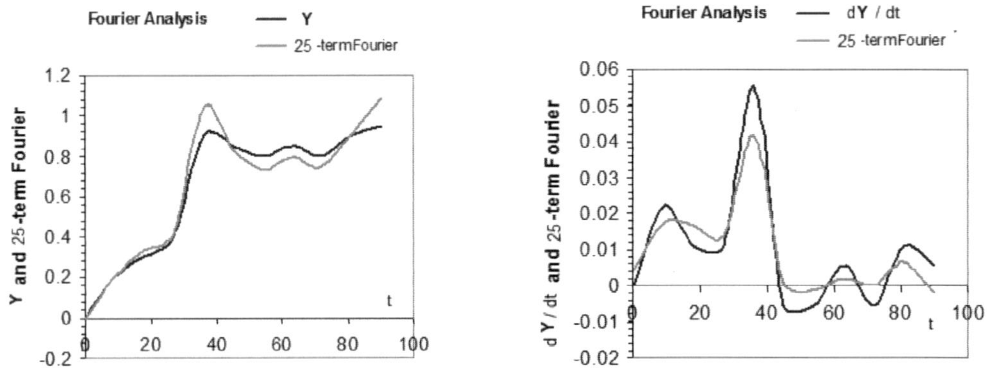

Figure 2. The black line is the variation in the time dimension (= Positivity / Negativity). The other is as Fourier fitting curve. On the right graph the dark line is the rate of change of Y in time t.

5. Initial Conditions

By using initial conditions [28, 31], once perceived homogeneity of the initial emotional field of students (which may contain behaviors of stagnation, indifference, anxiety, etc.), and a new emotional disposition occurs that drives toward the objectives presented. The type of initial condition depend on the level of complexity of content, how feasible is contextualize, the nature of the difficulty that the student has [32], etc., which gives it a high complexity [20, 33].

Initial conditions were applied at 36 minutes into the laboratory session 90 minutes because, in general, students are not adequately prepared their working guidelines. The teacher should invest time in introducing them to the activity of the day (time is important to the achievement of the objectives). The difficulties arising from the work of each team, the initial perception of the environment of emotional field indicate the type of initial condition that breaks the symmetry generating an environment of new emotional field that provides the impetus to start efforts toward the goal.

Was verified in this study the effectiveness of initial conditions such as:

demonstrative experiment limited to a fraction of time over the school period class, scheduled laboratory session within the course, subject to independent laboratory but it is the counterpart to the theoretical practice course with intersecting optimizing and improving content objectives learning, practical sessions and according professional profile is intended for the student, short videos (10 or 15 minutes) structured in a simple and accessible narrative oriented to specific thematic content, brief presentation of any specific item by a teacher researcher charismatic and empathetic (testimony), etc..

6. Results

6.1. Data analysis and graphics.

From the graphs constructed and based on the domain of behaviors experimentally verifiable, according to the ranges defined for X, Y, Z [10, 11, 25], follows:

1. Learning objectives are achieved if the variables of inquiry / persuasion balanced manifest. The experimental work presents a physics lab, for the inquiry, behaviors concatenated in a measurement process, a second behavior does not occur if you have not met before. Therefore it is essential that the activities related to each behavior that is contained in persuasion stimulate inquiry.

The graphs show that the domain of persuasion experimental sessions work necessarily entails a dissipation of effort generating a "damping" in the way $e^{-kf(t)}$ that drags the other variables (Y, Z).

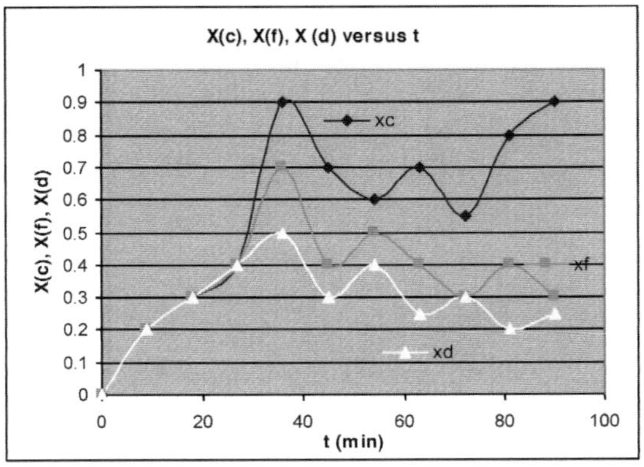

Figure 3. Each curve represents the behavior over time of the variable X (defined as quotient Inquiry / persuasion) for each dynamic: Weak Attractor (Xd), fixed (Xf) and chaotic (Xc).

When monitoring X (t), for different dynamic arises from its graphics a 'resonant' effect for teams that achieve learning objectives and resulting chaotic attractor dynamics type. This effect involute functional on X (t) is well described by exponential forms $e^{\lambda\, g(t)}$ of positive type and determines an increase in the time of X (t). The reinforcement or encouragement among students of the team, is based on the generation of emotional field, and positive elements such as humor, empathy, self-generation of confidence, etc..

2.Although the variable Y, emotions, involving the ratio of positivity and negativity with behaviors associated with numerator and denominator is what drives the process, if not properly channeled connectivity inter alumni team also tends to "damp "dissipating such a decay to a negative exponential function of time, $e^{-\theta\, h(t)}$.

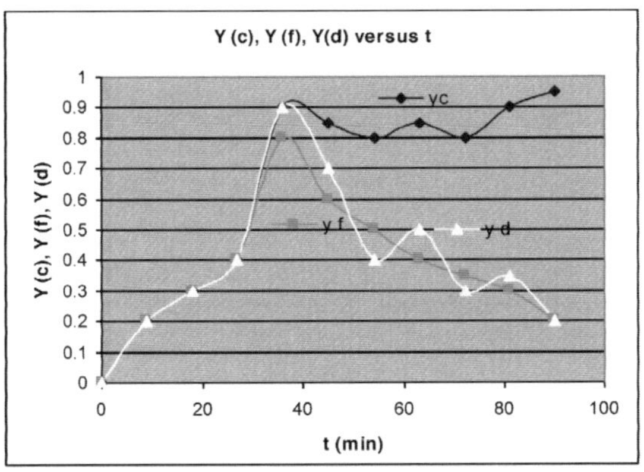

Figure 4. Each curve represents the behavior over time of the variable Y (defined as quotient Positivity/ Negativity) for each dynamic: Weak Attractor (Yd), fixed (Yf) and chaotic (Yc).

When generating the initial condition, induces a new emotional atmosphere what provides stimulus and energy to the educational goal. It is observed from the graphs that high performing teams are able to manage in the long term this "emotional" energy. Not so average performance teams and under which, despite having reached an "excited state" and associated energy, similar to the case of high performance not well "used" in time spreading inter and extra equipment seriously affecting the achievement educational goals (creating a climate where disorder prevails, work dispersion, jokes, little commitment to the rigor of measurements, etc..).

The positivity / negativity ratio is a major influence in the numerator. If very dominant positivity students too condescending teams between them become so neither facilitates the achievement of goals (Gottman in his study of proposed marriages lasting relationships for over 5.8 as appropriate [25].

3.Internal information / External information.

This variable has teams in laboratory work in physics a unique characteristic. From the observed, which exhibit more commitment to the objectives are oriented towards working within the team (internal information) and search for answers within. But this is not a general feature in the teams in this study, which is reflected in the final graphical results. The more data that is recorded is the external orientation, i.e. seek answers outside the team and in the particular teacher. While this statement is speculative, perhaps this has to do with an educational deformation tends to imitation or repetition (or rudely told to "copy" the work of the other teams or follow patterns of authority), in the context in which observations are made. Then the key factor in why internal information / external information are the denominator dominates for defined behaviors. This determines a damping effect on all experimental dynamic.

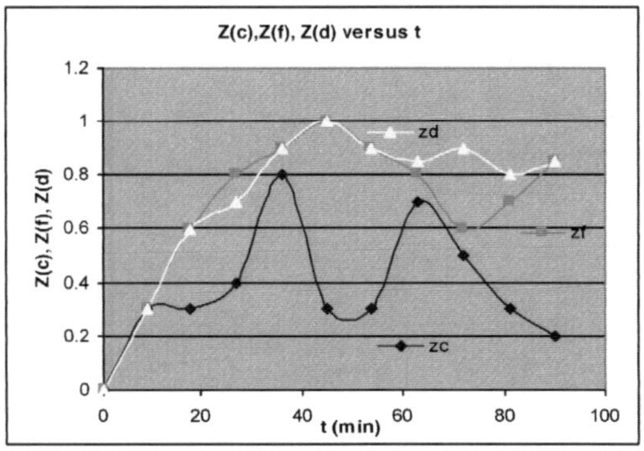

Figure 5. Each curve represents the behavior over time of the variable Z (defined as quotient Information Internal / External Information) for each dynamic: weak attractor (Zd), fixed (Zf) and chaotic (Zc).

For other lab sessions, under reference experimental - theoretical model implemented and shows the behavior of the four variables (X, Y, Z, t), can be prepared to perform the activity in order to reduce errors and increase efficiency. The teacher should read the "resume" (emotional and yield) of each student in order to rigorously select the suitable initial condition, a task that can implement any experienced teacher. This would allow, prior, identify and characterize the optimal initial condition and that time, anticipating the possible scene and conflict.

6.2. Graphical representation of the different dynamics.

The graph described by the variables X, Y and Z for high performance teams determines a dynamics called chaotic, as shown in figure [20, 34]:

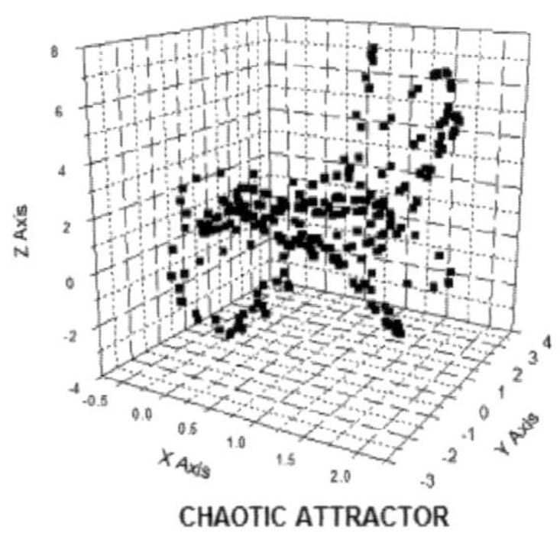

Figure 6. The graph is the three-dimensional representation of the columns of X, Y, Z, each of 1000 data - generated by its follow-up in time to different teams of students that satisfy the dynamics of chaotic attractor.

Average performance teams show numerical values for inquiry / persuasion (X) and positivity / negativity (Y) well below unity and decreasing. This means that in the case of variable X, the denominator are dominant teams incurring the persuasive type behaviors when in fact what is required in the measurement process is laboratory investigation of phenomena with capacity data recording instruments measure. Inside teams, an emotional field is generated with negativity domain behaviors apathy disorder, deny conflicts, etc. Not so for the internal information / external data showing a clear tendency to seek answers outside or inside the team.

The graph below shows how the variables X, Y, Z are combined to give rise to the dynamics of fixed point [20, 34]

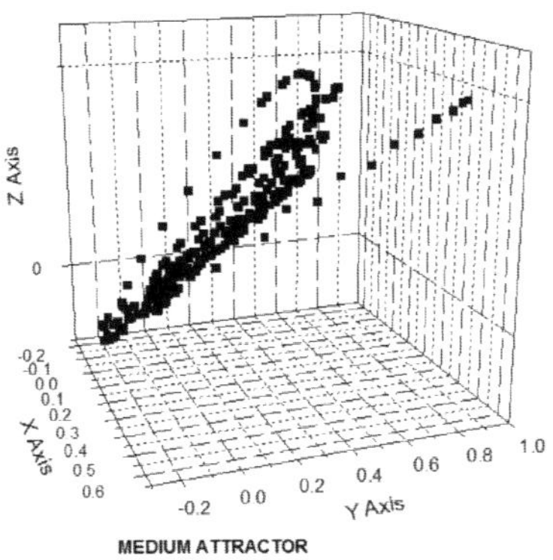

Figure 7. The graph is the three-dimensional representation of the columns of X, Y, Z, each of 1000 data - generated by its follow-up in time to different teams of students that satisfy the dynamics of fixed-point attractor.

If teams exhibit a low performance, working procedure is characterized by the search for solutions without validation analysis thereof. Learning is minimized because through copying results to other teams of students or improvising solutions is very hard to occur a meaningful learning as a team and as personal contribution of its components. The graph below shows how the variables X, Y, Z are combined to generates the dynamics of the weak point. [20, 34]:

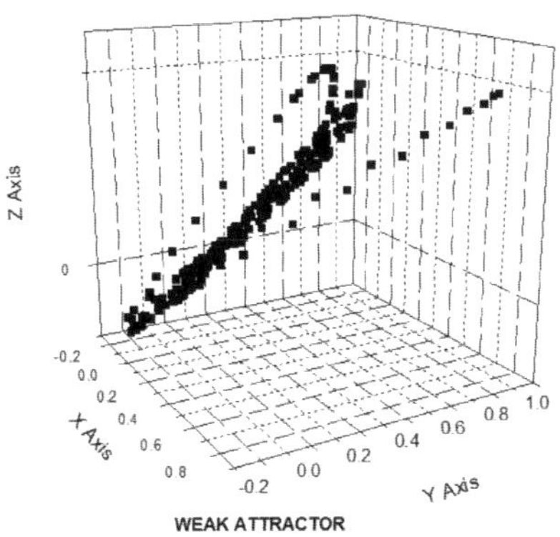

Figure 8. The graph is the three-dimensional representation of the columns of X, Y, Z, each of 1000 data - generated by its follow-up in time to different teams of students that satisfy the dynamics of weak fixed point attractor.

The projection of three-dimensional curves as each XY, XZ, and YZ allows getting in graphical form the centroids associated with the outbreaks of the attractors that keep trajectories within limits [10]. The numerical values indicate that each variable influence has on the other. This macroscopic view of nonlinear dynamics indicates the way in which relational links or connectivity (r) between members of the team is manifested.

7. Time Fourier series and Moving Averages

Using time series and moving averages of the data, an average value of the quotient convergence Positivity / Negativity, P / N is obtained for each case, which is a fundamental parameter for the characterization and classification of work dynamics:

$$\left\langle \left(\frac{P}{N} \right) \right\rangle_{WEAK} \rightarrow 0.375, \quad \left\langle \left(\frac{P}{N} \right) \right\rangle_{MEDIUM} \rightarrow 1.95, \quad \left\langle \left(\frac{P}{N} \right) \right\rangle_{CHAOTIC} \rightarrow 5.25$$

8. The cross-correlation function

Given two functions f (t) and g (t) the cross-correlation function [20] is given by:

$$C_{fg}(\tau) = f(t) * g(-t) = \int_{-\infty}^{\infty} f(t)\, g(\tau + t)\, dt$$

The cross-correlation function can be interpreted as a measure of the coincidence between two images, for various amounts of shift. It is possible to measure the ties through the discretized cross-correlation function applied to all time series data [35] that were generated from work, each team of students, for X, Y, Z. This function provides a measure of how strongly and how long it takes the particular conduct of a person, in time, to connect with the conduct of another person.

The graphs for each dynamic deliver an accurate descriptive picture from the pattern of behavior of teams from the correlation. The emotional field mediates the other variables, but, in itself, depends on the context of experiential reference (culture, values, social relations, economy, family, etc.) where students are inserted, and this framework and narrative, extra classroom and student teams, also plays a role in the achievements of the teams. If the educational institution has failed to structure emotional field conditions which prevail in the values and principles (honesty, nobility, generosity, solidarity, etc.) this will influence negatively the performance and "procedures" to achieve the goals by the students affected indicators. So excess positivity or negativity was reflected even in the teams 'best' performance. Tables of values follow:

Table 2. Tables of cross correlation values, C.

	Chaotic case	**Attractor mean Case**	**Attractor Weak Case**
Cxy	0,44	0,16	0,13
Cxz	0,24	0,28	0,23
Cyz	0,27	0,29	0,32

Since Cxy represents the influence between the variables of Inquiry / persuasion with Positivity / Negativity in the chaotic case "borders" 50% while in the remaining two cases is less than 20%. Similarly it observed in the groups of low and average yield, the greater the influence of negativity in positivity / negativity ratio and internal / external ratio Orientation, seeking a way to balance showing a tendency to get solutions by whatever means.

9. Graphical representation of the cross-correlation

Plotting the cross-correlation function in time the value of maximum travel (vertical axis of the coordinate system) arises for each graph XY (t), XZ ((t) and YZ (t). The calculation for the influence of Y (t) on X (t) was presented for the three dynamics studied:

Case 1.- Chaotic attractor

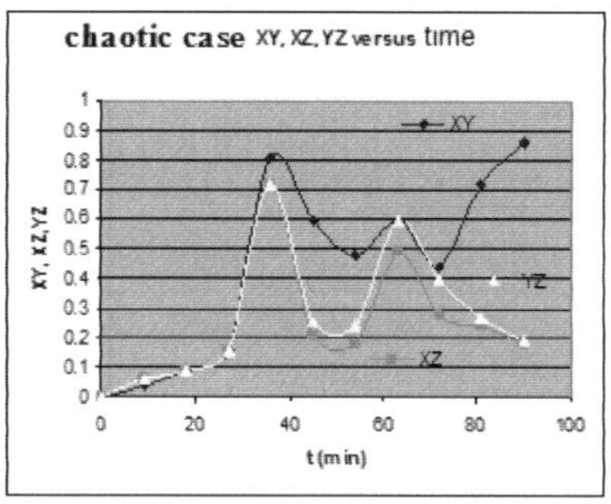

Graphics for cross-correlation with chaotic dynamics calculating the maximum connectivity value: 0.81 * 36 = 29.16 ~ 29

Case 2.- Medium tractor

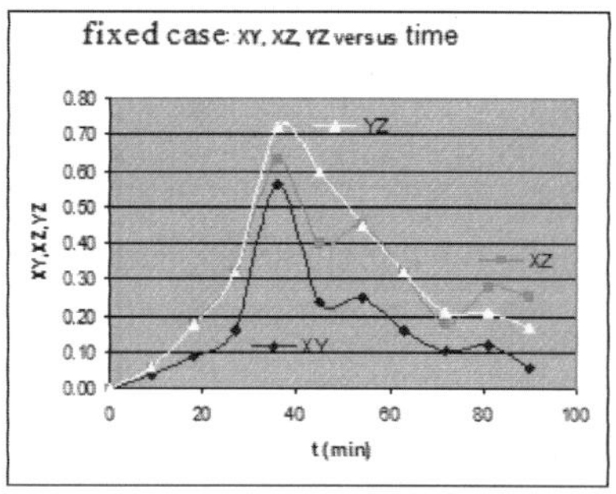

Graphics of the cross-correlation for the dynamics of the fixed point calculating the maximum connectivity value: 0.56 * 36 = 20.16 ~ 20

Case 3.- Weak attractor

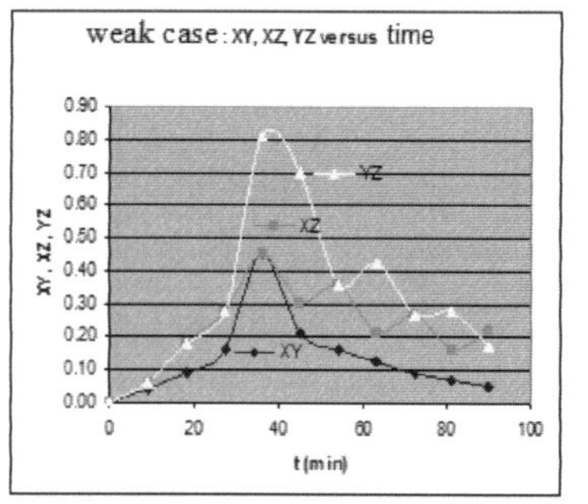

Cross-correlation graphs for dynamic weak point with calculating the maximum connectivity value: 0.45 * 36 = 16.2 ~ 16

Figure 9. Graphics of the cross-correlation for the dynamics.

11. Graphic Connectivity versus performance

Connectivity can be defined as the ability the team members show to expand their actions with the actions of others and to expand their own actions from the actions of others [36, 37]. So in particular for the relationship - influence of Y on X according to the different dynamics: Figure 10 gives the connectivity by calculating the centroid, cross correlation [23], the theoretical recursive method as MatLab [38] applied nonlinear model [34, 39, 40] and averaged value:

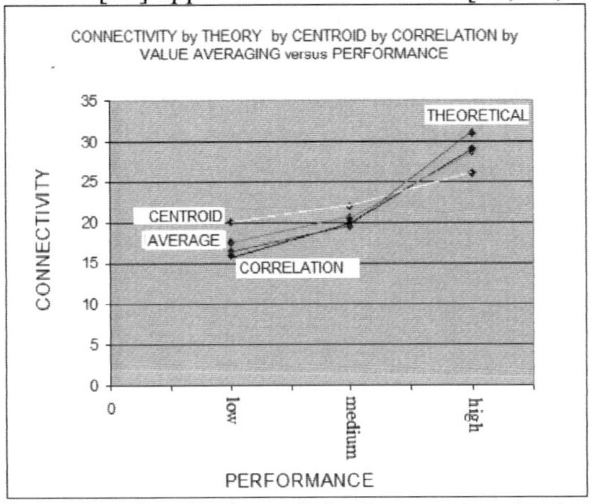

Figure 10. Graphic of Connectivity versus levels of school performance, the latter understood as significant learnings.

From the observation of the interaction [24] the dimensions of the model, the information work of teams of students [41] is extracted. According to the central hypothesis of this study is that to be mediated by the quotient Positivity / Negativity, the variable Y of emotions, is tested quantitatively, from the graphs in time, its influence and interdependence in Nonlinear way what directly affects the performance of the teams.

12. Conclusions

The approach used gives a quantitative numerical vision that allows qualitative conclusions regarding the observed performance of equipment and how emotions are crucial in achieving the learning objectives. Using time scales graduated at intervals of 9 minutes is appropriate because the emphasis is on how the process leads to the goal and the way in which the evolution of the observed variables (X, Y, Z) lead to meaningful learning This process is not instantaneous. It is unlikely that there is an understanding of concepts or instruments relating to physics in 1 second.

The transition between those low, medium and chaotic dynamics is given by the parameter r, connectivity, while its values for each case in the order 16, 20 and 29 respectively, which is verified by the experimental methods of the centroid and the correlation crusade confirming theoretical modeling of the nonlinear Lorenz equations. These numerical values are indicative of meaningful learning teams, revealing, in the order given, if it is low, medium or high. They confirm and justify that whoever drives the transformation and transition between dynamic is the evolution of the emotional field.

A more rigorous understanding of quality [42] in its relationship with significant learning as appears linked to the quotient Positivity / Negativity as demonstrated by the time series, the values of convergence are in good numbers defined (0.375 (weak), 1.95 (medium), 5.25 (chaotic)) indicating which is the dominant factor of the fraction. This allows building a domain of validity for the quality of the educational process which directly affects perception, integration and continuity of pupils in their learning and assessment made of the same. This is crucial in this study from the point of view of avoiding permanent abandonment or desertion of the various professions studying by young.

In the segment of students tested, many are imbued with an emotional dynamic that, most of the time; it feeds on some negativity and fatalism [43]. Yet their attitude is clearly aspirational in the sense of set goals to help them out of this emotional potential well.

References

[1] Capra F 1982 *El punto Crucial* (Buenos Aires: Estaciones)
[2] Connes A 1990 *Noncommutative Geometry* (San Diego: Academia Press)

[3] Stiglitz J 1987 The causes and consequences of the dependence of quality on Prices *Journal of Economic Literature* **25** p 1-48
[4] Chomsky N 2012 *Cómo funciona el mundo. Conversaciones con David Barsamian.* (Buenos Aires: Katz Editores)
[5] Balanzkat A., Blamire R and Kefala S 2006 The ICT Impact Report: A Review of Studies of ICT Impact on Schools in Europe Recuperado el 11 de noviembre de 2012 de http://insight.eun.org/ww/en/pub/insight/misc/specialreports/impact_study.htm
[6] Marqués P 2013 ¿Por qué no mejoran las notas de los alumnos con las TIC? Recuperado el 20 de Agosto de 2013 de http://peremarques.blogspot.com/
[7] Carr N 2011 *¿Que esta haciendo Internet con nuestras mentes? Superficiales* (México D.F.: Taurus)
[8] Rodríguez M 2010 *La teoría del aprendizaje significativo en la perspectiva de la psicología cognitiva* (Barcelona: Octaedro)
[9] Losada M 1999 The complex dynamics of high performance teams *Mathematical and Computer Modelling* **30** 9-10 p 179-192
[10] Losada M and Heaphy E 2004 The role of positivity and connectivity in the Performance of Business Teams *American Behavioral Scientist* **47** 6 p 740 – 765
[11] Losada M and Fredrickson L 2005 Positive Affect and Complex Dynamics of Human Flourishing *American Psychologist* **60** 7 p 678-86
[12] Grün E and Del Caño E 2003 *Ensayos sobre Sistémica y Cibernética* (Buenos Aires:Dunken)
[13] Laughlin R B 2007 Un *universo diferente. La reinvención de la física en la edad de la emergencia* (Buenos Aires: Katz)
[14] Mock T 9 Sep 2013 Climate Change Hill upset vital ocean chemical cycles *Journal Nature Climate Change* (University of East Anglia press release 64 / 13)
[15] Monzó J 2008 Pensamiento Sistémico. Teoría del caos: una mariposa en el reloj Recuperado el 07 de Octubre 2012 de http://www.blogger.com/profile/16346186836123661127.
[16] Pacheco P, Villagrán S, Guzmán C, Muñoz I, Quiroz E, Vásquez L and Donoso R 2011 Modelo de Dinámica de sistemas complejos para analizar las emociones, las formas relacionales y los aprendizajes significativos con proposición experimental para su comprobación Recuperado el 10 de Noviembre del 2012 de http://www.cimm.ucr.ac.cr/ocs/files/conferences/1/schedconfs/1papers/2535/submission/review/2535-6949-1-RV.pdf
[17] Pacheco P, Villagrán S, Guzmán C 2015 Estudio del campo emocional en aula y simulación de su evolución durante un proceso de enseñanza- aprendizaje para cursos de Ciencias *.Revista Estudios Pedagógicos* (aceptado para publicarse el primer semestre 2015)
[18] Canavos G C 1988 *Probabilidad y estadística* (Mexico D.F.: McGraW-Hill)
[19] Cronbach L J 1951 Coefficient alpha and the internal structure of tests *Psychometrika* **16** 3 p 297-334
[20] Sprott J C 2006 *Chaos and Time – Series Analysis* (New York: Oxford University Press)
[21] Maturana H 2001 *Emociones y Lenguaje en Educación y Política* (Santiago de Chile:Dolmen)
[22] Ibáñez N 2002 Las Emociones en el Aula *Estudios Pedagógicos* **28** p 31-45
[23] Pacheco P, Villagran S, Quiroz E 2013 Dinámica no lineal y rendimiento académico: verificación experimental e interpretación *Revista Internacional de Educación y Aprendizaje* **1** 1 p 49 - 73
[24] Bareman R and Gottman J M 1989 *Observación de la interacción: introducción al Análisis secuencia* (Madrid: Morata S.A)
[25] Gottman J 1994 *What predicts divorce?. The relationship between marital processes and marital outcomes* (New Jersey: Lawrence Erlbaum Associates)
[26] Baumeister R, Bratslavsky E, Finkenauer C, and Vohs K 2001 Bad is stronger than good *Review of General Psychology* **5** 4 p 323-370
[27] Cabero J 2001 *Tecnología Educativa,Diseño y Utilización de medios en la enseñanza* (Madrid:

Paidós)
[28] Zill D 1988 *Ecuaciones Diferenciales con Aplicaciones* (México D.F.: Ibero América)
[29] Goleman D 2005 *La Inteligencia Emocional*.(Barcelona:Cairos)
[30] Varela F, Thompson E and Rosch E 1997 *De cuerpo presente* (Barcelona: Gedisa)
[31] Faires J D and Burden R 2004 *Métodos Numéricos* (3ª Edición).(Madrid: Brooks / Cole, Division of Thomson Learning, Inc)
[32] Beilock S L, Gunderson E A, Ramirez G and Levine S C 2010 Female teachers´ math anxiety impacts girls´ math achievement *Proceedings of the National Academy of Sciences* **107** 5 p 1860 – 1863
[33] Schuster H G 1995 *Deterministic Chaos* (Weinheim: Wiley-VCH Verlag GMBH & CoKGaA)
[34] Kaspar F and Schuster H G 1986 Easily calculable measure for the complexity of spatiotemporal patterns *Physical Review* A **36** p 842–848
[35] Ziv J and Lempel A 1977 A Universal Algorithm for Sequential Data Compression *IEEE Transactions on Information Theory* **23** 3 p 337 - 343
[36] Cacioppo J T and Berntson G G 1999 The affect system: Architecture and operating characteristics *Current Directions in Psychological Science* **8** p 133-137
[37] Echeverría R 2005 *Ontología del Lenguaje* (Santiago de Chile: Lom)
[38] Nakamura S 1997 *Análisis Numérico y Visualización Grafica con Matlab* (Mexico: Pearson)
[39] Lorenz E 1976 Deterministic nonperiodic flow *Journal of Atmospheric Sciences* 20:69
[40] Grassberger P and Procaccia I 1982 Characterization of Strange Attractors *Physical Review Letters* **50** p 346 -349
[41] Treisman U 1992 Studying students studyng calculus: A look at the lives of minority mathematics students in collage *College Mathematics Journal* **23** 5 p 362 - 372.
[42] Álvarez-Tostado C 1997 Calidad de la Educación. Entre el eslogan y la utopía (Buenos Aires: Magisterio del Río de la Plata)
[43] Otten M and Jonas K 2014 Humiliation as an intense emotional experience: Evidence from the electro-encephalogram *Social Neuroscience* **9** 1 p 23 – 35

Journal of Physics: Conference Series **720** (2016) 012057 doi:10.1088/1742-6596/720/1/012057

Non university sources of science in Chile

Leopoldo Soto
Comisión Chilena de Energía Nuclear , Casilla 188-D, Santiago, Chile
and
Center for Research and Applications in Plasma Physics and Pulsed Power, P4, Chile

E-mail: lsoto@cchen.cl

Abstract. The following ideas are widely accepted in Chile with respect to scientific activity: a) is carried out mainly in universities and science is considered a naturally university activity, b) that was developed in a period of more than 150 years by isolated individual efforts, c) it was transformed into an institutionalized activity at the universities after the university reform movement at the end of the 1960 decade, d) the activity is finally institutionalized in the country with the creation of the "Comisión Nacional de Investigación Científica y Tecnológica, CONICYT (National Commission for Scientific and Technological Research) in 1967. This work presents preliminary findings showing that there are other institutional initiatives, different to the efforts from universities and directly dependent of the Chilean Government, in order to produce science and technology in Chile. This governmental initiatives start at the beginning of the Republic of Chile circa of 1810.

1. Introduction

The scientific activity produces more results than can be handled by universities. These results impact all areas of society: economy, health, education, natural resources, environment, freedom and welfare of citizens, international relations, defense, popular sovereignty and national sovereignty. This has been understood many years ago by the developed countries, where scientific activity is carried out in universities and state institutes, private centers, public and private productive sector and armed forces. Contrary to what happens in developed countries, in Chile more than 95% of the scientific research is conducted at universities, virtually without connection with state institutions, the productive sector and the armed forces, which makes practically impossible that the cycle Research + Development + Innovation can be completed.

The stories about the history of science in Chile consider mainly the contribution of individuals to scientific activity. Starting back with the contributions of Claudio Gay, a French scientist, who in 1830 was hired by the Government of Chile for a study the natural history of the area and everything that contributes to know the industry of the country, its commercial activities and management. In 1838 the Chilean government hires the Polish scientist Ignacio Domeyko to work as a teacher at the "Liceo de Coquimbo" (High School of Coquimbo). After that, between 1840 and 1846 Domeyko makes trips through the territory of Chile and describes its geology.

Henceforth, the various stories of the history of science in Chile refer to the contributions of individuals without relation with a policy or institutional interest in developing scientific activity in Chile (see for example reference [1]). In fact, scientific research is not considered an important activity in Chilean universities until after the university reform from the sixties of the twentieth century.

Thus, nowadays one may find the following ideas in the Chilean intellectual imaginary with respect to the scientific activity in Chile: a) is carried out mainly in universities and science in Chile is considered a naturally university activity, b) that was developed in a period of more than 150 years by isolated individual efforts, c) it was transformed into an institutionalized activity at the universities after the university reform movement at the end of the 1960 decade, d) the activity is finally institutionalized in the country with the creation of the "Comisión Nacional de Investigación Científica y Tecnológica, CONICYT (National Commission for Scientific and Technological Research) in 1967.

However, Chile has about 17 State Research Institutes. State Research Institutes are institutions for scientific and technological research created by the State outside the universities. Their goal was to increase the productivity of various economic and strategic sectors through the introduction of technological innovations and the creation knowledge. To do so, among his tasks it to included the development of basic and applied research, coordinating and dictating policies in this area, and to train specialists [2].

2. Research institutes created by the Chilean State.

Although the Bacteriological Institute was created in 1929, and the Institute of Experimental Medicine in 1937, the golden age of the State Research Institutes is the second half of the twentieth century, between 1957 and 1970. To get an idea, apart from the above, during that period the following institutes were created: Geological Survey (1957); Chilean Antarctic (1963); Fisheries Development (1963); Occupational Health and Air Pollution (1963); Agricultural Research (1964); Natural Resources Research (1964); Chilean Nuclear Energy Commission (1965); Forest (1965); Committee Harnessing Geothermal Energy (1967); National Hydraulic (1967); Center of Studies in Measurement and Quality Certification (1968); Technological Research (1968); Mining; and Metallurgical Research (1970).

From the mid-1970´s and 1980´s most of these institutions suffer a strong reduction in budget and self-financing policies are imposed to them [2]. During the subsequent governments, since 1989 to date, there are no policies oriented to the State Research Institutes. Government policies and instruments include incentives to universities and to the productive sector. None of these policies and instruments consider the particular reality of State Research Institutes. Thus, at present most of these are far from fulfilling one of their original objectives, mainly, to develop basic and applied research. In most of them their main activities are the provision of services and routine production activities. Even though, they have a small number of professional researchers, they have a lot of highly skilled and competent of other type of professionals and technicians. In addition, many of them have infrastructure and equipment that the universities do not have .

3. Others non-university initiatives previous to the State Research Institutes.

In a recent study entitled "Institutos de Investigación del Estado: Evolución y Perspectivas" (State Research Institutes: Evolution and Perspectives) presented at the "Encuentro Ciencia, Tecnología y

Journal of Physics: Conference Series **720** (2016) 012057 doi:10.1088/1742-6596/720/1/012057

Sociedad 2014" in Santiago, Chile [3], it is shown that there were other non-university initiatives, inclusive previous to the State Research Institutes, linked to government policies or public companies that allowed the creation of many organisms and institutions whose objectives were to conduct scientific and technological research.

A chronology regarding the creation of Chilean non-university scientific and technological institutions, it can be described as follows:

Stage I: First attempts, 1812-1939.
Stage II: Companies of the "Corporación de Fomento de la Producción, CORFO" (Chilean Economic Development Agency), 1939-1960.
Stage III: State Research Institutes of Chile, 1963-1973.
Stage IV: Decline and privatization, 1975-present.

<u>**Stage I**: First attempts, 1842-1939</u>
1811: "Fábricas y Maestranzas del Ejército, FAMAE" (Factories and Arsenals of the Chilean Army).
1817: Primer "Arsenal y Maestranza de la Armada", (First Factory and Arsenals of the Chilean Navy) at Port of Valparaiso.
1842: "Cuerpo de Ingenieros del Ejército" (Army Corps of Engineers).
1843: "Oficina de Estadísticas" (Statistics Office).
1874: "Oficina Hidrográfica de la Armada" (Navy Hydrographic Office).
1888: "Dirección de Obras Públicas" (Public Works Department). Laboratories are created: a) Railways and Telegraphs, b) Bridges, Roads and Hydraulic Constructions, c) Architecture
1891: "Oficina de Meteorología" (Weather Bureau, later Military Geographical Institute).
1892: "Instituto de Higiene" (Institute of Hygiene).
1896: "Arsenal y Maestranza de la Armada" at Port of Talcahuano (Factory and Arsenals of the Chilean Navy).
1898: "Taller Resistencia Materiales" (Resistance Workshop Materials). In association with University of Chile.
1908: "Instituto Sismológico" (Seismological Institute). Created after the Valparaiso earthquake of 1906.
1914: "Instituto Científico e Industrial del Salitre" (Scientific and Industrial Institute of Saltpetre).
1914: "Servicio Provisorio de Estudios Geológicos" (Interim Service of Geological Studies).
1919: "Servicio de Minas y Geología" (Service of Mines and Geology) former Interim Service of Geological Studies.
1922: "Instituto Geográfico Militar" (Military Geographical Institute, former Weather Bureau).
1925: "Cuerpo de Ingenieros de Minas" (Mining Engineers Corps).
1925: "Banco Central" (Central Bank).
1927: "Departamento de Navegación e Hidrografía (Department. Navigation and Hydrography), former Hydrographic Office.
1929: "Instituto Bacteriológico" (Bacteriological Institute) former Institute of Hygiene.
1930: "Ala de Mantenimiento de la Fuerza Aérea de Chile" (Maintenance Wing Air Force of Chile)
1930: "Departamento de Minas y Petróleo" (Department of Mines and Petroleum), former Mining Engineers Corps.
1937: "Instituto de Medicina Experimental" (Institute of Experimental Medicine).
1938: "Instituto de Investigación de Materiales" (Institute for Materials Research), former Resistance Workshop Materials is associated with Roads Laboratory.

Journal of Physics: Conference Series **720** (2016) 012057 doi:10.1088/1742-6596/720/1/012057

Stage II: Companies of the "Corporación de Fomento de la Producción, CORFO" (Chilean Economic Development Agency), 1939-1960.
1939: "Corporación de Fomento de la Producción, CORFO" (Chilean Economic Development Agency), 1939.
1943: "Empresa Nacional de Electricidad, ENDESA" (National Electricity Company).
1944: "Instituto Nacional de Investigaciones Tecnológicas y Normalización, INDITECNOR" (National Institute for Technological Research and Standardization).
1946: "Compañía de Aceros del Pacífico, CAP" (Pacific Steel Company).
1950: "Empresa Nacional del Petróleo, ENAP" (National Oil Company).
1952: "Servicio de Cooperación Técnica, SERCOTEC" (Technical Cooperation Service).
1953: "Dirección de Arquitectura" (Department of Architecture)
1957: "Instituto de Investigaciones Geológicas" (Institute of Geological Research"
1960: "Empresa Nacional de Minería, ENAMI"(Mining National Company)
1960: "Astilleros y Maestranzas de la Armada, ASMAR" (Astilleros and Arsenals of the Navy), former Factory and Arsenals of the Chilean Navy.
1960: "Ministerio de Agricultura" (Agriculture Ministry)

Stage III: State Research Institutes of Chile, 1960-1973.
1963: "Insituto Chileno Antártico" (Chilean Antarctic Institute)
1963: "Instituto de Fomento Pesquero" (Fisheries Development Institute)
1963: "Instituto de Higiene del Trabajo y Contaminación Atmosférica" (Institute of Occupational Health and Air Pollution)
1963: "Servicio Aerofotogramétrico de la Fuerza Aérea de Chile, SAF" (Photogrammetric Service of the Air Force of Chile).
1964: "Instituto Nacional de Investigación de Recursos Naturales, IREN" (National Research Institute of Natural Resources).
1964: "Instituto de Investigaciones Agropecuarias" (Institute of Agricultural Research).
1964: "Empresa Nacional de Telecomunicaciones, ENTEL" (National Telecommunications Company).
1964: "Comisión Chilena de Energía Nuclear", CCHEN (Chilean Nuclear Energy Commission).
1965: "Instituto Forestal" (Forest Institute).
1966: "Instituto Nacional de Capacitación, INACAP" (National Training Institute).
1967: "Comité de Aprovechamiento de Energía Geotérmica" (Committee of Geothermal Energy Utilization).
1967: "Servicio Agrícola y Ganadero, SAG" (Agriculture and Livestock Service)
1967: "Instituto Nacional de Hidráulica" (National Institute of Hydraulic).
1968: "Centro de Estudios, Medición y Certificación de Calidad Studies Center" (Measurement and Quality Certification).
1968: "Empresa Nacional de Computación,ECOM" (National Computing Company).
1968: "Instituto de Investigaciones Tecnológicas, INTEC" (Technological Research Institute).
1968: "Comisión Nacional de Investigación Científica y Tecnológica, CONICYT" (National Commission for Scientific and Technological Research).
1968: "Instituto Hidrográfico de la Armada, IHA" (Hydrographic Institute of the Navy).
1970: "Centro de Investigación Minera y Metalúrgica, CIMM" (Center for Mining and Metallurgical Research).
1970: "Instituto Nacional de Estadísticas, INE" (National Statistics Institute), former Statistics Office).
1973: "Coorporación Nacional Forestal, CONAF" (National Forest Corporation).

Stage IV: Decline and privatization, 1975-present.
1976: "Comisión Chilena del Cobre, COCHILCO" (Chilean Copper Commission).

1976: "Corporación Nacional del Cobre de Chile, CODELCO" (National Copper Corporation).
1984: "Empresa Nacional de Aeronáutica, ENAER" (National Aeronautics Company), former Maintenance Wing Air Force of Chile.
1985: "Centro de Informaciones de Recursos Naturales, CIREN" (Information Center of Natural Resources), former National Research Institute of Natural Resources.
1990: "Servicio Hidrográfico y Oceanográfico de la Armada, SHOA" (Hydrographic and Oceanographic Service of the Navy), former Navy Hydrographic Institute.

During this last period most of State Research Institutes suffer a strong reduction in budget and self-financing policies are imposed. In addition several State Companies are privatized.

4. Final remarks.

As was shown, the development of scientific and technological activity in Chile has older and beginnings more than the university sector and than the isolated individual contributions. In fact the above chronology shows that institutional efforts predate those by universities and the creation of the National Commission for Scientific and Technological Research, CONICYT.

These preliminary findings are part of a research on the history and institutionalization of scientific activity in Chile. This research in progress is motivated for the following questions: What are the origins and further development of scientific activity in Chile beyond the university environment? Why is there in Chile this "universitization" of science? Why the governmental efforts manifested in the creation of institutes and public companies were not intersected and there is not feedbacked with the university efforts? Is this a coincidence or a consequence of an ideological model?

It would be interesting also to compare the experiences of Latin American countries like Argentina, Mexico and Brazil.

5. Acknowledgments.

The author acknowledges the discussion with Claudio Gutierrez and Carlos Friedli. In Addition C. Gutierrez provided useful information. Support by PIA-CONICYT grant ACT 1115.

References

[1] Claudio Gutiérrez and Flavio Gutiérrez (2008) "Forjadores de la Ciencia en Chile. Problemas y Soluciones", Editorial RIL, Santiago.
[2] Enrique Dellacasa and Alvaro Guell (1990) "Evolución y financiamiento entre los años 1968 y 1988 de los institutos de investigación y desarrollo no universitarios, creados por el Estado", Arch. Biol. Med. Exp. 23, 65.
[3] Claudio Gutiérrez y Leopoldo Soto (2014) "Institutos de Investigación del Estado: Evolución y Perspectivas", Encuentro Ciencia, Tecnología y Sociedad CTS-Chile 2014, Book of Abstracts, p. 34. http://encuentroctschile.files.wordpress.com/2013/12/libro-de-resumenes-primer-encuentro-cts-chile.pdf

Hilda Cid: physicist, crystallographer, structural biologist

Gonzalo Gutiérrez
Departamento de Física, Facultad de Ciencias, Universidad de Chile
Casilla 653, Santiago, Chile
www.gnm.cl/gonzalo

E-mail: gonzalo@fisica.ciencias.uchile.cl

Abstract. A brief review of Dr. Hilda Cid, Chilean scientist who excelled in the field of crystallography, is presented. Dr. Cid was born in 1933 in Talcahuano, where she completed her primary and secondary education, graduating as a Teacher of mathematics and physics from the University of Chile in 1958. She continued graduate studies at MIT, USA, where obtained the PhD in 1964. Upon her return to Chile, she joined the Austral University in Valdivia, where she dedicated to characterize, by X-ray, biological material. During that time, she was actively involved in the social changes taking place in the university and in the country. Thus, following the bloody coup of 1973, she and her family were persecuted and forced to go into exile to Sweden. In the decade of the 1980 she returns to Chile, and settles at University of Concepción, where she teaches, conducts research and organizes a number of latino-american courses and symposiums. Knowing that Hilda Cid formed a real school, where her students -now distinguished professionals and researchers- cast their teachings through Chile and other countries, we ended the article with a reflection about why Hilda Cid, despite its obvious scientific achievements, remains ignored by most of the scientific establishment in Chile.

1. Introduction

Crystallography in Chile has a long history. Chile, as a mining country, have always required people and techniques in order to characterize the riches extracted from the earth. In the mid-nineteenth century, the State hired a chemical engineer of the French school, the Polish Ignacio Domeyko to teach mineralogy. The aim was to teach the new generations the latest hasten in exploration, prospecting, mining, as well as in the characterization of minerals.

Domeyko, educated in the Ecole de Mines in France, knew the state of the art on the subject, in particular the study of crystallography, which was in full bloom after the magnificent treaty by René Just Haüy was published in Paris in 1822 [1], a book which Domeyko knew well. Under the Government's mandate, Domeyko was sent to the northern city of Coquimbo and then Copiapó, creating the first School of Mines in the country. Crystallography was one of the main subjects in the school's program. There he introduced the concepts of crystallographic planes, and polymorphism symmetries, among others. Domeyko later moved to Santiago, where he started the area of materials science. Later he became rector of the main institution of higher education, the University of Chile.

The imprint of Domeyko had influence in Chile. The new generation of physicist in Chile reproduced in the late nineteenth century, soon after, the discovery by Roentgen, the X-rays.

Three months after the German scientist presented the x-ray picture of his wife's hand, Arturo Salazar and Luis Ladislao Zegers, obtained in Chile a photograph of the bones of their own hands: this radiograph was the second in America and the seventh in the world [2, 3].

Following these paths, in the early 1950s, when basic science began to professionalize at the University of Chile, crystallography was chosen as one of the key disciplines, along with nuclear physics and cosmic rays [3, 4]. The laboratory was formed in the Pedagogical Institute, led by Nahum Joel, who studied in Paris and England under professor J. B. Bernal [5]. Researchers were recruited among students of physics education, and among them two women: Hilda Cid and Isabel Garaycoechea. In the following we review some aspects of the multifaceted biography of Hilda Cid.

2. From Talcahuano to Concepción and then to Santiago

Hilda Cid Araneda was born in Talcahuano, a small port in southern Chile, on February 20, 1933, within a family of primary school teachers aware of the importance of education. Her father, Jose Cid Bastías, and her mother, Hilda Mendoza Aranda, formed a family in which their three children were university graduates. Daughter of public education, Hilda attended primary school at School No. 7 and 8 of Talcahuano, and her secondary education at the Liceo Fiscal the same city between 1944 and 1949. Outstanding student, as a teenager she began studying violin and then she took part of the Chamber Orchestra of the city of Concepción.

After successfully passing the Baccalaureate (the entrance exam to the University), in 1951 she enrolled in the Teacher's career in physics and mathematics at the University of Concepción, led by an early vocation for teaching. She worked as supervisor at the Laboratory of optics between the years 1954-55 and in 1955 won the University prize awarded to the best undergraduate student in physics and mathematics. This was an incentive to move to the Pedagogical Institute of the University of Chile, in the capital, Santiago, where she completed the requirements for obtaining the title of Teacher of Physics and Mathematics in 1958, with distinction (summa cum laude). Hilda moved to Santiago because there were opportunities to do research in physics. She worked in the laboratory of crystallography, while she was an assistant professor in the Optics lab. The research carried out between the years 1955 to 1958 allowed her to write her thesis degree entitled "Some optical methods in determining the crystal structures by X-ray", that was the basis for her firsts scientific papers. One of them was published in 1961 with Isabel Garaycochea, which is the first and (maybe still) the only paper published in the area of physics and materials science from Chile whose authors are two women [6]. Besides it scientific contribution, this paper is highly important for the history of exact sciences in Chile, as is among the firsts articles published from Chile.

3. From Santiago to Boston, nonstop

The passion for scientific investigation lead her to pursue postgraduate studies, being accepted into the Massachusetts Institute of Technology (MIT) in the USA. In September 1960, she undertook the trip to Boston, with two sons (Rodrigo, 4 years old; Jorge, 3 years old) and her husband. At MIT she continued her crystallography studies under Professor Martin J. Buerger, a world authority on that subject, who held the title of Institute Professor awarded in 1956 from MIT. Hilda Cid graduated of Master of Science in 1962, while pregnant with her third daughter, Rossana. Her thesis was about the determination, by X-ray, of the structure of a complex compound, the hexatitanate potassium [7]. Aware of the ability and talent of Hilda Cid, her advisor M. J. Buerger offered her to continue in MIT in order to obtain the Ph.D degree. MIT at that time was a redoubt formed almost only by men, both among its teachers and students. The doctoral program of the Geology department, at which M. J. Buerger worked, had graduated only 4 women between 1894 and 1964 (1931, 1939, 1948 and 1953) [8]. There was no customary for women to study science. An anecdote from the very Hilda Cid reveals the situation: when pregnant with her daughter Rossana and her belly growing remarkably, no one in the laboratory spoke a word of the situation. Only at the time of birth she told her advisor that would be missing a couple of days: "*I am going to give birth to a daughter*", to what Professor Buerger laconically replied: "*OK*". The situation was so unusual that neither the student nor the Professor knew how to proceed.

By working on her doctorate, in just two years she ended up her thesis, obtaining the PhD in 1964. She became the first Chilean woman to earn that degree in the area of the exact sciences [9]. Her thesis, entitled "Crystal structure of the turquoise group minerals," dealt with the determination of the atomic structure of a mineral called turquoise, which are phosphates of copper and aluminum. It gave her great knowledge and expertise in using X-rays to determine the structure of large and complex molecules. In this way, in 1964 she was able to determine the structure of the larger molecule resolved by this methodology to date: the atomic structure of terramycin [10].

4. From Boston to Valdivia

In 1964, upon receiving her doctorate, Hilda returned with their children to Chile, whereas her husband George Dresdner stayed on for another year in the USA to complete his Ph.D. in Biology. She joined the Lab of Crystallography at the University of Chile, now located at the School of Engineering, doing research, writing lectures notes and training new scientists, while waiting for her fourth son, Marcos. During the year 1965 she participated in the foundation of the Chilean Physical Society (*Sociedad Chilena de Física*, SOCHIFI). The decree of foundation of the Society appeared in the Official Gazette on November 9, 1965, and in Article 2 states that its purpose is 'to encourage scientific research in the field of physics and related sciences, disclosure of this discipline and contact people whose primary occupation is the practice of this science". The document details the first Directive: President: Dr. Jorge Zamudio Cristi; Secretary General: Dr. Hilda Cid-Dresdner; Treasurer: Engineer Egbert Schneider Hesse; Directors: Mr. Ruben Toro Valenzuela and Mr. Jorge Garcia Ruffinatti. This action is of prime importance: The SOCHIFI is the organization that would bring together the nascent Chilean physics community and organize, among other things, the bi-annual Symposia maintaining in contact Chileans physicists for the past forty years.

In 1967 she moved to the Austral University in Valdivia, eight hundred kilometers south of Santiago. This not only represented a change of city, but also a change of research topic. While completing her work on determining atomic structure of inorganic materials, she turned to biophysics, beginning to study the structure of macromolecules and biological material by means of X-ray diffraction. There were times of great effervescence in Chile, not only in the scientific field, but also at the university. There was a clamor for deep social changes in the country, which in the scientific community was expressed in the First National Congress of Scientists, one of whose objectives was to discuss a national science and technology program [11]. This Congress began with regional meetings in late 1971 and culminated in a major national meeting in July 1972. Hilda Cid participated actively in this process. In one of its resolutions, the Congress formally recognized the essential responsibility scientists and technologists had in devoting their research to solve the problems the country faced at various levels. The government, as expressed by socialist President Allende in his speech to the Parliament in 1973, was determined to implement the scientific community's mandate. However, it all ended in September 11, 1973 with a bloody military coup.

5. From Valdivia into exile to Uppsala

The coup changed violently the situation of Hilda Cid: she and her husband, who was the Dean of the Faculty of Science in Valdivia, were exonerated from the university, and persecuted. Finally, they were pushed into exile. In February 1974 the Dresdner-Cid family landed in Arlanda,

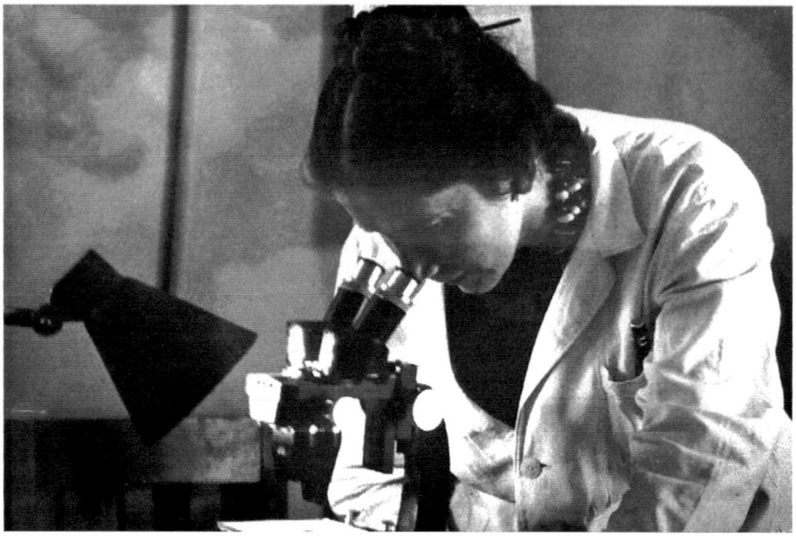

Figure 1. Hilda Cid ca. 1965, when she and a handful of persons founded the Chilean Physical Society.

Sweden, from where they travel to Uppsala, where they settled in a small apartment in the neighborhood of Flogsta. Hilda joined the Wallenberg Protein Laboratory at the University of Uppsala, where she devoted herself to the characterization by X-ray diffraction of biological molecules composed not only by the hundreds, but by the thousands of atoms. During her "swedish time" Hilda Cid published key articles on this topic. With her working group, she crystallized and characterized for the first time non-muscle actin. Also, they were able to determine the three-dimensional structure of carbonic anhydrase, an enzyme responsible for setting erythrocyte and convert carbon dioxide. Thus, she returned to the classic paradigm of molecular biology, which has brought so much success since the work of Watson and Crick, the close relationship between structure and function: "If you want to understand function, study structure" wrote Crick. Regarding the importance of her new research on protein tertiary structure, in an article published in 1980, she wonders if it's worth the effort it means. In what might be considered as her researcher philosophy testament, she responds: "*The question is whether it justifies the enormous effort of 5 years of work by a group of researchers who must bring together biologists, chemists, biochemists and crystallographers in determining the tertiary structure of a protein. The answer is yes. It is justified for some proteins. This means that the problem to be addressed should be relevant to justify the effort. In addition to the undeniable importance in the development of biological knowledge that has to determine the exact configuration of a protein molecule it must also be satisfied with the solution of a relevant problem.*" [12].

6. From Uppsala to Concepción

In the late seventies, having already become a prestigious researcher in the field of protein structure, she decided to return Chile from exile in 1979 to her alma mater, the University of Concepción. She arrived to the Department of Physiology, to settle later in Molecular Biology, until her retirement in 1996. She brought a microdensitometer, microscopes and software donated by the University of Uppsala, and installed them in the Laboratory of Molecular Biophysics. There she developed a fruitful research activity in the area of structure and protein crystallography, in addition to lecture at undergraduate and graduate level. Participated and organized numerous national and international courses and congresses and symposia, becoming a Latin American authority on the subject. During this time she married for the second time, forming a new family with Nelson Saavedra, professor at the University of Concepción. In parallel, driven by her strong commitment to social justice, she became actively involved in the pro-democracy and human rights movement that took place in the 1980s in Chile. In particular, she was instrumental in the organization of the Association of Academics of the University of Concepción, being elected President, and in supporting the student movement against the military intervention of universities. In this process, in the mid-eighties she was exonerated again from the university. The great pressure exerted by, and the solidarity of, the community, forced the authority to reconsider the call. That was the leadership of Hilda Cid.

7. The living legacy of Hilda Cid

As noted by Dr. Marta Bunster [13], her colleague and disciple of the University of Concepción, Hilda Cid formed a "*real school, where her students -now distinguished professionals and researchers- cast their teachings through Chile and other countries ...* ". And Dr. Bunster continued: "*It is admirable that Hilda Cid, with an initial training in physics, has successfully penetrated in the study of protein's structure-function, marking the beginning of these research in Chile. Dr. Cid is also a pioneer in promoting interdisciplinary scientific work, which she boosted in a visionary way, incorporating physical, mathematical and computational methods to their molecular biology research. Finally, an important feature that characterizes her entire research career is the persistent insistence on the need for science in Chile as an engine for development*".

Hilda Cid is an uncommon human being. A great female scientist; a great south American scientist; a committed social person. With so much scientific talent, when one searches Google, the only reference about her comes from the History of the 100 years of the Department Geology at MIT [8]. Little or nothing of her activities in Chile. Hilda Cid reunites not only a superior talent and a supreme human quality, but also features all that is negligible in our society today: she is female; she is provincial; she is the daughter of modest family; she was educated in public schools. Moreover, she did not follow a traditional career path: she sought the inter-discipline and changed research subject without complex: for physicists, she was a biologist; for biologists, she was a crystallographer. We could say that in a one-dimensional environment, she committed the sin of being expert in various fields, and not settle for lifelong keep singing the same ditty. To all of this, one should add her deep social commitment in difficult times in her country. The outcome is not difficult to guess: the silence, until today, of the establishment. It is time for the rest of us to honor the memory of Hilda Cid.

Acknowledgments

The author would like to thank Rossana Dresdner Cid and Claudio Gutiérrez for useful information, comments and assistance. This work was partially supported by grant Anillo CONICYT-PIA ACT-1115.

References

[1] Haüy R 1822, *Traité de Cristallographie* vol. 1, 2 (Paris: Bachelier et Huzard)
[2] Vogel E 2002 Historical background in the arising and development of Physics in Chile during the 20th century *Rev. Mex. Fis.* 48S3, 1-9;
[3] Gutiérrez C and Gutiérrez F 2006 Física: su trayectoria en Chile (1800-1960) *Revista Historia* 39:II, diciembre 2006, 477-496.
[4] Soto L et al 2008 Analysis and Projections of Physics in Chile *Journal of Physics: Conference Series* 134, 012052.

[5] Wittke O 2011, *Creación del Laboratorio de Cristalografía*, retrieved november 2014 http://www.cec.uchile.cl/cinetica/pcordero/recordando/Relatos/CreacionLab2.html
[6] Garaycochea I and Cid-Dresdner H 1961 On the setting of crystals for X-ray diffraction work *Acta Crystallogr.* **14**, 200-201.
[7] Cid-Dresdner H and Buerger MJ 1962 The Crystal structure of potassium hexa-titanate $K_2Ti_6O_{13}$, *Z. fur Kristallogr* **177** 411-430.
[8] Shrock Robert S 1982 *Geology at MIT 1865-1965* vol. 1 (Boston: The MIT Press)
[9] The next Chilean woman to obtain a PhD was Maria Esther Brandan, graduated in 1979 in Nuclear Physics at Wisconsin University. Hilda Cid is also the second Chilean to obtain a PhD in the area of physics (the first was Igor Saavedra, who obtain his PhD in Theoretical Physics from Manchester).
[10] Cid-Dresdner H 1965 The crystal structure of terramicina hydrochoride $C_22H_24N_2O_9HCl$, *Z. fur Kristallogr.* **121** 170-189.
[11] Power Margaret M 2011 The First National Congress of Scientists in Chile: The Popular Unity Government, Technology, Science, and Development, *SudHistoria* vol 2, nr. 2, January-June.
[12] Cid H 1980 Estructura de las proteínas *Revista Creces*, julio.
[13] Bunster M 1998 *Hilda Cid, reseña biográfica con motivo del Premio Municipal Concepción.*